Birkhäuser

Trends in Mathematics

Trends in Mathematics is a series devoted to the publication of volumes arising from conferences and lecture series focusing on a particular topic from any area of mathematics. Its aim is to make current developments available to the community as rapidly as possible without compromise to quality and to archive these for reference.

Proposals for volumes can be submitted using the Online Book Project Submission Form at our website www.birkhauser-science.com.

Material submitted for publication must be screened and prepared as follows:

All contributions should undergo a reviewing process similar to that carried out by journals and be checked for correct use of language which, as a rule, is English. Articles without proofs, or which do not contain any significantly new results, should be rejected. High quality survey papers, however, are welcome.

We expect the organizers to deliver manuscripts in a form that is essentially ready for direct reproduction. Any version of TEX is acceptable, but the entire collection of files must be in one particular dialect of TEX and unified according to simple instructions available from Birkhäuser.

Furthermore, in order to guarantee the timely appearance of the proceedings it is essential that the final version of the entire material be submitted no later than one year after the conference.

For further volumes:
http://www.springer.com/series/4961

Qamrul Hasan Ansari
Editor

Nonlinear Analysis

Approximation Theory, Optimization and Applications

Editor
Qamrul Hasan Ansari
Department of Mathematics
Aligarh Muslim University
Aligarh, Uttar Pradesh
India

ISBN 978-81-322-1882-1 ISBN 978-81-322-1883-8 (eBook)
DOI 10.1007/978-81-322-1883-8
Springer New Delhi Heidelberg New York Dordrecht London

Library of Congress Control Number: 2014939616

Printed on acid-free paper

Springer is part of Springer Science+Business Media (www.birkhauser-science.com)

In memory of Prof. S. P. Singh (d. 2013),
Professor of Mathematics (retd.),
Memorial University, St.John's, Canada

Preface

The approximation theory, optimization theory, theory of variational inequalities, and fixed point theory constitute some of the core topics of nonlinear analysis. These topics provide most elegant and powerful tools to solve the problems from diverse branches of science, social science, engineering, management, etc.

The theory of best approximation is applicable in a variety of problems arising in nonlinear functional analysis. The well-posedness and stability of minimization problems are topics of continuing interest in the literature on variational analysis and optimization. The variational inequality problem, complementarity problem, and fixed point problem are closely related to each other. However, they have their own applications within mathematics and in diverse areas of science, management, social sciences, engineering, etc. The split feasibility problem is a general form of the inverse problem which arises in phase retrievals and in medical image reconstruction. This book aims to provide the current, up-to-date, and comprehensive literature on different topics from approximation theory, variational inequalities, fixed point theory, optimization, complementarity problem, and split feasibility problem. Each chapter is self-contained and contributed by different authors. All chapters contain several examples and complete references on the topic.

Ky Fan's best approximation theorems, best proximity pair theorems, and best proximity point theorems have been studied in the literature when the fixed point equation $Tx = x$ does not admit a solution. "Best Proximity Points" contains some basic results on best proximity points of cyclic contractions and relatively nonexpansive maps. An application of a best proximity point theorem to a system of differential equations has been discussed here. Although, it is not possible to include all the available interesting results on best proximity points, an attempt has been made to introduce some results involving best proximity points and references of the related work have been indicated.

"Semi-continuity Properties of Metric Projections" presents some selected results regarding semi-continuity of metric projections onto closed subspaces of normed linear spaces. Though there are several significant results relevant to this topic, only a limited coverage of the results is undertaken, as an extensive survey is beyond our scope. This exposition is divided into three parts. The first one deals with results from finite dimensional normed linear spaces. The second one deals with results connecting semi-continuity of metric projection maps and duality maps. The third one deals with subspaces of finite codimension of infinite dimensional normed linear spaces.

The purpose of "Convergence of Slices, Geometric Aspects in Banach Spaces and Proximinality" is to discuss some notions of convergence of sequence of slices and relate these notions with certain geometric properties of Banach spaces and also to some known proximinality properties in best approximation theory. The results which are presented here are not new and in fact they are scattered in the literature in different formulations. The geometric and proximinality results discussed in this chapter are presented in terms of convergence of slices, and it is observed that several known results fit naturally in this framework. The presentation of the results in this framework not only unifies several results in the literature, but also it allows us to view the results as geometric results and understand some problems, which remain to be solved in this area. The chapter is in two parts. The first part begins from the classical works of Banach and Šmulian on the characterizations of smooth spaces and uniformly smooth spaces (or uniformly convex spaces) and present similar characterizations for other geometric properties including some recent results. Similarly, the second part begins from the classical results of James and Day on characterizations of reflexivity and strict convexity in terms of some proximinality properties of closed convex subsets and present similar characterizations for other proximinality properties including some recent results.

"Measures of Noncompactness and Well-Posed Minimization Problems" is devoted to present some facts concerning the theory of well-posed minimization problems. Some classical results obtained in the framework of that theory are presented but the focus here is mainly on the detailed presentation of the application of the theory of measures of noncompactness to investigations of the well-posedness of minimization problem.

"Well-Posedness, Regularization, and Viscosity Solutions of Minimization Problems" is divided into two parts. The first part surveys some classical notions for well-posedness of minimization problems. The main aim here is to synthesize some known results in approximation theory for best approximants, restricted Chebyshev centers and prox points from the perspective of well-posedness of these problems. The second part reviews Tikhonov regularization of ill-posed problems. This leads us to revisit the so-called viscosity methods for minimization problems using the modern approach of variational convergence. Lastly, some of these results are particularized to convex minimization problems, and also to ill-posed inverse problems.

In "Best Approximation in Nonlinear Functional Analysis," some results from fixed point theory, variational inequalities, and optimization theory are presented. At the end, convergence of approximating sequences and the sequence of iterative process are also given.

In "Hierarchical Minimization Problems and Applications," several iterative methods for solving fixed point problems, variational inequalities and zeros of monotone operators are presented. A generalized mixed equilibrium problem is considered. The hierarchical minimization problem over the set of intersection of fixed points of a mapping and the set of solutions of a generalized mixed equilibrium problem are considered. A new unified hybrid steepest-descent-like iterative algorithm for finding a common solution of a generalized mixed equilibrium problem and a common fixed point problem of uncountable family of nonexpansive mappings is presented and analyzed.

"Triple Hierarchical Variational Inequalities" is devoted to the theory of variational inequalities. A brief introduction of variational inequalities is given. The hierarchical variational inequalities are considered, and several iterative methods are presented. The triple hierarchical variational inequalities are discussed in detail along with several examples. Several solution methods are presented.

"Split Feasibility and Fixed Point Problems" is devoted to the theory of split feasibility problems and fixed point problems. The split feasibility problems and multisets split feasibility problems are described. Several solution methods, namely, CQ methods, are presented for these two problems. Mann-type iterative methods are given for finding the common solution of a split feasibility problem and a fixed point problem. Some methods and results are illustrated by examples.

The last chapter is devoted to the study of nonlinear complementarity problems in a Hilbert space. A notion of *-isotone is discussed in relation with solvability of nonlinear complementarity problems. The problem of finding nonzero solution of these problems is also presented.

We would like to thank our colleagues and friends who, through their encouragement and help, influenced the development of this book. In particular, we are grateful to Prof. Huzoor H. Khan and Prof. Satya Deo Tripathi who encouraged us (me and Prof. S. P. Singh) to hold the special session on Approximation Theory and Optimization in the Indian Mathematical Society Conference which was held at Banara Hindu University, Varanasi, India during January 12–15, 2012. Prof. S. P. Singh could not participate in this conference due to the illness. Most of the authors who contributed to this monograph presented their talks in this special session and agreed to be a part of this project.

We would like to convey our special thanks to Mr. Shamim Ahmad, Editor, Mathematics, Springer India for taking keen interest in publishing this book.

Last, but not the least, we would like to thank the members of our family for their infinite patience, encouragement, and forbearance.

Aligarh, India, February 2014 Qamrul Hasan Ansari

Contents

Contributors

M. Abbas Department of Mathematics and Applied Mathematics, University of Pretoria, Pretoria, South Africa

Qamrul Hasan Ansari Department of Mathematics, Aligarh Muslim University, Aligarh, India; Department of Mathematics and Statistics, King Fahd University of Petroleum and Minerals, Dhahran, Saudi Arabia

Józef Banaś Department of Mathematics, Rzeszów University of Technology, Rzeszów, Poland

Lu-Chuan Ceng Department of Mathematics, Shanghai Normal University, Shanghai, China; Scientific Computing Key Laboratory of Shanghai Universities, Shanghai, China

Himanshu Gupta Department of Mathematics, Aligarh Muslim University, Aligarh, India

V. Indumathi Department of Mathematics, Pondicherry University, Pondicherry, India

S. Z. Németh School of Mathematics, The University of Birmingham, Birmingham, UK

D. V. Pai Indian Institute of Technology Gandhinagar, VGEC Campus, Chandkheda, Ahmedabad, India

S. Rajesh Department of Mathematics, Indian Institute of Technology Madras, Chennai, India

Aisha Rehan Department of Mathematics, Aligarh Muslim University, Aligarh, India

D. R. Sahu Department of Mathematics, Banaras Hindu University, Varanasi, India

P. Shunmugaraj Department of Mathematics and Statistics, Indian Institute of Technology, Kanpur, India

S. P. Singh Department of Mathematics and Statistics, Memorial University, St. John's, NL, Canada

M. R. Singh Department of Physics and Astronomy, The University of Western Ontario, London, ON, Canada

P. Veeramani Department of Mathematics, Indian Institute of Technology Madras, Chennai, India

About the Editor

Qamrul Hasan Ansari is a professor at the Department of Mathematics, Aligarh Muslim University, Aligarh, India. He has been a visiting professor at several universities, namely, King Fahd University of Petroleum and Minerals, Dhahran, Saudi Arabia; National Sun Yat-sen University, Kaohsiung, Taiwan. He has visited several universities across the globe to deliver his talks. His main areas of research are variational analysis, optimization, convex analysis, and fixed-point theory. He has published more than 150 research papers in international journals. He is the author of two books, and editor of four contributed volumes. He is an associate editor of *Journal of Optimization Theory and Applications* and *Fixed Point Theory and Applications*. He has also served as an associate editor or a lead guest editor of several journals, namely, *Journal of Inequalities and Applications*, *Journal of Global Optimization*, *Positivity*, *Applicable Analysis*.

Best Proximity Points

P. Veeramani and S. Rajesh

Abstract Ky Fan's best approximation theorems, best proximity pair theorems, and best proximity point theorems have been studied in the literature when the fixed point equation $Tx = x$ does not admit a solution. This chapter contains some basic results on best proximity points of cyclic contractions and relatively nonexpansive maps. An application of a best proximity point theorem to a system of differential equations has been discussed. Though it is not possible to include all the available interesting results in best proximity points, an attempt has been made to introduce some results involving best proximity points and references of related work have been indicated.

Keywords Best approximant · Best proximity point · Best proximity pair · Cyclic contraction theorem · Best proximity point theorem · Relatively nonexpansive mappings · Set-valued maps · Upper semicontinuity for set-valued maps · Strictly normed spaces · Banach contraction theorem

1 Introduction

Consider the equation $Tx = x$, if the equation $Tx = x$ does not possess a solution, then it is contemplated to resolve the problem of finding an element x such that x is in proximity to Tx. In fact, the "Ky Fan's best approximation theorems" and "Best proximity pair theorems" are pertinent to be explored in this direction. In the setting of a metric space (X, d), if $T : A \to X$, then a best approximation theorem provides sufficient conditions that ascertain the existence of an element x_0, known as *best approximant*, such that

P. Veeramani (✉) · S. Rajesh
Department of Mathematics, Indian Institute of Technology Madras, Chennai 600036, India
e-mail: pvmani@iitm.ac.in

S. Rajesh
e-mail: srajeshiitmdt@gmail.com

Q. H. Ansari (ed.), *Nonlinear Analysis*, Trends in Mathematics,
DOI: 10.1007/978-81-322-1883-8_1,

$$d(x_0, Tx_0) = \text{dist}(Tx_0, A),$$

where $\text{dist}(A, B) := \inf\{d(x, y) : x \in A \text{ and } y \in B\}$ for any nonempty subsets A and B of X. Indeed, a classical best approximation theorem, due to Ky Fan [14], states that if K is a nonempty compact convex subset of a Banach space X and $T : K \to X$ is a single-valued continuous map, then there exists an element $x_0 \in K$ such that

$$d(x_0, Tx_0) = \text{dist}(Tx_0, K).$$

Later, this result has been generalized by many authors [2, 5, 6, 24–26, 32]. Despite the fact that the existence of an approximate solution is ensured by best approximation theorems, a natural question that arises in this direction is whether it is possible to guarantee the existence of an approximate solution that is optimal. In other words, if A and B are nonempty subsets of a normed linear space and $T : A \to B$ is a mapping, then the point to be mooted is whether one can find an element $x_0 \in A$ such that

$$d(x_0, Tx_0) = \min\{d(x, Tx) : x \in A\}.$$

An affirmative answer to this poser is provided by best proximity pair theorems. A best proximity pair theorem analyzes the conditions under which the optimization problem, namely

$$\min_{x \in A} d(x, Tx)$$

has a solution. Indeed, if T is a multifunction from A to B, then

$$d(x, Tx) \geq \text{dist}(A, B),$$

where $d(x, Tx) = \text{dist}(x, Tx) = \inf\{d(x, y) : y \in Tx\}$. So, the most optimal solution to the problem of minimizing the real-valued function $x \to d(x, Tx)$ over the domain A of the multifunction T will be the one for which the value $\text{dist}(A, B)$ is attained. In view of this standpoint, best proximity pair theorems are considered to expound the conditions that assert the existence of an element x_0 such that

$$d(x_0, Tx_0) = \text{dist}(A, B).$$

The pair (x_0, Tx_0) is called a *best proximity pair* of T and the point x_0 is called a *best proximity point* of T. If the mapping under consideration is a self-mapping, it may be observed that a best proximity pair theorem boils down to a fixed point theorem under certain suitable conditions. Because of the fact that

$$d(x, Tx) \geq \text{dist}(Tx, A) \geq \text{dist}(A, B), \quad \text{for all } x \in A,$$

an element x_0 satisfying the conclusion of a best proximity pair theorem is a best approximant but the refinement of the closeness between x_0 and its image Tx_0 is demanded in the case of best proximity pair theorems. For a detailed study on fixed point theory, one may refer [16, 17, 27, 30, 33].

Now, we will give some of the basic definitions which we use in this chapter.

Definition 1 Let X be a normed linear space over $\mathbb{F}$, where $\mathbb{F}$ is $\mathbb{R}$ or $\mathbb{C}$. The *closed unit ball* of X is defined as $\{x : x \in X,\ \|x\| \leq 1\}$ and is denoted by B_X. The *unit sphere* of X is defined as $\{x : x \in X,\ \|x\| = 1\}$ and is denoted by S_X.

Definition 2 A normed linear space is said to be a *Banach space* if the metric induced by the norm is a complete metric.

Definition 3 A normed linear space is said to be *rotund* or *strictly convex* or *strictly normed* if $\left\|\frac{x_1+x_2}{2}\right\| < 1$ whenever x_1 and x_2 are different points of S_X.

If the normed linear space X is strictly convex, then the norm $\|.\|$ of X is also called as *strictly convex norm*.

Example 1 [22] In $\mathbb{R}^n$, for $1 < p < \infty$, define $\|.\|_p$ by $\|x\|_p = \left\{\sum_{i=1}^n \mid x_i \mid^p\right\}^{\frac{1}{p}}$ where $x = (x_1, x_2, \ldots, x_n) \in \mathbb{R}^n$. Then $(\mathbb{R}^n, \|.\|_p)$ is strictly convex.

Definition 4 Let X be a nonzero normed linear space. Define a function $\delta_X : [0, 2] \to [0, 1]$ by the formula

$$\delta_X(\varepsilon) = \inf\left\{1 - \left\|\frac{1}{2}(x + y)\right\| : x, y \in S_X, \|x - y\| \geq \varepsilon\right\}$$

Then δ_X is called the *modulus of rotundity* or *modulus of convexity* of X. The space X is said to be *uniformly rotund* or *uniformly convex* if $\delta_X > 0$ whenever $0 < \varepsilon \leq 2$. Also note that, if X is uniformly convex, then X is strictly convex.

Remark 1 Suppose X is a strictly convex finite dimensional normed linear space. For $\varepsilon \in (0, 2]$, let $A_\varepsilon = \{(x, y) : x, y \in S_X$ and $\|x - y\| \geq \varepsilon\}$. Since norm is a continuous function, A_ε is a closed subset of the compact set $S_X \times S_X$. Hence, there exists $(x_0, y_0) \in A_\varepsilon$ such that $\delta_X(\varepsilon) = 1 - \|\frac{x_0+y_0}{2}\|$. Since X is strictly convex, $\left\|\frac{x_0+y_0}{2}\right\| < 1$. Therefore, $\delta_X(\varepsilon) > 0$, for $\varepsilon \in (0, 2]$. Hence, X is uniformly convex.

Example 2 [22] For $1 < p < \infty$, $l_p = \left\{x = \{x_n\}_{n\in\mathbb{N}} : \sum_{n\in\mathbb{N}} \mid x_n \mid^p < \infty\right\}$ is uniformly convex with respect to $\|.\|_p$, where $\|x\|_p = \left\{\sum_{n\in\mathbb{N}} \mid x_n \mid^p\right\}^{\frac{1}{p}}, x \in l_p$.

Definition 5 A nonempty subset K of a normed linear space X is said to be *boundedly compact* if $K \cap B[x, r]$ is compact for every $x \in X$ and $r > 0$, where $B[x, r]$ is the closed ball-centered at x and radius r.

Example 3 It is easy to see that every finite dimensional subspace of a normed linear space is boundedly compact.

If X is a normed linear space, then the notation X^* and the term "the dual space of X" always refer to the dual space of X with respect to the norm topology of X.

Definition 6 [22] Let X be a normed linear space. Then the topology for X induced by the collection

$$\mathfrak{S} = \left\{ f^{-1}(U) : f \in X^*, U \text{ is open in } \mathbb{F} \right\}, \quad \text{where } \mathbb{F} = \mathbb{R} \text{ or } \mathbb{C}$$

is called the *weak topology* of X or the X^* topology of X or the topology $\sigma(X, X^*)$.

Definition 7 Let X and Y be topological spaces. A *multivalued map* or *multifunction* or *set-valued map* T from X to Y denoted by $T : X \to 2^Y$ is defined to be a function which assigns to each element of $x \in X$ a nonempty subset Tx of Y. A fixed point of the multifunction $T : X \to 2^X$ will be a point $x \in X$ such that $x \in Tx$.

Example 4 Consider $X = \mathbb{R}^2$ with $\|.\|_\infty$ norm. Then the function $P_{B_X} : X \to B_X$ defined by $P_{B_X}(x) := \left\{ y \in B_X : \|x - y\| = \inf\limits_{z \in B_X} \|x - z\| \right\}$ is a multivalued function.

Definition 8 Let X and Y be topological spaces. Let $T : X \to 2^Y$ be a map. The map T is said to be *upper semi-continuous* (u.s.c) if $T^{-1}(A) := \{x \in X : Tx \cap A \neq \varnothing\}$ is closed in X whenever A is a closed subset of Y.

In case of $Y = \mathbb{R}$ and T is a single-valued map, we say that T is upper semi-continuous at $x \in X$, if $T(x) \geq \limsup\limits_{\alpha} T(x_\alpha) = \inf\limits_{\alpha} \sup\limits_{\alpha \leq \beta} T(x_\beta)$, whenever $\{x_\alpha : \alpha \in D\}$ is a net in X such that x_α converges to x, where D is a directed set and $f : D \to X$ is defined by $f(\alpha) = x_\alpha$, for $\alpha \in D$.

We say that T is *upper semi-continuous* on X if it is upper semi-continuous at each point of X.

Example 5 Let $X = \mathbb{R}^2$, $K = [0, 1] \times \{0\}$. Let $T : K \to 2^X$ be defined by $T(a, 0) = \{(0, 1)\}$; if $a \neq 0$ and $T(a, 0) =$ the line segment joining $(0, 1)$ and $(1, 0)$; if $a = 0$. Then T is upper semi-continuous.

Proposition 1 *Let A be a compact subset of a metric space (X, d). Then the metric projection $P_A : X \to 2^A$ defined as $P_A(x) = \{y \in A : d(x, y) = \text{dist}(x, A)\}$ is upper semi-continuous.*

Proof Let C be a nonempty closed subset of A. We claim that $P_A^{-1}(C)$ is a closed subset of X. Let $\{x_n\}$ be a sequence in $P_A^{-1}(C)$ such that x_n converges to x_0, for some $x_0 \in X$. Since $\{x_n\} \subseteq P_A^{-1}(C)$, for each $n \in \mathbb{N}$ there exists $y_n \in C$, such that $d(x_n, y_n) = \text{dist}(x_n, A)$.

As A is compact, $\{y_n\}$ has a subsequence, say $\{y_{n_k}\}$ such that y_{n_k} converges to y_0, for some $y_0 \in A$. Since the distance functions $d(.,.)$ and $\text{dist}(., A)$ are continuous, we have $d(x_{n_k}, y_{n_k})$ converges to $d(x_0, y_0)$ and $d(x_{n_k}, y_{n_k}) = \text{dist}(x_{n_k}, A)$ converges to $\text{dist}(x_0, A)$. Thus, $d(x_0, y_0) = \text{dist}(x_0, A)$. As C is a closed subset of the compact set A and $\{y_{n_k}\} \subseteq C$ such that y_{n_k} converges to y_0, $y_0 \in C$. Thus, there is a $y_0 \in C$ such that $d(x_0, y_0) = \text{dist}(x_0, A)$, that is, $x_0 \in P_A^{-1}(C)$. This proves our claim. □

Definition 9 Let X and Y be topological spaces. Let $T : X \to 2^Y$ be a map. The map T is said to be *lower semi-continuous* (l.s.c) if $T^{-1}(A) := \{x \in X : Tx \cap A \neq \varnothing\}$ is open in X whenever A is an open subset of Y.

In case of $Y = \mathbb{R}$ and T is a single-valued map, we say that T is lower semi-continuous at $x \in X$, if $T(x) \leq \liminf_{\alpha} T(x_\alpha) = \sup_{\alpha} \inf_{\alpha \leq \beta} T(x_\beta)$, whenever $\{x_\alpha : \alpha \in D\}$ is a net in X such that x_α converges to x, where D is a directed set and $f : D \to X$ is defined by $f(\alpha) = x_\alpha$, for $\alpha \in D$.

We say that T is *lower semi-continuous* on X if it is lower semi-continuous at each point of X. In this case, it is easy to see that T is l.s.c if and only if $-T = (-T)$ is u.s.c.

Proposition 2 *Suppose X is a topological space and $T : X \to \mathbb{R}$ is a single-valued map. Then the following statements are equivalent:*

(a) T is lower semi-continuous.
(b) $\{x \in X : T(x) > \alpha\}$ is open, for each $\alpha \in \mathbb{R}$.
(c) $\{x \in X : T(x) \leq \alpha\}$ is closed, for each $\alpha \in \mathbb{R}$.

Proof It is easy to see that (b) $\Leftrightarrow$ (c). Hence, it is enough to prove (a) $\Leftrightarrow$ (c).

Suppose $T : X \to \mathbb{R}$ is l.s.c. It is claimed that, for each $r \in \mathbb{R}$, $X_r = \{x \in X : T(x) \leq r\}$ is closed in X. Let $\{x_\alpha\}$ be a net in X_r such that x_α converges to $x_0 \in X$. Then $T(x_0) \leq \liminf_{\alpha} T(x_\alpha)$.

Since $x_\alpha \in X_r$, $T(x_\alpha) \leq r$, for all α. Thus, for each α, $\inf_{\alpha \leq \beta} T(x_\beta) \leq r$. Hence $T(x_0) \leq \liminf_{\alpha} T(x_\alpha) \leq r$. Therefore, $x_0 \in X_r$ and hence X_r is closed in X. This establishes (a) $\Rightarrow$ (c).

Conversely, suppose for each $r \in \mathbb{R}$, $X_r = \{x \in X : T(x) \leq r\}$ is closed in X. Then $X \backslash X_r$ is open in X. Let $\{x_\alpha\}$ be a net in X such that x_α converges to $x_0 \in X$.

Now, for every $\varepsilon > 0$, let $V_\varepsilon = \{x \in X : T(x) > T(x_0) - \varepsilon\}$. Then $x_0 \in V_\varepsilon$ and V_ε is open in X. Since x_α converges to x_0, there exists α_0 such that $x_\beta \in V_\varepsilon$, for all $\beta \geq \alpha_0$. Thus, $T(x_0) - \varepsilon \leq \inf_{\alpha_0 \leq \beta} T(x) \leq \liminf_{\alpha} T(x_\alpha)$ and hence $T(x_0) \leq \liminf_{\alpha} T(x_\alpha)$. This proves (c) $\Rightarrow$ (a). □

Remark 2 Suppose $T : X \to \mathbb{R}$ is a map on a topological space X. Then the following statements are equivalent:

(a) T is u.s.c.
(b) $\{x \in X : T(x) < \alpha\}$ is open in X, for each $\alpha \in \mathbb{R}$.
(c) $\{x \in X : T(x) \geq \alpha\}$ is closed in X, for each $\alpha \in \mathbb{R}$.

Suppose $T : X \to \mathbb{R}$ is a map on a topological space X. Then, it is easy to verify that T is continuous if and only if T is both lower semi-continuous and upper semi-continuous.

Proposition 3 *Let X be a normed linear space. Then the norm $\|.\|$ is weakly lower semi-continuous on X (that is, $\|.\|$ is l.s.c. with respect to the weak topology on X).*

Proof As norm is a continuous function, for each $\alpha \in \mathbb{R}$, the set $F_\alpha = \{x \in X : \|x\| \leq \alpha\}$ is a closed set in X. It is also clear that F_α is a convex set. Hence F_α is weakly closed. This proves that the norm function $\|.\|$ is weakly lower semi-continuous. □

Next, we prove a result which assures that a lower semi-continuous function defined on a weakly compact set attains its infimum.

Proposition 4 *Let K be a weakly compact subset of a Banach space X and let $f : K \to \mathbb{R}$ be a function such that f is weakly lower semi-continuous(that is, f is l.s.c with respect to the weak topology on X). Then there exists $x_0 \in K$ such that $f(x_0) = \inf_{x \in K} f(x)$.*

Proof Given that f is l.s.c. Then for each $\alpha \in \mathbb{R}$, $f^{-1}(\alpha, \infty)$ is a weakly open set in K and $K = \bigcup_{\alpha \in \mathbb{R}} f^{-1}(\alpha, \infty)$.

Since K is a weakly compact set, there exists $\alpha_1, \alpha_2, \ldots, \alpha_m$ such that $K \subseteq \bigcup_{i=1}^{m} f^{-1}(\alpha_i, \infty)$ and hence $K \subseteq f^{-1}(\alpha_0, \infty)$, where $\alpha_0 = \min\{\alpha_i : i = 1, 2, \ldots, m\}$.

Let $\beta = \inf_{x \in K} f(x)$. For $n \in \mathbb{N}$, there exists $x_n \in K$ such that $\beta \leq f(x_n) < \beta + \frac{1}{n}$.

Since K is weakly compact, every sequence in K has a subsequence, which converges weakly in K. Hence $\{x_n\} \subseteq K$ has a subsequence, say $\{x_{n_k}\}$ such that x_{n_k} converges weakly to x_0, for some $x_0 \in K$.

Since f is a lower semi-continuous function and x_{n_k} converges weakly to x_0, $f(x_0) \leq \liminf f(x_{n_k})$. But $\beta \leq f(x_{n_k}) \leq \beta + \frac{1}{n_k}$, hence $f(x_0) = \beta$. □

Proposition 5 *Let K be a nonempty weakly compact convex subset of a Banach space X and H be a nonempty bounded subset of X. Let $f : K \to \mathbb{R}$ be defined by $f(x) = r_x(H) = \sup\{\|x - y\| : y \in H\}$. Then f is a continuous function with respect to norm and f is a l.s.c function with respect to the weak topology on X.*

Proof Suppose $\{x_n\} \subset K$ such that x_n converges to x_0, for some $x_0 \in K$. Then for given $\varepsilon > 0$ there is a $N \in \mathbb{N}$ such that for $n \geq N$,

$$\begin{aligned}
\|x_n - x_0\| &< \varepsilon \\
|\ \|x_n - y\| - \|x_0 - y\|\ | \leq \|x_n - x_0\| &< \varepsilon \\
\|x_n - y\| - \varepsilon &< \|x_0 - y\| < \|x_n - y\| + \varepsilon \\
\sup_{y \in H}\{\|x_n - y\|\} - \varepsilon &\leq \sup_{y \in H} \|x_0 - y\| < \sup_{y \in H}\{\|x_n - y\|\} + \varepsilon \\
r_{x_n}(H) - \varepsilon &\leq r_{x_0}(H) \leq r_{x_n}(H) + \varepsilon \\
|\ r(x_n)(H) - r(x_0)(H)\ | &\leq \varepsilon \qquad (\text{Note } f(x) = r_x(H)).
\end{aligned}$$

Hence for $n \geq N$, we have $|\ f(x_n) - f(x_0)\ | \leq \varepsilon$. Thus, $f(x_n)$ converges to $f(x_0)$. This proves the continuity of $f(x) = r_{(x)}(H)$ with respect to the norm topology.

Since the norm $\|.\|$ is a convex function and K is a convex set, we get that f is a convex function. By the continuity of f, the set $F_\alpha = \{x \in K : f(x) \leq \alpha\}$ is a closed subset of K, for each $\alpha \in \mathbb{R}$. Also, it is easy to see that F_α is a convex subset of K. But for a convex set, weak closure and norm closure are the same. Thus, for every $\alpha \in \mathbb{R}$, F_α is weakly closed subset of K. Hence, f is a lower semi-continuous function with respect to the weak topology on X. □

2 Best Proximity Pair Theorems

Let A and B be nonempty subsets of a metric space X. We use the following notations in the sequel.

$$\begin{aligned}
\operatorname{dist}(A, B) &:= \inf\{d(a, b) : a \in A, b \in B\} \\
Prox(A, B) &:= \{(a, b) \in A \times B : d(a, b) = \operatorname{dist}(A, B)\} \\
A_0 &:= \{a \in A : d(a, b) = \operatorname{dist}(A, B) \text{ for some } b \in B\} \\
B_0 &:= \{b \in B : d(a, b) = \operatorname{dist}(A, B) \text{ for some } a \in A\}
\end{aligned}$$

Proposition 6 [3] *If A and B are nonempty subsets of a normed linear space X such that* $\operatorname{dist}(A, B) > 0$, *then $A_0 \subseteq Bd(A)$ and $B_0 \subseteq Bd(B)$ where $Bd(K)$ denotes the boundary of K for any $K \subseteq X$.*

Proof Let $x \in A_0$. Then, there exists $y \in B$ such that $d(x, y) = \operatorname{dist}(A, B)$. Since $\operatorname{dist}(A, B) > 0$, A and B are disjoint. Let $K = \{(1-t)x + ty : 0 \le t \le 1\}$. As K intersects both A and its complement $X \backslash A$, it must intersect the boundary of A. So, there exists $t_0 \in [0, 1]$ such that $z = (1-t_0)x + t_0 y \in Bd(A)$. To show $z = x$. It suffices to show $t_0 = 0$. Suppose $t_0 > 0$, then

$$\begin{aligned}
d(z, y) &= \|(1-t_0)x + t_0 y - y\| \\
&= (1-t_0)\|x - y\| \\
&= (1-t_0)\operatorname{dist}(A, B).
\end{aligned}$$

Thus, $d(z, y) < \operatorname{dist}(A, B)$, which is a contradiction. Hence $x = z$ and $A_0 \subseteq Bd(A)$. Similarly, we can prove $B_0 \subseteq Bd(B)$. □

Theorem 1 (Brouwer's Fixed Point Theorem) *Let B be the closed unit ball in $\mathbb{R}^n$. Then any continuous mapping $f : B \to B$ has at least one fixed point.*

Theorem 2 [13] *Let X be a Banach space and K be a nonempty compact convex subset of X. Let $\mathfrak{C}(K)$ be the family of all closed convex nonempty subsets of K. Then for any upper semi-continuous function $f : K \to \mathfrak{C}(K)$, there exists a point $x_0 \in K$ such that $x_0 \in f(x_0)$.*

The finite dimensional version of the above theorem is known as Kakutani's fixed point theorem for multifunctions.

Theorem 3 (Kakutani's Theorem) *Let X be a finite dimensional Banach space and K be a nonempty compact convex subset of X. Let $\mathfrak{C}(K)$ be the family of all closed convex nonempty subsets of K. Then for any upper semi-continuous function $f : K \to \mathfrak{C}(K)$, there exists a point $x_0 \in K$ such that $x_0 \in f(x_0)$.*

Remark 3 Let A and B be nonempty closed subsets of a normed linear space and $x \in A_0$. Then there exists $y \in B$ such that $d(x, y) = \text{dist}(A, B)$. This implies that $y \in B_0$ and $d(x, y) = \text{dist}(A, B) = \text{dist}(x, B) \leq \text{dist}(x, B_0) \leq d(x, y)$.

Theorem 4 [28] *Let A and B be nonempty compact convex subsets of a Banach space X and let $T : A \to B$ be a continuous function. Further, assume that $T(A_0) \subseteq B_0$. Then there exists an element $x \in A$ such that $d(x, Tx) = \text{dist}(A, B)$.*

Proof Consider the metric projection $P_A : X \to 2^A$ defined as $P_A(x) = \{a \in A : \|a - x\| = \text{dist}(x, A)\}$. As A is a nonempty compact convex set, for each $x \in X$, $P_A(x)$ is a nonempty closed, convex subset of A. By Proposition 1, P_A is an upper semi-continuous multivalued map.

Now, it is claimed that $P_A(Tx) \subseteq A_0$, for $x \in A_0$.

Let $y \in A$ be such that $y \in P_A(Tx)$. Then $\|Tx - y\| = \text{dist}(Tx, A)$. As $T(A_0) \subseteq B_0$, we have $Tx \in B_0$. There exists $a \in A$ such that $\|Tx - a\| = \text{dist}(A, B)$.

Now, $\text{dist}(Tx, A) = \|Tx - y\| \leq \|Tx - a\| = \text{dist}(A, B)$. Hence for $y \in A$ there exist $Tx \in B$, such that $\|Tx - y\| = \text{dist}(A, B)$. Thus, $y \in A_0$. Consequently, $P_A(Tx) \subseteq A_0$, for each $x \in A_0$.

Since A and B are compact sets, $A_0 \neq \emptyset$. Also as T is single valued, we have $P_A \circ T$ is a convex-valued multifunction. That is, for each $x \in A_0$, $P_A(Tx)$ is a closed convex subset of A_0.

By Theorem 2, there exists $x_0 \in A_0$ such that $x_0 \in P_A(Tx_0)$ and since $Tx_0 \in B_0$, we have $\|Tx_0 - a\| = \text{dist}(A, B)$, for some $a \in A$. But $\text{dist}(Tx_0, A) = \|x_0 - Tx_0\| \leq \|Tx_0 - a\|$. Thus $\|x_0 - Tx_0\| = \text{dist}(A, B)$. □

The above result has been further generalized by Kim et al., and Kim and Lee, for more details refer [18, 19]. In [4], Basha et al., obtained similar results.

3 Cyclic Contractions and Best Proximity Point Theorems

Definition 10 Let (X, d) be a metric space and $T : X \to X$ be a map such that $d(Tx, Ty) \leq kd(x, y)$, for every $x, y \in X$, where $k \in (0, 1)$. Then T is called a *contraction mapping* on X.

Notice that a contraction mapping is always continuous.

Theorem 5 (Cyclic Contraction Version of Banach Contraction Principle) [21] *Let A and B be nonempty closed subsets of a complete metric space (X, d), and $T : A \cup B \to A \cup B$ be a map satisfying:*

(i) $T(A) \subseteq B$ *and* $T(B) \subseteq A$;
(ii) for some $k \in (0, 1)$, $d(Tx, Ty) \leq kd(x, y)$, *for* $x \in A$, $y \in B$.

Then for any $x_0 \in A$, $x_n = T^n x_0 \to x$, *where* $x \in A \cap B$ *is such that* $Tx = x$. *Further* $d(x_n, x) \leq \frac{k^n}{1-k} d(x_1, x_0)$.

Proof Let $x_0 \in A$. Define the iterative sequence $\{x_n : n \in \mathbb{N} \cup \{0\}\}$ by $x_{n+1} = T(x_n)$, for $n \in \mathbb{N} \cup \{0\}$ (equivalently, $x_n = T^n(x_0)$, $n \in \mathbb{N}$). Let us prove $\{x_n\}$ is a Cauchy sequence. Now for $n \in \mathbb{N}$,

$$\begin{aligned} d(x_{n+1}, x_n) &\leq kd(x_n, x_{n-1}) \\ &\leq k^2 d(x_{n-1}, x_{n-2}) \\ d(x_{n+1}, x_n) &\leq k^n d(x_1, x_0) \end{aligned}$$

Therefore, for n and $m \in \mathbb{N}$, we have

$$\begin{aligned} d(x_n, x_{n+m}) &\leq d(x_n, x_{n+1}) + d(x_{n+1}, x_{n+2}) + ... + d(x_{n+m-1}, x_{n+m}) \\ &\leq k^n d(x_1, x_0) + k^{n+1} d(x_1, x_0) + \cdots + k^{n+m-1} d(x_1, x_0) \\ &= k^n d(x_1, x_0)[1 + k + k^2 + \cdots + k^{m-1}] \\ &< k^n d(x_1, x_0)[1 + k + k^2 + \cdots] \\ d(x_n, x_{n+m}) &\leq \frac{k^n}{1-k} d(x_1, x_0) \end{aligned}$$

As $k \in (0, 1)$, $\{x_n : n \in \mathbb{N} \cup \{0\}\}$ is a Cauchy sequence in the complete metric space X. Thus, $x_n \to x$, for some $x \in X$. Since every subsequence of a convergent sequence converges to the same limit, hence $x_{2n} \to x$ and $x_{2n-1} \to x$.

Notice that $\{x_{2n} : n \in \mathbb{N}\} \subseteq A$, $\{x_{2n-1} : n \in \mathbb{N}\} \subseteq B$, thus $x \in A \cap B$. Now

$$d(Tx, x_{2n}) \leq kd(x, x_{2n-1}).$$

Hence $x_{2n} \to Tx$. But the limit of a convergent sequence is unique, hence $Tx = x$.

Since $d(x_n, x_{n+m}) \leq \frac{k^n}{1-k} d(x_1, x_0)$, $d(x_n, x) = \lim_{m \to \infty} d(x_n, x_{n+m}) \leq \frac{k^n}{1-k} d(x_1, x_0)$. □

In case of $A = B = X$, the Banach contraction principle follows as a corollary to the above theorem.

Theorem 6 (Banach Contraction Principle) *Let* (X, d) *be a complete metric space and* $T : X \to X$ *be a contraction. Then* T *has a unique fixed point, say* x *in* X, *and for each* $x_0 \in X$ *the sequence of iterates* $\{x_n = T^n(x_0) : n \in \mathbb{N}\}$ *converges to the fixed point. Further* $d(x_n, x) \leq \frac{k^n}{1-k} d(x_1, x_0)$.

Suppose A and B are nonempty closed subsets of a metric space. Let T be a self-map on $A \cup B$ satisfying $T(A) \subseteq B$ and $T(B) \subseteq A$. It is to be noted that if T satisfies the contraction condition as given in Theorem 5, then $A \cap B \neq \varnothing$. In case of

$\text{dist}(A, B) > 0$, our aim is to find a contraction type condition which will guarantee the existence of a point $x \in A$ such that $d(x, Tx) = \text{dist}(A, B)$. Motivated by this, the authors in [9] introduced the following notion of cyclic contraction.

Definition 11 [9] Let A and B be nonempty subsets of a metric space X. A map $T : A \cup B \to A \cup B$ is said to be a *cyclic contraction map* if it satisfies:

(i) $T(A) \subseteq B$ and $T(B) \subseteq A$.
(ii) for some $k \in (0, 1)$, $d(Tx, Ty) \leq kd(x, y) + (1 - k)\text{dist}(A, B)$, for all $x \in A, y \in B$.

Since $\text{dist}(A, B) \leq d(x, y)$, for $x \in A$ and $y \in B$, $d(Tx, Ty) \leq d(x, y)$ for all $x \in A$, $y \in B$. Also note that the condition (ii) in the above definition can be written as $d(Tx, Ty) - \text{dist}(A, B) \leq k(d(x, y) - \text{dist}(A, B))$.

Example 6 In $(\mathbb{R}^2, \|.\|_2)$, let $A = \{(0, t) : 0 \leq t \leq 1\}$ and $B = \{(1, t) : 0 \leq t \leq 1\}$. Define $T : A \cup B \to A \cup B$ by $T(0, t) = \left(1, \frac{1-t}{2}\right)$ and $T(1, t) = \left(0, \frac{1-t}{2}\right)$. It is easy to see that $\text{dist}(A, B) = 1$, $T(A) \subseteq B$ and $T(B) \subseteq A$. Now, for $x = (0, t_1) \in A$ and $y = (1, t_2) \in B$,

$$
\begin{aligned}
\|Tx - Ty\|_2^2 &= \left\| \left(1, \frac{1 - t_1}{2}\right) - \left(0, \frac{1 - t_2}{2}\right) \right\|_2^2 \\
&= \left\| \left(1, \frac{t_2 - t_1}{2}\right) \right\|_2^2 \\
&= \frac{1}{4}\left(1 + (t_2 - t_1)^2\right) + \frac{1}{4} + \frac{1}{2} \\
&\leq \left\{\frac{1}{2}\sqrt{1 + (t_2 - t_1)^2}\right\}^2 + \frac{1}{4} + \frac{1}{2}\sqrt{1 + (t_2 - t_1)^2} \\
&= \left(\frac{1}{2}\sqrt{1 + (t_2 - t_1)^2} + \frac{1}{2}\right)^2 \\
&= \left(\frac{1}{2}\|(0, t_1) - (1, t_2)\|_2 + \frac{1}{2}\text{dist}(A, B)\right)^2 \\
\|Tx - Ty\|_2 &\leq \frac{1}{2}\|(0, t_1) - (1, t_2)\|_2 + \frac{1}{2}\text{dist}(A, B).
\end{aligned}
$$

Hence T is a cyclic contraction on $A \cup B$.

Proposition 7 [9] *Let A and B be nonempty subsets of a metric space X. Suppose $T : A \cup B \to A \cup B$ is a cyclic contraction map. Then for any $x_0 \in A \cup B$, $d(x_n, x_{n+1}) \to \text{dist}(A, B)$, where $x_{n+1} = Tx_n$, $n = 0, 1, 2, 3, \ldots$.*

Proof Fix $x_0 \in A \cup B$. Define $x_{n+1} = Tx_n$, where $n = 0, 1, 2, \ldots$. Now for $n \in \mathbb{N}$,

$$\begin{aligned} d(x_n, x_{n+1}) &= d(Tx_{n-1}, Tx_n) \\ &\leq kd(x_{n-1}, x_n) + (1-k)\text{dist}(A, B) \\ &\leq k[kd(x_{n-2}, x_{n-1}) + (1-k)\text{dist}(A, B)] + (1-k)\text{dist}(A, B) \\ &= k^2 d(x_{n-2}, x_{n-1}) + (1-k^2)\text{dist}(A, B). \end{aligned}$$

Hence $d(x_n, x_{n+1}) \leq k^n d(x_1, x_0) + (1-k^n)\text{dist}(A, B)$. Since $k \in (0, 1)$, we have $d(x_n, x_{n+1}) \to \text{dist}(A, B)$. □

Next, we give a simple existence result for a best proximity point.

Proposition 8 [9] *Let A and B be nonempty closed subsets of a complete metric space X. Let $T : A \cup B \to A \cup B$ be a cyclic contraction map and $x_0 \in A$. Define $x_{n+1} = Tx_n, n = 0, 1, 2, \ldots$. Suppose $\{x_{2n} : n \in \mathbb{N}\}$ has a convergent subsequence in A, then there exists $x \in A$ such that $d(x, Tx) = \text{dist}(A, B)$.*

Proof Let $\{x_{2n_k} : k \in \mathbb{N}\}$ be a subsequence of $\{x_{2n}\}$, which converges to some $x \in A$. Now

$$\text{dist}(A, B) \leq d(x, x_{2n_k - 1}) \leq d(x, x_{2n_k}) + d(x_{2n_k}, x_{2n_k - 1})$$

Since $d(x_{2n_k}, x_{2n_k-1}) \to \text{dist}(A, B)$ and $d(x, x_{2n_k}) \to 0$, $d(x, x_{2n_k-1}) \to \text{dist}(A, B)$. Also $\text{dist}(A, B) \leq d(x_{2n_k}, Tx) = d(Tx_{2n_k-1}, Tx) \leq d(x_{2n_k-1}, x)$. Thus, $d(x_{2n_k}, Tx) \to \text{dist}(A, B)$.

But, $\lim_{k\to\infty} d(x_{2n_k}, Tx) = d(x, Tx)$. Hence $d(x, Tx) = \text{dist}(A, B)$. □

The following proposition yields an existence result when one of the sets is boundedly compact.

Proposition 9 [9] *Let A and B be nonempty subsets of a metric space X. Suppose $T : A \cup B \to A \cup B$ is a cyclic contraction map. Then for $x_0 \in A \cup B$ and $x_{n+1} = Tx_n$, where $n = 0, 1, 2, \ldots$, the sequences $\{x_{2n}\}$ and $\{x_{2n+1}\}$ are bounded.*

Proof Suppose $x_0 \in A$. It is enough to prove either $\{x_{2n+1}\}$ or $\{x_{2n}\}$ is bounded. For, by Proposition 7, $d(x_{2n}, x_{2n+1})$ converges to $\text{dist}(A, B)$ and hence boundedness of one of the sequence will imply the boundedness of the other sequence.

Let us prove that $\{x_{2n+1}\}$ is bounded. That is to prove there exists $M > 0$, such that for all $n, m \in \mathbb{N}$, $d(x_{2n+1}, x_{2m+1}) \leq M$. Equivalently for every $y \in X$, there exist $M_y > 0$ such that $d(y, x_{2n+1}) < M_y$, for all $n \in \mathbb{N}$. But it is enough to prove this for some $x \in X$, then by triangle inequality the result follows for every $y \in X$.

It is claimed that there exists $M > 0$ such that $d(T^2x_0, x_{2m+1}) < M$ for all $m \in \mathbb{N}$.

Suppose $\{x_{2n+1}\}$ is not bounded. Let

$$M > \max\left\{\frac{2d(x_0, Tx_0)}{\frac{1}{k^2} - 1} + \text{dist}(A, B), d(T^2x_0, Tx_0)\right\}. \tag{1}$$

Then there exists $N \in \mathbb{N}$ such that $d\left(T^2x_0, T^{2N+1}(x_0)\right) > M$ and $d\left(T^2x_0, T^{2N-1}(x_0)\right) \leq M$. Now

$$\begin{aligned} M &< d\left(T^2x_0, T^{2N+1}(x_0)\right) \\ &\leq k\, d\left(Tx_0, T^{2N}(x_0)\right) + (1-k)\text{dist}(A, B) \\ &\leq k^2\, d\left(x_0, T^{2N-1}x_0\right) + (1-k^2)\text{dist}(A, B) \\ M - \text{dist}(A, B) &\leq k^2\left[d(x_0, T^{2N-1}x_0) - \text{dist}(A, B)\right] \end{aligned}$$

$$\begin{aligned} \frac{M - \text{dist}(A, B)}{k^2} + \text{dist}(A, B) &\leq d\left(x_0, T^{2N-1}x_0\right) \\ &\leq d(x_0, Tx_0) + d\left(Tx_0, T^2x_0\right) + d\left(T^2x_0, T^{2N-1}x_0\right) \\ \frac{M - \text{dist}(A, B)}{k^2} + \text{dist}(A, B) &\leq 2d(x_0, Tx_0) + M. \\ \left(\frac{1}{k^2} - 1\right)[M - \text{dist}(A, B)] &< 2d(x_0, Tx_0). \end{aligned}$$

Thus $M < \frac{2d(x_0, Tx_0)}{\frac{1}{k^2} - 1} + \text{dist}(A, B)$, which is a contradiction to (1). □

Theorem 7 [9] *Let A and B be nonempty closed subsets of a metric space* (X, d) *and let* $T : A \cup B \to A \cup B$ *be a cyclic contraction. If either A or B is boundedly compact, then there exists* $x \in A \cup B$ *such that* $d(x, Tx) = \text{dist}(A, B)$.

Proof Suppose A is boundedly compact. Fix $x_0 \in A$, let $x_n = Tx_{n-1}$, $n \in \mathbb{N}$. Now, by Proposition 7, $d(x_n, Tx_n) \to \text{dist}(A, B)$. Also from Proposition 9, the sequence $\{x_{2n} : n \in \mathbb{N}\} \subseteq A$ is bounded. Hence, the sequence $\{x_{2n} : n \in \mathbb{N}\}$ has a subsequence, say $\{x_{2n_k} : k \in \mathbb{N}\}$ which converges to $x_* \in A$. Thus, by Proposition 8, $\text{dist}(x_*, Tx_*) = \text{dist}(A, B)$. □

As every finite dimensional space is boundedly compact, we have the following result.

Corollary 1 [9] *Let A and B be nonempty closed subsets of a normed linear space X and let* $T : A \cup B \to A \cup B$ *be a cyclic contraction. If either the span of A or the span of B is a finite dimensional subspace of X, then there exists* $x \in A \cup B$ *such that* $d(x, Tx) = \text{dist}(A, B)$.

Proof Directly follows from the above theorem. □

Next we proceed to the result which gives the existence, uniqueness, and convergence of the iterative sequence to the unique best proximity point.

Lemma 1 [9] *Let A be a nonempty closed convex subset and B be a nonempty closed subset of a uniformly convex Banach space. Let* $\{x_n : n \in \mathbb{N}\}$ *and* $\{z_n : n \in \mathbb{N}\}$ *be sequences in A and* $\{y_n : n \in \mathbb{N}\}$ *be a sequence in B satisfying:*

(i) $\|z_n - y_n\| \to \text{dist}(A, B)$
(ii) *For every* $\varepsilon > 0$, *there exists* N_0 *such that for all* $m > n \geq N_0$, $\|x_m - y_n\| \leq \text{dist}(A, B) + \varepsilon$.

Then, for every $\varepsilon > 0$, *there exists* N_1 *such that for all* $m > n \geq N_1$, $\|x_m - z_n\| \leq \varepsilon$.

Proof Suppose there exists $\varepsilon_0 > 0$ such that for each $k \in \mathbb{N}$, there exists $m_k > n_k \geq k$, for which $\|x_{m_k} - z_{n_k}\| \geq \varepsilon_0$. Choose $\gamma \in (0, 1)$ such that $\varepsilon_0/\gamma > \text{dist}(A, B)$ and choose ε such that

$$0 < \varepsilon < \min\left\{\frac{\varepsilon_0}{\gamma} - \text{dist}(A, B), \frac{\text{dist}(A, B)\delta_X(\gamma)}{1 - \delta_X(\gamma)}\right\}.$$

As the modulus of convexity δ_X is strictly increasing and $\gamma < \frac{\varepsilon_0}{\text{dist}(A,B)+\varepsilon}$, we have $0 < \delta_X(\gamma) < \delta_X\left(\frac{\varepsilon_0}{\text{dist}(A,B)+\varepsilon}\right)$. Also from the choice of ε, we have

$$\begin{aligned}
\varepsilon &< \frac{\text{dist}(A, B)\delta_X(\gamma)}{1 - \delta_X(\gamma)} \\
(1 - \delta_X(\gamma))\varepsilon &< \text{dist}(A, B)\delta_X(\gamma) \\
&= [\delta_X(\gamma) - 1 + 1]\text{dist}(A, B) \\
(1 - \delta_X(\gamma))(\text{dist}(A, B) + \varepsilon) &< \text{dist}(A, B)
\end{aligned}$$

Now, by assumption (ii), for the chosen ε there exist N_1 such that for all $m_k > n_k \geq N_1$, $\|x_{m_k} - y_{n_k}\| \leq \text{dist}(A, B) + \varepsilon$. Also by assumption (i), there exists $N_2 \in \mathbb{N}$ such that for all $n_k \geq N_2$, $\|z_{n_k} - y_{n_k}\| \leq \text{dist}(A, B) + \varepsilon$. Let $N_0 = \max\{N_1, N_2\}$.

From the uniform convexity of X, for all $m_k > n_k \geq N_0$,

$$\begin{aligned}
\left\|\frac{x_{m_k} + z_{n_k}}{2} - y_{n_k}\right\| &\leq \left[1 - \delta_X\left(\frac{\varepsilon_0}{\text{dist}(A, B) + \varepsilon}\right)\right](\text{dist}(A, B) + \varepsilon) \\
&< [1 - \delta_X(\gamma)](\text{dist}(A, B) + \varepsilon) \\
&< \text{dist}(A, B).
\end{aligned}$$

As A is convex and $y_{n_k} \in B$, $\text{dist}(A, B) \leq \left\|\frac{x_{m_k}+z_{n_k}}{2} - y_{n_k}\right\|$. This gives a contradiction.

Hence for every $\varepsilon > 0$, there exists $N \in \mathbb{N}$ such that for all $m > n \geq N$, we have $\|x_m - z_n\| \leq \varepsilon$. □

Lemma 2 [9] *Let A be a nonempty closed convex subset and B be a nonempty closed subset of a uniformly convex Banach space X. Let $\{x_n\}$ and $\{z_n\}$ be sequences in A and $\{y_n\}$ be a sequence in B satisfying:*

(i) $\|x_n - y_n\| \to \text{dist}(A, B)$ *and*
(ii) $\|z_n - y_n\| \to \text{dist}(A, B)$.

Then $\|x_n - z_n\|$ *converges to zero.*

Proof Suppose $\|x_n - z_n\|$ does not converge to zero. Then there exists a $\varepsilon_0 > 0$ such that for every $k \in \mathbb{N}$, there exists $n_k \in \mathbb{N}$ such that $\|x_{n_k} - z_{n_k}\| \geq \varepsilon_0$. Choose $\gamma \in (0, 1)$ such that $\varepsilon_0/\gamma > \text{dist}(A, B)$ and choose ε such that

$$0 < \varepsilon < \min\left\{\frac{\varepsilon_0}{\gamma} - \text{dist}(A, B), \frac{\text{dist}(A, B)\delta_X(\gamma)}{1 - \delta_X(\gamma)}\right\}.$$

As the modulus of convexity δ_X is strictly increasing and $\gamma < \frac{\varepsilon_0}{\text{dist}(A,B)+\varepsilon}$, we have $0 < \delta_X(\gamma) < \delta_X\left(\frac{\varepsilon_0}{\text{dist}(A,B)+\varepsilon}\right)$. Also from the choice of ε, we have

$$\begin{aligned}
\varepsilon &< \frac{\text{dist}(A, B)\delta_X(\gamma)}{1 - \delta_X(\gamma)} \\
(1 - \delta_X(\gamma))\varepsilon &< \text{dist}(A, B)\delta_X(\gamma) \\
&= [\delta_X(\gamma) - 1 + 1]\text{dist}(A, B) \\
(1 - \delta_X(\gamma))(\text{dist}(A, B) + \varepsilon) &< \text{dist}(A, B)
\end{aligned}$$

From assumptions (i) and (ii), for the chosen ε there exists N_1 and $N_2 \in \mathbb{N}$ such that $\|x_n - y_n\| \leq \text{dist}(A, B) + \varepsilon$, for $n \geq N_1$ and $\|z_n - y_n\| \leq \text{dist}(A, B) + \varepsilon$, for $n \geq N_2$. Let $N_0 = \max\{N_1, N_2\}$. Then $\|x_{n_k} - y_{n_k}\| \leq \text{dist}(A, B) + \varepsilon$ and $\|z_{n_k} - y_{n_k}\| \leq \text{dist}(A, B) + \varepsilon$, for $n_k \geq N_0$.

Since X is uniformly convex, for $n_k \geq N_0$,

$$\begin{aligned}
\left\|\frac{x_{n_k} + z_{n_k}}{2} - y_{n_k}\right\| &\leq \left[1 - \delta_X(\frac{\varepsilon_0}{\text{dist}(A, B) + \varepsilon})\right](\text{dist}(A, B) + \varepsilon) \\
&\leq \left[1 - \delta_X(\gamma)\right](\text{dist}(A, B) + \varepsilon) \\
&< \text{dist}(A, B) \\
\left\|\frac{x_{n_k} + z_{n_k}}{2} - y_{n_k}\right\| &< \text{dist}(A, B)
\end{aligned}$$

As A is convex and $y_{n_k} \in B$, thus $\text{dist}(A, B) \leq \left\|\frac{x_{n_k}+z_{n_k}}{2} - y_{n_k}\right\|$. This gives a contradiction. Hence $\|x_n - z_n\| \to 0$. □

In view of the above lemmas, Suzuki et al. [29] introduced a notion of UC property and proved the existence of best proximity points. Espínola and Fernández-León [12] further generalized the UC property and shown the existence of best proximity points. Abkar and Gabeleh [1] proved the existence of best proximity points, even if the pair (A, B) does not satisfy the UC property. For more details refer [1, 12, 15, 23, 29] and the references there in.

Theorem 8 [7, 9] *Let A and B be nonempty closed convex subsets of a uniformly convex Banach space. Suppose $T : A \cup B \to A \cup B$ is a cyclic contraction map. Then T has a unique best proximity point in A (i.e., $\exists!\, x \in A$ such that $\|x - Tx\| =$*

dist(A, B)). *Further, if* $x_0 \in A$ *and* $x_{n+1} = Tx_n$, *then the sequence* $\{x_{2n}\}$ *converges to the best proximity point.*

Proof Fix $x_0 \in A$. Define the iterative sequence $\{x_n : n = 0, 1, 2, \ldots\}$ where $x_{n+1} = Tx_n, n = 0, 1, 2, \ldots$. By Proposition 7, $\|x_n - x_{n+1}\| = \|x_n - Tx_n\| \to \text{dist}(A, B)$ and $\|x_{n+1} - x_{n+2}\| = \left\|T^2(x_n) - Tx_n\right\| \to \text{dist}(A, B)$.

In particular $\|x_{2n} - x_{2n+1}\| = \|x_{2n} - Tx_{2n}\| \to \text{dist}(A, B)$ and $\|x_{2(n+1)} - x_{2n+1}\| = \left\|T^2(x_{2n}) - T(x_{2n})\right\| \to \text{dist}(A, B)$.

By Lemma 2, $\|x_{2n} - x_{2(n+1)}\| \to 0$. Similarly, we can show $\|Tx_{2n} - Tx_{2(n+1)}\| \to 0$ We want to prove $\{x_{2n}\}$ is a Cauchy sequence. It is enough to prove for every $\varepsilon > 0$ there exists N_0 such that for all $m > n \geq N_0$, $\|x_{2m} - Tx_{2n}\| \leq \text{dist}(A, B) + \varepsilon$. Then by Lemma 1, the result follows.

Suppose not, then there exists $\varepsilon > 0$ such that for all $k \in \mathbb{N}$, there exists $m_k > n_k \geq k$ for which $\|x_{2m_k} - Tx_{2n_k}\| \geq \text{dist}(A, B) + \varepsilon$, this m_k can be chosen such that it is the least integer greater than n_k to satisfy the above inequality. Now

$$\begin{aligned}
\text{dist}(A, B) + \varepsilon &\leq \|x_{2m_k} - Tx_{2n_k}\| \\
&\leq \|x_{2m_k} - x_{2(m_k-1)}\| + \|x_{2(m_k-1)} - Tx_{2n_k}\| \\
&\leq \|x_{2m_k} - x_{2(m_k-1)}\| + \text{dist}(A, B) + \varepsilon
\end{aligned}$$

Hence $\lim_{k\to\infty} \|x_{2m_k} - Tx_{2n_k}\| = \text{dist}(A, B) + \varepsilon$. Also,

$$\begin{aligned}
\|x_{2(m_k+1)} - Tx_{2(n_k+1)}\| &\leq k\|x_{2m_k+1} - Tx_{2n_k+1}\| + (1-k)\text{dist}(A, B) \\
&\leq k(k\|x_{2m_k} - Tx_{2n_k}\| + (1-k)\text{dist}(A, B)) + (1-k)\text{dist}(A, B) \\
\|x_{2(m_k+1)} - Tx_{2(n_k+1)}\| &\leq k^2\|x_{2m_k} - Tx_{2n_k}\| + (1-k^2)\text{dist}(A, B).
\end{aligned}$$

Since $\|x_{2n} - x_{2(n+1)}\| \to 0$, $\|Tx_{2n} - Tx_{2(n+1)}\| \to 0$, we have

$$\begin{aligned}
\|x_{2m_k} - Tx_{2n_k}\| &\leq \|x_{2m_k} - x_{2(m_k+1)}\| + \|x_{2(m_k+1)} - Tx_{2(n_k+1)}\| \\
&\quad + \|Tx_{2(n_k+1)} - Tx_{2n_k}\| \\
&\leq \|x_{2m_k} - x_{2(m_k+1)}\| + k^2\|x_{2m_k} - Tx_{2n_k}\| + (1-k^2)\text{dist}(A, B) \\
&\quad + \|Tx_{2(n_k+1)} - Tx_{2n_k}\|
\end{aligned}$$

Letting $n_k \to \infty$, we get

$$\text{dist}(A, B) + \varepsilon \leq k^2(\text{dist}(A, B) + \varepsilon) + (1-k^2)\text{dist}(A, B) = \text{dist}(A, B) + k^2\varepsilon$$

Since $k < 1$, the above inequality gives a contradiction. Thus, $\{x_{2n}\}$ is a Cauchy sequence in the closed subset A of the Banach space X, hence $x_{2n} \to x$, for some $x \in A$. From Proposition 8, it follows that $\|x - Tx\| = \text{dist}(A, B)$.

Suppose $x, y \in A$ are such that $x \neq y$ and $\|x - Tx\| = \text{dist}(A, B) = \|y - Ty\|$. Now $\left\|T^2x - Tx\right\| \leq \|x - Tx\| = \text{dist}(A, B)$, thus $\left\|T^2x - Tx\right\| = \text{dist}(A, B)$, similarly $\|T^2y - Ty\| = \text{dist}(A, B)$. Let $u_n = x$, $w_n = T^2x$ and $v_n = Tx, n \in \mathbb{N}$.

From Lemma 2, $T^2x = x$. Similarly, we can show $T^2y = y$. Therefore,

$$\|Tx - y\| = \left\|Tx - T^2y\right\| \le \|x - Ty\|,$$
$$\|Ty - x\| = \left\|Ty - T^2x\right\| \le \|y - Tx\|,$$

which implies $\|Ty - x\| = \|y - Tx\|$. Also note that, $\|y - Tx\| > \text{dist}(A, B)$.

Now,

$$\begin{aligned}\|Ty - x\| = \left\|Ty - T^2x\right\| &\le k\|y - Tx\| + (1-k)\text{dist}(A, B)\\ &< k\|y - Tx\| + (1-k)\|y - Tx\| = \|y - Tx\|.\end{aligned}$$

That is $\|Ty - x\| < \|y - Tx\|$, a contradiction. Hence $x = y$. □

Remark 4 In the above Theorem, if the convexity assumption is dropped, then the convergence and uniqueness is not guaranteed even in finite dimensional spaces. Consider $X = \mathbb{R}^4$, $A = \{e_1, e_3\}$ and $B = \{e_2, e_4\}$. Define $T(e_i) = e_{i+1}$, where $e_{4+i} = e_i$. It is easy to see T is cyclic contraction, but T does not have any best proximity point.

In [9], the authors have raised the following question, whether a best proximity point exists when A and B are nonempty closed and convex subsets of a reflexive Banach space? Some authors [1, 12, 29] have partially answered this question, for more details refer [1, 12, 29].

4 Relatively Nonexpansive Mappings and Best Proximity Point Theorems

We use the following notations in the sequel.

Let D and H be nonempty subset of a Banach space X. Define

(i) for $u \in X$, $\delta(u, D) = r_u(D) = \sup\{\|u - v\| : v \in D\}$ is called the radius of D relative to the point u;
(ii) $r(D) = \inf\{r_u(D) : u \in D\}$ is called the *Chebyshev radius* of D;
(iii) $C(D) = \{u \in D : r_u(D) = r(D)\}$ is the set of all *Chebyshev centers* of D;
(iv) $\delta(D) = \sup\{r_u(D) : u \in D\}$ is called the *diameter* of D;
(v) $\text{dist}(D, H) = \inf\{\|u - v\| : u \in D, v \in H\}$ is called the *distance* between D and H;
(vi) $\delta(D, H) = \sup\{\|u - v\| : u \in D, v \in H\}$.

Definition 12 [20] A convex subset K, with $\delta(K) > 0$, in a normed linear space X is said to have *normal structure*, if every bounded convex subset H of K with $\delta(H) > 0$ has a nondiametral point (i.e., there exists $x \in H$, such that $r_x(H) < \delta(H)$).

If every bounded convex subset with positive diameter has normal structure, then the space is said to have normal structure.

Example 7 It is easy to see that uniformly convex spaces have normal structure.

Definition 13 [20] Let K be a nonempty subset of a normed linear space X. A map $T : K \to X$ is said to be *nonexpansive*, if $\|Tx - Ty\| \leq \|x - y\|$, for all $x, y \in K$.

Definition 14 [31] Let K be a nonempty subset of a normed linear space X and $T : K \to K$ be a map. The set K is said to be a T—*regular set*, if $x \in K$, then $\frac{x+Tx}{2} \in K$.

Theorem 9 [31] *Let K be a nonempty weakly compact T—regular subset of a uniformly convex Banach space X and T be a nonexpansive map on K. Then T has a fixed point in K.*

Proof Let $\mathfrak{S} := \{F \subseteq K : F \text{ is nonempty and weakly closed}\}$ and $\mathfrak{F} := \{F \in \mathfrak{S} : T(F) \subseteq F \text{ and } F \text{ is } T\text{—}regular\}$. Clearly $\mathfrak{F}$ is nonempty, define $\leq$ on $\mathfrak{F}$ by $F_1 \leq F_2 \Leftrightarrow F_2 \subseteq F_1$. Then $(\mathfrak{F}, \leq)$ is a partially ordered set.

Suppose $\mathscr{T}$ is a totally ordered subset of $\mathfrak{F}$. Since $\mathscr{T}$ contains weakly compact subsets of K, it has finite intersection property. Thus, $F_0 = \bigcap_{F \in \mathcal{T}} F$ is a nonempty weakly closed subset of K. Notice that $T(F_0) \subseteq F_0$ and if $x \in F_0$, then $\frac{x+Tx}{2} \in F_0$. Thus $F_0 \in \mathfrak{F}$.

As every totally ordered subset of $\mathfrak{F}$ has a lower bound, by Zorn's lemma $\mathfrak{F}$ has a minimal element, say K_0. That is, K_0 is a minimal subset of K such that $T(K_0) \subseteq K_0$ and K_0 is T—regular.

It is claimed that K_0 is a singleton set.

Suppose $\delta(K_0) > 0$, this implies that for every $x \in K_0$, $Tx \neq x$. Let $F = \overline{co}(T(K_0)) \cap K_0$. Then, it is easy to see that F is a nonempty weakly closed T—*regular* subset of K_0 and $T(F) \subseteq F$. Hence the minimality of K_0 implies that $F = K_0$, that is $K_0 \subseteq \overline{co}(T(K_0))$.

It is easy to see that for every nonempty subset A of X,

$$r_x(A) = r_x(\overline{co}(A)), \text{ for } x \in X.$$

Hence,

$$r_x(T(K_0)) = r_x(\overline{co}(T(K_0))) = r_x(K_0), \quad \text{for all } x \in X. \tag{2}$$

Let $x_0 \in K_0$. Since $x_0 \neq Tx_0$, $\|x_0 - Tx_0\| = \varepsilon > 0$. Also as $x_0, Tx_0 \in K_0$, K_0 is T—regular and X is uniformly convex, there exists $\alpha \in (0, 1)$ such that $r_m(K_0) \leq \alpha R$, where $m = \frac{x_0+Tx_0}{2}$ and $R = \delta(K_0)$.

Let $F_0 = \{x \in K_0 : r_x(K_0) \leq \alpha R\}$. Then $m \in F_0$ and $F_0 \subsetneqq K_0$. As K_0 is weakly closed, Proposition 5 implies that F_0 is weakly closed. Now, it is claimed that $T(F_0) \subseteq F_0$. Let $x \in F_0$,

$$\begin{aligned} r_{Tx}(K_0) &= \sup\{\|Tx - z\| : z \in K_0\} \\ &= \sup\{\|Tx - Ty\| : y \in K_0\} \quad \text{(from(2))} \\ &\leq \sup\{\|x - y\| : y \in K_0\} \quad (T \text{ is nonexpansive}) \\ &\leq \alpha R. \end{aligned}$$

Therefore $Tx \in F_0$, whenever $x \in F_0$. It is easy to verify that $\frac{x+Tx}{2} \in F_0$, if $x \in F_0$. Hence F_0 is a T—regular set, $T(F_0) \subseteq F_0$ and $F_0 \subsetneqq K_0$. This contradicts the minimality of K_0. Thus $K_0 = \{x_0\}$, for some $x_0 \in K$ and $Tx_0 = x_0$. □

If K is convex, then K is always T—regular for any self-map on K. Hence as a corollary to the above theorem, we get the Browder-Kirk-Göhde fixed point theorem.

Theorem 10 [33] *Let K be a nonempty weakly compact convex subset of a uniformly convex Banach space X and let $T : K \to K$ be a nonexpansive map. Then T has a fixed point in K.*

The following theorem yields the existence of fixed points of nonexpansive mappings in Banach spaces.

Theorem 11 [20] *Let K be a nonempty weakly compact convex subset of a Banach space and $T : K \to K$ be a nonexpansive map. Suppose K has normal structure, then T has a fixed point in K.*

Definition 15 [8] Let A and B be nonempty subsets of a normed linear space X. A map $T : A \cup B \to X$ is said to be *relatively nonexpansive*, if $\|Tx - Ty\| \leq \|x - y\|$, for $x \in A$, $y \in B$.

Definition 16 Let A and B be nonempty subsets of a normed linear space such that $\text{dist}(A, B) > 0$ and $T : A \cup B \to A \cup B$ be a relatively nonexpansive map satisfying $T(A) \subseteq B$ and $T(B) \subseteq A$. A point $x \in A \cup B$ is said to be a *best proximity point* of T if $d(x, Tx) = \text{dist}(A, B)$.

Definition 17 [8] Let A and B be nonempty subsets of a normed linear space X. The pair (A, B) is said to be a *proximal pair*, if for every $(x, y) \in A \times B$ there exists $(x', y') \in A \times B$ such that $\|x - y'\| = \text{dist}(A, B) = \|x' - y\|$.

In [11], Espínola introduced a notion of proximinal pair, which we call as proximal pair. Espínola proved some interesting results about proximinal pair in strictly convex Banach space settings, for more details, see [11].

We say that the pair (A, B) has a property P, if both A and B have the property P.

Definition 18 [8] A convex pair (K_1, K_2) in a Banach space is said to have proximal normal structure, if $(H_1, H_2) \subseteq (K_1, K_2)$, is a bounded closed convex pair such that $\text{dist}(H_1, H_2) = \text{dist}(K_1, K_2)$ and $\delta(H_1, H_2) > \text{dist}(H_1, H_2)$, then there exists $(x_1, x_2) \in H_1 \times H_2$ such that $\delta(x_1, H_2) < \delta(H_1, H_2)$ and $\delta(x_2, H_1) < \delta(H_1, H_2)$.

A Banach space X is said to have *proximal normal structure*, if every convex pair in X has proximal normal structure. Note that the convex pair (K, K) has proximal normal structure if and only if K has normal structure.

Example 8 It is easy to see that every compact convex proximal pair has proximal normal structure.

Next we prove that, uniformly convex Banach spaces have proximal normal structure.

Proposition 10 *Every proximal pair (K_1, K_2) in a uniformly convex Banach space X has proximal normal structure.*

Proof Let $(H_1, H_2) \subseteq (K_1, K_2)$ be a nonempty bounded closed convex pair such that $d(H_1, H_2) = d(K_1, K_2)$ and $\delta(H_1, H_2) > d(H_1, H_2)$. It is claimed that there exists $(x_1, x_2) \in H_1 \times H_2$ such that $\delta(x_1, H_2) < \delta(H_1, H_2)$ and $\delta(x_2, H_1) < \delta(H_1, H_2)$.

Let $R = \delta(H_1, H_2)$ and $x, y \in H_1$ be such that $\|x - y\| \geq \varepsilon$, where $\varepsilon = \frac{\delta(H_1)}{2} > 0$.

Now, let $z \in H_2$, then $\|x - z\| \leq R$, $\|y - z\| \leq R$. Since $\|x - y\| = \varepsilon > 0$, by the uniform convexity of X, we have

$$\left\| z - \frac{x+y}{2} \right\| \leq R\left(1 - \delta_X\left(\frac{\varepsilon}{R}\right)\right)$$

$$\sup_{z \in H_2} \left\| z - \frac{x+y}{2} \right\| \leq R\left(1 - \delta_X\left(\frac{\varepsilon}{R}\right)\right)$$

$$< R \qquad \left(\text{Since } 0 < \left(1 - \delta_X\left(\frac{\varepsilon}{R}\right)\right) < 1\right).$$

Hence for $\frac{x+y}{2} \in H_1$, we get $\delta\left(\frac{x+y}{2}, H_2\right) < \delta(H_1, H_2)$. In a similar way, it can be shown that there exists $x \in H_2$, such that $\delta(x, H_1) < R$. □

Remark 5 Let A and B be nonempty subsets of a normed linear space X. A pair $(x, y) \in A \times B$ is said to be proximal in (A, B) if $d(x, y) = \|x - y\| = \text{dist}(A, B)$. We define $A_0 = \{x \in A : d(x, y') = \text{dist}(A, B) \text{ for some } y' \in B\}$ and $B_0 = \{y \in B : d(x', y) = \text{dist}(A, B) \text{ for some } x' \in A\}$. Then the pair (A_0, B_0) is a proximal pair obtained from (A, B).

Also it is easy to verify that, if A and B are nonempty weakly compact subsets, then A_0 and B_0 are nonempty weakly compact and $\text{dist}(A_0, B_0) = \text{dist}(A, B)$.

Lemma 3 *Let (A, B) be a nonempty weakly compact convex proximal pair in a Banach space. Let $T : A \cup B \to A \cup B$ be a relatively nonexpansive map such that $T(A) \subseteq B$ and $T(B) \subseteq A$. Further, suppose (K_1, K_2) is a nonempty weakly compact convex proximal pair, which is a subset of (A, B) such that* $\text{dist}(K_1, K_2) = \text{dist}(A, B) = d$ *and (K_1, K_2) is a minimal T invariant pair (i.e., there is no closed convex pair $(F_1, F_2) \subsetneqq (K_1, K_2)$ such that* $\text{dist}(F_1, F_2) = d$, *$T(F_1) \subseteq F_2$ and $T(F_2) \subseteq F_1$). Then*

(a) $\overline{co}(T(K_1)) = K_2$ *and* $\overline{co}(T(K_2)) = K_1$
(b) (K_1, K_2) *does not have proximal normal structure (i.e., for every* $(x, y) \in K_1 \times K_2$, *either* $r_x(K_2) = \delta(K_1, K_2)$ *or* $r_y(K_1) = \delta(K_1, K_2)$).

Proof Let $F_1 = \overline{co}(T(K_2))$ and $F_2 = \overline{co}(T(K_1))$. It is easy to see that (F_1, F_2) is a weakly compact convex subset of (K_1, K_2), and $\text{dist}(F_1, F_2) = d$. Now $T(F_1) \subseteq T(K_1) \subseteq F_2$, and similarly, $T(F_2) \subseteq T(K_2) \subseteq F_1$. But (K_1, K_2) is a minimal T invariant pair and $(F_1, F_2) \subseteq (K_1, K_2)$, hence $F_1 = K_1$ and $F_2 = K_2$. This establishes (a).

Let $\alpha = \inf_{y \in K_1} r_y(K_2)$, $\beta = \inf_{x \in K_2} r_x(K_1)$ and $r = \max\{\alpha, \beta\}$.

By Proposition 4 and 5, there exists $(x, y) \in K_1 \times K_2$ such that $r_x(K_2) = \alpha$ and $r_y(K_1) = \beta$. Since (K_1, K_2) is a proximal pair, there exists $(y', x') \in K_1 \times K_2$ such that $\|x - x'\| = \|y - y'\| = d$. Let $x_1 = \frac{x+y'}{2}$, $x_2 = \frac{y+x'}{2}$ and $R = \frac{(r+\delta(K_1,K_2))}{2}$.

Let $M_1 = \{x \in K_1 : r_x(K_2) \leq R\}$, and $M_2 = \{y \in K_2 : r_y(K_1) \leq R\}$. Then $x_1 \in M_1$, $x_2 \in M_2$ and $\|x_1 - x_2\| = d$ and the pair (M_1, M_2) is a nonempty closed convex subset of (K_1, K_2) such that $\text{dist}(M_1, M_2) = d$.

Since $K_1 = \overline{co}(T(K_2))$, for $x \in M_1$,

$$\begin{aligned} r_{Tx}(K_1) &= \sup\{\|Tx - y\| : y \in K_1\} \\ &= \sup\{\|Tx - Tz\| : z \in K_2\} \\ &\leq \sup\{\|x - z\| : z \in K_2\} \\ &\leq R. \end{aligned}$$

Thus $Tx \in M_2$, if $x \in M_1$. Hence $T(M_1) \subseteq M_2$ and in a similar way it follows that $T(M_2) \subseteq M_1$. Since (K_1, K_2) is minimal, $(M_1, M_2) = (K_1, K_2)$.

Now $\delta(K_1, K_2) = \sup_{x \in M_1 = K_1} r_x(K_2) = \sup_{y \in M_2 = K_2} r_y(K_1)$. But for $(x, y) \in M_1 \times M_2$, $r_x(K_2) \leq R$ and $r_y(K_1) \leq R$. This implies that $\delta(K_1, K_2) \leq R$ and hence $r = \delta(K_1, K_2)$. That is either $\alpha = \delta(K_1, K_2)$ or $\beta = \delta(K_1, K_2)$. Thus for every $(x, y) \in K_1 \times K_2$, either $r_x(K_2) = \delta(K_1, K_2)$ or $r_y(K_1) = \delta(K_1, K_2)$. This proves (b). □

Lemma 4 *Let* (A, B) *be a nonempty weakly compact convex proximal pair in a Banach space. Let* $T : A \cup B \to A \cup B$ *be a relatively nonexpansive map such that* $T(A) \subseteq A$ *and* $T(B) \subseteq B$. *Further, suppose* (K_1, K_2) *is a nonempty weakly compact convex proximal pair, which is a subset of* (A, B) *such that* $\text{dist}(K_1, K_2) = \text{dist}(A, B) = d$ *and* (K_1, K_2) *is a minimal* T *invariant pair (i.e., there is no closed convex pair* $(F_1, F_2) \subsetneq (K_1, K_2)$ *such that* $\text{dist}(F_1, F_2) = d$, $T(F_1) \subseteq F_1$ *and* $T(F_2) \subseteq F_2$). *Then*

(a) $\overline{co}(T(K_1)) = K_1$ *and* $\overline{co}(T(K_2)) = K_2$.
(b) (K_1, K_2) *does not have proximal normal structure (i.e., for every* $(x, y) \in K_1 \times K_2$, *either* $r_x(K_2) = \delta(K_1, K_2)$ *or* $r_y(K_1) = \delta(K_1, K_2)$).

Proof The proof follows exactly the same way as that of Lemma 3. □

The following theorem gives a sufficient condition for the existence of a best proximity point for a relatively nonexpansive map.

Theorem 12 [8] *Let (A, B) be a nonempty weakly compact convex pair in a Banach space. Let $T : A \cup B \to A \cup B$ be a relatively nonexpansive map such that $T(A) \subseteq B$ and $T(B) \subseteq A$. Suppose (A, B) has proximal normal structure, then there exists $(x, y) \in A \times B$ such that $\|x - Tx\| = \|y - Ty\| = \text{dist}(A, B)$.*

Proof Let $\text{dist}(A, B) = d$. Suppose (A_0, B_0) is the proximal pair obtained from (A, B). Then $\text{dist}(A_0, B_0) = d$, $T(A_0) \subseteq B_0$ and $T(B_0) \subseteq A_0$. For, $(x, y) \in A_0 \times B_0$ is such that $d(x, y) = d$, then $d(Tx, Ty) \leq d(x, y) \leq d$. Thus, $(Ty, Tx) \in A_0 \times B_0$.

Let $\mathfrak{S} = \{(H_1, H_2) \subseteq (A_0, B_0) : (H_1, H_2)$ is closed, convex and $d(H_1, H_2) = d\}$ and $\mathfrak{F} = \{(H_1, H_2) \in \mathfrak{S} : T(H_1) \subseteq H_2$ and $T(H_2) \subseteq H_1\}$. Then $(A_0, B_0) \in \mathfrak{F}$.

Define $\leq$ on $\mathfrak{F}$ by $(K_1, K_2) \leq (H_1, H_2) \Leftrightarrow (K_1, K_2) \subseteq (H_1, H_2)$. Then $(\mathfrak{F}, \leq)$ is a partially ordered set.

Let $\mathcal{T}$ be a totally ordered subset of $\mathfrak{F}$. As (A_0, B_0) is a weakly compact pair, then $\mathcal{T}$ contains weakly compact subsets of (A_0, B_0) and it is totally ordered. Hence $\mathcal{T}$ has finite intersection property. Thus, $F_1 = \bigcap_{(H_1, H_2) \in \mathcal{T}} H_1$ and $F_2 = \bigcap_{(H_1, H_2) \in \mathcal{T}} H_2$ are nonempty weakly compact convex subsets of A_0 and B_0, respectively and $T(F_1) \subseteq F_2$, $T(F_2) \subseteq F_1$.

Also for every $(H_1, H_2) \in \mathcal{T}$, let $(x_{H_1}, y_{H_2}) \in H_1 \times H_2$ be such that $\|x_{H_1} - y_{H_2}\| = d$. Then $\{x_{H_1} : (H_1, H_2) \in \mathcal{T}\}$ and $\{y_{H_2} : (H_1, H_2) \in \mathcal{T}\}$ are nets in the weakly compact subsets A_0 and B_0 respectively. It is possible to choose weakly convergent subnets $\{x_\alpha\}$ and $\{y_\alpha\}$ (with the same indices) such that weak-$\lim_\alpha x_\alpha = x_0$, for some $x_0 \in A_0$ and weak-$\lim_\alpha y_\alpha = y_0$, for some $y_0 \in B_0$. Then clearly $x_0 \in F_1$ and $y_0 \in F_2$.

By weak lower semicontinuity of the norm,

$$\|x_0 - y_0\| \leq \text{dist}(A_0, B_0) = d;$$

hence, $\text{dist}(F_1, F_2) = d$. Thus, $(F_1, F_2) \in \mathfrak{F}$.

That is every totally ordered subset of $\mathfrak{F}$ has a lower bound. Hence by Zorn's lemma $\mathfrak{F}$ has a miniimal element, say (K_1, K_2). Note that $\text{dist}(K_1, K_2) = d$.

It is claimed that $\delta(K_1, K_2) = d$. Suppose $\delta(K_1, K_2) > d$. Now from Lemma 3, the pair (K_1, K_2) satisfies the following:

(i) $\overline{co}(T(K_2)) = K_1$ and $\overline{co}(T(K_1)) = K_2$.
(ii) (K_1, K_2) does not have proximal normal structure.

But (K_1, K_2) has proximal normal structure. Hence $\delta(K_1, K_2) = d$. That is for every $x \in K_1 \cup K_2$, $\|x - Tx\| = d$. □

Theorem 13 [8] *Let (A, B) be a nonempty weakly compact convex pair in a strictly convex Banach space. Let $T : A \cup B \to A \cup B$ be a relatively nonexpansive* *map such that $T(A) \subseteq A$ and $T(B) \subseteq B$. Suppose (A, B) has proximal normal structure, then there exists $(x, y) \in A \times B$ such that $x = Tx$, $y = Ty$, and $\|Tx - Ty\| = \|x - y\| = \operatorname{dist}(A, B)$.*

Proof It is easy to see that $T(A_0) \subseteq A_0$ and $T(B_0) \subseteq B_0$, where (A_0, B_0) is the proximal pair obtained from the pair (A, B).

Let $\mathfrak{S} = \{(H_1, H_2) \subseteq (A_0, B_0) : (H_1, H_2)$ is closed, convex and $\operatorname{dist}(H_1, H_2) = d\}$ and $\mathfrak{F} = \{(H_1, H_2) \in \mathfrak{S} : T(H_1) \subseteq H_1$ and $T(H_2) \subseteq H_2\}$. Then $(A_0, B_0) \in \mathfrak{F}$.

Define $\leq$ on $\mathfrak{F}$ by $(K_1, K_2) \leq (H_1, H_2) \Leftrightarrow (K_1, K_2) \subseteq (H_1, H_2)$. Then $(\mathfrak{F}, \leq)$ is a partially ordered set. Let $\mathcal{T}$ be a totally ordered subset of $\mathfrak{F}$.

Now $\mathcal{T}$ contains weakly compact subsets of (A_0, B_0) and it is totally ordered. Hence, $\mathcal{T}$ has finite intersection property. Thus, $F_1 = \bigcap\limits_{(H_1, H_2) \in \mathcal{T}} H_1$ and $F_2 = \bigcap\limits_{(H_1, H_2) \in \mathcal{T}} H_2$ are nonempty weakly compact convex subsets of A_0 and B_0, respectively. Now, it is easy to verify that $T(F_1) \subseteq F_1$, $T(F_2) \subseteq F_2$ and $\operatorname{dist}(F_1, F_2) = d$. Thus $(F_1, F_2) \in \mathfrak{F}$.

That is every totally ordered subset of $\mathfrak{F}$ has a lower bound. Hence by Zorn's lemma $\mathfrak{F}$ has a minimal element, say (K_1, K_2). Note that $\operatorname{dist}(K_1, K_2) = d$.

It is claimed that $\delta(K_1, K_2) = d$.

Suppose $\delta(K_1, K_2) > d$. Now from Lemma 4, the pair (K_1, K_2) satisfies the following:

(i) $\overline{co}(T(K_1)) = K_1$ and $\overline{co}(T(K_2)) = K_2$.
(ii) (K_1, K_2) does not have proximal normal structure.

But (K_1, K_2) has proximal normal structure. Hence $\delta(K_1, K_2) = d$. Also the strict convexity of X implies that K_1 and K_2 are singleton sets. Hence for every $x \in K_1 \cup K_2$, $Tx = x$. □

Next, we show that Kransnosel'skíi's iteration process yields a convergence result if X is uniformly convex and the relatively nonexpansive map $T : A \cup B \to A \cup B$ satisfies $T(A) \subseteq A$ and $T(B) \subseteq B$. We assume that (A_0, B_0) is the proximal pair obtained from (A, B).

Theorem 14 [8] *Let A and B be nonempty bounded closed convex subsets of a uniformly convex Banach space. Suppose $T : A \cup B \to A \cup B$ is a relatively nonexpansive map such that $T(A) \subseteq A$ and $T(B) \subseteq B$. Let $x_0 \in A_0$. Define $x_{n+1} = \frac{(x_n + Tx_n)}{2}$, $n = 1, 2, \ldots$. Then $\lim\limits_{n} \|x_n - Tx_n\| = 0$.*

Moreover, if $\overline{T(A)}$ is a compact set, then $\{x_n\}$ converges to a fixed point of T.

Proof By Theorem 13, there exists $y \in B_0$ such that $Ty = y$. Since

$$\begin{aligned}\|x_{n+1} - y\| &= \left\| \frac{(x_n + Tx_n)}{2} - \frac{(y + Ty)}{2} \right\| \\ &\leq \frac{\|x_n - y\|}{2} + \frac{\|Tx_n - Ty\|}{2} \\ &\leq \frac{\|x_n - y\|}{2} + \frac{\|x_n - y\|}{2} \\ \|x_{n+1} - y\| &\leq \|x_n - y\|\end{aligned}$$

$\{\|x_n - y\|\}$ is a nonincreasing sequence. Hence

$$\lim_n \|x_n - y\| \text{ exists, say } \lim \|x_n - y\| = r \geq 0. \tag{3}$$

Suppose $r = 0$, then $\{x_n\}$ converges to the fixed point y. Also

$$\begin{aligned}\|Tx_n - y\| &\leq \|2x_{n+1} - x_n - y\| \\ &\leq 2\|x_{n+1} - y\| + \|x_n - y\| \\ \|Tx_n - y\| &\to 0.\end{aligned}$$

Hence $\|x_n - Tx_n\| \to 0$.

Now, consider the case $r > 0$. Suppose $\|x_n - Tx_n\|$ does not converge to zero. Then there exists an $\varepsilon_0 > 0$ and a subsequence $\{x_{n_k}\}$ of $\{x_n\}$ such that $\|x_{n_k} - Tx_{n_k}\| \geq \varepsilon_0$, for all $k \in \mathbb{N}$.

Choose $\gamma \in (0, 1)$ such that $\varepsilon_0/\gamma > r$ and choose ε such that

$$0 < \varepsilon < \min\left\{\frac{\varepsilon_0}{\gamma} - r, \frac{r\delta_X(\gamma)}{1 - \delta_X(\gamma)}\right\}.$$

As X is uniformly convex, the modulus of convexity $\delta_X(.)$ is strictly increasing. Since $0 < \gamma < \frac{\varepsilon_0}{r+\varepsilon}$, $0 < \delta_X(\gamma) < \delta_X\left(\frac{\varepsilon_0}{r+\varepsilon}\right)$. Also from the choice of ε,

$$\begin{aligned}\varepsilon &< \frac{r\delta_X(\gamma)}{1 - \delta_X(\gamma)} \\ (1 - \delta_X(\gamma))\varepsilon &< r\delta_X(\gamma) \\ &= [\delta_X(\gamma) - 1 + 1]r \\ (1 - \delta_X(\gamma))(r + \varepsilon) &< r\end{aligned}$$

Therefore, $[1 - \delta_X(\frac{\varepsilon_0}{r+\varepsilon})](r + \varepsilon) < r$. As $\|x_n - y\| \to r$, choose $N \in \mathbb{N}$ such that $\|x_n - y\| \leq r + \varepsilon$, for $n \geq N$. Also as $Ty = y$, hence for $n \geq N$, $\|Tx_n - y\| = \|Tx_n - Ty\| \leq \|x_n - y\| \leq r + \varepsilon$. Therefore, for every $n_k \geq N$, $\|x_{n_k} - y\| \leq r + \varepsilon$, $\|Tx_{n_k} - y\| \leq r + \varepsilon$ and $\|x_{n_k} - Tx_{n_k}\| \geq \varepsilon_0$. Hence from the uniform convexity of X,

$$\begin{aligned}\|y - x_{n_k+1}\| &= \left\| y - \frac{x_{n_k} + Tx_{n_k}}{2} \right\| \\ &\leq \left(1 - \delta_X\left(\frac{\varepsilon_0}{r+\varepsilon}\right)\right)(r+\varepsilon) \\ &< r\end{aligned}$$

This contradicts to $\|x_n - y\| \geq r$, for $n \in \mathbb{N}$. Hence $\|x_n - Tx_n\| \to 0$.

If $\overline{T(A)}$ is a compact set, then $\{Tx_n\}$ has a subsequence $\{Tx_{n_k}\}$ which converges to a point $z \in \overline{T(A)}$. Also as $\|x_{n_k} - Tx_{n_k}\| \to 0$, $x_{n_k} \to z$.

Let $d = \text{dist}(A, B)$. Choose $w \in B_0$ such that $\|z - w\| = d$. Since $\|x_{n_k} - w\| \to \|z - w\|$, $\|Tx_{n_k} - Tw\| \to \|z - Tw\|$ and $\|Tx_{n_k} - Tw\| \leq \|x_{n_k} - w\|$, thus $\|z - Tw\| = d$.

As X is a strictly convex space, $Tw = w$. Similarly $d \leq \|Tz - w\| = \|Tz - Tw\| \leq \|z - w\| = d$, thus $Tz = z$.

Since $\|x_{n_k} - w\| \to d$ and $\{\|x_n - w\|\}$ is nonincreasing, $\|x_n - w\| \to d$. Thus, by Lemma 2, $\|x_n - z\| \to 0$. □

Proposition 11 *Suppose A is a nonempty closed convex subset of a real Hilbert space X. For any $x \in X$, let $P_A x$ denote the unique point of A for which $\|x - P_A x\| = \text{dist}(x, A)$, where $\text{dist}(x, A) = \inf\{\|x - y\| : y \in A\}$. Then for any $z \in A$,*

$$\langle z - P_A(x), P_A(x) - x \rangle \geq 0. \tag{4}$$

Proof Let $y = P_A(x)$. For any $z \in A \backslash \{y\}$, we have $\|x - y\| < \|x - z\|$. Fix $z \in A \backslash \{y\}$, and let $w = (1 - r)y + rz$, for $0 < r < 1$. Now

$$\begin{aligned}\|x - w\|^2 &= \|x - y + r(y - z)\|^2 \\ &= \langle x - y + r(y - z), x - y + r(y - z) \rangle \\ &= \langle x - y, x - y \rangle + r\langle x - y, y - z \rangle \\ &\quad + r\langle y - z, x - y \rangle + r^2\langle y - z, y - z \rangle \\ &= \|x - y\|^2 + r^2\|y - z\|^2 + 2r\langle x - y, y - z \rangle \\ \|x - w\|^2 - \|x - y\|^2 - r^2\|y - z\|^2 &= 2r\langle x - y, y - z \rangle\end{aligned}$$

As $w \neq y$, $\varepsilon = \|x - w\|^2 - \|x - y\|^2 > 0$. Suppose $\varepsilon > \|y - z\|^2$, then clearly $0 < 2r\langle x - y, y - z \rangle$, and hence, $0 < \langle x - y, y - z \rangle = \langle z - y, y - x \rangle$. Suppose $\varepsilon \leq \|y - z\|^2$, then choose $r \in (0, 1)$ such that $r^2\|y - z\|^2 < \varepsilon$ this implies that $0 < 2r\langle x - y, y - z \rangle$, and hence, $0 < \langle x - y, y - z \rangle = \langle z - y, y - x \rangle$. □

The following observation provides an example of a relatively nonexpansive mapping.

Theorem 15 [8] *Let A and B be nonempty closed and convex subsets of a real Hilbert space X. Let $P : A \cup B \to A \cup B$ be the restriction of P_B on A and the restriction of P_A on B. Then $P(A) \subseteq B$, $P(B) \subseteq A$ and $\|Px - Py\| \leq \|x - y\|$, for all $x \in A$ and $y \in B$.*

Proof Suppose $x \in A$ and $y \in B$. Then from Eq. (4) in Proposition 11,

$$\langle y - P_B(x), P_B(x) - x\rangle \geq 0, \quad \langle x - P_A(y), P_A(y) - y\rangle \geq 0$$

Adding the above two terms, we have

$$\langle y - P_B(x), P_B(x) - x\rangle - \langle x - P_A(y), y - P_A(y)\rangle \geq 0.$$

Note that

$$\langle x_1, y_1\rangle - \langle x_2, y_2\rangle = \langle x_1, y_1 - y_2\rangle + \langle x_1 - x_2, y_2\rangle = \langle x_1 - x_2, y_1\rangle + \langle x_2, y_1 - y_2\rangle. \quad (5)$$

Let $x_1 = y - P_B(x)$, $y_1 = P_B(x) - x$, $x_2 = x - P_A(y)$ and $y_2 = y - P_A(y)$. Now as $\langle x_1, y_1\rangle - \langle x_2, y_2\rangle \geq 0$ and from (5), we have

$$\langle y - P_B(x), P_B(x) + P_A(y) - (x + y)\rangle + \langle y - x + P_A(y) - P_B(x), y - P_A(y)\rangle \geq 0.$$

Similarly,

$$\langle y - x + P_A(y) - P_B(x), P_B(x) - x\rangle + \langle x - P_A(y), P_B(x) + P_A(y) - (x + y)\rangle \geq 0.$$

Adding the above two inequality, we get

$$\begin{aligned} &\langle (P_A(y) + P_B(x)) - (x + y), (x + y) - (P_B(x) + P_A(y))\rangle \\ &+\langle y - x + P_A(y) - P_B(x), y - x + P_B(x) - P_A(y)\rangle \geq 0. \end{aligned} \quad (6)$$

Consider the second-term in the above Eq. (6),

$$\begin{aligned} &\langle y - x + P_A(y) - P_B(x), y - x + P_B(x) - P_A(y)\rangle \\ &\quad = \langle y - x, y - x\rangle + \langle P_A(y) - P_B(x), P_B(x) - P_A(y)\rangle \\ &\qquad +\langle y - x, P_B(x) - P_A(y)\rangle + \langle P_A(y) - P_B(x), y - x\rangle \\ &\quad = \|y - x\|^2 - \langle P_A(y) - P_B(x), P_A(y) - P_B(x)\rangle \\ &\qquad +\langle y - x, P_B(x) - P_A(y)\rangle - \langle y - x, P_B(x) - P_A(y)\rangle \\ &\quad = \|y - x\|^2 - \|P_A(y) - P_B(x)\|^2. \end{aligned}$$

From the first-term of Eq. (6),

$$\begin{aligned} &\langle (P_A(y) + P_B(x)) - (x + y), (x + y) - (P_B(x) + P_A(y))\rangle \\ &\quad = -\langle ((x + y) - (P_A(y) + P_B(x)), (x + y) - (P_B(x) + P_A(y))\rangle \\ &\quad = -\|(x + y) - (P_A(y) + P_B(x)\|^2. \end{aligned}$$

Thus, $\|y - x\|^2 - \|P_A(y) - P_B(x)\|^2 - \|(x + y) - (P_A(y) + P_B(x))\|^2 \geq 0$. Therefore,

$$\begin{aligned}\|P_A(y) - P_B(x)\|^2 &\leq \|y - x\|^2 - \|(x + y) - (P_A(y) + P_B(x))\|^2 \\ &\leq \|y - x\|^2 \\ \|P_A(y) - P_B(x)\|^2 &\leq \|y - x\|^2.\end{aligned}$$

Hence, $\|P_A(y) - P_B(x)\| \leq \|y - x\|$. □

Remark 6 Suppose A and B are closed convex subsets of a real Hilbert space and $T : A \cup B \to A \cup B$ is a relatively nonexpansive map satisfying $T(A) \subseteq A$ and $T(B) \subseteq B$. Define $U : A \cup B \to A \cup B$ by setting $Ux = P_B Tx$, if $x \in A$ and $Uy = P_A Ty$, if $y \in B$. Then by Theorem 15, U is a relatively non-expansive map. By Theorem 12, there exists $x_0 \in A_0$ such that $\|x_0 - Ux_0\| = \|x_0 - P_B Tx_0\| = \text{dist}(A, B)$. Since $\|x_0 - P_B x_0\| = \text{dist}(A, B)$, and T is relatively nonexpansive $\text{dist}(A, B) \leq \|Tx_0 - TP_B x_0\| \leq \|x_0 - P_B x_0\| = \text{dist}(A, B)$. But $\|Tx_0 - P_B Tx_0\| = \text{dist}(A, B)$. As the best approximant is unique, we have $P_B Tx_0 = TP_B x_0$. Thus $\|x_0 - P_B Tx_0\| = \|Tx_0 - P_B Tx_0\| = \text{dist}(A, B)$ again by the uniqueness of the best approximant, $x_0 = Tx_0$.

Theorem 16 [8] *Let A and B be nonempty bounded closed convex subsets of a real Hilbert space such that $A = A_0$ and $B = B_0$. Suppose $T : A \cup B \to A \cup B$ is a relatively nonexpansive map satisfying $T(A) \subseteq A$ and $T(B) \subseteq B$. Then T is nonexpansive on $A \cup B$.*

Proof From the above observation, we have

$$T(P_B(u)) = P_B(Tu), \quad \text{for all } u \in A. \tag{7}$$

We claim that for $x \in A$, $P_A P_B(x) = x$. Since best approximation is unique and $\|P_A P_B(x) - P_B(x)\| = \text{dist}(A, B) = \|x - P_B(x)\|$, thus $P_A P_B(x) = x$. Similarly, for $y \in B$, $P_B P_A(y) = y$. Therefore for $x \in A$ and $y \in B$, $\|x - y\| = \|P_A P_B(x) - P_B P_A(y)\|$. From Theorem 15,

$$\begin{aligned}\|P_A P_B(x) - P_B P_A(y)\| &\leq \|P_B(x) - P_A(y)\| \\ &\leq \|x - y\|.\end{aligned}$$

Hence $\|P_A(y) - P_B(x)\| = \|x - y\|$. From Eq. (6) in Theorem 15, we get $\|x + y - (P_B(x) + P_A(y))\| = 0$. This implies that

$$x - P_A(y) = P_B(x) - y. \tag{8}$$

Let $u, v \in A$ and $x = Tu$ and $y = P_B(Tv)$. From Eq. (8), $Tu - Tv = P_B(Tu) - P_B(Tv)$ and by Parallelogram law,

$$\begin{aligned}2(\|Tu - Tv\|^2 + \|Tv - P_B(Tv)\|^2) &= \|Tv - P_B(Tu)\|^2 + \|Tu - P_B(Tv)\|^2\\ &\leq \|Tv - T(P_Bu)\|^2 + \|Tu - T(P_Bv)\|^2 \text{ (By (7))}\\ &\leq \|v - P_Bu\|^2 + \|u - P_Bv\|^2\\ &= 2\left(\|u - v\|^2 + \|v - P_Bv\|^2\right).\end{aligned}$$

Since $\|Tv - P_BTv\| = \|v - P_Bv\| = \text{dist}(A, B)$, therefore, $\|Tu - Tv\| \leq \|u - v\|$. □

Theorem 16 in conjunction with the fixed point theorem for nonexpansive mappings immediately ensures the existence of a fixed point x_0 of T in A and the unique point $y_0 \in B$ which is nearest to x_0 satisfies $Ty_0 = y_0$ and $\|x_0 - y_0\| = \text{dist}(A, B)$.

5 Applications of Best Proximity Point Theorems

Let $S = \left\{(x, y) \in \mathbb{R}^2 :| x - x_0 |\leq a, | y - y_0 |\leq b\right\}$, for some $a, b > 0$ and $(x_0, y_0) \in \mathbb{R}^2$. Suppose (x, y_1) and (x, y_2) are two points in S, let $f(x, y)$ and $g(x, y)$ be real-valued functions defined on S.

Consider the following system of differential equations.

$$\frac{\mathrm{d}y}{\mathrm{d}x} = f(x, y) \quad y(x_0) = y_2, \tag{9}$$

$$\frac{\mathrm{d}y}{\mathrm{d}x} = g(x, y) \quad y(x_0) = y_1. \tag{10}$$

Clearly, it does have a solution when $f = g$ and $y_1 = y_2$. Suppose $y_1 \neq y_2$ and $f \neq g$. Define $C_a = \{y \in C[x_0 - a, x_0 + a] :| y(x_0) - y_0 |\leq b\}$, $A = \{y \in C_a : y(x_0) = y_2\}$ and $B = \{y \in C_a : y(x_0) = y_1\}$. Then for any $y \in A$ and $z \in B$, $\|y - z\| \geq| y_1 - y_2 |$ and $\text{dist}(A, B) =| y_1 - y_2 |$.

Let $T : A \cup B \to X$ be defined as

$$T(y(x)) = y_1 + \int_{x_0}^{x} g(t, y(t))\mathrm{d}t, \quad y \in A,$$

$$T(z(x)) = y_2 + \int_{x_0}^{x} f(t, z(t))\mathrm{d}t, \quad z \in B.$$

It is easy to see that $T(A) \subseteq B$ and $T(B) \subseteq A$. Then, under what conditions on f and g does there exist $w \in A \cup B$ such that $d(w, Tw) = \text{dist}(A, B)$? If such a function w exists on an interval containing x_0 in $[x_0 - a, x_0 + a]$, then the pair (w, Tw) is called an optimum solution for the system of differential equations given in (9) and (10).

Also note that, if ϕ_1 is a solution of (9) and ϕ_2 is a solution of (10), then the pair (ϕ_1, ϕ_2) need not form an optimum solution.

We use the following notation in the sequel. Let $a \geq 0$, define $\phi_a : \mathbb{R} \times \mathbb{R} \to \mathbb{R}^+$ by $\phi_a(x, y) = \frac{|x|}{|y|}$, when $y \neq 0$ and $\phi_a(x, y) = a$, if $y = 0$.

Theorem 17 [10] *Let S, T, A, and B be as defined above and $y_1 < y_2$. Suppose f and g are continuous functions on S satisfying:*

(i) $| f(x, z) - g(x, y) | \leq K(| y - z | - | y_1 - y_2 |)$, *for some* $K > 0$, *whenever* $| y - z | \geq | y_1 - y_2 |$.
(ii) $f(x, z) \geq g(x, y)$, *if* $x \leq x_0$ *and* $f(x, z) \leq g(x, y)$, *if* $x \geq x_0$, *whenever* $| y - z | \leq | y_1 - y_2 |$.

Then, for any $\beta < \min\left\{a, \frac{b-|y_1-y_0|}{M}, \frac{b-|y_2-y_0|}{M}, \frac{1}{K}, \phi_a(y_1 - y_2, N)\right\}$, *there exists* $w \in A \cup B$ *such that* $d(w, Tw) = \operatorname{dist}(A, B)$, *where* $A, B \subseteq C_\beta = \{y \in C[x_0 - \beta, x_0 + \beta] : | y(x_0) - y_0 | \leq b\}$, *$M$ is the bound for both f and g and*

$$N = \sup\{| f(x, z) - g(x, y) | : | y - z | \leq | y_1 - y_2 |\},$$

that is, (w, Tw) is an optimum solution for the system of differential equations given in (9) and (10).

Proof Let $y \in A$, then $Ty(x_0) = y_1$, also

$$\begin{aligned}
| Ty(x) - y_0 | &= | y_1 - y_0 + \int_{x_0}^{x} g(t, y(t))\mathrm{d}t | \\
&\leq | y_1 - y_0 | + \int_{x_0}^{x} | g(t, y(t)) | \,\mathrm{d}t \\
&\leq | y_1 - y_0 | + M \int_{x_0}^{x} \mathrm{d}t \\
&= | y_1 - y_0 | + M | x - x_0 | \\
&\leq | y_1 - y_0 | + \beta M \\
&\leq b.
\end{aligned}$$

Hence, $T(A) \subseteq B$. Similarly $T(B) \subseteq A$.

To prove that T is a cyclic contraction take $y \in A$, $z \in B$, and assume $x \geq x_0$,

$$|Ty(x) - Tz(x)| = \left| y_2 - y_1 + \int_{x_0}^{x} (f(t, z(t)) - g(t, y(t)))\mathrm{d}t. \right| \tag{11}$$

Now,

$$\int_{x_0}^{x} (f(t, z(t)) - g(t, y(t)))\mathrm{d}t = \int_{[x_0,x]} (f(t, z(t)) - g(t, y(t)))\mathrm{d}t,$$

where $\int_{[x_0,x]}(f(t, z(t)) - g(t, y(t)))\mathrm{d}t$ is the Lebesgue integral of $(f(t, z(t)) - g(t, y(t)))$ over the interval $[x_0, x]$. Now, let

$$C_1 = \{t \in [x_0, x_0 + \beta] : \mid y(t) - z(t) \mid > \mid y_1 - y_2 \mid\},$$
$$C_2 = \{t \in [x_0, x_0 + \beta] : \mid y(t) - z(t) \mid \leq \mid y_1 - y_2 \mid\}.$$

Since y and z are continuous functions, we have both C_1 and C_2 are disjoint measurable sets. Therefore,

$$\int_{x_0}^{x} (f(t, z(t)) - g(t, y(t)))\mathrm{d}t = \int_{C_1} (f(t, z(t)) - g(t, y(t)))\mathrm{d}t + \int_{C_2} (f(t, z(t)) - g(t, y(t)))\mathrm{d}t.$$

Hence from (11),

$$\begin{aligned} \mid Ty(x) - Tz(x) \mid &= \left| y_2 - y_1 + \int_{C_1} (f(t, z(t)) - g(t, y(t)))\mathrm{d}t \right. \\ &\quad \left. + \int_{C_2} (f(t, z(t)) - g(t, y(t)))\mathrm{d}t \right| \\ &\leq \left| y_2 - y_1 + \int_{C_2} (f(t, z(t)) - g(t, y(t)))\mathrm{d}t \right| \\ &\quad + \left| \int_{C_1} (f(t, z(t)) - g(t, y(t)))\mathrm{d}t \right|. \end{aligned}$$

In C_2, $\mid y(t) - z(t) \mid \leq \mid y_1 - y_2 \mid$, for $x \geq x_0$ by condition(ii), we get $f(t, z(t)) \leq g(t, y(t))$, so $\int_{C_2}(f(t, z(t)) - g(t, y(t))) \leq 0$ and

$$\left|\int_{C_2} (f(t, z(t)) - g(t, y(t)))\right| \leq \int_{C_2} | (f(t, z(t)) - g(t, y(t))) | \mathrm{d}t$$
$$\leq N \int_{C_2} \mathrm{d}t \leq N \mid x - x_0 \mid$$
$$\leq N\beta <\mid y_1 - y_2 \mid.$$

Therefore,

$$\mid Ty(x) - Tz(x) \mid \leq \mid y_2 - y_1 \mid + \int_{C_1} \mid (f(t, z(t)) - g(t, y(t))) \mid \mathrm{d}t$$
$$\leq \mid y_2 - y_1 \mid + \int_{C_1} K(\mid y(t) - z(t) \mid - \mid y_1 - y_2 \mid)\mathrm{d}t$$
$$\leq \mid y_2 - y_1 \mid + K\beta \max_{t \in [x_0 - \beta, x_0 + \beta]} (\mid y(t) - z(t) \mid - \mid y_1 - y_2 \mid)$$
$$\leq \mid y_2 - y_1 \mid + K\beta(\|y - z\| - \mid y_1 - y_2 \mid)$$
$$= K\beta\|y - z\| + (1 - K\beta) \mid y_1 - y_2 \mid$$
$$\mid Ty(x) - Tz(x) \mid \leq K\beta\|y - z\| + (1 - K\beta) \mid y_1 - y_2 \mid.$$

As $K\beta < 1$, the map T is a cyclic contraction. A similar proof can be given for the case $x \leq x_0$.

Now for any $y \in A$,

$$\mid Ty(x) \mid = \left| y_1 + \int_{x_0}^{x} g(t, y(t))\mathrm{d}t \right|$$
$$\leq \mid y_1 \mid + \int_{x_0}^{x} \mid g(t, y(t)) \mid \mathrm{d}t$$
$$\mid Ty(x) \mid \leq \mid y_1 \mid + M\beta.$$

Hence, the family $\{Ty\}_{y \in A}$ is uniformly bounded. Let $x_1, x_2 \in [x_0 - \beta, x_0 + \beta]$,

$$\mid Ty(x_1) - Ty(x_2) \mid \leq \left| y_1 + \int_{x_1}^{x} g(t, y(t))\mathrm{d}t - (y_1 + \int_{x_2}^{x} g(t, y(t))\mathrm{d}t) \right|$$
$$= \left| \int_{x_1}^{x_2} g(t, y(t))\mathrm{d}t \right| \leq \int_{x_1}^{x_2} |g(t, y(t))| \,\mathrm{d}t$$
$$\mid Ty(x_1) - Ty(x_2) \mid \leq M \mid x_1 - x_2 \mid.$$

Since this holds for any $y \in A$, $\{Ty\}_{y \in A}$ is a family of equicontinuous functions. Therefore, by Arzela-Ascoli's theorem, $T(A)$ lies in a compact subset of B. Similarly, $T(B)$ lies in a compact subset of A. Hence by Proposition 8, there exists $w \in C_\beta$ such that $d(w, Tw) = \text{dist}(A, B)$. □

Acknowledgments The first author dedicates this work to Prof. D. V. Pai and Prof. W. A. Kirk for their guidance and inspiration. We also would like to thank the referees for their constructive comments and valuable suggestions.

References

1. Abkar, A., Gabeleh, M.: Best proximity points for asymptotic cyclic contraction mappings. Nonlinear Anal. **74**, 7261–7268 (2011)
2. Basha, S.S., Veeramani, P.: Best approximations and best proximity pairs. Acta Sci. Math. (Szeged) **63**, 289–300 (1997)
3. Basha, S.S., Veeramani, P.: Best proximity pair theorems for multifunctions with open fibers. J. Approx. Theory **103**, 119–129 (2000)
4. Basha, S.S., Veeramani, P., Pai, D.V.: Best proximity pair theorems. Indian J. Pure Appl. Math. **32**, 1237–1246 (2001)
5. Beer, G., Pai, D.V.: The prox map. J. Math. Anal. Appl. **156**, 428–443 (1991)
6. Beer, G., Pai, D.V.: Proximal maps, prox maps and coincidence points. Numer. Funct. Anal. Optim. **11**, 429–448 (1990)
7. Eldred, A.A.: Existence and convergence of best proximity points. Ph. D. thesis, Indian Institute of Technology-Madras, Chennei (2006)
8. Eldred, A.A., Kirk, W.A., Veeramani, P.: Proximal normal structure and relatively nonexpansive mappings. Studia Math. **171**, 283–293 (2005)
9. Eldred, A.A., Veeramani, P.: Existence and convergence of best proximity points. J. Math. Anal. Appl. **323**, 1001–1006 (2006)
10. Eldred, A.A., Veeramani, P.: On best proximity pair solutions with applications to differential equations. J. Indian Math. Soc. (N.S.) 2007. Special volume on the occasion of the centenary year of IMS (1907–2007), 51–62 (2008)
11. Espínola, R.: A new approach to relatively nonexpansive mappings. Proc. Am. Math. Soc. **136**, 1987–1995 (2008)
12. Espínola, R., Fernández-León, A.: On best proximity points in metric and Banach spaces. Can. J. Math. **63**, 533–550 (2011)
13. Fan, K.: Fixed-point and minimax theorems in locally convex topological linear spaces. Proc. Nat. Acad. Sci. U.S.A. **38**, 121–126 (1952)
14. Fan, K.: Extensions of two fixed point theorems of F. E. Browder. Math. Z. **112**, 234–240 (1969)
15. Fernández-León, A.: Existence and uniqueness of best proximity points in geodesic metric spaces. Nonlinear Anal. **73**, 915–921 (2010)
16. Goebel, K., Kirk, W.A.: Topics in metric fixed point theory. Cambridge Studies in Advanced Mathematics. Cambridge University Press, Cambridge (1990)
17. Khamsi, M.A., Kirk, W.A.: An Introduction to Metric Spaces and Fixed Point Theory. Wiley-Interscience, New York (2001)
18. Kim, W.K., Kum, S., Lee, K.H.: Semicontinuity of the solution multifunctions of the parametric generalized operator equilibrium problem. Nonlinear Anal. **71**, 2182–2187 (2009)
19. Kim, W.K., Lee, K.H.: Existence of best proximity pairs and equilibrium pairs. J. Math. Anal. Appl. **316**, 433–466 (2006)

20. Kirk, W.A.: A fixed point theorem for mappings which do not increase distances. Am. Math. Mon. **72**, 1004–1006 (1965)
21. Kirk, W.A., Srinivasan, P.S., Veeramani, P.: Fixed points for mappings satisfying cyclical contractive conditions. Fixed Point Theory **4**, 79–89 (2003)
22. Megginson, R.E.: An Introduction to Banach Space Theory. Springer-Verlag, New York (1998)
23. Piatek, B.: On cyclic Meir-Keeler contractions in metric spaces. Nonlinear Anal. **74**, 35–40 (2011)
24. Prolla, J.B.: Fixed point theorems for set valued mappings and existence of best approximations. Numer. Funct. Anal. Optim. **5**, 449–455 (1982/1983)
25. Reich, S.: Approximate selections, best approximations, fixed points and invariant sets. J. Math. Anal. Appl. **62**, 104–113 (1978)
26. Sehgal, S.M., Singh, S.P.: A generalization to multifunctions of Fan's best approximation theorem. Proc. Am. Math. Soc. **102**, 534–537 (1988)
27. Singh, S.P., Watson, B., Srivastava, P.: Fixed Point Theory and Best Approximation: The KKM-Map Principle. Kluwer Academic Publishers, New York (1997)
28. Srinivasan, P.S., Veeramani, P.: On existence of equilibrium pair for constrained generalized games. Fixed Point Theory Appl. **1**, 21–29 (2004)
29. Suzuki, T., Kikkawa, M., Vetro, C.: The existence of best proximity points in metric spaces with the property UC. Nonlinear Anal. **71**, 2918–2926 (2009)
30. Takahashi, W.: Nonlinear Functional Analysis: Fixed Point Theory and Its Applications. Yokohama Publishers Inc, Yokohama (2000)
31. Veeramani, P.: On some fixed point theorems on uniformly convex Banach spaces. J. Math. Anal. Appl. **167**, 160–166 (1992)
32. Vetrivel, V., Veeramani, P., Bhattacharyya, P.: Some extensions of Fan's best approximation theorem. Numer. Funct. Anal. Optim. **13**, 397–402 (1992)
33. Zeidler, E.: Nonlinear Functional Analysis and Its Applications - I: Fixed Point Theorems. Springer, New York (1986)

Semi-continuity Properties of Metric Projections

V. Indumathi

Abstract This chapter presents some selected results regarding semi-continuity of metric projections onto closed subspaces of normed linear spaces. Though there are several significant results relevant to this topic, only a limited coverage of the results is undertaken, as an extensive survey is beyond our scope. This exposition is divided into three parts. The first one deals with results from finite dimensional normed linear spaces. The second one deals with results connecting semi-continuity of metric projection maps and duality maps. The third one deals with subspaces of finite codimension of infinite dimensional normed linear spaces.

Keywords Proximinal set · Strongly proximinal set · Best approximation · Upper and lower semicontinuity for set-valued maps · Pre-duality maps · Metric projection · Strongly subdifferentiable maps · Quasi-polyhedral points

1 Metric Projections onto Finite Dimensional Spaces

In this survey article, we present some selected results regarding semi-continuity of metric projections onto closed subspaces of normed linear spaces. Though there are several significant results relevant to this topic, we undertake only a limited coverage of the results as an extensive survey is beyond our scope.

This exposition is divided into three parts. The first one deals with results from finite dimensional normed linear spaces. The second one deals with results connecting semi-continuity of metric projection maps and duality maps.

The third one deals with subspaces of finite co-dimension of infinite dimensional normed linear spaces.

V. Indumathi (✉)
Department of Mathematics, Pondicherry University, Pondicherry 605014, India
e-mail: pdyindumath@gmail.com

Q. H. Ansari (ed.), *Nonlinear Analysis*, Trends in Mathematics,
DOI: 10.1007/978-81-322-1883-8_2,

Throughout we consider only real normed linear spaces and we assume all subspaces are closed.

If X is a normed linear space, X^* will denote the dual of X, B_X the closed unit ball, $\{x \in X : \|x\| \leq 1\}$ and S_X the unit sphere $\{x \in X : \|x\| = 1\}$, of X. If x is in X and $r > 0$ then the open and closed balls with center x and radius r are denoted by

$$B(x, r) = \{z \in X : \|x - z\| < r\},$$

and

$$B[x, r] = \{z \in X : \|x - z\| \leq r\},$$

respectively. Further, if $A \subseteq X$, $x \in X$ and $\varepsilon > 0$, then we set

$$B(A, \varepsilon) = \{x \in X : d(x, A) < \varepsilon\},$$

$$d(x, A) = \inf\{\|x - a\| : a \in A\}, \quad \text{for } x \in X,$$

and

$$P_A(x) = \{a \in A : \|x - a\| = d(x, A)\}.$$

Further, we set

$$A^{\perp} = \{f \in X^* : f \equiv 0 \text{ on A}\}.$$

We now have

Definition 1 Let $A \subseteq X$. Then A is said to be *proximinal* in X if $P_A(x)$ is nonempty for each $x \in X$. Any element in $P_A(x)$ is called *a nearest element to x from A* or *a best approximation to x from A*. The set A is said to be *Chebyshev* if $P_A(x)$ is a singleton set for all $x \in X$.

The set valued map P_A defined on X, is called the metric projection from X onto A. The following stronger notion of proximinality figures in an essential way, often in our discussion. Note that if $d = d(x, A)$ then

$$P_A(x) = B[x, d] \cap A.$$

For $\delta > 0$, set

$$\begin{aligned} P_A(x, \delta) &= \{y \in A : \|x - y\| \leq d(x, A) + \delta\} \\ &= B[x, d + \delta] \cap A. \end{aligned}$$

We now have the following definition from Godefroy and Indumathi [14].

Definition 2 Let X be a normed linear space. A proximinal subset A of X is said to be *strongly proximinal* at x in X, if given $\varepsilon > 0$ there exists $\delta > 0$ such that

$$d(y, P_A(x)) < \varepsilon \text{ for all } y \in P_A(x, \delta)$$

or equivalently

$$P_A(x, \delta) \subseteq B(P_A(x), \varepsilon).$$

If A is strongly proximinal at each x in X, then we say A is strongly proximinal in X.

Let $F : X \to Y$ be a set-valued map. We say F is *lower semi-continuous* (l.s.c.) at x_0 of X if for any open set U of X such that $U \cap F(x_0) \neq \emptyset$, the set $\{x \in X : F(x) \cap U \neq \emptyset\}$ is a neighbourhood of x_0 and F is *upper semi-continuous* (u.s.c.) at x_0 of X if for any open set U of X such that $F(x_0) \subseteq U$, the set $\{x \in X : F(x) \subseteq U\}$ is a neighbourhood of x_0.

The set-valued map F is said to be *Hausdorff lower semi-continuous* (H.l.s.c.) at x_0 of X if given $\varepsilon > 0$ there exists $\delta > 0$ such that $x \in B(x_0, \delta)$ implies $F(x_0) \subseteq B(F(x), \varepsilon)$.

We say F is *Hausdorff upper semi-continuous* (H.u.s.c.) at x_0 in X if for any $\varepsilon > 0$, the set $\{x \in X : F(x) \subseteq B(F(x_0), \varepsilon)\}$ is a neighbourhood of x_0.

The set valued map F is would be called *Hausdorff semi-continuous* if it is both H.u.s.c. and H.l.s.c.

We observe that

$$F \ \ H.l.s.c. \Rightarrow F \ \ l.s.c, \text{ while } F \ \ u.s.c \Rightarrow F \ \ H.u.s.c.$$

Our discussion would involve the above semi-continuity concepts with reference to metric projections.

It is easily verified that if Y is strongly proximinal then the metric projection is H.u.s.c.

Remark 1 A well-known fact, that can be proved using the usual compactness arguement, is that any finite dimensional subspace Y of a normed linear space X is strongly proximinal and hence the metric projection P_Y is upper Hausdorff semi-continuous. We observe that for a single-valued map, all the above four notions of semi-continuity coincide with the usual notion of continuity of a single-valued map. Thus if Y is a finite dimensional Chebychev subspace of X, then P_Y is continuous.

A single-valued map f on X is said to be a selection for F if $f(x) \in F(x)$ for each x in X. The set valued map F is said to have a continuous selection if it has a selection that is continuous. Among the semi-continuity properties of the metric projection, l.s.c. gains prominence because of the following important theorem of Michael.

Theorem 1 (Michael Selection Theorem) [21] *If X is a paracompact, Hausdorff topological space, Y is a Banach space and $F : X \to 2^Y$ is a nonempty closed convex set valued and lower semi-continuous mapping, then F has a continuous selection; that is, there exists a continuous $s : X \to Y$ such that $s(x) \in F(x)$*

for each x in X. In particular if Y is a subspace of a normed linear space X with $P_Y(x) \neq \phi$ for all $x \in X$ and P_Y lower semi-continuous on X then P_Y has a continuous selection.

However, l.s.c. is not a necessary condition for the existence of a continuous selection as the following example of Deutsch and Kenderov from [8] shows.

Example 1 [8] Let B be the convex hull of the circle $\Gamma = \{(x_1, x_2, 0) : x_1^2+x_2^2 = 1\}$ and the two points (0, 0, 1) and (0, 0, −1) in $\mathbf{R}^3$. (B is double cone formed by placing the two cones with vertices (0, 0, 1) and (0, 0, −1) in such a way that, their common circular base coincides.) Then B is a closed convex, symmetric set with nonempty interior. Let X be the normed linear space $\mathbf{R}^3$, with the norm for which B is the closed unit ball.

Let $Y = sp(1, 0, 1)$. Then Y is the line L through (0, 0, 0) and (1, 0, 1), which is parallel to the line segment l , lying on the unit sphere, joining (−1, 0, 0) in Γ and the vertex (0, 0, 1). If $x = (x_1, x_2, x_3)$ and $x_2 \neq 0$ then $P_Y(x) = \{(x_3, 0, x_3)\}$.

If $x_2 = 0$, then $P_Y(x)$ is a line segment of nonzero length containing the point $(x_3, 0, x_3)$. It is clear that $f(x) = (x_3, 0, x_3)$ is a continuous selection for P_Y but P_Y is not l.s.c.

An important weaker notion than l.s.c, that is also a necessary condition for the existence of continuous selections, is that of approximate lower semi-continuity (a.l.s.c.).

Definition 3 [7, 8] We say F is *approximate lower semi-continuous* (a.l.s.c) at x_0 if for each $\varepsilon > 0$ there exists $\delta > 0$ such that

$$\bigcap \{B(F(x), \varepsilon) : x \in B(x_0, \delta)\} \neq \phi.$$

Weaker notions than a.l.s.c. can be naturally defined as follows. Let k be a positive integer ≥ 2. The following notion from [8] is weaker than a.l.s.c.

The set valued map F is said to be $k-l.s.c$ at x_0 if given $\varepsilon > 0$ there exists $\delta > 0$ such that $\cap_{i=1}^k B(F(x_i), \varepsilon) \neq \phi$ for every choice of $k-$points in $B(x_0, \delta)$. It follows from Helly's theorem that if Y is a subspace of finite dimension n and F is closed convex valued then

$$F \text{ is a.l.s.c} \Leftrightarrow F \text{ is (n+1)−l.s.c.} \tag{1}$$

Clearly

$$\begin{aligned} F \text{ l.s.c at } x_0 &\Rightarrow F \text{ is a.l.s.c at } x_0 \\ &\Rightarrow F \text{ is k−l.s.c at } x_0 \text{ for } k \geq 2. \end{aligned}$$

We refer the reader to the papers [5, 6] of Brown, for a thorough discussion about a.l.s.c. of a set valued map and its derived maps, presenting a lucid and overall perspective of these concepts in relation to the existence of continuous selections.

Let $\varepsilon > 0$. The set valued map F is said to have an ε-approximate continuous selection if there is a continuous map $s_\varepsilon : X \to Y$ such that $s_\varepsilon(x) \in B(F(x), \varepsilon)$ for each $x \in X$.

A parallel result to Michael selection theorem, for a.l.s.c, was proved in [8].

Theorem 2 [8] *Let X be a paracompact space and Y be a normed linear space. Let $F : X \to 2^Y$ have closed,convex images. Then F is a.l.s.c if and only if for each $\varepsilon > 0$, F has a continuous ε-approximate selection.*

Examples of a.l.s.c. and u.s.c. maps with no continuous selections have long been known. Zhivkov [27] constructed an example of a space X of dimension five- and a three-dimensional subspace Y of X such that the metric projection P_Y is a.l.s.c. but does not have a continuous selection. However, Deutsch and kenderov [8] showed that if dim $Y = 1$ then

$$\begin{aligned} P_Y \text{ has a continuous selection} &\Leftrightarrow P_Y \text{is a.l.s.c.} \\ &\Leftrightarrow P_Y \; is \; 2.l.s.c. \end{aligned}$$

Brown [2] and Deutsch and Kenderov [8] have independently constructed examples of one dimensional subspace Y of a three-dimensional space X such that P_Y does not have a continuous selection, which in turn implies P_Y is not 2.l.s.c. We observe that by Remark 1 and Corollary 1, given later in Sect. 2, P_Y is u.s.c.

Example 2 [2, 7] Let X be $\mathbf{R}^3$ with norm generated by the unit ball $B = co\ (l \cup D \cup -D)$, where l is the line segment joining $(1, 0, 0)$ and $(-1, 0, 0)$ and D is the semicircle

$$\{(1, y, z) : y \geq 0,\ z\ \geq 0 \text{ and } y^2 + z^2 = 1\}.$$

If Y is the one dimensional subspace sp(1,0,0) and z is a point that moves on the circle $C = \{0, y, z) : y^2 + z^2 = 1\}$, we have $P_Y(z) = \{(-1, 0, 0)\}$ if $z > 0$ and $P_Y(z) = \{(1, 0, 0)\}$ if $z < 0$. It is clear that P_Y cannot have a continuous selection.

As observed earlier, a.l.s.c. does not guarantee existence of continuous selections for metric projections. However, we have surprising positive results when the space $C(Q)$ is considered and the following results of Wu Li and T. Fisher are impressive.

Theorem 3 (Li [19]) *Let Q be a compact Hausdorff topological space, $C(Q)$ the space of real valued continuous maps defined on Q with supnorm. If Y is a finite dimensional subspace of $C(Q)$ then the metric projection P_Y has a continuous selection if and only if P_Y is a.l.s.c.*

Theorem 4 (Fisher [18]) *Let Q be a compact Hausdorff topological space, $C(Q)$ the space of real valued continuous maps defined on Q with supnorm. If Y is a finite dimensional subspace of $C(Q)$ then the metric projection P_Y has a continuous selection if and only if P_Y is 2−l.s.c.*

The proofs of the above two theorems are elaborate and technical in nature. However, the proof of Fisher, via optimization techniques, leads to a stronger conclusion, viz, the sufficiency of 2- l.s.c. for the existence of the continuous selection. We refer the reader to the papers [5, 6] of Brown, for detailed discussion and comparison of the proofs of Wu Li and Fisher.

The above results are extended to $X = C_0(T)$, the space of real continuous functions which vanish at infinity on a locally compact Hausdorff space T, in a later work of Wu Li. He also proved the equivalence of a.l.s.c and existence of continuous selections for finite dimensional subspaces of $L_1(T, \mu)$.

Theorem 5 [20] *Let (T, μ) be a positive measure space and $X = L_1(T, \mu)$, the space of real integrable functions on (T, μ) equipped with the usual norm. If Y is a finite dimensional subspace of X then, P_Y has a continuous selection if and only if P_Y is a.l.s.c.*

We now discuss some geometric conditions that play a vital role in the semicontinuity of metric projections. We need the definition of polyhedral spaces in the discussion now and later.

Definition 4 A finite dimensional normed linear space X is said to be *polyhedral* if extreme points of B_X is a finite set. A normed linear space is called polyhedral if every one of its finite dimensional subspace is polyhedral.

In Brown [3], defined property P for a normed linear space (If x and z in X satisfy $\|x + z\| \leq \|x\|$, then there exist positive constants δ and η such that $\|y + \eta z\| \leq \|y\|$ if y is in $B(x, \delta)$) and showed that normed linear spaces with property P are precisely those spaces in which metric projections onto all finite dimensional subspaces are l.s.c. In [1], the equivalence of Property P to metric projection onto every one dimensional subspace being l.s.c, was shown.

Strictly convex spaces and finite dimensional, polyhedral normed linear spaces are examples of spaces with property (P). We recall that Singer [23] if a normed linear space is strictly convex, then every proximinal subspace of X is Chebyshev. Thus if X is strictly convex, every finite dimensional subspace of X is Chebyshev and by Remark 1 above, the metric projection onto every finite dimensional space is single-valued and continuous and equivalently, l.s.c.

A normed linear space X is said to have property (CS1) Brown et al. [7] whenever Y is a one dimensional subspace of X then P_Y has a continuous selection or equivalently P_Y is 2−a.l.s.c. Clearly property (P) implies property (CS1).

In Brown et al. [7], an example of a three-dimensional space with property (CS1) but not having property (P) was given. Further it was shown in that paper that a normed linear space X has property (CS1) if and only if the metric projection P_Y is a.l.s.c for every finite dimensional subspace Y of X.

We end this section with two comments. Consider the class of those Banach spaces X, for which the following hold: If Y any finite dimensional subspace of X, then P_Y has a continuous selection if and only if P_Y is a.l.s.c. We note that the above class includes $C(Q)$ and $L_1(T,\nu)$. It would be desirable to have more examples of nonstrictly convex spaces in this class.

Given a Banach space X, identifying some finite dimensional subspaces Y of X with dim $Y \geq 2$, for which P_Y has a continuous selection if P_Y a.l.s.c, would be an interesting problem. Using the notion of derived maps of Brown [6], the subspaces for which the derived map of the metric projection is l.s.c, would have the above property.

2 Pre-duality Maps and Metric Projections

In this section, we present mostly results from [9] connecting semi-continuity properties of metric projections and the pre-duality maps.

We need the following facts about upper semi-continuity of set-valued maps later, for proving Theorem 8. The fact below is from [25].

Fact 1 [25] *Let X and Y be normed linear spaces and $F : X \to 2^Y$ is a set valued map with nonempty closed convex and bounded images. Assume that F is positively homogeneous: that is, $F(\alpha x) = \alpha x$ for $\alpha \geq 0$ and x in X. Then F is u.s.c if and only if F is H.u.s.c and $F(x)$ is compact for each $x \in X$.*

Proof Assume F is H.u.s.c and $F(x)$ is compact for each $x \in X$. Fix $x_0 \in X$. Suppose F is not u.s.c at x_0. Then there exists a sequence $\{x_n\}$ in X converging to x_0, a neighborhood U of $F(x_0)$ and a sequence $\{y_n\}$ in Y such that $y_n \in F(x_n) \setminus U$ for all $n \geq 1$. Since F is H.u.s.c at x_0, $d(y_n, F(x_0)) \to 0$ as $n \to \infty$. Then there exists $\{z_n\} \subseteq F(x_0)$ such that $\|z_n - y_n\| < \frac{1}{n}$ for all $n \geq 1$. Now $F(x_0)$ compact implies $\{z_n\}$ has a convergent subsequence that converge to $z_0 \in F(x_0)$ and hence $\{y_n\}$ has a convergent subsequence converging to z_0. Since $z_0 \in U$ this implies $y_n \in U$ for all large enough n. This contradicts $y_n \notin U$ for all $n \geq 1$. Hence, F is u.s.c.

Conversely assume F is u.s.c. Clearly F is H.u.s.c. We only show that $F(x)$ is compact for each x in X. Fix x_0 in X and assume $\{y_n\} \subseteq F(x_0)$ has no convergent subsequence. Let $\beta_n = \sup\{\beta y_n : \beta y_n \in F(x_0)\}$. Then $\beta_n \geq 1$ and since $F(x_0)$ is closed, $\beta_n y_n \in F(x_0)$. Note that $\lambda\beta_n y_n \notin F(x_0)$ for $\lambda > 1$, for each $n \geq 1$. Let $\{\lambda_n\}$ be a sequence of scalars such that $\lambda_n > 1$ for all n. Clearly, the sequence $\{\lambda_n\beta_n y_n\}$ lies outside the set $F(x_0)$. We claim that the sequence $\{\lambda_n\beta_n y_n\}$ does not have a convergent subsequence.

Select $M > 0$ such that $\sup\limits_{x\in F(x_0)} \|x\| \leq M$. Since $\{y_n\}$ does not have a convergent subsequence, without loss of generality we can and do assume $\min\{\|y_n\| : n \geq 1\} = \delta > 0$. Then

$$\sup_{n\geq 1} \beta_n \leq \frac{M}{\delta} \text{ and } 0 < \frac{\delta}{\lambda_n M} \leq \frac{1}{\lambda_n \beta_n} \leq 1, \quad \text{for all } n \geq 1.$$

Assume $\{\lambda_n \beta_n y_n\}$ has a convergent subsequence, say $\{\lambda_{n_k} \beta_{n_k} y_{n_k}\}$. Then by the above, the sequence of scalars $\left\{\frac{1}{\lambda_{n_k} \beta_{n_k}}\right\}$ has a convergent subsequence that converges to a limit in the open interval $(0, 1)$. This would imply the sequence $\{y_{n_k}\}$ has a convergent subsequence and contradict our assumption. Hence we conclude $\{\lambda_n \beta_n y_n\}$ does not have a convergent subsequence.

Let $w_n = \beta_n y_n$ and $x_n = \frac{n+1}{n} x_0$ for $n \geq 1$. Then $\frac{n+1}{n} w_n \in F(x_n)$ for all $n \geq 1$. Clearly (x_n) converges to x_0 and $A = \left\{\frac{n+1}{n} w_n : n \geq 1\right\}$ is a closed set, since the sequence $\left\{\frac{n+1}{n} w_n\right\}$ does not have a convergent subsequence. Clearly $A \cap F(x_n)$ is nonempty for all $n \geq 1$, while $A \cap F(x_0)$ is an empty set. This contradicts upper semi-continuity of F at x_0 and $F(x_0)$ is compact. □

The following corollary of the above theorem is immediate.

Corollary 1 *Let X be a Banach space and Y a proximinal subspace of X. Then the metric projection P_Y is u.s.c. if and only if P_Y is H.u.s.c. and $P_Y(x)$ is compact for each x in X.*

Let X be a normed linear space and Y be a proximinal subspace of X. Set

$$D_Y = \{x \in X : d(x, Y) = 1\}.$$

Then it is easy to check that the metric projection P_Y is H.u.s.c. (l.s.c.) on X if and only if P_Y is H.u.s.c. (l.s.c.) on the set D_Y.

The following theorem of Morris [22] is needed for proving Theorem 8 below.

Theorem 6 [22] *Let X be a normed linear space and Y be a proximinal subspace of finite codimension in X. Then P_Y is u.s.c if and only if $P_Y^{-1}\{0\}$ is boundedly compact.*

Proof Assume $P_Y^{-1}\{0\}$ is boundedly compact. Fix x in D_Y. Note that $x - P_Y(x) \subseteq P_Y^{-1}\{0\}$, is a bounded set and therefore is compact. By Corollary 1, we only have to show that P_Y is H.u.s.c.

If P_Y is not H.u.s.c at x_0, there exists $\{x_n\} \subseteq D_Y$ and $\varepsilon > 0$ such that $\{x_n\}$ converges to x and a sequence $\{y_n\}$ with $y_n \in P_Y(x_n)$ for all $n \geq 1$ and

$$d(y_n, P_Y(x)) \geq \varepsilon, \quad \text{for all } n \geq 1. \tag{2}$$

Now $x_n - y_n$ is in $P_Y^{-1}\{0\}$ and $\|x_n - y_n\| \leq 1$, for each $n \geq 1$. So $\{x_n - y_n\}$ has a convergent subsequence, say, $x_{n_k} - y_{n_k}$ that converges to some $z \in P_Y^{-1}\{0\}$. This implies $\{y_{n_k}\}$ converges to $z - x$ in Y. Now

$$\left\|x_{n_k} - y_{n_k}\right\| \to \|x - (z - x)\| = \|z\| = d(z, Y) = d(x, Y).$$

So $z - x \in P_Y(x)$ and this contradicts (2).

Conversely assume that P_Y is u.s.c. Then P_Y is u.H.s.c and $P_Y(x)$ is compact for each x in X. Let $\{x_n\}$ be any sequence in $P_Y^{-1}\{0\} \cap S_X$. Since X/Y and hence

$Y^\perp$ is finite dimensional, and the sequence $\{x_n + Y\}$ is bounded, it has a convergent subsequence that converges to some $x + Y$ in X/Y. W.l.o.g. we can and do assume that the sequence $\{x_n + Y\}$ converges to $x + Y$, that is, $\lim_{n\to\infty} \|x_n - x + Y\| = 0$.

Select y_n in Y such that $\lim_{n\to\infty} \|x_n - x - y_n\| = 0$. Then the sequence $\{x_n - y_n\}$ converges to x and

$$-y_n \in P_Y(x_n) - y_n = P_Y(x_n - y_n), \quad \text{for all } n \geq 1.$$

Since P_Y is u.s.c at x, $d(-y_n, P_Y(x)) \to 0$ as $n \to \infty$. So there exists $\{z_n\} \subseteq P_Y(x)$ such that $\|y_n + z_n\| \to 0$ as $n \to \infty$. Now $\{z_n\}$ has a convergent subsequence that converges to $z \in P_Y(x)$, as $P_Y(x)$ is compact. This implies $\{y_n\}$ has a convergent subsequence, say $\{y_{n_k}\}$, that converges to z and $\{x_{n_k}\} = \{x_{n_k} - y_{n_k} + y_{n_k}\}$ converges to $x - z \in P_Y^{-1}\{0\}$. This implies (x_n) has a convergent subsequence that converges to an element of $P_Y^{-1}\{0\}$ and this completes the proof. □

For $x \in X$, set

$$J_{X^*}(x) = \{f \in X^* : \|f\| = 1 \text{ and } f(x) = \|x\|\}.$$

Note that $J_{X^*}(x)$ is a nonempty subset, by the Hahn-Banach theorem. The map $x \to J_{X^*}(x),\ x \in X$ is called the *duality map* on X. For f in X^*, define

$$J_X(f) = \{x \in S_X : f(x) = \|f\|\}.$$

Note $J_X(f)$ can be empty. If it is nonempty, we recall that f is said to be a *norm attaining functional*. We will denote the class of all norm attaining functionals on X by $NA(X)$. The set-valued map $f \to J_X(f)$ from X^* into X is called the *pre-duality map* on X^*.

Remark 2 Let X be a normed linear space and Y be a subspace of X. If x is in X it is well known that [23]

$$d(x, Y) = \{\max\{f(x) : f \in Y^\perp \text{ and } \|f\| = 1\}.$$

Thus for f in $S_{Y^\perp}$ and x is in $J_X(f)$, we have

$$1 = \|x\| \geq d(x, y) = 1$$

and hence equality holds. Thus, x is in $P_Y^{-1}\{0\} \cap S_X$. Conversely, if x is in $P_Y^{-1}\{0\} \cap S_X$, then x is in $J_X(f)$ for some f in $S_{Y^\perp}$.

Remark 3 Let Y be a proximinal subspace of finite codimension in a normed linear space X. Then an useful corollary of a characterization, of proximinal subspaces of finite codimension of Garkavi (see Godefroy and Indumathi [14]), implies that $Y^\perp$ is contained in $NA(X)$. In other words, $J_X(f)$ is nonempty for each f in $Y^\perp$.

We now have the following result from [9], relating continuity properties of the metric projection with that of the pre-duality map.

Theorem 7 [9] *Let X be a normed linear space and Y be a proximinal subspace of finite codimension in X. Then the following assertions are equivalent.*

(a) P_Y *is u.s.c.*
(b) P_Y *is H.u.s.c and* $P_Y(x)$ *is compact for each* x *in* X.
(c) $P_Y^{-1}\{0\}$ *is boundedly compact.*
(d) $J_X|_{S_{Y^\perp}}$ *is H.u.s.c and* $J_X|_{S_{Y^\perp}}(f)$ *is compact for each* f *in* $S_{Y^\perp}$.
(e) $J_X|_{S_{Y^\perp}}$ *is u.s.c.*
(f) $J_X|_{Y^\perp}$ *is u.s.c.*

Proof (a) $\Leftrightarrow$ (b) $\Leftrightarrow$ (c) follows from the above two theorems.
(c) $\Rightarrow$ (d). Note that $J_X(f)$ is a closed subset of $P_Y^{-1}\{0\}$ for f in $S_{Y^\perp}$, so is compact. Pick f in $S_{Y^\perp}$. If J_X is not H.u.s.c at f then there exist $\varepsilon > 0$ and sequences $\{f_n\} \subseteq S_{Y^\perp}$ converging to f and $\{x_n\} \subseteq J_X(f_n)$ such that

$$d(x_n, J_X(f)) \geq \varepsilon, \quad \text{for all } n \geq 1. \tag{3}$$

Now $\{x_n\} \subseteq P_Y^{-1}\{0\} \cap S_X$ for all $n \geq 1$. Hence $\{x_n\}$ has a convergent subsequence $\{x_{n_k}\}$ converging to, say x_0. Clearly

$$\|x_0\| = 1 = f(x_0) = \lim_{k\to\infty} f_{n_k}(x_{n_k}).$$

So, $x_0 \in J_X(f)$. This contradicts (3).
(d) $\Leftrightarrow$ (e) Follows from Fact 1.
(e) $\Leftrightarrow$ (f). This is so since $J_x(\alpha f) = \alpha J_X(f)$ for $\alpha > 0$ and f in X^*.
(d) $\Rightarrow$ (c). Suppose $P_Y^{-1}\{0\}$ is not boundedly compact. Then there exists a sequence $\{x_n\} \subseteq P_Y^{-1}\{0\} \cap S_X$, that has no convergent subsequence. Pick f_n in $S_{Y^\perp} \cap J_{X^*}(x_n)$ for each $n \geq 1$. Since codim $Y = \dim Y^\perp < \infty$, without loss of generality assume $\{f_n\}$ converges to $f \in S_{Y^\perp}$. Now (d) implies $d(x_n, J_X(f)) \to 0$ as $n \to \infty$. So there exists $\{v_n\} \subseteq J_X(f)$ such that $\|x_n - v_n\| < \frac{1}{n}$ for all $n \geq 1$. Now $\{v_n\}$ has a convergent subsequence, so $\{x_n\}$ also has a convergent subsequence. This gives a contradiction and completes the proof. □

We observe that u.s.c can not be replaced by l.s.c in Theorem 7, as shown by the following example from [9]. Let $X = \mathbb{R}^3$ with l_∞-norm and $Y = \{(\alpha, 0, 0) : \alpha \in \mathbb{R}\}$. Then P_Y is l.s.c since X is a finite dimensional polyhedral space (See comments at the end of Sect. 1) but it is easy to verify that $J_X|_{S_{Y^\perp}}$ is not l.s.c in this case. In general if $3 \leq \dim X < \infty$ and Y be a subspace of X with $\dim Y \leq \dim X - 2$ then P_Y is l.s.c but $J_X|_{S_{Y^\perp}}$ is not. However, the implication in the reverse direction holds and we describe the details below. We need some facts in the sequel. All the results given below in this section are from [9].

Theorem 8 *Let X be reflexive. If $J_X|_{S_{X^*}}$ is l.s.c then X is strictly convex.*

Proof Pick f_0 in S_{X^*}. Suppose $J_X(f_0)$ has two distinct elements. Since X is reflexive, by a result of Lindenstrauss there exists $\{f_n\} \subseteq S_{X^*}$ such that $\|.\|_{X^*}$ is smooth at f_n and $\{f_n\} \to f_0$. Now J_X is l.s.c at f_0, $J_X(f_n)$ is a singleton set for each n but $J_X(f_0)$ is not a singleton set, gives a contradiction. □

Fact 2 *Let X be Banach and Y be a factor reflexive subspace. Then $C = \{f \in S_{Y^\perp} : J_X(f) - J_X(f) \subseteq Y\}$ is dense in $S_{Y^\perp}$. Further if J_X is l.s.c then $C = S_{Y^\perp}$.*

Proof Since X/Y is reflexive, by a result of Lindenstrauss

$$\{f \in S_{(X/Y)^*} : f \text{ is a smooth point}\}$$

is dense in $S_{(X/Y)^*}$. Recall that $(X/Y)^* \simeq Y^\perp$ and note that if $f \in S_{Y^\perp}$ is a smooth point with $f(x) = f(z) = 1$ then $x + Y = z + Y$ or equivalently $x - z \in Y$. Hence C is dense in $S_{Y^\perp}$.

Now assume J_X is l.s.c. Since $X \setminus Y$ is open, the set

$$U = \{f \in S_{Y^\perp} : [J_X(f) - J_X(f)] \cap (X \setminus Y) \neq \emptyset\}$$

is open, as the map $f \to (J_X(f) - J_X(f))$ is l.s.c on X^*. Clearly $C \cap U = \emptyset$, U is open in $S_{Y^\perp}$, and C is dense in $S_{Y^\perp}$ gives a contradiction if U is nonempty. Hence $U = \emptyset$. □

Corollary 2 *If X is Banach and Y is a factor reflexive subspace of X then for every f in $S_{Y^\perp}$ and x in $J_X(f)$ we have $J_X(f) \subseteq x + Y$ and $x - P_Y(x) = J_X(f)$.*

Proof Since x is in $J_X(f)$ and by Remark 2, $\|x\| = \|x + Y\| = 1$. Let z be in $J_X(f)$. Then by Fact 2, $z - x \in Y$ and so $z = x - y$ for some y in Y. Clearly, $\|x - y\| = \|z\| = 1 = d(x, Y)$ and $y \in P_Y(x)$. So, $z \in x - P_Y(x)$ and $J_X(f) \subseteq x - P_Y(x)$.

Now for any y in $P_Y(x)$, we have

$$f(x - y) = f(x) = 1 \quad \text{and} \quad \|x - y\| = \|x + Y\| = 1.$$

Thus $x - P_Y(x) \subseteq J_X(f)$. □

We can now prove the main result. For a normed linear space X and x in X, we denote by $\hat{x}$, the image of x under the canonical embedding of X into X^{**}.

Theorem 9 *Let X be a Banach space and Y be a proximinal subspace of finite codimension in X. If $J_X|_{S_{Y^\perp}}$ is l.s.c., then P_Y is l.s.c.*

Proof We first observe that it is enough to prove P_Y is l.s.c. on $D_Y = \{x \in X : d(x, Y) = 1\}$. Pick any x in X with $d(x, Y) = \|x + Y\| = 1$. It suffices to show that P_Y is l.s.c. at x or equivalently $I - P_Y$ is l.s.c. at x. Let $\phi = \widehat{x}|_{Y^\perp}$. Then $\phi \in S_{(Y^\perp)^*}$. If $f \in J_{Y^\perp}(\phi)$ then $f(x) = \widehat{x}(f) = \phi(f) = 1$ and $x \in J_X(f)$. By the above corollary $x - P_Y(x) = J_X(f)$. Thus we have

$$J_X(f) = x - P_Y(x), \quad \text{for all } f \in J_{Y^\perp}(\phi). \tag{4}$$

Hence to show $I - P_Y$ is l.s.c. at x, it is enough to prove the following: Given y_0 in $P_Y(x)$ and $\varepsilon > 0$, there exists $\eta > 0$ such that if $z \in D_Y$ and $\|x - z\| < \eta$ then we have $B(x - y_0, \varepsilon) \cap (z - P_Y(z)) \neq \emptyset$.

Pick any f in $J_{Y^\perp}(\phi)$. Since J_X is l.s.c. at f and (4) holds, there exists $\delta_f > 0$ such that $g \in S_{Y^\perp}$, $\|f - g\| < \delta_f$ implies $B(x - y_0, \varepsilon) \cap J_X(g) \neq \emptyset$. Since $J_{Y^\perp}(\phi) \subseteq S_{Y^\perp}$ is closed and compact, the open cover $\left\{B\left(f, \frac{\delta_f}{2}\right) : f \in S_{Y^\perp}\right\}$ has a finite subcover, say, $\left\{B\left(f_i, \frac{\delta_{f_i}}{2}\right) : 1 \leq i \leq k\right\}$. If

$$0 < 2\delta < \min\{\delta_{f_i} : 1 \leq i \leq k\},$$

then for any $f \in J_{Y^\perp}(\phi)$ and $g \in S_{Y^\perp}$ satisfying $\|f - g\| < \delta$ we have

$$B(x - y_0, \varepsilon) \cap J_X(g) \neq \emptyset. \tag{5}$$

Now dim $Y^\perp < \infty$. Using usual compactness arguments, it is easily shown that the map $J_{Y^\perp}$ is H.u.s.c. on $(Y^\perp)^*$. In particular, $J_{Y^\perp}$ is H.u.s.c. at ϕ. So there exists $\eta > 0$ such that $\psi \in S_{(Y^\perp)^*}$, $\|\phi - \psi\| < \eta$ and $g \in J_{Y^\perp}(\psi)$ implies $\|f - g\| < \delta$ for some f in $J_{Y^\perp}(\phi)$. Consequently, (5) holds.

Now pick any $z \in D_Y$ satisfying $\|x - z\| < \eta$. Let $\psi = \widehat{z}|_{Y^\perp}$. Then $\psi \in S_{(Y^\perp)^*}$, $\|\phi - \psi\| < \eta$. Pick any $g \in J_{Y^\perp}(\psi)$. We have $J_X(g) = z - P_Y(z)$ and this with (5) implies

$$B(x - y_0, \varepsilon) \cap (z - P_Y(z)) \neq \emptyset.$$

□

Note that P_Y and J_X are single valued if X is strictly convex. Hence, we have the following Corollary of the above theorem.

Corollary 3 *Let X be reflexive and assume $J_X|_{S_{X^*}}$ is l.s.c. Then every closed linear subspace Y of finite codimension has a continuous metric projection.*

3 Metric Projection onto Subspaces of Finite Codimension

In this section,we list some recent results which derive continuity properties of metric projections onto subspaces of finite codimension, using "polyhedral" related geometric conditions. However, we begin with a well known result from [12], giving a sufficient condition for continuity of metric projection in reflexive, strictly convex spaces and then describe a striking negative result of P. D. Morris regarding continuity of metric projections onto Chebeychev subspaces of finite codimension in the space $C(Q)$.

Theorem 10 (Glicksberg [12]) *Let X be reflexive and strictly convex Banach space and every $f \in X^*$ is Fréchet smooth. Then the metric projection onto every closed subspace of X is continuous.*

Proof Let Y be a closed subspace of X. Then Y is Chebyshev. Pick x in D_Y. Then there exists $f \in S_{Y^\perp}$ such that $f(x) = 1 = d(x, Y) = \left\|\widehat{x}|_{Y^\perp}\right\|$. Let z in X^{**} be the unique norm preserving extension of $\widehat{x}|_{Y^\perp}$ to X^*. As X is reflexive, z is in X and $\{x - z\} = P_Y(x)$, equivalently $Q_Y(x) = \{z\}$. We will show that Q_Y is continuous at x.

Since X is reflexive and $\|.\|_{X^*}$ is Fréchet smooth at f, given $\varepsilon > 0$ there exists $\delta > 0$ such that $w \in S_X$ and $f(w) > 1 - \delta$ imply $\|z - w\| < \delta$. Let $u \in D_Y$ and assume $\|x - u\| < \delta$. If $\{v\} = Q_Y(u)$, then $f(v) = f(u) > 1 - \delta$ and therefore $\|z - v\| < \varepsilon$ and Q_Y is continuous at x. □

Let H be a hyperplane or a subspace of codimension 1 in X. It is well known that $H = \ker f$, for some f in X^* and H is proximinal in X if and only if the set $J_X(f)$ is nonempty.

We recall that if X is a Banach space, then using the famous James Theorem, we have X is reflexive if and only if every hyperplane is proximinal.

Also, $P_H(x) = \left\{x - \frac{f(x)}{\|f\|} J_X(f)\right\}$ for any x in X and it can be thus easily shown that P_H is Hausdorff metric continuous.

The situation is dramatically different if we consider proximinal subspaces of codimension ≥ 2. Below, we describe a striking negative result of P. D. Morris, which says that if $X = C(Q)$ and Y is a Chebychev subspace of codimension ≥ 2, then P_Y is not continuous on X.

We need the following facts about Chebyshev subspaces in the sequel.

Fact 3 [15] *Let Y be a Chebyshev subspace of finite codimension of a normed linear space X. If the metric projection P_Y is continuous on X, then $S_{X/Y}$ is homeomorphic to $S_X \cap P_Y^{-1}\{0\}$.*

Proof Let $W : X \to X/Y$ be the quotient map, I the identity map on X and $V : X/Y \to P_Y^{-1}\{0\}$ given by $V(x + Y) = x - P_Y(x)$, $x \in X$. It is easy to check that V is a well defined bijective map on X/Y. We now observe that the following diagram commutes.

Since W is open and continuous, V is continuous on X/Y if P_Y is continuous. Note that the inverse V^{-1} of V, is the restriction of the quotient map W to the set $P_Y^{-1}\{0\}$ and hence continuous. Thus, V is a homeomorphism if and only if P_Y is continuous. Further, V is an isometry and therefore, $V(S_{X/Y})$ is the set $S_X \cap P_Y^{-1}\{0\}$. This completes the proof. □

The characterizations, given below, of semi-Chebyshev subspaces is from [23].

Proposition 1 *Let Y be a subspace of X. Then Y is semi-Chebyshev if and only if there do not exist $f \in S_{Y^\perp}$, $x \in X$ and $y_0 \in Y$ such that $f(x_0) = \|x_0\| = \|x_0 - y_0\|$.*

The following result [23, Theorem 2.1] for a subspace of codimension n shows that semi-Chebyshevity of Y restricts the dimension of the set $J_X(f)$, for each nonzero f in $Y^{\perp}$. If A is a set, aff A denotes the affine hull of A and rel.int A denotes the relative interior of A (interior of A with respect to aff A).

Proposition 2 *Let X be a Banach space and Y be a subspace of codimension n in X. Then Y is semi-Chebyshev implies that for every f in $Y^{\perp} \setminus \{0\}$ the set $J_X(f)$ is of dimension $r \leq n-1$.*

Proof Assume that there exists f in $Y^{\perp} \setminus \{0\}$ with $J_X(f)$ having dimension $\geq n$. Then $J_X(f)$ contains $n+1$ affinely linearly independent elements, say, $\{w_1, w_2, \ldots, w_{n+1}\}$. Let $A = co\{w_1, w_2, \ldots, w_{n+1}\}$. Then A is a compact and convex subset of $J_X(f)$ and dim $A = n$. Recall $rel.int A \neq \emptyset$ and pick x_0 in $rel.int A$. Then $0 \in rel.int(A - x_0)$. We have $Y_1 = sp(A - x_0) = aff(A) - x_0$. Then dim $Y_1 = n$ and $0 \in int(A - x_0)$, considered as a subset of Y_1. Thus every z in Y_1 is a positive multiple of an element in $A - x_0$.

Now if $Y_1 \cap Y = \{0\}$ then $X = Y \oplus Y_1$, since codim $Y = n =$ dim Y_1. Now $f \equiv 0$ on $A - x_0$ and therefore $f \equiv 0$ on Y_1 and $f \equiv 0$ on Y. This is a contradiction to $\|f\| = 1$. Hence there exists y_0 in $Y_1 \cap Y \setminus \{0\}$. Choose $\delta > 0$ small so that $-\delta y_0 \in A - x_0$. That is, $x_0 - \delta y_0 \in A$. Note that $\delta y_0 \neq 0$ and $x_0 - \delta y_0 \in A \subseteq J_X(f)$. Since $x_0 \in J_X(f)$, using Proposition 1 we get a contradiction to Y being Chebyshev. □

We now apply the above result to the space $C(Q)$. Let Y be a proximinal subspace of finite codimension in $C(Q)$. Assume $\mu \in S_{Y^{\perp}}$ and $Q \setminus S(\mu)$ has r points say $\{q_1, \ldots, q_r\}$. Define x and x_i, $1 \leq i \leq r$ in $C(Q)$, with norm one and satisfying

$$x(t) = \begin{cases} 1 & \text{if } t \in S(\mu^+), \\ -1 & \text{if } t \in S(\mu^-), \\ 0 & \text{Otherwise}, \end{cases}$$

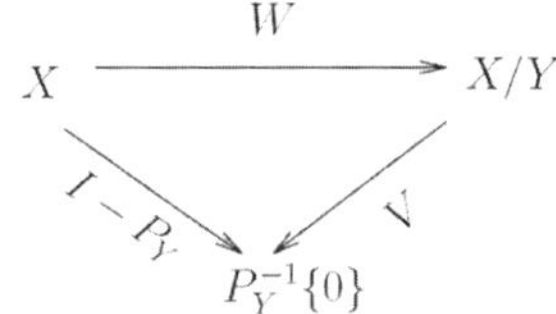

and

$$x_i(t) = \begin{cases} 1 & \text{if } t \in S(\mu^+) \cup \{q_i\}, \\ -1 & \text{if } t \in S(\mu^-), \\ 0 & \text{Otherwise}. \end{cases}$$

Then $\{x, x, \ldots, x_r\} \subseteq J_{C(Q)}(\mu)$ is a linearly independent set. If Y is Chebyshev then $r \leq$ dim $J_{C(Q)}(\mu) \leq n-1$. Hence $Q \setminus S(\mu)$ has at most $n-2$ points. Since this set is open, it is contained in the set of isolated points of Q.

We are now in a position to prove the result of Morris mentioned earlier.

Theorem 11 (Morris [22]) *Let Y be a Chebyshev subspace of finite codimension in $C(Q)$. Then P_Y is continuous if and only if $codim\ Y = 1$.*

Proof Since Q is infinite compact, Q has a limit point, say q_0. For any $\mu \in Y^{\perp}$, $Q \setminus S(\mu)$ consist of isolated points. So $q_0 \in S(\mu)$. Since $\mu \in Y^{\perp}$ was chosen arbitrarily $q_0 \in \cap\{S(\mu) : \mu \in Y^{\perp}\}$.

Pick x in $P_Y^{-1}\{0\} \cap S_X$. Then there exists a $\mu \in S_{Y^{\perp}}$ such that $\mu(x) = 1$. Hence $|x(q_0)| = 1$. Set $A = \{x \in P_Y^{-1}\{0\} \cap S_X : x(q_0) = 1\}$. Thus both A and $-A$ are non empty closed sets, $A \cap -A = \phi$. Further $P_Y^{-1}\{0\} \cap S_X = A \cup -A$ is disconnected.

Now assume P_Y is continuous. Then by Fact 3, $S_X \cap P_Y^{-1}\{0\}$ is homeomorphic to $S_{X/Y}$ and so $S_{X/Y}$ is disconnected. This implies $\dim X/Y = 1$. □

It was thought that negative results like the one above, may not occur in "nice" spaces. For instance, it was conjectured that if X is reflexive and strictly convex, the scenario may be different and the metric projections onto subspaces of X would be continuous. However, in [4], Brown constructed an example of a proximinal subspace of codimension 2 in a reflexive, strictly convex space with a discontinuous metric projection. However, some of the recent results show that there are indeed a large class of spaces, for which the metric projection onto subspaces of finite codimension have strong continuity properties. We describe them below.

We recall that in a finite dimensional polyhedral space X, the metric projection P_Y is l.s.c. for every subspace Y of X. The following series of results show that polyhedral condition plays a crucial role in the continuity properties of metric projections onto proximinal subspaces of finite codimension too. As far as we are aware, the first result of this kind was given in [14]. Recall that $NA(X)$ denotes the set of norm attaining functionals on X and by $NA_1(X)$, we denote those functionals in $NA(X)$ of norm one.

Theorem 12 [14] *Let X be a Banach space and Y be a subspace of finite codimension in X with $Y^{\perp}$ polyhedral. Assume that $Y^{\perp} \subseteq NA(X)$. Then Y is proximinal and the metric projection P_Y has a continuous selection.*

The geometric notion of *QP-points* [26] was utilized in the same paper to get a sufficient condition for strong proximinality and hence for H.u.s.c. of metric projections. To describe the relevant results from this paper, we need the following two definitions that are central to this part of the discussion.

Definition 5 Let X be a Banach space and $F : X \to \mathbb{R}$ be a convex function. We say F is *strongly subdifferentiable* (SSD) at $x \in X$, if the one sided limit $\lim\limits_{t \to 0^+} \frac{F(x+ty)-F(x)}{t}$ exists uniformly for $y \in S_X$.

Definition 6 [26] Let X be a Banach space. An element x in S_X is called a *Quasi-polyhedral point (QP-point)* if there exists $\delta > 0$ such that $J_{X^*}(y) \subseteq J_{X^*}(x)$ for every y in $B(x, \delta) \cap S_X$. If every element of S_X is a *QP-point* of X, then X is said to be a *QP-space*.

We now have the following result linking the *QP-points* and SSD points.

Lemma 1 [14] *Let X be a Banach space and x in S_X be a QP-point. Then the norm of X is SSD at x.*

Proof Since x is a *QP-point*, there exists $\delta > 0$ such that if $\omega \in B(x, 2\delta) \cap S_X$ then $J_{X^*}(\omega) \subseteq J_{X^*}(x)$. Select any $y \in S_X$ and fix $0 < t < \delta$. If $\omega = \frac{x+ty}{\|x+ty\|}$ then $\omega \in S_X$ and $\|x - \omega\| < 2\delta$. Thus $J_{X^*}(x + ty) = J_{X^*}(\omega) \subseteq J_{X^*}(x)$.

Pick any s such that $0 < s < t$ and let $\lambda = s/t$. Then $x+sy = \lambda(x+ty)+(1-\lambda)x$. Now, for any f in $J_{X^*}(x + ty)$,

$$\|x + sy\| \geq f(x + sy) = \lambda\|x + ty\| + (1 - \lambda)\|x\| \geq \|x + sy\|.$$

Hence $f(x + sy) = \|x + sy\|$ and

$$\frac{\|x + sy\| - \|x\|}{s} = \frac{f(x + sy) - f(x)}{s} = f(y).$$

It is now clear that for all y in S_X, we have

$$\lim_{s\to0^+} \frac{\|x + sy\| - \|x\|}{s} = \frac{\|x + ty\| - \|x\|}{t} = \frac{f(x + ty) - f(x)}{t} = f(y)$$

and the norm of X is SSD at x. □

The following useful characterization of SSD points of the norm of the dual space leads to a characterization of strongly proximinal hyperplanes.

Theorem 13 [14] *Let X be a Banach space and $f \in S_{X^*}$. Then the norm of X^* is SSD at f if and only if $f \in NA_1(X)$ and given $\varepsilon > 0$ there exists $\delta_\varepsilon > 0$, such that for $x \in B_X$ satisfying $f(x) > 1 - \delta_\varepsilon$, we have $d(x, J_X(f)) < \varepsilon$.*

Corollary 4 [14] *Let X be a Banach space and $f \in X^*$. Then $H = \ker f$ is strongly proximinal in X if and only if the norm of X^* is SSD at f.*

We now give a characterization of *QP-points*, that helps to visualize *QP-points* on the unit sphere.

Fact 4 *Let X be a Banach space and $x \in S_X$. Then x is a QP-point of X if and only if there exists $\varepsilon > 0$ such that if $y \in S_X$ and $\|x - y\| < \varepsilon$, then the line segment $[x, y]$ lies on the sphere S_X.*

Proof Since x is a *QP-point*, there exists $\varepsilon > 0$ such that $J_{X^*}(y) \subseteq J_{X^*}(x)$ for all $y \in S_X \cap B(x, \varepsilon)$. Select any $y \in S_X \cap B(x, \varepsilon)$ and $f \in J_{X^*}(y)$. Then $f \in J_{X^*}(x)$ and if $\omega \in [x, y]$ then $f(x) = f(y) = f(\omega) = 1$. Since $\|\omega\| \leq 1$, this implies $\|\omega\| = 1$ and $[x, y] \subseteq S_X$.

For the converse, let $x \in S_X$ and select $\varepsilon > 0$ so that the given condition holds. Let $\alpha = \varepsilon/2$ and $z \in S_X \cap B(x, \alpha)$. Considering the 2-dimensional subspace generated by x and z and using the given condition, we can easily get $y \in B(x, \varepsilon) \cap S_X$ such that $z \in (x, y)$. So there exists $\lambda, 0 < \lambda < 1$, such that $z = \lambda x + (1 - \lambda)y$. If $f \in J_{X^*}(z)$ then

$$f(z) = 1, \ f(x) \leq 1 \text{ and } f(y) \leq 1,$$

and so $1 = f(z) = \lambda f(x) + (1 - \lambda) f(y) \leq 1$. This implies $f(x) = f(y) = 1$ and $f \in J_{X^*}(x)$. Thus $J_{X^*}(z) \subseteq J_{X^*}(x)$ for all $z \in S_X \cap B(x, \alpha)$ and x is a *QP-point* of X. □

For a norm attaining functional f in X^*, the functional f being a *QP-point* can be characterized in terms of the sets $J_X(.)$, instead of the sets $J_{X^{**}}(.)$, as the following result shows.

Fact 5 *Let X be a Banach space and $f \in NA_1(X)$. Then f is a QP-point of X^* if there exists $\alpha > 0$ such that $J_X(g) \subseteq J_X(f)$, for all $g \in B(f, \alpha) \cap NA_1(X)$.*

Proof The necessity follows the above Fact. To prove sufficiency, let $f \in NA_1(X)$ satisfy the condition of the lemma. If $g \in NA_1(X)$ and $\|f - g\| < \varepsilon$, then by assumption $J_X(g) \subseteq J_X(f)$. Pick any z in $J_X(g)$. Then $f(z) = g(z) = 1$ and so $(f + g)(z) = 2$. Since $\|z\| = 1$, this implies $\|f + g\| = 2$. Hence

$$\|f + g\| = 2 \ \text{ for all } g \in B(f, \varepsilon) \cap NA_1(X).$$

By the Bishop-Phelps theorem, the set $B(f, \varepsilon) \cap NA_1(X)$, is dense in $B(f, \varepsilon) \cap S_{X^*}$ and this with the continuity of the norm function yields

$$\|f + g\| = 2 \text{ for all } g \in B(f, \varepsilon) \cap S_{X^*}.$$

It is now easy to verify the above equality implies $[f, g] \subseteq S_{X^*}$ if $g \in B(f, \varepsilon/2) \cap S_{X^*}$. By Fact 4, f is a *QP-point* of X^*. □

We now fix some notation, used hereafter. Let X be a normed linear space and $\{f_1 \dots f_n\} \subseteq X^*$. We define subsets $J_X(f_1, \dots f_i)$ for $1 \leq i \leq n$ inductively as follows.

$$J_X(f_1) = \{x \in B_X : f_1(x) = \|f_1\|\}.$$

Having defined $J_X(f_1)$ we define

$$J_X(f_1, \dots f_i) = \{x \in J_X(f_1 \dots f_{i-1}) : f_i(x) = \sup\{f_i(y) : y \in J_X(f_1 \dots f_{i-1})\}\}$$

for $2 \leq i \leq n$. Note that $J_X(f_1) \neq \emptyset \Leftrightarrow f_1 \in NA(X)$. The sets $J_X(f_1 \dots f_i)$ can be empty and if nonempty, they are faces of B_X.

However, if X is finite dimensional then the sets $J_X(f_1 \dots f_i)$ are nonempty for $1 \leq i \leq n$. Further if $\dim X = n$ and $(f_1 \dots f_n)$ is a basis of X^*, then $J_X(f_1 \dots f_n)$ is a singleton set. We set

$$\alpha_i = \sup\{f_i(x) : x \in J_X(f_1, \ldots, f_{i-1})\}, \quad \text{for } 2 \le i \le n.$$

Clearly,

$$\begin{aligned} J_X(f_1 \ldots f_i) &= \{x \in J_X(f_1 \ldots f_{i-1}) : f_i(x) = \alpha_i\} \\ &= \bigcap_{j=1}^{i} \left\{x \in B_X : f_j(x) = \alpha_j\right\} \end{aligned}$$

for $2 \le i \le n$. Further if $x_0 \in J_X(f_1 \ldots f_i)$ then,

$$J_X(f_1 \ldots f_i) = \left\{x \in B_X : f_j(x) = f_j(x_0) \text{ for } 1 \le j \le i\right\}, \quad \text{for } 1 \le i \le n.$$

Theorem 14 *Let X be a Banach space and Y be a proximinal subspace of finite codimension n in X. Then Y is strongly proximinal if and only if for every basis $f_1, \ldots, f_n$ of $Y^\perp$*

$$\lim_{\varepsilon \to 0} \sup\{d(y, J_X(f_1 \ldots f_i)) : y \in J_X(f_1 \ldots, f_i, \varepsilon)\} = 0$$

for $1 \le i \le n$.

Remark 4 It can be shown that [14] if X is a Banach space and Y is a subspace of finite codimension in X such that each functional in $Y^\perp \cap S_{X^*}$ is a *QP-point* of X^*, then Y is proximinal in X.

We are now in a position to prove the following theorem.

Theorem 15 [14] *Let X be a Banach space such that every f in $NA(X) \cap S_{X^*}$ is a QP-point of X^*. Then every subspace Y of finite codimension $Y^\perp \subseteq NA(X)$ is strongly proximinal and P_Y is H.u.s.c.*

Proof By the above Remark, Y is proximinal in X. By Lemma 1, every $f \in NA_1(X)$ is a SSD point of X^*.

Let $(f_1, \ldots, f_n)$ be a basis of $Y^\perp$. We now show that we can select positive scalars λ_i, $1 \le i \le n$, such that

$$J_X(f_1, \ldots, f_i) = J_X\left(\sum_{j=1}^{i} \lambda_j f_j\right), \quad \text{for } 1 \le i \le n. \tag{6}$$

We use induction on n. We take $\lambda_1 = 1$ and note that the case $n = 1$ is trivial. Inductively assume that $\lambda_j > 0$ for $1 \le j \le i-1$ have been chosen so that if $g_{i-1} = \sum_{j=1}^{i-1} \lambda_j f_j$ then $J_X(g_{i-1}) = J_X(f_1, f_2, \ldots, f_{i-1})$. Now $g_{i-1} \in Y^\perp$ and so is a *QP-point* of X^*, by assumption. Using Fact 5, choose $\lambda_i > 0$ small enough so that

$$J_X(g_{i-1} + \lambda_i f_i) \subseteq J_X(g_{i-1}).$$

By induction assumption,

$$J_X(g_{i-1}, f_i) = J_X(f_1, f_2, \ldots, f_{i-1}, f_i).$$

We have $J_X(g_{i-1} + \lambda_i f_i) \subseteq J_X(g_{i-1})$ and $\lambda_i > 0$. It is now easy to verify that $J_X(g_{i-1}, f_i) = J_X(g_{i-1} + \lambda_i f_i)$ and we have

$$J_X\left(\sum_{j=1}^{i} \lambda_j f_j\right) = J_X(g_{i-1} + \lambda_i f_i) = J_X(g_{i-1}, f_i) = J_X(f_1, f_2, \ldots, f_{i-1}, f_i). \quad (7)$$

This completes the induction and (6) holds. It now follows that

$$x \in J_X\left(\sum_{j=1}^{i} \lambda_j f_j\right) \Rightarrow f_i(x) = \alpha_i \text{ for } 1 \leq i \leq n,$$

where $\alpha_1 = \|f_1\|$ and $\alpha_i = \sup\{f_i(y) : y \in J_X(f_1, f_2, \ldots, f_{i-1})\}$, for $2 \leq i \leq n$.

We now proceed to show that the condition of Theorem 13 holds for the basis $(f_1, f_2, \cdots, f_n)$. Recall that

$$J_X(f_1 \ldots, f_i, \varepsilon) = \bigcap_{j=1}^{i} \{x \in B_X : f_j(x) > \alpha_j - \varepsilon\}.$$

Now $\sum_{j=1}^{i} \lambda_j f_j \in Y^{\perp} \subseteq NA(X) \subseteq$ *QP-points* of X^*, for $1 \leq i \leq n$. Thus the norm of X^* is SSD at $\sum_{j=1}^{i} \lambda_j f_j$ for $1 \leq i \leq n$. So by Theorem 13

$$\lim_{\varepsilon \to 0} \sup\left\{d\left(y, J_X\left(\sum_{j=1}^{i} \lambda_j f_j\right)\right) : y \in J_X\left(\sum_{j=1}^{i} \lambda_j f_j, \varepsilon\right)\right\} = 0$$

for $1 \leq i \leq n$. It is easy to check that this with (7) implies

$$\lim_{\varepsilon \to 0} \sup\{d(y, J_X(f_1 \ldots f_i)) : y \in J_X(f_1 \ldots, f_i, \varepsilon)\} = 0$$

for $1 \leq i \leq n$. By Theorem 14, Y is strongly proximinal in X. □

It is a natural question to ask whether a stronger conclusion in Theorem 12 is possible. More precisely, can we conclude P_Y is l.s.c. under the conditions of Theorem 12?

The proof of Theorem 12 is rather short and essentially makes use of a property of finite dimensional polyhedral space that allows continuous selection for measures supported on extreme points. However, an imitation of the same proof did not seem to carry further. Relatively long and elaborate proofs were used to show that P_Y is l.s.c. in Theorem 12 if

(i) X is a subspace of c_0. Indumathi [17]
(ii) X is a separable Banach space with Property (*) (V. P. Fonf and J. Lindenstrauss, Pre-print 2003. See also Fonf et al. [13])

where Property (*) as in Definition 9, below.

We observe that (ii) is a generalization of the earlier result (i). However, it was shown in [16] that no additional condition on the Banach space X is , in fact , needed. More precisely,

Theorem 16 [16] *Let X be a Banach space and Y be a proximinal subspace of finite codimension in X with $Y^\perp$ polyhedral. Then P_Y is l.s.c.*

We need some definitions and preliminary results to prove the above theorem. Let X be a Banach space, Y be a closed subspace of finite codimension in X. Set

$$Q_Y(x) = x - P_Y(x), x \in X$$

and for a finite subset $\{f_1, \ldots, f_k\}$ of $Y^\perp$, let

$$Q_{f_1,\ldots,f_k}(x) = \bigcap_{i=1}^{k} \{y \in B_X : f_i(y) = f_i(x)\}.$$

Clearly, $Q_{f_1,\ldots,f_k}(x)$ is either empty or convex and the domain of the set valued map $Q_{f_1,\ldots,f_k}$ will be taken as the set $D_Y = \{x \in X : d(x, Y) = 1\}$ in the sequel. Note that if $\{f_1, \ldots, f_k\} \subseteq Y^\perp$ then $Q_{f_1,\ldots,f_k}(x) \subseteq Q_Y(x)$ and equality holds if $\{f_1, \ldots, f_k\}$ is a basis of $Y^\perp$. Further if Y is proximinal, $Q_Y(x)$ is nonempty for each x and hence $Q_{f_1,\ldots,f_k}(x)$ is nonempty in this case. For $k > 1$ and $x \in D_Y$, define

$$\alpha_{x,k} = \inf\{f_k(z) : z \in Q_{f_1,\ldots,f_{k-1}}(x)\}$$

and

$$\beta_{x,k} = \sup\{f_k(z) : z \in Q_{f_1,\ldots,f_{k-1}}(x)\}.$$

We now have the following Proposition from [17].

Proposition 3 *Let X be a Banach space, Y be proximinal in X and $x \in D_Y$. Assume that there exists a finite subset $\{f_1, \ldots, f_{k+1}\}$, $1 \leq k < n$, of $Y^\perp$ such that the map $Q_{f_1,\ldots,f_k}$ is H.l.s.c at x and further*

$$\alpha_{x,k+1} < f_{k+1}(x) < \beta_{x,k+1}.$$

Then $Q_{f_1,\ldots,f_{k+1}}$ *is H.l.s.c at* x.

Proof Let $2\eta = \min\{\beta_{x,k+1} - f_{k+1}(x), f_{k+1}(x) - \alpha_{x,k+1}\}$. Then $\eta > 0$. Since $Q_{f_1,\ldots,f_k}$ is H.l.s.c at x, given $\varepsilon > 0$, there exists $\delta > 0$ such that for any z in $Q_{f_1,\ldots,f_k}(x)$ and y in D_Y with $\|x - y\| < \delta$, there exists w in $Q_{f_1,\ldots,f_k}(y)$ such that $\|z - w\| < \frac{\eta\varepsilon}{8}$. Without loss of generality we assume that $0 < \delta < \frac{\eta\varepsilon}{8}$, $0 < \varepsilon < 1$, and $\|f_i\| = 1$ for $1 \leq i \leq n$. Now, if $y \in D_Y$ and $\|x - y\| < \delta$, it follows easily that

$$\beta_{y,k+1} > \beta_{x,k+1} - \frac{\eta}{8} \ , \ \alpha_{y,k+1} < \alpha_{x,k+1} + \frac{\eta}{8} \tag{8}$$

$$\alpha_{y,k+1} < f_{k+1}(y) < \beta_{y,k+1}. \tag{9}$$

Fix $z \in Q_{f_1,\ldots,f_{k+1}}(x)$. We have to show that there exists v in $Q_{f_1,\ldots,f_{k+1}}(y)$ such that $\|z - v\| < \varepsilon$.

Since $Q_{f_1,\ldots,f_{k+1}}(x) \subseteq Q_{f_1,\ldots,f_k}(x)$, there exists w in $Q_{f_1,\ldots,f_k}(y)$ such that $\|z - w\| < \frac{\eta\varepsilon}{8}$. We have

$$f_{k+1}(z) = f_{k+1}(x), \quad \|w - z\| < \frac{\eta}{8}, \quad \|x - y\| < \frac{\eta\varepsilon}{8} < \frac{\eta}{8}.$$

This together with (8) and (9) implies

$$\begin{aligned}\beta_{y,k+1} - f_{k+1}(w) &= \beta_{y,k+1} - \beta_{x,k+1} + \beta_{x,k+1} - f_{k+1}(x)\\ &\quad + f_{k+1}(x) - f_{k+1}(z) + f_{k+1}(z) - f_{k+1}(w) > 2\eta - \frac{\eta}{8} + \frac{\eta}{8} > \eta.\end{aligned} \tag{10}$$

Similarly we can show that

$$f_{k+1}(w) - \alpha_{y,k+1} > \eta. \tag{11}$$

Also,

$$\begin{aligned}|f_{k+1}(y) - f_{k+1}(w)| &\leq |f_{k+1}(w) - f_{k+1}(z)| + |f_{k+1}(z) - f_{k+1}(x)|\\ &\quad + |f_{k+1}(x) - f_{k+1}(y)|\\ &< \frac{\eta\varepsilon}{8} + \frac{\eta\varepsilon}{8} = \frac{\eta\varepsilon}{4} < \frac{\eta}{4}.\end{aligned} \tag{12}$$

If $f_{k+1}(w) = f_{k+1}(y)$, then $w \in Q_{k+1}(y)$ and $\|w - z\| < \varepsilon$. Take $v = w$ in this case. Otherwise, we slightly perturb w to get an element of $Q_{f_1,\ldots,f_{k+1}}(y)$ as follows. Note that using (10)–(12), we can get w_1 in $Q_{f_1,\ldots,f_k}(y)$ such that

$$|f_{k+1}(w_1) - f_{k+1}(w)| > \eta, \tag{13}$$

and $f_{k+1}(y)$ lies in between $f_{k+1}(w)$ and $f_{k+1}(w_1)$. Choose $0 < \lambda < 1$ such that

$$f_{k+1}(\lambda w + (1 - \lambda)w_1) = f_{k+1}(y)$$

and take $v = \lambda w + (1-\lambda) w_1$. Since w and w_1 are in $Q_{f_1,\ldots,f_k}(y)$, v is in $Q_{f_1,\ldots,f_{k+1}}(y)$. Also,

$$(1-\lambda)[f_{k+1}(w_1) - f_{k+1}(w)] = f_{k+1}(y) - f_{k+1}(w).$$

This together with (12) and (13) gives

$$1 - \lambda < \frac{\eta\varepsilon}{4\eta} = \frac{\varepsilon}{4}.$$

Hence

$$\|w - v\| = (1-\lambda)\,\|w - w_1\| \leq 2(1-\lambda) < \frac{2\varepsilon}{4} = \frac{\varepsilon}{2},$$
$$\|z - v\| \leq \|z - w\| + \|w - v\| < \frac{\varepsilon}{2} + \frac{\varepsilon}{2} = \varepsilon.$$

Hence the proof is complete. □

Definition 7 Let Y be a proximinal subspace of codimension n in a Banach space X and x, an element of D_Y. We say x is a *k-corner point*, $1 \leq k \leq n$, with respect to a linearly independent set of functionals $\{f_1, \ldots, f_k\}$ in $Y^{\perp}$ if $Q_{f_1,\ldots,f_k}(x) = \cap_{i=1}^{k} J_X(f_i)$.

We now require some well known facts about finite dimensional convex sets. Let E be a finte dimensional normed linear space. For $C \subseteq A \subseteq E$, where C is convex and A is affine, by "interior of C with respect to A", we mean the interior of C, considered as a subset of the affine space A. The set of all extreme points of C would be denoted by $extC$. A subset D of C is called *extremal* if D contains an interior point of a line segment l in C then D contains l. Clearly, a singleton extremal set is an extreme point.

We now state the following result from [16].

Lemma 2 *Let X be a Banach space, Y be a proximinal subspace of finite codimension n in X with $Y^{\perp}$ polyhedral. For each x_0 in D_Y, there is a basis $\{f_1, \ldots, f_n\}$ of $Y^{\perp}$ such that $Q_{f_1,\ldots,f_n}(x_0) = \bigcap_{i=1}^{k} J_X(f_i)$ either with $k = n$, or with $1 \leq k \leq n$. In the later case we have*

$$\alpha_{x_0,j} < f_j(x_0) < \beta_{x_0,j}, \quad \text{for } k+1 \leq j \leq n.$$

The following theorem is an immediate from Proposition 3 and Lemma 2.

Theorem 17 [16] *Let X be a Banach space, Y a proximinal subspace of finite codimension n in X with $Y^{\perp}$ polyhedral. Assume that, whenever x in D_Y is a k-corner point with respect to a set of linearly independent functionals $\{f_1, \ldots, f_k\}$ in $Y^{\perp}$ for some positive integer k, $1 \leq k \leq n$, then the map $Q_{f_1,\ldots,f_k}$ is H.l.s.c at x. Then the metric projection P_Y is H.l.s.c on X.*

The result below shows that the conditions of the above theorem hold under natural assumptions. More specifically, we have

Lemma 3 *Let X be a Banach space and Y be a proximinal subspace of finite codimension in X, with* $Y^{\perp}$ *polyhedral. If* x_0*, in* D_Y*, is a k-corner point with respect to a set of linearly independent functionals* $\{f_1, \ldots, f_k\}$ *in* $Y^{\perp}$*, then the set valued map* $Q_{f_1,\ldots,f_k}$ *is H.l.s.c at* x_0*.*

It is now easily seen that Theorem 16 follows from Theorem 17 and Lemma 3 above. Hence to complete the proof of Theorem 16, it suffices to prove Lemma 3.

To prove Lemma 3, we need the result quoted below. By $(\mathbb{R}^+)$, we denote the set of non-negative real numbers.

Proposition 4 *Let E be a finite dimensional polyhedral space and let* $ext\ B_E = \{e_1, \ldots, e_m\}$*. Then there exists a continuous map* $A : B_E \to (\mathbb{R}^+)^m$ *such that, if* $A(x) = (\mu_i(x))_{i=1}^m$*, then*

$$\sum_{i=1}^{m} \mu_i(x) = 1, \quad and \quad x = \sum_{i=1}^{m} \mu_i(x) e_i$$

for all x in B_E*.*

We now continue the proof of Lemma 3.

Proof The finite dimensional space $Y^{\perp}$ is polyhedral and so is its dual, $(Y^{\perp})^*$. Thus $B_{(Y^{\perp})^*}$ has only finite number of extreme points. As

$$S_{(Y^{\perp})^*} = \{\phi_x : x \in D_Y\},$$

there exists a finite subset $\{x_1, \ldots, x_m\}$ of D_Y such that

$$ext\ B_{(Y^{\perp})^*} = \{\phi_{x_1}, \ldots, \phi_{x_m}\}.$$

Let x be in D_Y. Then ϕ_x is in $S_{(Y^{\perp})^*}$. Taking $E = (Y^{\perp})^*$ in Proposition 4, let $A(\phi_x) = (\mu_i(\phi_x))_{i=1}^m$. Since the map $x \to \phi_x$ is continuous, the map $x \to A(\phi_x))$ is continuous from D_Y into $(\mathbb{R}^+)^m$. We abbreviate, $\mu_i(\phi_x)$ as $\mu_i(x)$ for $1 \leq i \leq m$. Then $\sum_{i=1}^m \mu_i(x) = 1$ and

$$\phi(x) = \sum_{i=1}^{m} \mu_i(x) \phi_{x_i}. \tag{14}$$

By assumption, x_0 is in D_Y and is a k-corner point with respect to a set of linearly independent functionals $\{f_1, \ldots, f_k\}$ in $Y^{\perp}$. We need to show that the set valued map $Q_{f_1,\ldots,f_k}$ is H.l.s.c at x_0. For this purpose, we first define a set valued map T_k on D_Y as follows:

$$T_k(x) = \sum_{i=1}^{m} \mu_i(x) Q_{f_1,\dots,f_k}(x_i), \quad \text{for } x \in D_Y.$$

We now claim that T_k is H.l.s.c on D_Y.

To see this, fix x in D_Y and $\varepsilon > 0$. Since the map $x \to (\mu_i(x))_{i=1}^m$ is continuous from D_Y into $(\mathbb{R}^+)^m$, there exists $\delta > 0$ such that for z in $D_Y \cap B(x, \delta)$, we have

$$\sum_{i=1}^{m} |\mu_i(z) - \mu_i(x)| < \varepsilon.$$

For any v in $T_k(x)$, there is a v_i in $Q_{f_1,\dots,f_k}(x_i)$, for $1 \le i \le m$, such that $v = \sum_{i=1}^m \mu_i(x) v_i$. Let $w = \sum_{i=1}^m \mu_i(z) v_i$. Clearly w is in $T_k(z)$ and $\|v - w\| < \varepsilon$ as $\|v_i\| \le 1$, for $1 \le i \le m$. It now follows that

$$T_k(x) \subseteq T_k(z) + B(0, \varepsilon),$$

for any z in $B(x, \delta)$. Hence the map T_k is H.l.s.c at x and hence, on D_Y.

We now proceed to show that $Q_{f_1,\dots,f_k}$ is l.s.c at x_0. In order to do this, we first show that

$$T_k(x_0) = Q_{f_1,\dots,f_k}(x_0) \text{ and } T_k(x) \subseteq Q_{f_1,\dots,f_k}(x), \quad \text{for all } x \in D_Y.$$

To begin with, note that

$$Q_{f_1,\dots,f_k}(x_0) = \cap_{j=1}^m J_X(f_j). \tag{15}$$

Now for x in D_Y, by Eq. (14), we have

$$f_j(x) = \sum_{i=1}^{m} \mu_i(x) f_j(x_i), \quad \text{for } 1 \le j \le k.$$

Select any z in $T_k(x)$. Then there are elements $z_i \in Q_{f_1,\dots,f_k}(x_i)$ for $1 \le i \le m$, such that $z = \sum_{i=1}^m \mu_i(x) z_i$. Note that $\|z\| \le 1$, as $\|z_i\| \le 1$ for $1 \le i \le m$. Also for $1 \le j \le k$,

$$f_j(z) = \sum_{i=1}^{m} \mu_i(x) f_j(z_i) = \sum_{i=1}^{m} \mu_i(x) f_j(x_i) = f_j(x).$$

Since $\|z\| \le 1$, this implies z is in $Q_{f_1,\dots,f_k}(x)$ and we have

$$T_k(x) \subseteq Q_{f_1,\dots,f_k}(x), \quad \forall x \in D_Y.$$

Now $\phi_{x_0} = \sum_{i=1}^m \mu_i(x_0) \phi_{x_i}$, and by Eq. (15),

$$f_j(x_0) = \phi_{x_0}(f_j) = \|f_j\| = \sum_{i=1}^{m} \mu_i(x_0)\phi_{x_i}(f_j), \quad \text{for } 1 \le j \le k.$$

Since ϕ_{x_i} are norm one elements of $(Y^{\perp})^*$, we must have

$$f_j(x_i) = \phi_{x_i}(f_j) = \|f_j\|, \quad \text{for } 1 \le j \le k,$$

whenever $\mu_i(x_0) \ne 0$. Hence

$$Q_{f_1,\dots,f_k}(x_i) = \bigcap_{j=1}^{k} J_X(f_j) = Q_{f_1,\dots,f_k}(x_0),$$

whenever $\mu_i(x_0) \ne 0$, $1 \le i \le m$. It is now easy to see that

$$T_k(x_0) = \sum_{i=1}^{m} \mu_i(x_0) Q_{f_1,\dots,f_k}(x_i) = Q_{f_1,\dots,f_k}(x_0).$$

Thus

$$T_k(x_0) = Q_{f_1,\dots,f_k}(x_0) \text{ and } T_k(x) \subseteq Q_{f_1,\dots,f_k}(x), \quad \text{for all } x \in D_Y. \tag{16}$$

Since the map T_k is H.l.s.c at x_0, it now easily follows from Eq. (16) that the map $Q_{f_1,\dots,f_k}$ is also H.l.s.c at x_0. This completes the proof of the Lemma. □

Dutta and Narayana [10] proved that if Y is a strongly proximinal subspace of finite codimension in $C(Q)$ then P_Y is Hausdorff metric continuous. Here too, polyhedral condition plays an important role. They in fact show that $Y^{\perp}$ is polyhedral in this case and use it to prove their conclusion.

Dutta and Shanmugaraj [11] quantified strong proximinality through

$$\varepsilon(x, t) = \inf\{r > 0 : P_Y(x, t) \subseteq P_Y(x) + rB_Y\}$$

for $x \in X \backslash Y$ and $t \ge 0$. They have proved that if Y is a strongly proximinal subspace of finite codimension, P_Y is Hausdorff semi-continuous at x in X if and only if $\varepsilon(t)$ is continuous at x for every $t > 0$.

Recently, in 2011, a long and comprehensive paper "Best Approximation in polyhedral spaces" by Fonf et al. [13] presents significant results, linking geometric properties of a Banach space X with that of the continuity properties metric projection onto subspaces of finite codimension. We need the following definitions to state the results.

Definition 8 [13] A set $\mathscr{B} \subseteq S_{X^*}$ is a *boundary* for X if for each x in X there exists $f \in \mathscr{B}$ with $f(x) = \|x\|$.

Definition 9 [13] A Banach Space X satisfies a *property* $(*)$ if there exists a boundary $\mathscr{B} \subseteq S_{X^*}$ such that $\mathscr{B}' \cap NA(X) \neq \emptyset$, where $\mathscr{B}'$ is the set of all w^*-accumulation points of $\mathscr{B}$.

Definition 10 [13] A Banach Space X satisfies a *property* (Δ) if there exists a boundary $\mathscr{B} \subseteq S_{X^*}$ such that the set

$$\{f \in \mathscr{B} : f(x) = 1\} = J_{X^*}(x) \cap \mathscr{B}$$

is finite for each $x \in S_X$.

It is known that if X is a QP-space then X is polyhedral. The result below from [13] explains the relation between the above geometric conditions.

Fact 6 [13] *Let X be a Banach space. Then*

$$\begin{aligned} X \text{ has Property } (*) &\Rightarrow X \text{ is } QP \text{ with } \Delta. \\ &\Leftrightarrow X \text{ is polyhedral with } \Delta. \end{aligned}$$

Definition 11 Let X be a Banach space and Y be a closed subspace of X. Then the *effective domain* of P_Y, denoted by $\operatorname{dom} P_Y$, is the set $\{x \in X : P_Y(x) \neq \phi\}$.

Theorem 18 [13] *Let Y be a closed subspace of X. Then*

(a) *If X is polyhedral with (Δ), then P_Y is H.l.s.c on* $\operatorname{dom} P_Y$. *In particular, P_Y restricted to* $\operatorname{dom} P_Y$ *admits a continuous selection by Michael's selection theorem.*
(b) *If X is polyhedral with (Δ), P_Y is not necessarily H.u.s.c on* $\operatorname{dom} P_Y$, *even when Y is proximinal with a finite codimension.*
(c) *If X satisfies $(*)$, then P_Y is Hausdorff continuous on* $\operatorname{dom} P_Y$.

Remark 5 We would like to mention here that Theorem 5.1 of [13], which says that a proximinal subspace Y of a Banach space X is strongly proximinal if and only if the metric projection P_Y is H.u.s.c, is incorrect. While it is easy to prove that strong proximinality of Y implies P_Y is H.u.s.c., the implication in the reverse direction is not true.

To see this, let X be a Banach space and $H = \ker f$, where f is in $NA(X)$. Then H is a proximinal hyperplane and it is easily shown that P_H is H.u.s.c. However, many examples of proximinal hyperplanes which are not strongly proximinal are known [14]. Thus the conclusion of Theorem 5.1 of [13] does not hold.

Before concluding the article, we make two observations. The notion of a.l.s.c. has not been discussed much in the context of metric projections onto subspaces of infinite dimension and in particular, onto subspaces finite codimension. It is desirable to characterize the class of Banach spaces X such that for every proximinal subspace Y of finite codimension in X, the metric projection map P_Y is a.l.s.c.

We are not aware of any characterization of a proximinal subspace of finite codimension Y in $C(Q)$ with the metric projection P_Y having a continuous selection or P_Y is a.l.s.c. It would be desirable to obtain some results in that direction.

References

1. Blatter, J., Morris, P.D., Wulbert, D.E.: Continuity of the set-valued metric projection. Math. Ann. **178**, 12–24 (1968)
2. Brown, A.L.: Some problems of linear analysis. Ph.D Thesis, University of Cambridge (1961)
3. Brown, A.L.: Best n- dimensional approximation to sets of functions. Proc. London Math. Soc. **14**, 577–594 (1964)
4. Brown, A.L.: A rotund reflexive space having a subspace of codimension two with a discontinuous metric projection. Michigan Math. J. **21**, 145–151 (1974)
5. Brown, A.L.: Set valued mappings, continuous selections, and metric projections. J. Approx. Theory **57**, 48–68 (1989)
6. Brown, A.L.: The derived mappings and the order of a set-valued mapping between topological spaces. Set-valued Anal. **5**, 195–208 (1997)
7. Brown, A.L., Deutsch, F., Indumathi, V., Kenderov, P.S.: Lower semicontinuity concepts, continuous selections and set valued metric projections. J. Approx. Theory **115**, 120–143 (2002)
8. Deutsch, F., Kenderov, P.: Continuous selections and approximate selection for set valued mappings and Applications to Metric projections. SIAM J. Math. Anal. **14**, 185–194 (1983)
9. Deutsch, F., Pollul, W., Singer, I.: On set-valued metric projections, Hahn- Banach extension mapsand spherical image maps. Duke Math. J. **40**, 355–370 (1973)
10. Dutta, S., Narayana, D.: Strong proximinality and continuity of metric projection in C(K). Colloq. Math. **109**, 119–128 (2007)
11. Dutta, S., Shanmugaraj, P.: Modulus of strong proximinality and continuity of metric projection. Set-valued Variational Anal. **19**, 271–28 (2011)
12. Fan, K., Glicksberg, D.: Some geometric properties of the spheres in a normed linear space. Duke Math. J. **25**, 553–568 (1958)
13. Fonf, V.P., Lindenstrauss, J., Vesely, L.: Best approximation in polyhedral Banach spaces. J. Approx. Theory **163**, 1748–1771 (2011)
14. Godefroy, G., Indumathi, V.: Strong proximinality and polyhedral spaces. Rev. Mat. Complut. **14**, 105–125 (2001)
15. Holmes, R.B.: On the continuity of best approximation operators. In: Symposium on Infinite Dimensional Topology. Princeton University Press, Princeton (1972)
16. Indumathi, V.: Metric projections and polyhedral spaces. Set-Valued Anal. **15**, 239–250 (2007)
17. Indumathi, V.: Metric projections of closed subspaces of c_0 onto subspaces of finite codimension. Colloq. Math. **99**, 231–252 (2004)
18. Fischer, T.: A continuity condition for the existence of a continuous selection for a set-valued mapping. J. Approx. Theory **49**, 340–345 (1987)
19. Li, W.: The characterization of continuous selections for metric projections in $C(X)$. Scientia Sinica A (English) **31**, 1039–1052 (1988)
20. Li, W.: Various continuities of metric projections. In: Nevai, P., Pinkus, A. (eds.) $L_1(T, \mu)$ in Progress in Approximation Theory, pp. 583–607. Academic Press, Boston (1991)
21. Michael, E.: Continuous selections, I. Ann. Math. **63**, 361–382 (1956)
22. Morris, P.D.: Metric projections onto subspaces of finite codimension. Duke Math. J. **35**, 799–808 (1968)
23. Singer, I.: On Best Approximation in Normed Linear Spaces by Elements of Linear Subspaces. Springer, New York (1970)
24. Singer, I.: The Theory of Best Approximation and Functional Analysis. CBMS, SIAM, Philadelphia (1974)
25. Singer, I.: Set-valued metric projections. In: Butzer, P.L., Kahane, J.P., Sz-Nagy, B. (eds.) Linear Operators and Approximation, pp. 217–233. Birkhauser (1972)
26. Wegmann, R.: Some properties of the peak set mapping. J. Approx. Theory **8**, 262–284 (1973)
27. Zhivkov, N.V.: A characterisation of reflexive spaces by means of continuousapproximate selections for metric projections. J. Approx. Theory **56**, 59–71 (1989)

Convergence of Slices, Geometric Aspects in Banach Spaces and Proximinality

P. Shunmugaraj

Abstract Some geometric properties of Banach spaces and proximinality properties in best approximation theory are characterized in terms of convergence of slices. The paper begins with some basic geometric properties of Banach spaces involving slices and their geometric interpretations. Two notions of convergence of sequence sets, called Vietoris convergence and Hausdorff convergence, with their characterizations are presented. It is observed that geometric properties such as uniform convexity, strong convexity, Radon-Riesz property, and strong subdifferentiability of the norm can be characterized in terms of the convergence of slices with respect to the notions mentioned above. Proximinality properties such as approximative compactness and strong proximinality of closed convex subsets of a Banach space are also characterized in terms of convergence of slices.

Keywords Slices · Convergence of slices · Proximinality · Strict convex normed spaces · Smooth normed spaces · Uniform convex spaces · Radon-Riesz property · Strong subdifferentiability · Convergence of sequences of sets · Vietoris convergence · Hausdorff convergence · Measure of noncompactness · Continuity of set-valued maps · Duality mappings · Preduality mappings

1 Introduction

The purpose of this chapter is to discuss some notions of convergence of sequence of slices and relate these with certain geometric properties of Banach spaces and also to some known proximinality properties in best approximation theory. The results which are presented here are not new and in fact they are scattered in the literature in different formulations. We present these results in terms of convergence of slices and

P. Shunmugaraj (✉)
Department of Mathematics and Statistics, Indian Institute of Technology, Kanpur, India
e-mail: psraj@iitk.ac.in

Q. H. Ansari (ed.), *Nonlinear Analysis*, Trends in Mathematics,
DOI: 10.1007/978-81-322-1883-8_3,

we observe that several known results fit naturally in this framework. The presentation of the results in this framework not only unifies several results in the literature, it also allows us to view the results as geometric results and understand some problems which remain to be solved in this area.

The chapter is in two parts. In the first part, we begin from the classical works of Banach and Šmulian on the characterizations of smooth spaces and uniformly smooth spaces (or uniformly convex spaces) and present similar characterizations for other geometric properties including some recent results. Similarly, in the second part, we begin from the classical results of James and Day on characterizations of reflexivity and strict convexity in terms of some proximinality properties of closed convex subsets and present similar characterizations for other proximinality properties including some recent results. To be more specific, let us now define the slices formally.

Let X be a real Banach space and X^* its dual. We denote the closed unit ball and the unit sphere of X by B_X and S_X respectively. For $x^* \in S_{X^*}$ and $0 \leq \delta < 1$, we define

$$S(X, x^*, \delta) = \left\{x \in B_X : x^*(x) \geq 1 - \delta\right\}.$$

The set $S(X, x^*, \delta)$ is called a *slice* of B_X defined by x^* and δ. The geometric interpretation of a slice is given in the Sect. 2 (see Fig. 2). Similarly for $x \in S_X$ and $0 \leq \delta < 1$, we define the slice $S(X^*, x, \delta)$ of B_{X^*}, defined by x and δ, as follows

$$S(X^*, x, \delta) = \left\{x^* \in B_{X^*} : x^*(x) \geq 1 - \delta\right\}.$$

In the first part of the chapter, it is observed that geometric properties such as uniform convexity, strong convexity, Radon-Riesz property, and strong subdifferentiability of the norm can be characterized in terms of the convergence of sequence of slices

$$S(X, x^*, 1/n), \quad S(X^*, x, 1/n) \text{ and } S(X^{**}, x^*, 1/n) \text{ as } n \to \infty.$$

Observe that these are sequences of sets in X, X^* and X^{**}, respectively. A sequence of slices is shown in Fig. 7.

In the second part, proximinality properties such as strong proximinality and approximative compactness in best approximation theory are characterized in terms of convergence of slices. To be more specific, we need some notions from best approximation theory.

Let C be a nonempty, closed, and convex subset in a Banach space X and $x \in X$. For $\delta \geq 0$, consider the following set

$$P_C(x, \delta) = \{y \in C : \|x - y\| \leq d(x, C) + \delta\},$$

where $d(x, C)$ denotes the distance between x and C. The set $P_C(x, 0)$ is called the set of *best approximations* to x in C and the set $P_C(x, \delta)$, for $\delta > 0$, is called the set of *nearly best approximations* to x in C. The set $P_C(x, 0)$ could be empty but

$P_C(x, \delta) \neq \emptyset$, for $\delta > 0$. The geometric interpretation of the set $P_C(x, \delta)$ is given Sect. 5 (see Fig. 10).

We will see that the proximinality properties mentioned above can be characterized in terms of convergence of sequence of sets $(P_C(x, 1/n))$ as $n \to \infty$. In the second part of the paper, we will relate the convergence of slices with the convergence of the sets $P_C(x, 1/n)$. This will illustrate that the slice convergence connects the geometry of Banach spaces and the theory of best approximation.

The chapter is organized as follows. In Sect. 2, we recall some basic results involving slices and present their geometric interpretations. We use two notions of convergence of sequence of sets, called *Vietoris convergence* and *Hausdorff convergence* which are presented with their characterizations in Sect. 3. The characterizations of the geometric properties of Banach spaces in terms of convergence of slices are discussed in Sect. 4. Section 5 is devoted to relate proximinality properties and geometric properties of Banach spaces through the convergence of slices.

2 Preliminaries

In this section, we present the geometric interpretation of the slices and some basic results in Functional Analysis involving slices which are required in the sequel. The geometric visualization presented here would help, at least for the beginners, to appreciate the results presented in this and the subsequent sections. Throughout the section we assume that X is a real Banach space.

In Functional Analysis, there are several results which involve elements of X^* or S_{X^*}. Many such results are geometric in nature and their proofs also use the geometric visualizations. To visualize the results geometrically, the nonzero elements of X^* are associated with hyperplanes.

A set $H = \{x \in X : x^*(x) = c\}$ for some $x^* \in S_{X^*}$ and $c \in \mathbb{R}$ is called a *hyperplane* (see Fig. 1). If $x^* \in S_{X^*}$, we associate this element of the dual space with a hyperplane

$$\left\{x \in X : x^*(x) = c\right\}, \quad \text{for some } \; c \in \mathbb{R}\backslash\{0\}.$$

So, the elements of dual spaces are geometrically visualized as hyperplanes in X. For example if $X = \mathbb{R}^3$ then the hyperplanes are planes in $\mathbb{R}^3$.

2.1 Distance Formula and Slices

The formula appearing in the following result is a generalization of the distance formula in $\mathbb{R}^3$. The formula helps us to locate a hyperplane or, at least, to know the distance between the origin and a hyperplane. The formula is illustrated in Fig. 1.

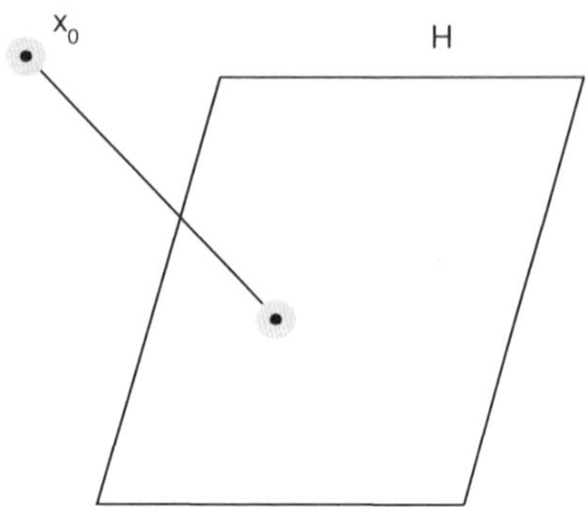

Fig. 1 Distance between the point x_0 and hyperplane $H := \{x \in X : x^*(x) = c\}$

Theorem 1 (Ascoli's Formula) *Let* $x^* \in X^* \backslash \{0\}$, $x_0 \in X$ *and*

$$H = \left\{x \in X : x^*(x) = c\right\}, \quad \text{for some } c \in \mathbb{R}.$$

Then,

$$d(x_0, H) = \frac{|x^*(x_0) - c|}{\|x^*\|}.$$

Proof For any $h \in H$, $|x^*(x_0) - x^*(h)| \leq |x^*(x_0) - c| \leq \|x^*\| \, \|x_0 - h\|$. This implies that $|x^*(x_0) - c| \leq \|x^*\| d(x_0, H)$.

To prove $\|x^*\| d(x_0, H) \leq |x^*(x_0) - c|$, it is sufficient to show that

$$\alpha d(x_0, H) < |x^*(x_0) - c|,$$

for all α such that $0 < \alpha < \|x^*\|$. For $0 < \alpha < \|x^*\|$, find $y \in S_X$ such that $|x^*(y)| > \alpha$. Define

$$h = x_0 - \left(\frac{x^*(x_0) - c}{x^*(y)}\right) y.$$

It is easy to verify that $h \in H$ and $\alpha \|x_0 - h\| < |x^*(x_0) - c|$. This shows that $\alpha d(x_0, H) < |x^*(x_0) - c|$. □

We use the formula given above as follows. Let $x^* \in S_{X^*}$, $H_0 = \{x \in X : x^*(x) = 1\}$ and $H_n = \left\{x \in X : x^*(x) = 1 - \frac{1}{n}\right\}$. Then from the previous formula we get that $d(0, H_0) = 1$ and $d(0, H_n) = 1 - \frac{1}{n}$. We can also verify that

$$d(B_X, H_0) = \inf \{\|x - y\| : x \in B_X \text{ and } y \in H_0\} = 0$$

and $d(H_0, H_n) = \frac{1}{n}$.

We can easily visualize that a hyperplane $H = \{x \in X : x^*(x) = c\}$ divides the space X into two parts $H^+ = \{x \in X : x^*(x) \geq c\}$ and $H^- = \{x \in X : x^*(x) \leq c\}$. Note that $H = H^+ \cap H^-$ and the half space H^+ or H^- lies on one side of H.

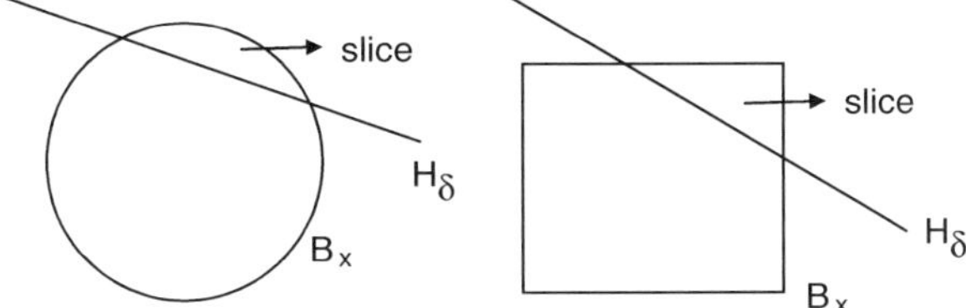

Fig. 2 Slice $S(X, x^*, \delta)$ formed by slicing B_X by the hyperplane $H_\delta = \{x \in X : x^*(x) = 1 - \delta\}$

A *slice* $S(X, x^*, \delta)$, $0 \leq \delta < 1$, $x^* \in S_{X^*}$, is the intersection of the unit ball B_X and the half space $\{x \in X : x^*(x) \geq 1 - \delta\}$. We cut the unit ball into two pieces with the hyperplane $\{x \in X : x^*(x) = 1 - \delta\}$ and take the slice lying in $\{x \in X : x^*(x) \geq 1 - \delta\}$. The slice $S(X, x^*, \delta)$ is illustrated in Fig. 2.

In case $0 < \delta < 1$, the slice $S(X, x^*, \delta)$ is nonempty. Observe that if we take

$$Y = \{x \in X : x^*(x) = 0\}$$

and $x_0 \in S(X, x^*, \delta)$ then $d(x_0, Y) \geq 1 - \delta$. Does this resemble Riesz Lemma?

Let us discuss the slice for the case $\delta = 0$. It can be easily verified that if $x^* \in S_{X^*}$, then the hyperplane $H = \{x \in X : x^*(x) = 1\}$ does not intersect the interior of the unit ball B_X. If $S(X, x^*, 0) \neq \emptyset$, then geometrically it is clear in this case that the hyperplane H touches the unit ball at all points of $S(X, x^*, 0)$. Therefore if $S(X, x^*, 0) \neq \emptyset$ we say that the hyperplane $\{x \in X : x^*(x) = 1\}$, defined by $x^* \in S_{X^*}$, is a *supporting hyperplane* supporting B_X at every point of $S(X, x^*, 0)$ and x^* is a *support functional*. Since the set $S(X, x^*, 0)$ is the intersection of S_X and a hyperplane, it is sometimes called a face of B_X. See Figs. 3 and 4 for the illustration of supporting hyperplanes and faces.

Remark 1 Observe that $S(X^*, x, \delta)$, $x \in X$, is a slice of the ball B_{X^*} generated by the element $x \in X^{**}$ and $S(X, x^*, \delta) = S(X^{**}, x^*, \delta) \cap X$.

Let see the geometric interpretation of some well-known results in Functional Analysis which would help us to visualize several other geometric results.

The following geometric result is an immediate consequence of the Hahn-Banach extension theorem.

Theorem 2 *For every $x_0 \in S_X$, there exists $x^* \in S_{X^*}$ such that $x^*(x_0) = 1$.*

Theorem 2 says that for a given point x_0 in S_X we can always find a supporting hyperplane $\{x \in X : x^*(x_0) = 1\}$, defined by some $x^* \in S_{X^*}$, supporting B_X at x_0 and hence $S(X^*, x, 0)$ is always nonempty for all $x \in S_X$. The following question is natural.

Question 1 For given $x^* \in S_{X^*}$, is it possible to find an element $x \in S_X$ such that $x^*(x) = 1$?

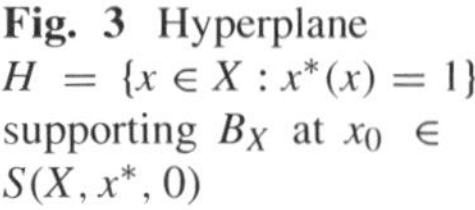

Fig. 3 Hyperplane $H = \{x \in X : x^*(x) = 1\}$ supporting B_X at $x_0 \in S(X, x^*, 0)$

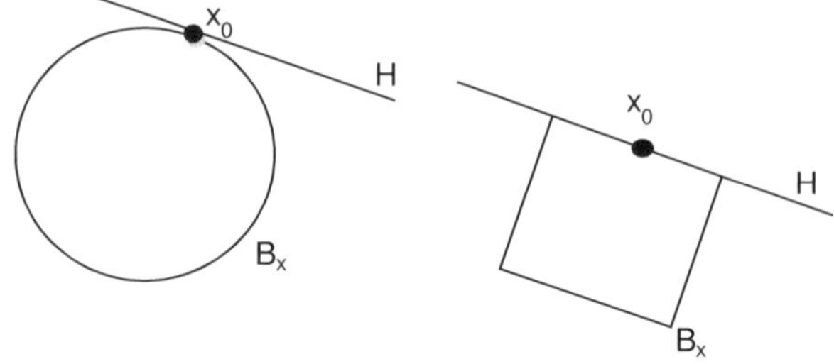

Fig. 4 Multiple hyperplanes supporting B_X at $x \in S(X, x^*, 0)$

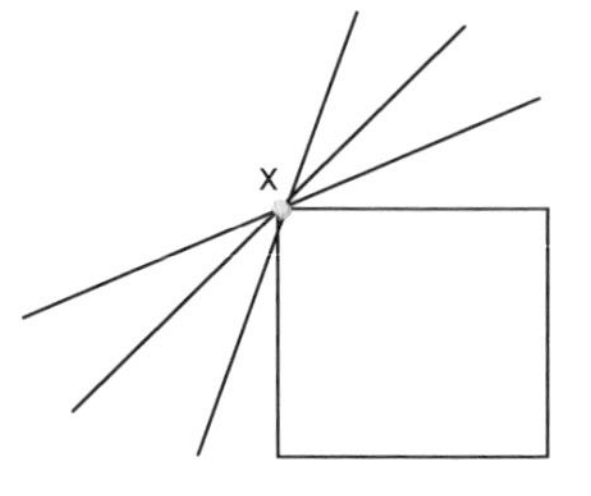

The question can be geometrically interpreted as follows. Suppose we are given a hyperplane $H = \{x \in X : x^*(x) = 1\}$. Does H support B_X at some point? The same question can also be posed as follows. Is it necessary that every $x^* \in S_{X^*}$ is a support functional or $S(X, x^*, 0) \neq \emptyset$ for every $x^* \in S_{X^*}$? The following simple example illustrates that $S(X, x^*, 0)$ could be empty for some $x^* \in S_{X^*}$.

Example 1 Let $X = \{x \in (C[0, 1], \| \|_\infty) : x(0) = x(1) = 0\}$ and define $x^*(x) = \int_0^1 x(t)dt$. It can be verified that $x^* \in S_{X^*}$ and $x^*(x) < 1$ for all $x \in B_X$. Therefore $S(X, x^*, 0) = \emptyset$.

Observe that for $x^* \in S_{X^*}$, $S(X, x^*, 0) = B_X \cap H_0$ where $H_0 = \{x \in X : x^*(x) = 1\}$. The above example illustrates that although $d(B_X, H_0) = 0$ and $d(0, H_0) = 1$, the hyperplane H_0 may not touch the unit ball B_X, that is, it may happen that the slice $S(X, x^*, 0)$ could be empty for some x^*. So the following question is also natural.

Question 2 Under what condition on X, every $x^* \in S_{X^*}$ is a support functional?

It follows from the Banach-Alaoglu theorem that if X is reflexive then B_X is weakly compact and hence every $x^* \in S_{X^*}$ is a support functional. What about the converse? This question remained open for about thirty years and finally it was settled by James [27].

Theorem 3 (James Theorem) *X is reflexive if and only if every $x^* \in S_{X^*}$ is a support functional.*

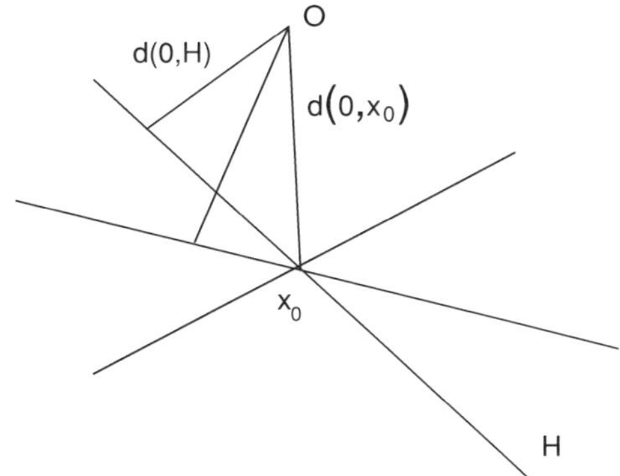

Fig. 5 Norm of x_0 is equal to the supremum of the distances between 0 and the hyperplanes passing through x_0

In terms of slices the James theorem can be stated as follows: X is reflexive if and only $S(X, x^*, 0) \neq \emptyset$ for all $x^* \in S_{X^*}$. We refer to [29] for the proof of the James theorem.

We discussed some geometric results above and their geometric interpretations by associating each $x^* \in S_{X^*}$ with a hyperplane. We will consider one more result involving elements from S_{X^*} and see its geometric interpretation.

The following result is an immediate consequence of the Hahn-Banach theorem.

Theorem 4 *Let $x_0 \in X$ such that $x_0 \neq 0$. Then $\|x_0\| = \sup\{|x^*(x_0)| : x^* \in S_{X^*}\}$.*

The above result is a kind of a duality result. The geometric interpretation will explain this. For $x^* \in S_{X^*}$, consider the hyperplane $H_{x^*} = \{x \in X : x^*(x) = x^*(x_0)\}$. Then by Ascoli's formula $d(0, H_{x^*}) = |x^*(x_0)|$. Now Theorem 4 can be rewritten as follows:

$$\|0 - x_0\| = \sup\{d(0, H_{x^*}) : x^* \in S_{X^*}\}.$$

The above observation says the following: Consider all hyperplanes passing through the point x_0 and find the supremum of the distances between 0 and these hyperplanes which is same as the minimum distance between the points 0 and x_0. This geometric interpretation is shown in Fig. 5.

We considered the origin and a point x_0 above and wrote $\|0 - x_0\|$ as max of $d(0, H_{x^*})$. One can ask the following question. Is it possible to write the distance between a point x_0 and a closed convex subset C as a maximum of the distances between x_0 and all hyperplanes separating x_0 and C? This is explained in Fig. 6. This is possible and this is one of the duality results in minimum norm problems [28].

We will use the basic results in Functional Analysis such as the Hahn-Banach separation theorem and Hanh-Banach extension theorem in general forms, Goldstine theorem and Eberlian-Šmulian theorem. For the statements of these theorems and their proofs, we refer to [29].

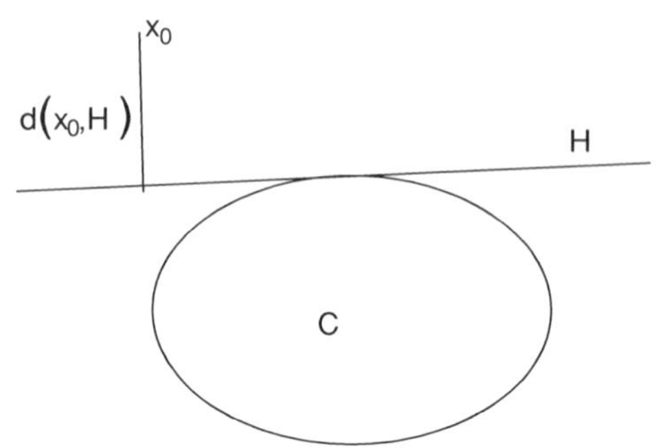

Fig. 6 Distance between a point x_0 and the set C is equal to the supremum of the distances between x_0 and the separating hyperplanes

2.1.1 Notes and Remarks

Generally a hyperlane in X is defined as a translate of a maximal proper subspace of X or a translate of a subspace of codimension one [4, 24, 28]. It can be shown that a set H is a closed hyperplane in X (with respect to the definition mentioned above) if and only if there exist $x^* \in S_{X^*}$ and $c \in \mathbb{R}$ such that $H = \{x \in X : x^*(x) = c\}$.

Figure 6 is used in [28] for illustration of the minimum norm problems and a similar figure can be seen on the cover page of the journal "Numerical Functional Analysis and Optimization".

2.2 Strict Convexity and Smoothness

We define two basic geometric properties of Banach spaces called "strictly convex" and "smooth" in terms of faces of B_X and B_{X^*}, respectively. We show that the notion of smoothness is associated with a notion of differentiability of the norm. We will also see the relation between strict convexity and smoothness.

The space X is called *strictly convex* if for any $x^* \in S_{X^*}$, the face $S(X, x^*, 0)$ is a singleton whenever it is nonempty. So if X is strictly convex then every supporting hyperplane can support or say touch B_X at only one point (see Figs. 3 and 4).

Theorem 5 *The space X is strictly convex if and only if S_X does not contain a line segment.*

Proof Let X be strictly convex. Suppose $x, y \in S_X, x \neq y$ and the line segment $[x, y]$ containing x and y is contained in S_X. By the Hahn-Banach separation theorem there exists a hyperplane $H_0 = \{x \in X : x^*(x) = 1\}$, $x^* \in S_{X^*}$, separating $[x, y]$ and B_X. It is easy to very that $[x, y] \subseteq S(X, x^*, 0)$ which is a contradiction. The converse is obvious from the definition. □

The space X is said to be *smooth* if for any $x \in S_X$, the face $S(X^*, x, 0)$ in B_{X^*} is a singleton. Note that the smoothness of X is defined by the faces of B_{X^*} whereas strict convexity is defined by the faces of B_X. If X is smooth then for every point $x \in S_X$, there exits a unique supporting hyperplane which supports B_X at x. One might ask whether the notion of smoothness corresponds to any differentiability.

We will first relate the Gâteaux differentiability of the norm and the support functionals. Note that the smoothness is related to the support functionals.

Let $x \in S_X$. Suppose the norm $\|\cdot\|$ of X is Gâteaux differentiable at x and f is its derivative. Then, for $h \in S_X$,

$$|f(h)| = \left|\lim_{t\to 0} \frac{\|x+th\| - \|x\|}{t}\right| \leq \lim_{t\to 0} \frac{\|th\|}{|t|} = 1.$$

and

$$f(x) = \lim_{t\to 0^+} \frac{\|x+tx\| - \|x\|}{t} = \|x\| = 1.$$

This shows that if the norm $\|\cdot\|$ of X is Gâteaux differentiable at $x \in S_X$, then its derivative f is in $S(X^*, x, 0)$ or f is a support functional which supports B_X at x. The natural questions are:

1. Is every element of $S(X^*, x, 0)$ associated with some kind of differentiability of the norm of X?
2. Suppose $S(X^*, x, 0)$ is a singleton, can we say that the norm of X is Gâteaux differentiable at x?

We will address these questions below. The results of this subsection and their proofs will be used in Sect. 4. Let us first express the set $S(X^*, x, 0)$ in different forms which will reveal the relation between the set $S(X^*, x, 0)$ and some kind of differentiability of the norm.

Lemma 1 *Let $x_0 \in S_X$. Then*

$$\begin{aligned} S(X^*, x_0, 0) &= \left\{x^* \in X^* : x^*(x) - x^*(x_0) \leq \|x\| - \|x_0\| \quad \forall\, x \in X\right\} \\ &= \left\{x^* \in X^* : x^*(h) \leq \|x_0 + h\| - \|x_0\| \quad \forall\, h \in X\right\}. \end{aligned}$$

Proof If $x^* \in S(X^*, x_0, 0)$ then it is clear that $x^*(x) - x^*(x_0) \leq \|x\| - 1$ for all $x \in X$.

Conversely, suppose

$$x^*(x) - x^*(x_0) \leq \|x\| - \|x_0\|$$

for all $x \in X$. If we take $x = x_0 + y$ then $x^*(y) \leq \|x_0 + y\| - \|x_0\| \leq \|y\|$ for any $y \in X$. This shows that $\|x^*\| \leq 1$. If $x = 0$, $x^*(x_0) \geq \|x_0\|$ and if $x = 2x_0$ then $x^*(x_0) \leq \|x_0\|$. This proves that $x^* \in S(X^*, x_0, 0)$. □

In Convex Analysis (see [16, 35]), the elements in the set

$$\{x^* \in X^* : x^*(x) - x^*(x_0) \leq \|x\| - \|x_0\| \quad \forall\, x \in X\}$$

are called *subdifferentials* of the convex function $\|\cdot\|$ at x_0.

The second form of the set $S(X^*, x_0, 0)$ that we are going to write will give a better picture about the relation between the set and the differentiability of the norm function. We need some definitions and basic results.

It is known [29, p. 483] that, for a fixed $x \in S_X$ the function $t \to \frac{\|x+th\|-\|x\|}{t}$ is increasing on $\mathbb{R}\backslash\{0\}$ for any $h \in X$ and hence the one sided limits (at x in the direction h)

$$d_+(x, h) = \lim_{t\to 0^+} \frac{\|x + th\| - \|x\|}{t},$$

and

$$d_-(x, h) = \lim_{t\to 0^-} \frac{\|x + th\| - \|x\|}{t}$$

exist. Moreover, the map $d_+(x, \cdot)$ is positive homogeneous and finitely subadditive.

The following lemma is a consequence of Lemma 1.

Lemma 2 *Let $x \in S_X$. Then*

$$S(X^*, x, 0) = \left\{x^* \in S_{X^*} : d_-(x, h) \leq x^*(h) \leq d_+(x, h) \quad \forall\, h \in X\right\}.$$

Proof It is easy to see from Lemma 1 that

$$S(X^*, x, 0) \subseteq \{x^* \in S_{X^*} : d_-(x, h) \leq x^*(h) \leq d_+(x, h) \quad \forall\, h \in X\}.$$

If $x^* \in S_{X^*}$ satisfies the condition $d_-(x, h) \leq x^*(h) \leq d_+(x, h) \quad \forall\, h \in X$, then

$$1 = d_-(x, x) \leq x^*(x) \leq d_+(x, x) = 1.$$

This shows that $x^* \in S(X^*, x, 0)$. □

It is clear from Lemma 2 that if the norm of X is Gâteaux differentiable at $x \in S_X$ then $S(X^*, x, 0)$ is a singleton. In fact, the converse is also true which is stated in the following result which is due to Banach. We will also use the proof of the following theorem in Sect. 4.

Theorem 6 *Let $x \in S_X$. Then $S(X^*, x, 0)$ is a singleton if and only if the norm $\|\cdot\|$ of X is Gâteaux differentiable at x.*

Proof Suppose that the norm $\|\cdot\|$ of X is not Gâteaux differentiable at x. Then there exist $h \in X$ and $\alpha \in \mathbb{R}$ such that $d_-(x, h) < \alpha < d_+(x, h)$. Define $M = \text{span}\{h\}$ and a linear functional g on M by $g(\lambda h) = \lambda\alpha$ for all $\lambda \in \mathbb{R}$. By using the fact that $d_+(x, \cdot)$ is positive homogeneous and $d_-(x, h) = -d_+(x, -h)$

we get that $g(y) \leq d_+(x, y)$ for all $y \in M$. By the Hahn-Banach extension theorem [29, p. 73] there is a linear functional x^*_α on X such that $x^*_\alpha(y) \leq d_+(x, y)$ for all $y \in X$. Moreover, for all $y \in X$,

$$d_-(x, y) = -d_+(x, -y) \leq -x^*_\alpha(-y) = x^*_\alpha(y).$$

To prove $x^*_\alpha \in S_{X^*}$, observe that

$$x^*_\alpha(y) \leq d_+(x, y) \leq \frac{1}{1}\|x + 1y\| - \|x\| \leq \|y\|,$$

for all $y \in X$ and $1 = d_-(x, x) \leq x^*_\alpha(x) \leq d_+(x, x) = 1$. This proves that $x^*_\alpha \in S_{X^*}$ and $S(X^*, x, 0)$ is not a singleton. □

Corollary 1 *The space X is smooth if and only if the norm $\|\cdot\|$ of X is Gâteaux differentiable on S_X.*

The following result is a consequence of the fact that $X \subseteq X^{**}$.

Theorem 7 *The space X is smooth if X^* is strictly convex. Similarly, X is strictly convex if X^* is smooth.*

Proof Suppose X^* is strictly convex. Then $S(X^*, x^{**}, 0)$ is at most a singleton for every $x^{**} \in S_{X^{**}}$. This implies that $S(X^*, x, 0)$ is a singleton for every $x \in S_X$ as $S_X \subseteq S_{X^{**}}$. This proves that X is smooth.

Suppose X^* is smooth. Then for every $x^* \in S_{X^*}$, $S(X^{**}, x^*, 0)$ is a singleton. This implies that $S(X, x^*, 0) = S(X^{**}, x^*, 0) \cap X$ which is either empty or a singleton. This proves the result. □

The proof of the following result is immediate from Theorem 7.

Theorem 8 *Suppose X is reflexive. Then X is smooth if and only if X^* is strictly convex and X is strictly convex if and only if X^* is smooth*

We define a notion called uniformly convex which is a stronger notion compared to strictly convex. We do not define this notions in terms of slices; however, it will be characterized in terms of slices in Sect. 4.

Suppose X is strictly convex. We have seen in Theorem 5 that if $x, y \in S_X, x \neq y$, then $1 - \left\|\frac{x+y}{2}\right\| > 0$. But this quantity need not be uniformly bounded from below.

We say that X is *uniformly convex* if for every $\varepsilon > 0$ $(0 < \varepsilon \leq 2)$ there is a $\delta > 0$ such that for all $x, y \in B_X$ with $\|x - y\| \geq \varepsilon$ we have $1 - \left\|\frac{x+y}{2}\right\| > \delta$.

Remark 2 For examples of strictly convex spaces, smooth spaces, and uniformly convex spaces, we refer to [13, 24, 29]. We will discuss some other geometric properties which are stronger than strict convexity and weaker than uniform convexity. However, we will not present examples for spaces satisfying such properties and we refer to [29] for examples.

3 Convergence of Sequence of Sets

Since the slices are sets, for studying the convergence of sequence of slices we need to know the convergence of sequence sets which is discussed in this section.

There are several notions of convergence of sequence of sets but in this section, we will discuss only two notions of convergence called Vietoris and Hausdorff convergence. For other notions of convergence of sequence of sets, we refer to [4, 31]. Since the continuity of the set-valued mappings can be defined in terms of the convergence of sequence of sets, we also touch upon the definitions of the continuity of the set-valued mappings at the end of this section.

In this section, we assume that X and Y are metric spaces. Let $CL(X)$ denote the set of all nonempty closed subsets of X. The set of all nonempty closed and bounded subsets will be denoted by $CLB(X)$. Throughout the section, we assume that $\{C_n\}$ is a sequence in $CL(X)$ and $C_0 \in CL(X)$. For $C \in CL(X)$ and $\epsilon > 0$ we write $B_\epsilon(C)$ for $\{x \in X : d(x, C) < \epsilon\}$. Whenever we write $\{n_k\}$, it is understood that it is a strictly increasing sequence in $\mathbb{N}$.

3.1 Definitions of Vietoris and Hausdorff Convergence

We present the definitions of the convergence of sequence of sets in the sense of Vietoris and Haudorff and some examples.

3.1.1 Vietoris Convergence

Each notion of convergence of sequence of sets has two parts, called upper and lower parts, which are analogous to limsup and liminf of a sequence of real numbers.

Let us first define the upper part of the Vietoris convergence, called V^+ convergence.

We say that $C_n \xrightarrow{V^+} C_0$ if whenever $C_0 \subseteq V$ for an open set V of X, then $C_n \subseteq V$ eventually. The intuitive idea of this notion of convergence is that if C_0 is small enough to be contained in an open set V then C_n's are also small enough to be contained in V.

The lower part of the Vietoris convergence, called V^-, is defined as follows.

We say that $C_n \xrightarrow{V^-} C_0$ if $C_0 \cap V \neq \emptyset$ for an open set V of X, then $C_n \cap V \neq \emptyset$ eventually. The intuitive idea of this convergence is that if C_0 is big enough to hit an open set V then C_n's are also big enough to hit V.

The sequence (C_n) converges to C_0 in the Vietoris sense, denoted by $C_n \xrightarrow{V} C_0$, if $C_n \xrightarrow{V^+} C_0$ and $C_n \xrightarrow{V^-} C_0$. The Vietoris convergence is a topological convergence, in the sense that there is a topology τ_V on $CL(X)$, called *Vietoris topology*, such that

$C_n \xrightarrow{V} C_0$ if only if $C_n \longrightarrow C_0$ in the topology τ_V. We refer to the book [4] for more details about the Vietoris topology.

3.1.2 Hausdorff Convergence

We first define the upper part of the Hausdorff convergence, called H^+ convergence. We say that $C_n \xrightarrow{H^+} C_0$ if for every $\varepsilon > 0$, $C_n \subseteq B_\varepsilon(C_0)$ eventually.

We say that $C_n \xrightarrow{H^-} C_0$ if for every $\varepsilon > 0$, $C_0 \subseteq B_\varepsilon(C_n)$ eventually.

The sequence (C_n) converges to C_0 in the Hausdorff sense, denoted by $C_n \xrightarrow{H} C_0$, if $C_n \xrightarrow{H^+} C_0$ and $C_n \xrightarrow{H^-} C_0$.

The Hausdorff convergence is also a topological convergence. In fact, if $C_n, C_0 \in CLB(X)$ for all n, then $C_n \xrightarrow{H} C_0$ if and only if $C_n \longrightarrow C_0$ in the Hausdorff metric H on $CLB(X)$. The well-known *Hausdorff metric H* is defined below

For $A, B \in CL(X)$, we define $h(A, B) = \sup\{d(a, B) : a \in A\}$ and

$$H(A, B) = \max\{h(A, B), h(B, A)\}.$$

It is easily seen that $h(A, B) < \varepsilon$ if and only if $A \subset B_\varepsilon(B)$. This shows that $h(C_n, C_0) \to 0$ if and only if $C_n \xrightarrow{H^+} C_0$. Similarly, $h(C_0, C_n) \to 0$ if and only if $C_n \xrightarrow{H^-} C_0$.

Note that if A is unbounded then $h(A, B)$ could be infinite. It can be easily verified that H is an infinite-valued pseudo-metric on $CL(X)$ and it is a metric on $CLB(X)$. We refer to [4] for more details on the Hausdorff metric and infinite-valued Hausdorff seudometric.

Remark 3 It is clear from the definitions that $C_n \xrightarrow{V^+} C_0$ implies $C_n \xrightarrow{H^+} C_0$. On the other hand, it can be easily derived from the definitions that $C_n \xrightarrow{H^-} C_0$ implies $C_n \xrightarrow{V^-} C_0$. The converses need not be true which are illustrated in the following examples.

Example 2 (a) Let $C_0 = \{(x, 0) \in \mathbb{R}^2 : x \in \mathbb{R}\}$ and $C_n = \left\{\left(x, \frac{1}{n}\right) : x \in \mathbb{R}\right\}$. Then $C_n \xrightarrow{H^+} C_0$ but $C_n \stackrel{V^+}{\not\longrightarrow} C_0$.

(b) Let $C_0 = \mathbb{R}$ and $C_n = [-n, n]$, $n \in \mathbb{N}$. Then $C_n \xrightarrow{V^-} C_0$ but $C_n \stackrel{H^-}{\not\longrightarrow} C_0$.

(c) Let $C_0 = \{(x, 0) \in \mathbb{R}^2 : x \in \mathbb{R}\}$ and $C_n = \left\{t\left(1, \frac{1}{n}\right) : t \in \mathbb{R}\right\}$ for all $n \in \mathbb{N}$. Then $C_n \xrightarrow{V^-} C_0$ but $C_n \stackrel{H^-}{\not\longrightarrow} C_0$.

Remark 4 It is clear from the definition that Vietoris convergence can be defined for a sequence of subsets of a topological space. Hausdorff convergence can also be defined for sequence of subsets of certain nonmetrizable spaces. For example, in a normed linear space, we say that $C_n \xrightarrow{H_w^+} C_0$ if $C_n \subseteq C_0 + N$ for every weak neighborhood N of 0. Similarly, we can define H_w^-, V_w^+, V_w^-, H_{w*}^-, V_{w*}^+ and V_{w*}^- convergence. These convergence notions are used for studying the convergence of slices. However, in this chapter we will present only the Vietoris and Hausdorff convergence of slices with respect to the metric associated with the given norm.

3.2 Characterizations of Convergence of Sequence of Sets

We now characterize the convergence of sequence of sets in terms of the sequence of elements from the sequence of sets. We will use these characterizations in the later sections for the convergence of slices and of sets of nearly best approximations.

Let us see a characterization of the upper part of the Hausdorff convergence.

Theorem 9 *Consider the following statements.*

(a) $C_n \xrightarrow{H^+} C_0$.
(b) *If* (x_n) *is such that* $x_n \in C_n$ *for every* n *then* $d(x_n, C_0) \to 0$.
(c) *If* (x_k) *is such that* $x_k \in C_{n_k}$ *for every* k *and* $x_k \to x_0$ *for some* x_0 *then* $x_0 \in C_0$.

Then (a) $\Leftrightarrow$ *(b)* $\Rightarrow$*(c).*

Proof This can be easily derived from the definition. □

Some characterizations of the upper part of the Vietoris convergence are derived in the following result.

Theorem 10 *Consider the following statements.*

(a) $C_n \xrightarrow{V^+} C_0$.
(b) *Every sequence* (x_{n_k}) *such that* $x_{n_k} \in C_{n_k} \backslash C_0, k = 1, 2, \ldots$ *has a convergent subsequence converging to some element* $x_0 \in C_0$.
(c) $C_n \xrightarrow{H^+} C_0$ *and every sequence* (x_{n_k}) *such that* $x_{n_k} \in C_{n_k} \backslash C_0, k = 1, 2, \ldots$ *has a convergent subsequence.*
(d) $C_n \xrightarrow{H^+} C_0$.
(e) *Every sequence* (x_{n_k}) *such that* $x_{n_k} \in C_{n_k}, k = 1, 2, \ldots$ *has a convergent subsequence converging to some element* $x_0 \in C_0$.

Then (a) $\Leftrightarrow$ *(b)* $\Leftrightarrow$ *(c)* $\Rightarrow$ *(d)* $\Leftarrow$ *(e). If* C_0 *is compact then all five statements are equivalent.*

Proof (a) $\Rightarrow$ (b): Suppose $C_n \xrightarrow{V^+} C_0$ and let $x_{n_k} \in C_{n_k} \backslash C_0, k = 1, 2, \ldots$. We will first show that it has a convergent subsequence. If not, then the set $F = \{x_{n_k} : k = 1, 2, \ldots\}$ is closed and $C_0 \subseteq F^c$ but $C_n \nsubseteq F^c$ eventually which is a contradiction. Suppose $x_{n_k} \to x_0$ for some x_0. If x_0 does not belong to C_0, then the set $F = \{x_{n_k} : k = 1, 2, \ldots\} \cup \{x_0\}$ is closed and $C_0 \subseteq F^c$ but $C_n \nsubseteq F^c$ eventually which is a contradiction.

(b) $\Rightarrow$ (a): Let us assume that there exists an open set V and a sequence n_k such that $C_0 \subseteq V$ but $C_{n_k} \nsubseteq V$ for every k. Choose x_{n_k} such that $x_{n_k} \in C_{n_k} \backslash V$ for every k. By (b) there exists a subsequence of (x_{n_k}) converging to some $x_0 \in V$ which is a contradiction.

(a) $\Rightarrow$ (c): This follows easily from (a) and (b).

(c) $\Rightarrow$ (d): This is obvious.

(e) $\Rightarrow$ (d): Since (e) implies (b) and (b) is equivalent to (a), (e) implies (d).

Suppose that C_0 is compact and (d) holds. To prove (c), let (x_{n_k}) be such that

$$x_{n_k} \in C_{n_k} \backslash C_0, \ k = 1, 2, \ldots.$$

Since $C_n \xrightarrow{H^+} C_0$, by Theorem 9, $d(x_{n_k}, C_0) \to 0$. Therefore there exists $y_{n_k} \in C_0$ such that $d(x_{n_k}, y_{n_k}) \to 0$. Since (y_{n_k}) has a convergent subsequence, (x_{n_k}) has a convergent subsequence. This proves (d)$\Rightarrow$ (c) under the assumption that C_0 is compact.
If C_0 is compact then the proof of (d) $\Rightarrow$ (e) follows from (b). □

Let us make some observations on the convergence V^+ and H^+. Suppose that (x_n) is a sequence such that $x_n \in C_n$, for every n. Then, in case of $C_n \xrightarrow{H^+} C_0$, the sequence (x_n) is close to the set C_0. On the other hand when $C_n \xrightarrow{V^+} C_0$, the sequence (x_n) is not only close to the set C_0 but it has a convergent subsequence if the sequence (x_n) is not eventually in C_0. Observe that a kind of compactness argument is involved when we deal with V^+ but which is missing when we deal with H^+. This observation leads to the following characterization.

We need two definitions.

For a sequence (C_n) in $CL(X)$, we define $\overline{\lim}\, C_n$ as follows:

$$\overline{\lim}\, C_n = \{x \in X : \text{there exists a sequence } (x_k) \text{ such that } x_k \in C_{n_k} \text{ and } x_k \to x\}.$$

3.2.1 Measure of Noncompactness

Let A be a nonempty bounded subset of X. Then the *Hausdorff index of noncompactness* $\alpha(A)$ of A is defined as follows [2]:

$$\alpha(A) = \inf\{\epsilon > 0 : A \subseteq B_\epsilon(F) \text{ for some finite set } F \subseteq X\}.$$

It is clear that $\alpha(A) = 0$ if and only if A is totally bounded.

In the following result, we write F_n for $\bigcup_{i=n}^{\infty}(C_i \backslash C_0)$ for a given sequence (C_n) in $CL(X)$ and $C_0 \in CL(X)$.

The proof of the following theorem is similar to the proof of [9, Lemma 1].

Theorem 11 *Let X be a complete metric space. Then the following statements are equivalent.*

(a) $C_n \xrightarrow{V^+} C_0$.
(b) $\overline{\lim}\, C_n = C_0$ *and* $\alpha(F_n) \to 0$.

Proof (a) $\Rightarrow$ (b): Let $C_n \xrightarrow{V^+} C_0$. Then Theorem 10 implies that $\overline{\lim}\, C_n = C_0$. Suppose $\lim \alpha(F_n) \neq 0$. As (F_n) is a descending sequence, there exists $\epsilon > 0$ such that $\alpha(F_n) > 2\epsilon$ for every n. Choose some $x_1 \in F_1$ and find $x_2 \in F_2$ such that $d(x_1, x_2) > \epsilon$. Find $x_3 \in F_3$ such that $d(x_1, x_3) > \epsilon$ and $d(x_2, x_3) > \epsilon$. Repeat the process and find a sequence (x_n) such that $x_n \in F_n$ for every n and $d(x_i, x_j) > \epsilon$ when $i \neq j$. Observe that (x_n) does not have a convergent subsequence which contradicts (a) $\Rightarrow$ (b) of Theorem 10
(b) $\Rightarrow$ (a): We will use (b) $\Rightarrow$ (a) of Theorem 10. Suppose that (x_k) is a sequence such that $x_k \in C_{n_k} \backslash C_0$, $k = 1, 2, \ldots$. Then $x_k \in F_{n_k}$ for every k and by (b), $\alpha(F_{n_k}) \to 0$. This implies that $\alpha(\{x_1, x_2, \ldots\}) = 0$. Therefore (x_k) has a Cauchy subsequence and hence (x_k) has a convergent subsequence converging to some x_0. Since $\overline{\lim}\, C_n = C_0$, $x_0 \in C_0$. Therefore (a) follows from (b) $\Rightarrow$ (a) of Theorem 10. □

We will now present the characterizations for the lower parts of the Vietoris and Hausdorff convergence without proofs. The proofs can be easily derived from the definitions.

Theorem 12 *The following statements are equivalent.*

(a) $C_n \xrightarrow{V^-} C_0$.
(b) *For every* $x_0 \in C_0$ *there exists* $x_n \in C_n$, $n \geq 1$, *such that* $x_n \to x_0$.

Theorem 13 *The following statements are equivalent.*

(a) $C_n \xrightarrow{H^-} C_0$.
(b) *For any sequence* (x_n) *in* C_0 *there exists* $y_n \in C_n$, $n \geq 1$, *such that* $d(x_n, y_n) \to 0$.

It is clear from Theorems 12 and 13 that $C_n \xrightarrow{H^-} C_0 \Rightarrow C_n \xrightarrow{V^-} C_0$. The reverse implication need not be true in general which is illustrated in Example 2.

The following result follows from Theorems 12 and 13.

Theorem 14 *If C_0 is compact then $C_n \xrightarrow{V^-} C_0 \Leftrightarrow C_n \xrightarrow{H^-} C_0$.*

3.3 Convergence of Sequence of Nested Sets

In this subsection, we assume that the sequence (C_n) satisfies the condition that $C_{n+1} \subseteq C_n$ for all n. It is clear from the definitions that in this case $C_n \xrightarrow{V^+} C_0 \Leftrightarrow C_n \xrightarrow{V} C_0$. Similarly, $C_n \xrightarrow{H^+} C_0 \Leftrightarrow C_n \xrightarrow{H} C_0$. Observe that the sequence of slices $\left(S\left(X, x^*, \frac{1}{n}\right)\right)$ is a nested sequence.

Theorem 15 *Suppose that X is complete and (C_n) is a sequence in $CL(X)$ satisfying the condition $C_{n+1} \subseteq C_n$ for all n. Let $C_0 = \bigcap_{n\geq 1} C_n$. Consider the following statements.*

(a) *C_0 is nonempty and $C_n \xrightarrow{V} C_0$.*
(b) *Every sequence (x_n) such that $x_n \in C_n \backslash C_0, n = 1, 2, \ldots$ has a convergent subsequence converging to some element $x_0 \in C_0$.*
(c) *Every sequence (x_n) such that $x_n \in C_n, n = 1, 2, \ldots$ has a convergent subsequence converging to some element $x_0 \in C_0$.*
(d) $\alpha(C_n) \to 0$.
(e) *$diam(C_n) \to 0$.*
(f) *There exists x_0 such that $C_0 = \{x_0\}$ and every sequence (x_n) satisfying $x_n \in C_n,\ n \in \mathbb{N}$, has a convergent subsequence converging to x_0.*

Then (a) $\Leftrightarrow$ (b) $\Leftarrow$ (c) $\Leftrightarrow$ (d) $\Leftarrow$ (e) $\Leftrightarrow$ (f). If C_0 is compact, then (b) $\Rightarrow$ (c). If C_0 is a singleton then, (d) $\Rightarrow$ (e).

Proof (a) $\Leftrightarrow$ (b): Since the sequence is nested, the equivalence follows from (a) $\Leftrightarrow$ (b) of Theorem 10.

(c) $\Rightarrow$ (b): This is obvious.

(c) $\Rightarrow$ (d): Observe that (c) implies that C_0 is nonempty and compact. Note that, by Theorem 10, (c) implies that $C_n \xrightarrow{V} C_0$. Therefore, (d) follows from Theorem 11.

(d) $\Rightarrow$ (c): Observe that (d) implies that C_0 is nonempty and compact. Moreover, if $x_k \in C_{n_k}$ and $x_k \to x_0$, then $x_0 \in C_{n_k}$ for every k. Therefore, $x_0 \in C_0$. This proves that $\overline{\lim} C_n = C_0$. Therefore, (c) follows from (b) $\Rightarrow$ (a) of Theorem 11 and (a) $\Rightarrow$ (e) of Theorem 10.

(e) $\Rightarrow$ (d): Since $\alpha(C_n) \leq \operatorname{diam}(C_n)$, the implication is obvious.

(e) $\Rightarrow$ (f): Since X is complete and $\operatorname{diam}(C_n) \to 0$, $C_0 = \{x_0\}$ for some x_0. Let (x_n) be a sequence such that $x_n \in C_n,\ n \in \mathbb{N}$. Then, by (e), the sequence (x_n) is Cauchy and hence it converges to x_0.

(f) $\Rightarrow$ (e): By (b) $\Rightarrow$ (a), $C_n \xrightarrow{V} C_0$. Since C_0 is a singleton, (e) follows.

If C_0 is compact, then (b) $\Rightarrow$ (c) follows from the fact that the sequence is nested and (e) $\Rightarrow$ (b) of Theorem 10.

If C_0 is a singleton, then (a) $\Rightarrow$ (e). This proves (d) $\Rightarrow$ (e). □

3.4 Continuity of Set-Valued Mappings

In this section, we define semicontinuities of set-valued mappings in terms of convergence of sequence of sets. We need these definitions for Sect. 4.

Let $F : X \to CL(Y)$ and $x \in X$. The set-valued map F is said to be *upper semicontinuous* (in short, usc) at x if $F(x_n) \xrightarrow{V^+} F(x)$ whenever a sequence (x_n) in X converges to x. If F is usc at every point of x, we say that F is usc (on X). The map F is called *lower semicontinuous* (in short, lsc) at x if $F(x_n) \xrightarrow{V^-} F(x)$ whenever $x_n \to x$. The map F is *continuous* at x if it is lsc and usc at x. If F is continuous at every point of x, we say that F is continuous (on X).

Hausdofff semicontinuities are defined as follows.

The map F is *Hausdorff upper semicontinuous* (in short, Husc) at x if $F(x_n) \xrightarrow{H^+} F(x)$ whenever $x_n \to x$. Similarly F is *Hausdorff lower semicontinuous* (in short, Hlsc) at x if $F(x_n) \xrightarrow{H^-} F(x)$ whenever $x_n \to x$.

Since the semicontinuities of F are defined in terms of convergence of sets, for every result presented in Sect. 3.2, there is a counterpart for F. For example, we can use Theorem 10 to relate the usc of F at a point x with the behavior of the sequence of elements from $F(x_n)$ when $x_n \to x$.

It is clear from the definitions that usc implies Husc and Hlsc implies lsc. The converses need not be true which are illustrated in the following examples. The following examples are modifications of the examples presented in Example 2.

Example 3 (a) Let $F : \mathbb{R} \to CL(\mathbb{R})$ be defined by $F(x) = \left[-\frac{1}{|x|}, \frac{1}{|x|}\right]$ if $x \neq 0$ and $F(0) = \mathbb{R}$. The function is not Hlsc at $x = 0$ and it is continuous on $\mathbb{R}$.

(b) Let $F : \mathbb{R} \to CL(\mathbb{R}^2)$ be defined by $F(x) = \left\{(x, y) \in \mathbb{R}^2 : 0 \leq y \leq \frac{1}{|x|}\right\}$ if $x \neq 0$ and $F(0) = \left\{(0, y) \in \mathbb{R}^2 : y \geq 0\right\}$. The function is Husc but not usc.

(c) Let $F : \mathbb{R} \to CL(\mathbb{R}^2)$ be defined by $F(x) = \left\{(t, xt) \in \mathbb{R}^2 : t \in \mathbb{R}\right\}$. Then F is lsc but not Hlsc.

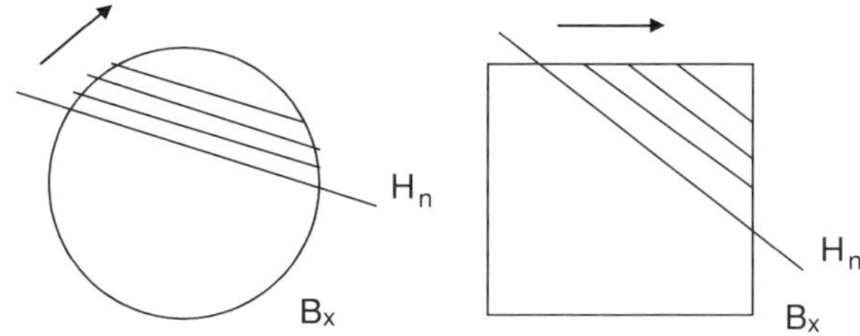

Fig. 7 Sequence of slices $S(X, x^*, \frac{1}{n})$ of B_X formed by the hyperplanes $H_n = \{x \in X : x^*(x) = 1 - \frac{1}{n}\}$

4 Convergence of Slices and Geometry of Banach Spaces

In this section, some known geometric properties of Banach spaces are characterized in terms of convergence of sequence of slices

$$S\left(X, x^*, 1/n\right),\ S\left(X^*, x, 1/n\right) \text{ and } S\left(X^{**}, x^*, 1/n\right) \quad \text{as } n \to \infty.$$

A sequence of slices is illustrated in Fig. 7.

We first present two results of Šmulian characterizing smooth and uniformly convex spaces. Throughout this section, we assume that X is a real Banach space.

4.1 Characterization of Strictly Convex and Smooth Spaces

We present characterizations of strictly convex and smooth spaces in terms of sequence of slices. The following result is due to Šmulian [32].

Theorem 16 *The following two statements are equivalent.*

(a) *X is smooth.*
(b) *Let $x \in S_X$ and $x_n^* \in S\left(X^*, x, \frac{1}{n}\right)$ for every n. Then (x_n^*) is a w^*-convergent sequence.*

Proof (a) $\Rightarrow$ (b): Suppose $x \in S_X$ and $x_n^* \in S\left(X^*, x, \frac{1}{n}\right)$ for every n. Since X is smooth, $S(X^*, x, 0)$ is a singleton, say $\{x_0^*\}$. Since $x_n^*(x) \to 1 = x_0^*(x)$, if (x_n^*) is w^*-convergent then $x_n^* \xrightarrow{w^*} x_0^*$. In fact every w^*-convergent subnet of (x_n^*) converges to x_0^* in the w^*-topology. So we claim that $x_n^* \xrightarrow{w^*} x_0^*$. Suppose that there exist ϵ and $y \in S_X$ such that $\left|(x_{n_k}^* - x_0^*)(y)\right| \geq \epsilon$ for some subsequence n_k. Since B_{X^*} is w^*-compact, $(x_{n_k}^* - x_0^*)$ has a subnet converging to 0 in the w^*-topology which is a contradiction.

(b) $\Rightarrow$ (a): Suppose that $S(X^*, x, 0)$ has two distinct elements x_0^* and y_0^*. Then the sequence $(x_n^*) = (x_0^*, y_0^*, x_0^*, y_0^*, \ldots)$ satisfies the condition that $x_n^* \in S\left(X^*, x, \frac{1}{n}\right)$ for every n but (x_n^*) is not a w^*-convergent sequence which is a contradiction. $\square$

In the following result, we present a characterization of strictly convex spaces which is analogous to the previous result.

Corollary 2 *Consider the following statements.*

(a) *X is strictly convex.*
(b) *Let $x^* \in S_{X^*}$ and $x_n \in S\left(X, x^*, \frac{1}{n}\right)$ for every n. Then every weakly convergent subsequence of (x_n) converges weakly to the same limit.*
(c) *Let $x^* \in S_{X^*}$ and $x_n^{**} \in S\left(X^{**}, x^*, \frac{1}{n}\right)$ for every n. Then (x_n^{**}) is a w^*-convergent sequence.*
(d) *The norm of X^* is Gâteaux differentiable on S_{X^*}; that is, X^* is smooth.*

Then (a) $\Leftrightarrow$ (b) $\Leftarrow$ (c) $\Leftrightarrow$ (d). If X is reflexive then all four statements are equivalent.

Proof (a) $\Rightarrow$ (b): Let X be strictly convex. Suppose $x^* \in S_{X^*}$ and $x_n \in S\left(X, x^*, \frac{1}{n}\right)$ for every n. If (x_n) has a weakly convergent subsequence converging weakly to some x_0, then $x_0 \in S(X, x^*, 0)$. Since $S(X, x^*, 0)$ is a singleton, (b) follows.

(b) $\Rightarrow$ (a): Suppose for some $x^* \in S_{X^*}$, $S(X, x^*, 0)$ has two distinct elements x_0 and y_0. Then the sequence $(x_n) = (x_0, y_0, x_0, y_0, \ldots)$ satisfies the condition that $x_n \in S\left(X, x^*, \frac{1}{n}\right)$ for every n but (x_n) has two subsequential limits.

The proof of (c) $\Leftrightarrow$ (d) follows from Theorem 16 and (d) $\Rightarrow$ (a) is proved in Theorem 7.

If X is reflexive, by Theorem 8, (a) implies (d). Therefore all four statements are equivalent. □

Observe, from the previous two results, that the smoothness of X is characterized by the slices of B_{X^*} whereas the strict convexity of X is characterized by the slices of B_X or $B_{X^{**}}$.

4.2 Characterizations of Uniformly Convex Spaces

The following result which characterizes uniformly convex spaces is due to Šmulian [33].

Theorem 17 *The following statements are equivalent.*

(a) *X is uniformly convex.*
(b) *$diam\left(S\left(X, x^*, \frac{1}{n}\right)\right) \to 0$ uniformly for all $x^* \in S_{X^*}$.*
(c) *$diam\left(S\left(X^{**}, x^*, \frac{1}{n}\right)\right) \to 0$ uniformly for all $x^* \in S_{X^*}$.*
(d) *The norm of X^* is uniformly Fréchet differentiable on S_{X^*}.*

Proof (a) $\Rightarrow$ (b): Let X be uniformly convex and $0 < \epsilon \leq 2$. Since X is uniformly convex, there is a $\delta > 0$ such that for every $x, y \in B_X$ with $\|x - y\| \geq \epsilon$

we have $\left\|\frac{x+y}{2}\right\| < 1-\delta$. We show that $\text{diam}\left(S\left(X, x^*, \frac{\delta}{2}\right)\right) < \epsilon$ for all $x^* \in S_{X^*}$ which proves (b). Suppose that for some $x^* \in S_{X^*}$ there exist $x, y \in S\left(X, x^*, \frac{\delta}{2}\right)$ such that $\|x-y\| \geq \epsilon$. Then $x^*\left(\frac{x+y}{2}\right) \leq \left\|\frac{x+y}{2}\right\| < 1-\delta$ which contradicts that $\frac{x+y}{2} \in S\left(X, x^*, \frac{\delta}{2}\right)$.

(b) $\Rightarrow$ (c): Since X is Banach and $\text{diam}\left(S\left(X, x^*, \frac{1}{n}\right)\right) \to 0$ for all $x^* \in S_{X^*}$,

$$S(X, x^*, 0) = \bigcap \left\{S\left(X, x^*, \frac{1}{n}\right) : n \in \mathbb{N}\right\} \neq \emptyset, \quad \text{for all } x^* \in S_{X^*}.$$

Therefore by the James theorem, X is reflexive. This proves (c).

The proof of (c) $\Rightarrow$ (b) is obvious as $S\left(X, x^*, \frac{1}{n}\right) \subseteq S\left(X^{**}, x^*, \frac{1}{n}\right)$.

(b) $\Rightarrow$ (a): First observe that (b) implies that X is strictly convex. Suppose $\epsilon > 0$. Then by (b), there exists a $\delta > 0$ such that $\text{diam}(S(X, x^*, \delta)) < \epsilon$ for all $x^* \in S_{X^*}$. We claim that for all $x, y \in S_X$ such that $\|x-y\| \geq \epsilon$ we have $\left\|\frac{x+y}{2}\right\| < 1-\frac{\delta}{2}$. Suppose that there exist $x, y \in S_X$ such that $\|x-y\| \geq \epsilon$ but $\left\|\frac{x+y}{2}\right\| \geq 1-\frac{\delta}{2}$. Find $x^* \in S_{X^*}$ such that $x^*\left(\frac{x+y}{2}\right) = \left\|\frac{x+y}{2}\right\|$. Therefore, by assumption, $x^*\left(\frac{x+y}{2}\right) = \left\|\frac{x+y}{2}\right\| > 1-\frac{\delta}{2}$. This implies that $x^*(x) > 1-\delta$ and $x^*(y) > 1-\delta$. Hence $x, y \in S(X, x^*, \delta)$. Therefore $\|x-y\| < \epsilon$ which is a contradiction.

The proof of (a) $\Leftrightarrow$ (d) is involved and we refer to [29] for the proof. □

The following result, called Milman-Pettis Theorem, can be obtained as a consequence of the previous result.

Theorem 18 *Every uniformly convex Banach space is reflexive.*

Proof This follows from the proof of (b) $\Rightarrow$ (c) of Theorem 17. □

In Theorem 17 and Corollary 2, we characterized the geometric properties uniform convexity and strict convexity, respectively. Several geometric properties which are weaker than uniform convexity and stronger than strict convexity have been introduced in the literature. Our aim is to present characterizations, similar to the ones presented in Theorem 17 or Corollary 2, to the other geometric properties.

4.3 Strong Convexity and Its Characterizations

Theorem 17 motivates the introduction of the following geometric property which is weaker than uniform convexity.

We say that the space X is *strongly convex* if

$$\operatorname{diam}\left(S\left(X, x^*, \frac{1}{n}\right)\right) \to 0, \quad \text{for all } x^* \in S_{X^*}.$$

The notion of strong convexity was actually introduced in a different formulation in [14] which is explained at the end of this section.

We use the term strongly convex which is also used in [36]. However, the term strongly rotund is used for strongly convex in [29].

It is clear from the proof of (b) $\Rightarrow$ (c) of Theorem 17 that every strongly convex space is strictly convex and reflexive. But a space which is strictly convex and reflexive need not be strongly convex. So we need some additional property for characterizing the strong convexity. The following result helps to get the required additional property.

We say that X has the *Radon-Riesz property* if the relative weak and norm topologies coincide on the unit sphere S_X of X.

The Radon-Riesz property is also called *Kadec-Klee property* [12, 29].

In the rest of the section, whenever we write $S\left(X, x^*, \frac{1}{n}\right) \to S(X, x^*, 0)$ in the sense of V or H, it is understood that the limiting set $S(X, x^*, 0)$ is nonempty.

The following result is from [7, 8].

Theorem 19 *The following statements are equivalent.*

(a) *For every* $x^* \in S_{X^*}$, $S\left(X, x^*, \frac{1}{n}\right) \xrightarrow{V} S(X, x^*, 0)$ *and* $S(X, x^*, 0)$ *is compact.*
(b) *X is reflexive and has the Radon-Riesz property.*

Proof (a) $\Rightarrow$ (b): Since $S(X, x^*, 0)$ is nonempty for all $x^* \in S_{X^*}$, X is reflexive.

Let $x_n, x \in S_X$, for every n and $x_n \xrightarrow{w} x$. Find $x^* \in S_{X^*}$ such that $x^*(x) = 1$. Then $x^*(x_n) \to x^*(x) = 1$ and hence $x_n \in S(X, x^*, \delta_n)$ for some $\delta_n \to 0$. Without loss of generality, we assume that $\delta_n = \frac{1}{n}$. Since $S(X, x^*, 0)$ is compact and

$$S\left(X, x^*, \frac{1}{n}\right) \xrightarrow{V} S(X, x^*, 0),$$

by Theorem 17, (x_n) has a convergent subsequence converging to x. In fact, we can show that every subsequence of (x_n) has a convergent subsequence converging to x. Therefore, $x_n \to x$. This proves that X has the Radon-Riesz property.

(b) $\Rightarrow$ (a): Let $x_n \in S\left(X, x^*, \frac{1}{n}\right)$ for all n. By reflexivity of X, there exists a subsequence (x_{n_k}) of (x_n) such that $x_{n_k} \xrightarrow{w} x$ for some $x \in B_X$. Since

$$x^*(x_{n_k}) \to x^*(x),$$

$x^*(x) = 1$. Therefore $\|x\| = 1$ and by the weak lower semicontinuity of the norm, $\|x_{n_k}\| \to 1$. Therefore, by the Radon-Riesz property, $x_{n_k} \to x$. Since $x \in S(X, x^*, 0)$,

$$S(X, x^*, 0) \neq \emptyset.$$

Since $S(X, x^*, 0) \subseteq S(X, x^*, \frac{1}{n})$ for all n, the above argument shows that $S(X, x^*, 0)$ is compact. Moreover, by Theorem 10, $S\left(X, x^*, \frac{1}{n}\right) \xrightarrow{V} S(X, x^*, 0)$. □

We will now characterize strongly convex spaces.

Theorem 20 *The following statements are equivalent.*

(a) *X is strongly convex.*
(b) *X is reflexive, strictly convex and has the Radon-Riesz property.*

Proof (a) $\Rightarrow$ (b): From the proof of (b) $\Rightarrow$ (c) of Theorem 17 we get that X is reflexive and strictly convex. By Theorem 15, $\text{diam}\left(S\left(X, x^*, \frac{1}{n}\right)\right) \to 0$ implies that $S\left(X, x^*, \frac{1}{n}\right) \xrightarrow{V} S(X, x^*, 0)$. Therefore, by Theorem 19, X has the Radon-Riesz property.

(b) $\Rightarrow$ (a): By Theorem 19, for every $x^* \in S_{X^*}$, $S\left(X, x^*, \frac{1}{n}\right) \xrightarrow{V} S(X, x^*, 0)$ and $S(X, x^*, 0)$ is nonempty. Since X is strictly convex, $S(X, x^*, 0)$ is a singleton for every $x^* \in S_{X^*}$. Now (a) follows from Theorem 15. □

The following characterization of strong convexity is analogous to Corollary 2 and Theorem 17. We basically drop the condition "uniformly" from Theorem 17 for obtaining the following result.

Theorem 21 *The following statements are equivalent.*

(a) *X is strictly convex, reflexive and has the Radon-Riesz property.*
(b) *$diam\left(S\left(X, x^*, \frac{1}{n}\right)\right) \to 0$ for all $x^* \in S_{X^*}$.*
(c) *$diam\left(S\left(X^{**}, x^*, \frac{1}{n}\right)\right) \to 0$ for all $x^* \in S_{X^*}$.*
(d) *The norm of X^* is Fréchet differentiable on S_{X^*}.*

Proof The proof of (a) $\Leftrightarrow$ (b) follows from Theorem 20 and proof of (b) $\Leftrightarrow$ (c) follows from the proof of (b) $\Leftrightarrow$ (c) of Theorem 17. We refer to [29] for the proof of (a) $\Leftrightarrow$ (d). □

The implications (a) $\Leftrightarrow$ (b) of Theorem 21 are due to [14]. The implication (c) $\Leftrightarrow$ (d) of Theorem 21 is due to Šmulian [33].

In the following characterization of strong convexity, we see that the strong convexity property is equivalent to a property which was introduced by Ky Fan and Glicksberg. The following result is due to Ky Fan and Glicksberg [14].

Theorem 22 *The following statements are equivalent.*

(a) *X is strongly convex.*
(b) *For every nonempty convex subset C of X, $diam(C \cap tB_X) \to 0$ as $t \downarrow d(0, C)$.*

Proof (a) $\Rightarrow$ (b): Let C be a nonempty convex subset of X. Suppose $d(0, C) = 0$ and $t \downarrow d(0, C)$. Since $(C \cap tB_X) \subseteq tB_X$, $\text{diam}(C \cap tB_X) \to 0$.

Let $d(0, C) > 0$. In this case we can assume that $d(0, C) = 1$. Because, if $k > 0$ and A is a nonempty subset of X, then $d(0, kA) = kd(0, A)$ and

$$\text{diam}\,(kA \cap ktB_X) = \text{diam}\,(k(A \cap tB_X)) = k\ \text{diam}(A \cap tB_X).$$

Let $d(0, C) = 1$. Find $x^* \in S_{X^*}$ such that the hyperplane $H = \{x \in X : x^*(x) = 1\}$ separates B_X and C. We can assume that $C \subseteq H^+ = \{x \in X : x^*(x) \geq 1\}$. We show that $\text{diam}\big(H^+ \cap (1 + \frac{1}{n})B_X\big) \to 0$ as $n \to 0$ which proves (b). Let

$$x_n \in H^+ \cap \left(1 + \frac{1}{n}\right) B_X, \quad \text{for all } n.$$

Then,

$$y_n = \left(1 - \frac{1}{n^2}\right) \frac{x_n}{\|x_n\|} \in S\left(X, x^*, \frac{1}{n}\right) \quad \text{and} \quad \|x_n - y_n\| \to 0.$$

Since $\text{diam}\left(S\left(X, x^*, \frac{1}{n}\right)\right) \to 0$, by Theorem 15, (y_n) converges to $S(X, x^*, 0)$ which is a singleton. Hence (x_n) converges to $S(X, x^*, 0)$. Therefore, by Theorem 15, $\text{diam}\left(H^+ \cap \left(1 + \frac{1}{n}\right) B_X\right) \to 0$.

(b) $\Rightarrow$ (a): This proof is exactly similar to the proof given above. Let $x^* \in S_{X^*}$ and $x_n \in S\left(X, x^*, \frac{1}{n}\right)$. Consider $C = H^+ := \{x \in X : x^*(x) \geq 1\}$. Then

$$y_n = \left(1 - \frac{1}{n}\right)^{-1} \frac{x_n}{\|x_n\|} \in H^+ \cap \left(1 - \frac{1}{n}\right)^{-1} B_X \quad \text{and} \quad \|x_n - y_n\| \to 0.$$

Repeat the steps of the proof of (a) $\Rightarrow$ (b). □

The following corollary is an immediate consequence of Theorem 22 and the definition of nearly best approximation given in the introduction.

Corollary 3 *Let X be a strongly convex and C a nonempty closed convex subset of X. Then diam* $\left(P_C\left(0, \frac{1}{n}\right)\right) \to 0$.

In Sect. 5, we will show that the converse of the previous corollary is also true.

4.4 Reflexive Space with Radon-Riesz Property

In this subsection we discuss a property which is weaker than strong convexity.

We need the following lemma which will also be used in Sect. 5.

The following result is from [8].

Lemma 3 *Let* $x^* \in S_{X^*}$. *If* $S\left(X, x^*, \frac{1}{n}\right) \xrightarrow{V} S(X, x^*, 0)$, *then* $S(X, x^*, 0)$ *is compact.*

Proof Let (x_n) be a sequence in $S(X, x^*, 0)$ for some $x^* \in S_{X^*}$. Choose a sequence (y_n) in X such that $\|y_n\| < 1$ for all $n \in \mathbb{N}$ and $\|x_n - y_n\| \to 0$. Since $x^*(x_n - y_n) \to 0$, $x^*(y_n) \to 1$. Therefore there exists a subsequence (y_{n_k}) of (y_n) such that

$$y_{n_k} \in S\left(X, x^*, \frac{1}{k}\right) \setminus S(X, x^*, 0), \quad \text{for all } k \in \mathbb{N}.$$

By Theorem 10, there exists a subsequence $(y_{n_{k_j}})$ of (y_{n_k}) converging to some element $y \in S(X, x^*, 0)$. Therefore, $x_{n_{k_j}} \to y$ which proves that $S(X, x^*, 0)$ is compact. □

Lemma 3 says that the convergence of slices in the Vietoris sense forces the face contained in the slices to be compact. This does not happen when we deal with the Hausdorff convergence of slices. This illustrates as to how strong the Vietoris convergence of slices is compared to the Hausdorff convergence.

We will relax the condition "strict convexity" in Theorem 21 to get a characterization for the space which is reflexive and has the Radon-Riesz property.

We need the following lemma proved in [17].

Lemma 4 *Let* $x^* \in S_{X^*}$ *and* $0 < \delta < 1$. *Then* $S(X^{**}, x^*, \delta)$ *is contained in the weak* closure of* $S(X, x^*, \delta)$ *(in* X^{**}*).*

Proof Let $x_0^{**} \in S(X^{**}, x^*, \delta)$ and N be a weak* neighborhood of x_0^{**} in X^{**}. Now N_1 defined by

$$N_1 = N \cap \{x^{**} \in X^{**} : x^{**}(x^*) > 1 - \delta\}$$

is also a weak* neighborhood of x_0^{**}. Note that, by the Goldstine Theorem [29], B_X is weak* dense in $B_{X^{**}}$. Therefore, $N_1 \cap B_X \neq \emptyset$ and hence $N_1 \cap S(X, x^*, \delta) \neq \emptyset$. This proves the lemma. □

We will now prove the main result of this subsection.

Theorem 23 *The following statements are equivalent.*

(a) *X reflexive and has the Radon-Riesz property.*
(b) *For every* $x^* \in S_{X^*}$, $S\left(X, x^*, \frac{1}{n}\right) \xrightarrow{V} S(X, x^*, 0)$.
(c) *For every* $x^* \in S_{X^*}$, $S\left(X^{**}, x^*, \frac{1}{n}\right) \xrightarrow{V} S(X^{**}, x^*, 0)$.

Proof (a) ⇒ (b) follows from Theorem 19.

(b) ⇒ (a): By Lemma 3, (b) implies that $S(X, x^*, 0)$ is compact. Therefore the proof of (b) ⇒ (a) follows from (a) ⇒ (b) of Theorem 19.

As proved in Lemma 3, we can also show that (c) implies that $S(X^{**}, x^*, 0)$ is compact for every $x^* \in S_{X^*}$.

(c) $\Rightarrow$ (b): Let (n_k) be an increasing sequence of integers and

$$x_{n_k} \in S\left(X, x^*, \frac{1}{n_k}\right) \backslash S(X, x^*, 0), \quad \text{for } k = 1, 2, \ldots.$$

Then $x_{n_k} \in S(X^{**}, x^*, \frac{1}{n_k})$ for $k = 1, 2, \ldots$ Therefore (c) implies, by Theorem 10, that the sequence (x_{n_k}) has a subsequence converging to some point $x \in S(X^{**}, x^*, 0)$. Since $(x_{n_k}) \in X$, $x \in S(X, x^*, 0)$. Hence by Theorem 10, $S\left(X, x^*, \frac{1}{n}\right) \xrightarrow{V} S(X, x^*, 0)$.

(b) $\Rightarrow$ (c): Let $\epsilon > 0$. Then (b) implies that there exists $m \in \mathbb{N}$ such that

$$S\left(X, x^*, \frac{1}{m}\right) \subseteq S(X, x^*, 0) + \epsilon B_X \subseteq S(X, x^*, 0) + \epsilon B_{X^{**}}.$$

Observe that $S(X, x^*, 0) + \epsilon B_{X^{**}}$ is a w^*-closed subset of X^{**}. Since, by Lemma 4, $S\left(X^{**}, x^*, \frac{1}{m}\right)$ is contained in the w^*-closure of $S(X, x^*, \frac{1}{m})$, we have

$$S\left(X^{**}, x^*, \frac{1}{m}\right) \subseteq S(X, x^*, 0) + \epsilon B_{X^{**}} \subseteq S(X^{**}, x^*, 0) + \epsilon B_{X^{**}}.$$

This shows that

$$S\left(X^{**}, x^*, \frac{1}{n}\right) \xrightarrow{H} S(X^{**}, x^*, 0)$$

and $S(X^{**}, x^*, 0) \subseteq S(X, x^*, 0) + \epsilon B_{X^{**}}$. The compactness of $S(X, x^*, 0)$ implies that $\alpha(S(X^{**}, x^*, 0) \leq 2\epsilon$ where α denotes the Hausdorff index of noncompactness (see Sect. 2). Since ϵ is arbitrary, $\alpha(S(X^{**}, x^*, 0) = 0$. Therefore $S(X^{**}, x^*, 0)$ is compact and hence $S\left(X^{**}, x^*, \frac{1}{n}\right) \xrightarrow{V} S(X^{**}, x^*, 0)$. □

The implication (b) $\Rightarrow$ (c) of Theorem 23 is proved in [18].

Theorem 23 can also be stated as follows because of Lemma 3 and Theorem 10.

Theorem 24 *The following statements are equivalent.*

(a) *X reflexive and has the Radon-Riesz property.*
(b) *For every* $x^* \in S_{X^*}$, $S\left(X, x^*, \frac{1}{n}\right) \xrightarrow{H} S(X, x^*, 0)$ *and* $S(X, x^*, 0)$ *is compact.*
(c) *For every* $x^* \in S_{X^*}$, $S\left(X^{**}, x^*, \frac{1}{n}\right) \xrightarrow{H} S(X^{**}, x^*, 0)$ *and* $S(X^{**}, x^*, 0)$ *is compact.*

Remark 5 (a) If we compare Theorem 23 with Theorems 17 and 21, the fourth condition regarding a differentiability of the dual norm is missing in Theorem 23.

One may ask the following question: Is there any kind of differentiability of the dual norm which can characterize the space which is reflexive with Radon-Riesz property? We will give a partial answer to this question in the next section.

(b) A reflexive space with Radon-Riesz property is also called Efimov-Stechkin property in [29, 34] and drop property in [18].

4.5 Space with SSD Dual Norm

We relaxed the condition "uniformly" from Theorem 17 and stated Theorem 21. Similarly, the condition strict convexity was relaxed from Theorem 21 to come to Theorem 23. Now we relax the compactness of $S(X, x^*, 0)$ in Theorem 24. So we want to see a characterization for the space in which

$$S\left(X, x^*, \frac{1}{n}\right) \xrightarrow{H} S(X, x^*, 0),$$

whenever $S(X, x^*, 0)$ is nonempty. A characterization of this space, analogous to Theorem 21, was achieved in two papers [15, 20] which will be discussed in this subsection.

We need a definition and two lemmas

We say that the norm of X is *strongly subdifferentiable* (in short ssd) at $x \in S_X$ if the one sided limit

$$\lim_{t \to 0^+} \frac{\|x + th\| - \|x\|}{t}$$

exists uniformly for $h \in S_X$.

Lemma 5 *Let $x \in S_X$. Then for $h \in S_X$, there exists $x^* \in S(X^*, x, 0)$ such that $x^*(h) = d_+(x, h)$.*

Proof It follows from Lemma 2 that

$$d_-(x, h) \leq x^*(h) \leq d_+(x, h), \quad \text{for all } x^* \in S(X^*, x, 0).$$

If $d_-(x, h) < \alpha < d_+(x, h)$ for some $\alpha \in \mathbb{R}$, then from the proof of Theorem 6, it follows that there exists $x^*_\alpha \in S(X^*, x, 0)$ such that

$$x^*_\alpha(h) = \alpha < d_+(x, h).$$

This proves that

$$d_+(x, h) = sup\{x^*(h) : x^* \in S(X^*, x, 0)\}.$$

Since $S(X^*, x, 0)$ is weak* compact there exits $x^* \in S(X^*, x, 0)$ such that $x^*(h) = d_+(x, h)$. □

In the following lemma we discuss the convergence of faces.

Lemma 6 *Let $x \in S_X$ and $S\left(X^*, x, \frac{1}{n}\right) \xrightarrow{H} S(X^*, x, 0)$. Then for $\epsilon > 0$, there exists a $\delta > 0$ such that $S(X^*, y, 0) \subseteq B_\epsilon(S(X^*, x, 0))$ whenever $\|x - y\| < \delta$ and $y \in S_X$; that is, $S(X^*, x_n, 0) \xrightarrow{H} S(X^*, x, 0)$ whenever $x_n \to x$ and $x_n \in S_X$.*

Proof Let $S\left(X^*, x, \frac{1}{n}\right) \xrightarrow{H} S(X^*, x, 0)$ and $\epsilon > 0$. Then there exists $m \in \mathbb{N}$ such that $S\left(X^*, x, \frac{1}{m}\right) \subseteq B_\epsilon(S(X^*, x, 0)$. Suppose $\|x - y\| < \frac{1}{m}$ and $y^* \in S(X^*, y, 0)$. Then,

$$y^*(x) = y^*(y) - y^*(y - x) > 1 - \frac{1}{m}.$$

Therefore, $y^* \in S\left(X^*, x, \frac{1}{m}\right) \subseteq B_\epsilon(S(X^*, x, 0))$. □

Interestingly the converse of the previous lemma is also true and is discussed in Sect. 4.6. The convergence of faces is illustrated in Fig. 8.

We now present a characterization of the convergence of slices in the Hausdorff sense.

Theorem 25 *Let $x \in S_X$. Then the following statements are equivalent.*

(a) $S\left(X^*, x, \frac{1}{n}\right) \xrightarrow{H} S(X^*, x, 0)$.
(b) *The norm of X is ssd at x.*

Proof (a) $\Rightarrow$ (b): Let $\epsilon > 0$. By Lemma 6, there exists a $\delta > 0$ such that $0 < \delta < 1$ and

$$S(X^*, y, 0) \subseteq B_\epsilon(S(X^*, x, 0))$$

whenever $y \in S_X$ and $\|x - y\| < \delta$. We claim that for all $h \in S_X$,

$$\frac{\|x + th\| - \|x\|}{t} - d_+(x, h) < \epsilon, \quad \text{for all } t \text{ satisfying } 0 < t < \frac{\delta}{4}$$

which proves that the norm of X is ssd at x. Let $0 < t < \frac{\delta}{4}$ and $h \in S_X$. Find $x_t^* \in S_{X^*}$ such that $x_t^*(x + th) = \|x + th\|$. Now

$$\|x + th\| \geq \|x\| - t\|h\| \geq 1 - \frac{\delta}{4} > \frac{1}{2}.$$

Let $x_t = \frac{x+th}{\|x+th\|}$ and observe that $x_t^* \in S(X^*, x_t, 0)$ and

$$\|x_t - x\| \leq \frac{1}{2}[(1 - \|x + th\|)\|x\| + t\|h\|] \leq \frac{1}{2}\left[\frac{\delta}{4} + t\|h\|\right] < \delta.$$

Therefore, by Lemma 6, there exists $x^* \in S(X^*, x, 0)$ such that $\|x_t^* - x^*\| < \epsilon$. Since, by Lemma 2, $x^*(h) \leq d_+(x, h)$,

$$\frac{\|x+th\|-1}{t} - d_+(x,h) \le \frac{x_t^*(x+th)-x_t^*(x)}{t} - x^*(h) \le x_t^*(h) - x^*(h) < \epsilon.$$

(b) $\Rightarrow$ (a): Suppose that (a) is not true. Then there exists $\epsilon > 0$ and a sequence (x_n^*) such that $x_n^* \in S(X^*, x, \frac{1}{n})$ and $d(x_n^*, S(X^*, x, 0)) > \epsilon$ for all $n \in \mathbb{N}$. We will find a sequence (z_n) in S_X and a real positive sequence (t_n) converging to 0 such that

$$\frac{\|x+t_n z_n\| - \|x\|}{t_n} - d_+(x, z_n) \not\to 0 \quad \text{as } t_n \to 0,$$

which contradicts (b).

Note that $S(X^*, x, 0)$ is a weak* compact convex subset of X^* and

$$B_\epsilon(x_n^*) \cap S(X^*, x, 0) = \emptyset, \quad \text{for all } n \in \mathbb{N}.$$

Therefore, by the separation theorem in X^* with weak topology [29] there exists $z_n \in S_X$, for each n, such that

$$x_n^*(z_n) - \epsilon \ge \sup\{x^*(z_n) : x^* \in S(X^*, x, 0)\}.$$

By Lemma 5, there exists $z_n^* \in S(X^*, x, 0)$ such that $z_n^*(z_n) = d_+(x, z_n)$ for all n. Therefore, for $t > 0$,

$$\frac{\|x+tz_n\| - \|x\|}{t} - d_+(x, z_n) \ge \frac{x_n^*(x+tz_n) - z_n^*(x)}{t} - z_n^*(z_n) \ge \frac{(x_n^* - z_n^*)(x)}{t} + \epsilon.$$

Observe that $(z_n^* - x_n^*)(x) \to 0$ because $x_n^* \in S\left(X^*, x, \frac{1}{n}\right)$ and $z_n^* \in S(X^*, x, 0)$. Therefore, if we define

$$t_n = \frac{2(z_n^* - x_n^*)(x)}{\epsilon},$$

then $t_n \ge 0$, $t_n \to 0$ and

$$\frac{\|x+t_n z_n\| - \|x\|}{t_n} - d_+(x, z_n) \ge -\frac{\epsilon}{2} + \epsilon = \frac{\epsilon}{2}.$$

This proves the claim and hence it proves (a). □

Theorem 25 is proved in [15] and the proof of (b) $\Rightarrow$ (a) is adopted from [26].

We now present a characterization of the space whose dual norm is ssd. We need the following Lemma.

Lemma 7 *Let $\epsilon > 0$, $y^{**} \in B_{X^{**}}$ and $x^* \in S_{X^*}$. If $B_\epsilon(y^{**}) \cap S(X^{**}, x^*, 0) \ne \emptyset$ then $B_\epsilon(y^{**}) \cap S\left(X, x^*, \frac{1}{n}\right) \ne \emptyset$ for every $n \in \mathbb{N}$.*

Proof Suppose

$$x^{**} \in B_\epsilon(y^{**}) \cap S(X^{**}, x^*, 0).$$

By the Goldstine theorem there exists a net $\{x_\alpha\}_{\alpha \in I}$ in B_X such that $x_\alpha \to x^{**}$ in the weak* topology. Let

$$C = co\{x_\alpha : \alpha \in I\}$$

and $n \in \mathbb{N}$. Without loss of generality, we assume that $C \subseteq S\left(X, x^*, \frac{1}{n}\right)$. We claim that $C \cap B_\epsilon(y^{**}) \neq \emptyset$ which will prove the result. Suppose $C \cap B_\epsilon(y^{**}) = \emptyset$. Then by the separation theorem [29, Theorem 1.8.5], there exists $h^* \in S_{X^*}$ such that

$$h^*(y^{**}) - \epsilon \geq h^*(c), \quad \text{for all } c \in C.$$

This implies that $h^*(y^{**} - x^{**}) \geq \epsilon > \|y^{**} - x^{**}\|$ which is a contradiction. □

We now state the main result of this subsection.

Theorem 26 *The following statements are equivalent.*

(a) *For every* $x^* \in S_{X^*}$, $S\left(X, x^*, \frac{1}{n}\right) \xrightarrow{H} S(X, x^*, 0)$.
(b) *For every* $x^* \in S_{X^*}$, $S\left(X^{**}, x^*, \frac{1}{n}\right) \xrightarrow{H} S(X^{**}, x^*, 0)$.
(c) *The norm of* X^* *is ssd on* S_{X^*}.

Proof The proof of (a) ⇒ (b) follows from the first part of the proof of (b) ⇒ (c) of Theorem 23. The equivalance of (b) and (c) follows from Theorem 25.

(b) ⇒ (a): Let $x^* \in S_{X^*}$. Suppose that for $\epsilon > 0$ there exists a $\delta > 0$ such that

$$S(X^{**}, x^*, \delta) \subseteq B_{\frac{\epsilon}{2}}(S(X^{**}, x^*, 0)).$$

We will show that $S(X, x^*, 0) \neq \emptyset$ and $S(X, x^*, \delta) \subseteq B_\epsilon(S(X, x^*, 0))$.

Let $x \in S(X, x^*, \delta)$. Since $B_{\frac{\epsilon}{2}}(x) \cap S(X^{**}, x^*, 0) \neq \emptyset$, by Lemma 7,

$$B_{\frac{\epsilon}{2}}(x) \cap S\left(X, x^*, \frac{1}{n}\right) \neq \emptyset, \quad \text{for all } n,$$

and hence by (b),

$$B_{\frac{\epsilon}{2}}(x) \cap S\left(X, x^*, \frac{1}{n}\right) \subseteq B_{\frac{\epsilon}{4}}(S(X^{**}, x^*, 0)$$

eventually. Therefore, find $x_1 \in B_X$ such that

$$\|x - x_1\| < \frac{\epsilon}{2} \text{ and } d(x_1, S(X^{**}, x^*, 0) < \frac{\epsilon}{2^2}.$$

Repeating the same steps, construct inductively a sequence (x_n) in B_X such that

$$\|x_{n-1} - x_n\| < \frac{\epsilon}{2^n} \quad \text{and} \quad d(x_n, S(X^{**}, x^*, 0)) < \frac{\epsilon}{2^{n+1}}.$$

Therefore (x_n) converges to some $x_0 \in X \cap S(X^{**}, x^*, 0) = S(X, x^*, 0)$. This proves that $S(X, x^*, 0) \neq \emptyset$. Since for any n,

$$\|x - x_n\| \leq \|x - x_1\| + \|x_1 - x_2\| + \cdots + \|x_{n-1} - x_n\| < \frac{\epsilon}{2} + \frac{\epsilon}{2^2} + \cdots + \frac{\epsilon}{2^n},$$

we get $\|x - x_0\| < \epsilon$. This shows that $S(X, x^*, \delta) \subseteq B_\epsilon(S(X, x^*, 0))$. □

The following result is a consequence of Theorem 26.

Corollary 4 *If the norm of X^* is ssd on S_{X^*} then X is reflexive.*

Proof Since any one of the three statements in Theorem 26 implies that

$$S(X, x^*, 0) \neq \emptyset$$

for every $x^* \in S_{X^*}$, the space X is reflexive. □

In view of Theorem 26, we can restate Theorem 24 as follows.

Theorem 27 *The following statements are equivalent.*

(a) *X reflexive and has the Radon-Riesz property.*
(b) *For every $x^* \in S_{X^*}$, $S\left(X, x^*, \frac{1}{n}\right) \xrightarrow{H} S(X, x^*, 0)$ and $S(X, x^*, 0)$ is compact.*
(c) *For every $x^* \in S_{X^*}$, $S\left(X^{**}, x^*, \frac{1}{n}\right) \xrightarrow{H} S(X^{**}, x^*, 0)$ and $S(X^{**}, x^*, 0)$ is compact.*
(d) *The norm of X^* is ssd on S_{X^*} and $S(X, x^*, 0)$ is compact for all $x^* \in S_{X^*}$.*

In Theorem 27, we obtained a characterization similar to Theorem 21. However, the statement (d), in particular the compactness condition in (d), looks bit superficial. It would be interesting if the statement (d) in Theorem 27 is characterized in terms of a differentiability of the norm of X^* which is stronger than ssd.

Compared to Theorem 17 and Theorem 21, the characterization given in Theorem 26 looks incomplete. It is clear from Theorem 26 that if the norm of X^* is ssd on S_{X^*} then X is reflexive. However, the dual norm of a reflexive space need not be ssd on S_{X^*}. It would be interesting to see as to what additional geometric property is required to characterize the ssd of the dual norm on S_{X^*}. From Theorem 27 one can see that such an additional geometric property has to be weaker than the Radon-Riesz property.

4.6 Duality and Preduality Mappings

The set-valued mappings

$$D : S_X \longrightarrow 2^{S_{X^*}} \text{ defined by } D(x) = S(X^*, x, 0),$$

and

$$PD : S_{X^*} \longrightarrow 2^{S_X} \text{ defined by } PD(x^*) = S(X, x^*, 0)$$

are called, respectively, the *duality* and *preduality mappings* for X. Note that the map D is a nonempty valued map whereas the map PD could be empty valued. We have seen in Sect. 2, that X is smooth if and only if D is single valued. Moreover, X is strictly convex and reflexive if and only if the map PD is single valued.

The geometrical implications of the semicontinuities, especially the usc and Husc, of the duality and preduality mappings have been studied by several authors [3, 15, 17, 18, 20, 21, 25]. In this subsection, we relate the convergence of slices and the semicontinuities of the mappings D and PD.

Let $x \in S_X$. As per the definition,

D is *usc* at x if $S(X^*, x_n, 0) \xrightarrow{V^+} S(X^*, x, 0)$ whenever (x_n) in S_X converges to x, and

D is *Husc* at x if $S(X^*, x_n, 0) \xrightarrow{H^+} S(X^*, x, 0)$ whenever (x_n) in S_X converges to x. Similarly we define the usc and Husc of PD.

In Lemma 6, we noticed that

$$\left\{S\left(X^*, x, \frac{1}{n}\right) \xrightarrow{H^+} S(X^*, x, 0)\right\} \Rightarrow \left\{S(X^*, x_n, 0) \xrightarrow{H^+} S(X^*, x, 0) \text{ if } x_n \to x\right\}.$$

This implication says that the convergence of slices in the H^+ sense implies the Husc of the duality mapping D. In other words, it says that the convergence of slices in X^* in the H^+ sense implies the convergence of the faces in X^* in the H^+ sense. We will also see that the statements in the above implications are equivalent. We will also discuss a similar equivalence for the V^+ convergence of faces and slices in X^*. Further, we will take up the equivalence of convergence of the slices and of faces in X. These will illustrate the geometrical implications of the semicontinuities of the mappings D and PD. See Fig. 8.

We need the following result called *Bishop-Phelps-Bollobas Theorem*. This result is a generalization of the famous Bishop-Phelps theorem [5] which says that the collection of support functionals of the unit ball B_X is dense in S_{X^*}. The proof of the following result is involved, and we refer to [6] for its proof.

Theorem 28 *For ϵ such that $0 < \epsilon < 1$, let $x \in S_X$ and $x^* \in S_{X^*}$ be such that*

$$|x^*(x) - 1| < \frac{\epsilon^2}{4}.$$

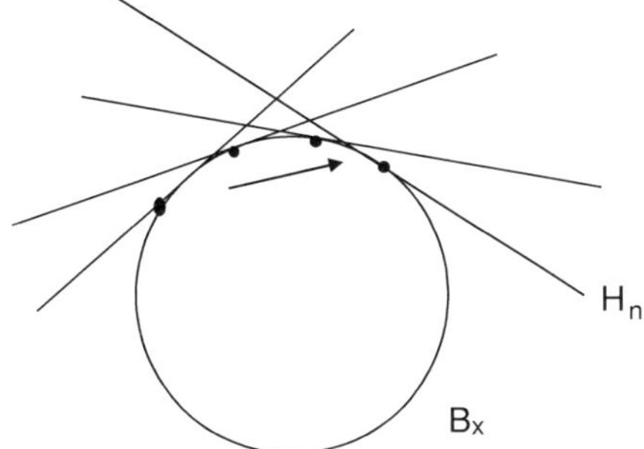

Fig. 8 Sequence of faces $S(X, x_n^*, 0)$ of B_X formed by the hyperplanes $H_n = \{x \in X : x_n^*(x) = 1\}$

Then there exist $y \in S_X$ *and* $y^* \in S_{X^*}$ *such that*

$$y^*(y) = 1, \quad \|x^* - y^*\| \leq \epsilon \text{ and } \|x - y\| \leq \epsilon.$$

The following result reveals the relation between the convergence of slices and of faces in X^*.

Theorem 29 *Let* $x \in S_X$. *Consider the following statements.*

(a) $S(X^*, x_n, 0) \xrightarrow{H^+} S(X^*, x, 0)$ *whenever* (x_n) *in* S_X *converges to* x.
(b) $S(X^*, x, \frac{1}{n}) \xrightarrow{H^+} S(X^*, x, 0)$.
(c) $S(X^*, x, \frac{1}{n}) \xrightarrow{V^+} S(X^*, x, 0)$.
(d) $S(X^*, x_n, 0) \xrightarrow{V^+} S(X^*, x, 0)$ *whenever* (x_n) *in* S_X *converges to* x.

Then (a) $\Leftrightarrow$ (b) $\Leftarrow$ (c) $\Rightarrow$ (d). *If* $S(X^*, x, 0)$ *is compact then the four statements are equivalent.*

Proof (a) $\Rightarrow$ (b): We will use Theorem 28 and Theorem 10 to prove this implication. Let $x_n^* \in S(X^*, x, \frac{1}{n})$ for all n. By Theorem 28, there exists $y_n \in S_X$ and $y_n^* \in S(X^*, y_n, 0)$ such that $y_n \to x$ and $\|x_n^* - y_n^*\| \to 0$. Since $y_n \to x$, by (a), there exists $z_n^* \in S(X^*, x, 0)$ such that $\|y_n^* - z_n^*\| \to 0$. Consequently, $\|x_n^* - z_n^*\| \to 0$. This proves (b).

(b) $\Rightarrow$ (a): This is proved in Lemma 6.

(c) $\Rightarrow$ (b): This is obvious.

(c) $\Rightarrow$ (d): Lemma 3 implies that $S(X^*, x, 0)$ is compact. Since $S(X^*, x, 0)$ is compact, $(c) \Rightarrow (d)$ follows from the implication (b) $\Rightarrow$ (a).

If $S(X^*, x, 0)$ is compact, then it follows from Theorem 10 that all four statements are equivalent. $\square$

The implication (a) $\Rightarrow$ (b) is due to [17].

The following corollary reveals the geometrical implication of the Husc of the duality mapping.

Corollary 5 *Let $x \in S_X$. Then the following statements are equivalent.*

(a) *The norm of X is ssd at x.*
(b) *The duality mapping D is Husc at x.*

Proof The equivalence follows immediately from Theorem 25 and (a) $\Leftrightarrow$ (b) of Theorem 29. □

Let us now discuss the relation between the convergence of slices and faces in X.

Theorem 30 *Let X be reflexive and $x^* \in S_{X^*}$. Consider the following statements.*

(a) $S(X, x_n^*, 0) \xrightarrow{H^+} S(X, x^*, 0)$ *whenever* (x_n^*) *in* S_{X^*} *converges to* x^*.
(b) $S\left(X, x^*, \frac{1}{n}\right) \xrightarrow{H^+} S(X, x^*, 0)$.
(c) $S\left(X, x^*, \frac{1}{n}\right) \xrightarrow{V^+} S(X, x^*, 0)$.
(d) $S(X, x_n^*, 0) \xrightarrow{V^+} S(X, x^*, 0)$ *whenever* (x_n^*) *in* S_{X^*} *converges to* x^*.

Then (a) $\Leftrightarrow$ (b) $\Leftarrow$ (c) $\Rightarrow$ (d). If $S(X, x^*, 0)$ is compact then all statements are equivalent.

Proof The theorem follows immediately from Theorem 29 and the fact that $X = X^{**}$. □

4.7 Notes and Remarks

In this section, we studied V^+ and H^+ convergence of slices and of faces and their geometric implications. We refer to [7, 8, 17, 18, 21] for the convergence of slices and of faces in the sense of V_w^+, H_w^+, $V_{w^*}^+$, $H_{w^*}^+$ and their geometric implications. A characterization of the Wijsman convergence [4, 31] of slices is obtained in [8]. It will be interesting to relate other geometric properties and the other notions of convergence of sets.

It is observed in [1, 8, 19, 22, 23] that the results stated in this section have their counterparts for the set of subdifferentials and the set of ϵ-subdifferentials of convex function and its conjugate.

5 Slices and Proximinality

In this section, we assume that X is a real Banach space. Let C be a nonempty, closed, and convex subset of X and $x \in X$. For $\delta \geq 0$, recall that the set

$$P_C(x, \delta) = \{y \in C : \|x - y\| \leq d(x, C) + \delta\}.$$

is called the *set of nearly best approximations* to x in C and the set $P_C(x, 0)$ is called the set of *best approximations* to x in C. The set $P_C(x, 0)$ could be empty but the set $P_C(x, \delta)$ is always nonempty when $\delta > 0$. For simplicity, we will denote $P_C(x, 0)$ as $P_C(x)$.

In this section, we will relate the geometric properties of Banach spaces discussed in the previous section with some properties in best approximation theory. As a consequence, we will see that the slices and their convergence determine some properties in best approximation theory. In fact, we have already seen in Corollary 3 that there is a relation between the convergence of the sets $P_C\left(x, \frac{1}{n}\right)$ and the convergence of slices $S\left(X, x^*, \frac{1}{n}\right)$ as $n \to \infty$ (see Fig. 11).

5.1 Slices and Set of Nearly Best Approximations

We will make some observations which relate the faces and slices with the set of best approximations and the set of nearly best approximations.

For $c > 0$, we denote $\overline{B}_c(0) = \{x \in X : \|x\| \leq c\}$.

Throughout this section, we will assume that C is a nonempty closed convex subset of X. Suppose $d(0, C) = 1$ and $H = \{x \in X : x^*(x) = 1\}$, where $x^* \in S_{X^*}$, is a hyperplane separating the unit ball B_X and the set C. Then,

$$P_C(0) = B_X \cap C \subseteq P_H(0) = B_X \cap H = S(X, x^*, 0) \tag{1}$$

and in fact

$$P_C(0) = S(X, x^*, 0) \cap C$$

(see Fig. 9). Observe that if $P_C(0) \neq \emptyset$ then $S(X, x^*, 0) \neq \emptyset$ and if $S(X, x^*, 0) = \emptyset$ then $P_C(0) = \emptyset$. Moreover, every face is a set of best approximations to 0 in a hyperplane and

$$P_C\left(0, \frac{1}{n}\right) = \overline{B}_{\left(1+\frac{1}{n}\right)}(0) \cap C$$

(see Fig. 10).

The following result, which is illustrated in Fig. 11, reveals the relationship between the slices and sets of nearly best approximations.

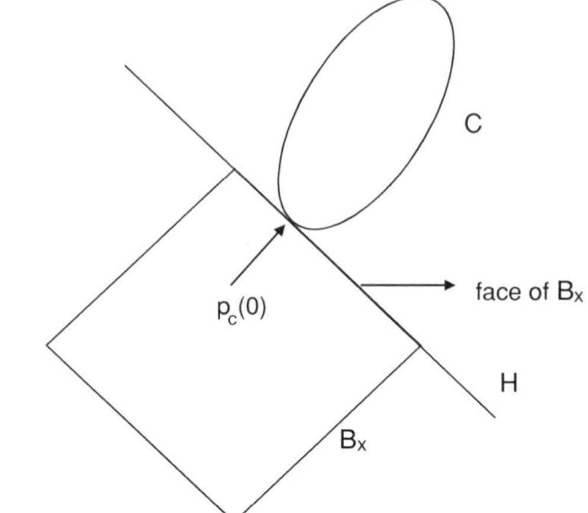

Fig. 9 Separation of B_X and C by the hyperplane H and $P_C(0)$ as a subset of the face $S(X, x^*, 0)$

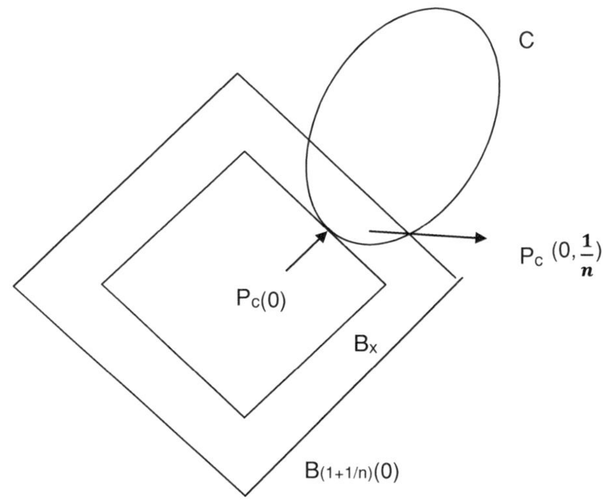

Fig. 10 Set of nearly best approximations $P_C(0, \frac{1}{n})$ as the intersection of C and the ball $B_{1+\frac{1}{n}}(0)$

We denote by $CC(X)$ the collection of all nonempty closed convex subsets of X and by $CH(X)$ the collection of all hyperplanes H of the form

$$H = \{x \in X : x^*(x) = \alpha\}, \quad \text{for some real } \alpha \geq 0 \text{ and } x^* \in S_{X^*}.$$

The set $\{H \in CH(X) : d(0, H) = 1\}$ is denoted by $CH_1(X)$. It follows from Ascoli's formula (Theorem 1) that if $H \in CH_1(X)$ then

$$H = \{x \in X : x^*(x) = 1\}, \quad \text{for some } x^* \in S_{X^*}.$$

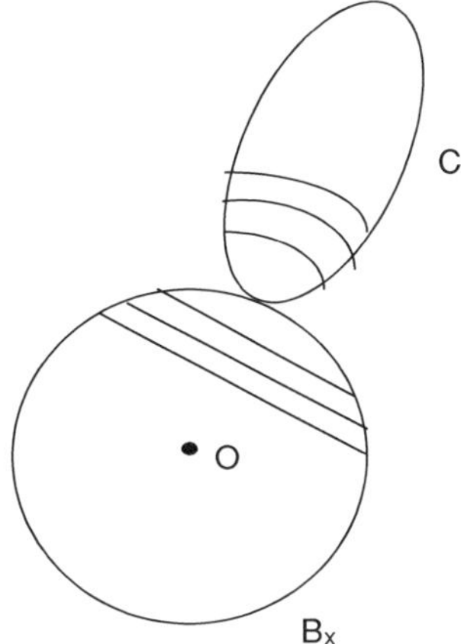

Fig. 11 Sequence of slices $S(X, x^*, \frac{1}{n})$ and sequence of sets of nearly best approximations $P_C(0, \frac{1}{n})$

Lemma 8 *Suppose $d(0, C) = 1$ and for $x^* \in S_{X^*}$, $H = \{x \in X : x^*(x) = 1\}$ is a hyperplane separating the unit ball B_X and the set C. Then for every sequence (x_n) satisfying*

(a) $x_n \in S\left(X, x^*, \frac{1}{n}\right)$ *for all n, there exist sequences $\{y_n\}$ and $\{\delta_n\}$ such that $\delta_n \to 0$, $y_n \in P_H(0, \delta_n)$ and $\|x_n - y_n\| \to 0$.*
(b) $x_n \in P_H\left(0, \frac{1}{n}\right)$ *for all n, there exist sequences $\{y_n\}$ and $\{\delta_n\}$ such that $\delta_n \to 0$, $y_n \in S(X, x^*, \delta_n)$ and $\|x_n - y_n\| \to 0$.*
(b) $x_n \in P_C\left(0, \frac{1}{n}\right)$ *for all n, there exist sequences $\{y_n\}$ and $\{\delta_n\}$ such that $\delta_n \to 0$, $y_n \in P_H(0, \delta_n)$ and $\|x_n - y_n\| \to 0$.*

Proof To prove the first statement, let $x_n \in S(X, x^*, \frac{1}{n})$ for all $n > 1$. Let $y_n = \frac{x_n}{x^*(x_n)}$ and $\delta_n = \frac{1}{n-1}$ for $n > 1$. Then $y_n \in P_H(0, \delta_n)$ and $\|x_n - y_n\| \to 0$.

If $x_n \in P_H(0, \frac{1}{n})$ for all $n > 1$, let $y_n = \frac{x_n}{\|x_n\|}$ and $\delta_n = \frac{1}{n+1}$. Then $y_n \in S(X^*, x^*, \delta_n)$ and $\|x_n - y_n\| \to 0$ which proves the second statement. If $x_n \in P_C(0, \frac{1}{n})$ for all n, let $y_n = \frac{x_n}{x^*(x_n)}$ and $\delta_n = \frac{1}{n}$ for n. Then $y_n \in P_H(0, \delta_n)$ and $\|x_n - y_n\| \to 0$. □

5.2 *Reflexivity and Strict Convexity*

We will present some characterizations of reflexivity and strict convexity in terms of some properties in best approximation theory.

We need some basic definitions from the theory of best approximation.

We say that the set C is *proximinal* if $P_C(x) \neq \emptyset$ for every $x \in X$. The set C is called *Chebyshev* if it is proximinal and $P_C(x)$ is a singleton for all $x \in X$.

We will first see under what conditions on X, C is proximinal or Chebyshev. We need the following facts which are easy to verify.

Fact 1. For $x_0 \in X$, $y \in P_C(x_0)$ if and only if $y - x_0 \in P_{-x_0+C}(0)$.

Fact 2. If $d(0, C) > 0$ then $d\left(0, \frac{C}{d(0,C)}\right) = 1$ and $\frac{1}{d(0,C)}C$ is a closed convex subset.

The following result which is an immediate consequence of the James theorem characterizes the reflexivity.

Theorem 31 *The following statements are equivalent.*

(a) *X is reflexive.*
(b) *Every $C \in CC(X)$ is proximinal.*
(c) *Every $H \in CH(X)$ is proximinal.*
(d) *Every $H \in CH_1(X)$ is proximinal.*

Proof (a) $\Rightarrow$ (b): Let X be reflexive. If $x \in X$ and C is a nonempty closed convex subset, then the set $P_C(x, \frac{1}{n})$ is weakly compact for each n. Therefore

$$\bigcap_{n=1}^{\infty} P_C\left(x, \frac{1}{n}\right) = P_C(x) \neq \emptyset.$$

(b) $\Rightarrow$ (c) $\Rightarrow$ (d): This is obvious.

(d) $\Rightarrow$ (a): Let $x^* \in S_{X^*}$ and $H = \{x \in X : x^*(x) = 1\} \in CH_1(X)$. Since

$$P_H(0) = S(X, x^*, 0) \neq \emptyset,$$

by the James Theorem, X is reflexive. □

The following characterization of strict convexity is a consequence of the fact that the unit sphere of a strictly convex space does not contain line segments.

Theorem 32 *The following statements are equivalent.*

(a) *X is strictly convex.*
(b) *$P_C(x)$ is at most a singleton for every $C \in CC(X)$ and $x \in X$.*
(c) *$P_H(x)$ is at most a singleton for every $H \in CH(X)$ and $x \in X$.*
(d) *$P_H(x)$ is at most a singleton for every $H \in CH_1(X)$ and $x \in X$.*

Proof (a) $\Rightarrow$ (b): Let X be strictly convex. This implies that $S(X, x^*, 0)$ is at most a singleton for every $x^* \in S_{X^*}$. Let $x \in X$ and C be a closed convex subset of X. Because of Facts 1 and 2 given above we can assume that $x = 0$ and $d(0, C) = 1$. Since, by equation (1), $P_C(0) \subseteq S(X, x^*, 0)$ for some $x^* \in S_{X^*}$, (b) follows.

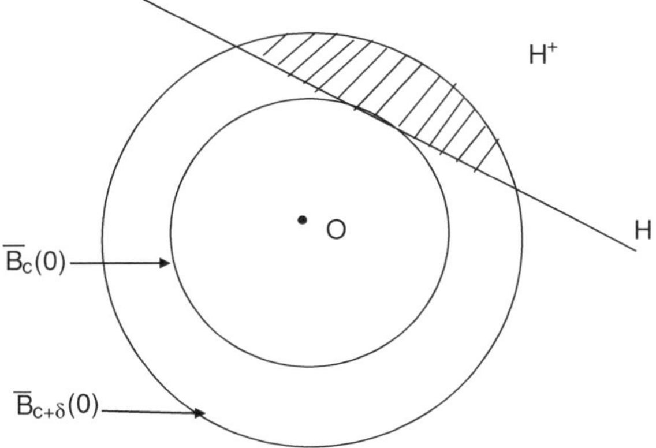

Fig. 12 The set $H^+(x^*, c) \cap \overline{B}_{c+\delta}(0)$ as the intersection of the ball $\overline{B}_{c+\delta}(0)$ and the half space H^+

(b) $\Rightarrow$ (c) $\Rightarrow$ (d): This is obvious.

(d) $\Rightarrow$ (a): Let $x^* \in S_{X^*}$. Then for $H = \{x \in X : x^*(x) = 1\}$, $P_H(0) = S(X, x^*, 0)$ and hence, by (d), $S(X, x^*, 0)$ is at most a singleton. □

The following result is an immediate consequence of the previous two results.

Theorem 33 *The following statements are equivalent.*

(a) *X is reflexive and strictly convex.*
(b) *Every $C \in CC(X)$ is Chebyshev.*
(c) *Every $H \in CH(X)$ is Chebyshev.*
(d) *Every $H \in CH_1(X)$ is Chebyshev.*

5.3 Characterizations of Uniform Convexity

In Sect. 4, we characterized the uniformly convex spaces in terms of convergence of slices of X and X^{**}. In this section, we will characterize the notion of uniform convexity in terms of convergence of sets of nearly best approximations.

For given $x^* \in S_{X^*}$ and a positive real number c, we denote

$$H^+(x^*, c) = \{x \in X : x^*(x) \geq c\}.$$

Lemma 9 *Let X be a uniformly convex space and c be a positive real number. Then, for $\epsilon > 0$ there exists a $\delta > 0$ such that*

$$\text{diam}\left(H^+(x^*, c) \cap \overline{B}_{c+\delta}(0)\right) < \epsilon$$

uniformly for every $x^* \in S_{X^*}$ (see Fig. 12).

Proof Let $x^* \in S_{X^*}$ and $\epsilon > 0$. In the proof of (a) $\Rightarrow$ (b) of Theorem 17, we have already seen that there exists a $\delta > 0$ such that $\delta < 1$ and

$$\operatorname{diam}(S(X, x^*, \delta)) < \epsilon. \tag{2}$$

Case I Let $c \geq 1$. We claim that

$$\operatorname{diam}\left(H^+(x^*, c) \cap \overline{B}_{c+\delta}(0)\right) < (1+c)\epsilon. \tag{3}$$

Suppose $x, y \in H^+(x^*, c) \cap \overline{B}_{c+\delta}(0)$. Define $\overline{x} = \frac{x}{c+\delta}$ and $\overline{y} = \frac{y}{c+\delta}$. Then $\overline{x}, \overline{y} \in B_X$ and

$$x^*(\overline{x}) = \frac{x^*(x)}{c+\delta} \geq \frac{c}{c+\delta} = 1 - \frac{\delta}{c+\delta} > 1 - \delta.$$

Similarly $x^*(\overline{y}) > 1 - \delta$. Therefore, by (2), $\|\overline{x} - \overline{y}\| < \epsilon$. This implies that

$$\|x - y\| < (c+\delta)\epsilon < (1+c)\epsilon.$$

Case II Let $0 < c < 1$. We claim that

$$\operatorname{diam}\left(H^+(x^*, c) \cap \overline{B}_{c+c\delta}(0)\right) < 2\epsilon.$$

Suppose $x, y \in H^+(x^*, c) \cap \overline{B}_{c+c\delta}(0)$. Define $\overline{x} = \frac{x}{c}$ and $\overline{y} = \frac{y}{c}$. Then $\overline{x}, \overline{y} \in \overline{B}_{1+\delta}(0)$ and

$$x^*(\overline{x}) = \frac{x^*(x)}{c} \geq 1$$

Similarly $x^*(\overline{y}) \geq 1$. Therefore

$$\overline{x},\ \overline{y} \in H^+(x^*, 1) \cap \overline{B}_{1+\delta}(0).$$

Hence by (3), $\|\overline{x} - \overline{y}\| < 2\epsilon$. This implies that $\|x - y\| < c2\epsilon < 2\epsilon$. □

We will now state the main result.

Theorem 34 *The following statements are equivalent.*

(a) *X is uniformly convex.*
(b) *$diam\left(S\left(X, x^*, \frac{1}{n}\right)\right) \to 0$ uniformly for all $x^* \in S_{X^*}$.*
(c) *$diam\left(P_H\left(0, \frac{1}{n}\right)\right) \to 0$ uniformly for all $H \in CH_1(X)$.*
(d) *For $\alpha \in \mathbb{R}^+$, $diam\left(P_H\left(x, \frac{1}{n}\right)\right) \to 0$ uniformly for all $x \in X$ and $H \in CH(X)$ such that $d(x, H) = \alpha$.*
(e) *For $\alpha \in \mathbb{R}^+$, $diam\left(P_C\left(x, \frac{1}{n}\right)\right) \to 0$ uniformly for all $x \in X$ and $C \in CC(X)$ such that $d(x, C) = \alpha$.*

Proof (a) $\Leftrightarrow$ (b): This follows from Theorem 17.

(a) $\Rightarrow$ (e): Let $\alpha > 0$ and $\epsilon > 0$. Suppose that $x \in X$ and $C \in CC(X)$ is such that $d(x, C) = \alpha$. By Facts 1 and 2 we can assume, without loss of generality, that $x = 0$. Find $x^* \in S_{X^*}$ such that the hyperplane

$$H = \{x \in X : x^*(x) = \alpha\}$$

separates $\overline{B}_\alpha(0)$ and C. By Lemma 9, there exists a $\delta > 0$ such that

$$\text{diam}(H^+(x^*, \alpha) \cap \overline{B}_{\alpha+\delta}(0)) < \epsilon.$$

Since

$$P_C(0, \delta) \subseteq (H^+(x^*, \alpha) \cap \overline{B}_{\alpha+\delta}(0)),$$

$\text{diam}(P_C(0, \delta)) < \epsilon$. This proves (e).

(e) $\Rightarrow$ (d) $\Rightarrow$ (c): This is obvious.

(c) $\Rightarrow$ (b): Suppose $\epsilon > 0$. By (c), there exists a $\delta > 0$ such that $\delta < \epsilon$ and

$$\text{diam}(P_H(0, \delta)) < \epsilon, \tag{4}$$

for any $H = \{x \in X : x^*(x) = 1\}$, $x^* \in S_{X^*}$. We claim that

$$\text{diam}(S(X, x^*, \delta)) < 3\epsilon.$$

Let $x, y \in S(X, x^*, \delta)$. Then define $\overline{x} = P_H(x)$ and $\overline{y} = P_H(y)$. Now $\|x - \overline{x}\| < \delta < \epsilon$ and $\|y - \overline{y}\| < \epsilon$. Therefore $\|\overline{x}\| \leq \|x - \overline{x}\| + \|x\| \leq \delta + 1$ and $\|\overline{y}\| \leq 1 + \delta$. Hence $\overline{x}$ and $\overline{y}$ are in $P_H(0, \delta)$ and therefore, by (4), $\|\overline{x} - \overline{y}\| < \epsilon$. Consequently

$$\|x - y\| \leq \|x - \overline{x}\| + \|\overline{x} - \overline{y}\| + \|y - \overline{y}\| \leq 2\delta + \epsilon < 3\epsilon.$$

This proves (b). □

5.4 Characterizations of Strong Convexity

We present some characterizations of the strong convexity property in terms of the convergence of sets of nearly best approximations.

We need some definitions from the theory of best approximation.

Let $C \in CC(X)$ and $x \in X$. A sequence (x_n) in C is called a *minimizing sequence* for $x \in X$ in C if

$$\|x - x_n\| \to d(x, C).$$

The set C is called *approximatively compact* if every minimizing sequence in C has a convergent subsequence.

A proximinal set C is said to be *strongly proximinal* at x if $d(x_n, P_C(x)) \to 0$ whenever (x_n) is a minimizing sequence for x in C. If C is strongly proximinal at every $x \in X$ then we say that C is *strongly proximinal*.

From the definition it is clear that every approximatively compact set C is proximinal and $P_C(x)$ is compact for every x.

We will now see the behavior of the sets $P_C(x, \frac{1}{n})$, $n \in \mathbb{N}$, in case C is approximatively compact or strongly proximinal.

Proposition 1 *Let $C \in CC(X)$. Consider the following statements.*

(a) *C is approximatively compact.*
(b) *For every $x \in X$, $P_C(x, \frac{1}{n}) \xrightarrow{H} P_C(x)$ and $P_C(x)$ is compact.*
(c) *For every $x \in X$, $P_C(x, \frac{1}{n}) \xrightarrow{H} P_C(x)$.*
(d) *C is strongly proximinal.*

Then (a) $\Leftrightarrow$ (b) $\Rightarrow$ (c) $\Leftrightarrow$ (d).

Proof (a) $\Rightarrow$ (b): Suppose C is approximatively compact and $x \in X$. Then $P_C(x)$ is compact. We will use Theorem 10. Suppose $y_n \in P_C\left(x, \frac{1}{n}\right)$, $n \in \mathbb{N}$. Then (y_n) is a minimizing sequence and hence it has a convergent subsequence converging to an element of $P_C(x)$. This proves (b).

(b) $\Rightarrow$ (a): Let (y_n) be a minimizing sequence in C for $x \in X$. Then there exists a subsequence (y_{n_k}) such that $y_{n_k} \in P_C\left(x, \frac{1}{k}\right)$ for all $k \in \mathbb{N}$. Therefore, by Theorem 10, (y_{n_k}) has a convergent subsequence.

(b) $\Rightarrow$ (c): This is obvious.

(c) $\Leftrightarrow$ (d): This follows easily from the definitions. □

It is clear from the previous result that approximative compactness is stronger than strong proximinality. Moreover, Proposition 1 motivates us to relate the properties approximative compactness and strong proximinality with geometrical properties of X as the convergence of sets $P_C\left(x, \frac{1}{n}\right)$ is related to the convergence of slices (see Lemma 8 and Fig. 11).

We will first characterize the strong convexity property in terms of approximative compactness. We relax the condition "uniformly" from Theorem 34 for the following result.

Theorem 35 *The following statements are equivalent.*

(a) *X is strongly convex.*
(b) *$diam\big(S\left(X, x^*, \frac{1}{n}\right)\big) \to 0$ for all $x^* \in S_{X^*}$.*
(c) *$diam\big(P_H\left(0, \frac{1}{n}\right)\big) \to 0$ for all $H \in CH_1(X)$.*
(d) *$diam\big(P_H\left(x, \frac{1}{n}\right)\big) \to 0$ for all $x \in X$ and $H \in CH(X)$.*

(e) *diam*$\left(P_C\left(x, \frac{1}{n}\right)\right) \to 0$ *for all* $x \in X$ *and* $C \in CC(X)$.
(f) *Every* $C \in CC(X)$ *is Chebyshev and approximatively compact.*

Proof The proof of the equivalence of (a) and (b) is given in Theorem 21.

(b) $\Rightarrow$ (c): We will use Lemma 8 and Theorem 15. Let $H = \{x \in X : x^*(x) = 1\}$ for some $x^* \in S_{X^*}$. Then $d(0, H) = 1$. Suppose $x_n \in P_H(0, \frac{1}{n})$ for all n. Then by Lemma 8, there exist sequences (y_n) and (δ_n) such that

$$\delta_n \to 0, y_n \in S(X^*, x^*, \delta_n) \text{ and } \|x_n - y_n\| \to 0.$$

Condition (b) implies that diam $\left(S\left(X, x^*, \frac{1}{n}\right)\right) \to 0$ and $S(X, x^*, 0) = \{x_0\}$ for some $x_0 \in S_X$. Therefore $y_n \to x_0$ and hence $x_n \to x_0$. Now (c) follows from Theorem 15.

(c) $\Rightarrow$ (d): This follows from Facts 1 and 2.

(d) $\Rightarrow$ (e): Let $C \in CC(X)$ and $x \in X$. By Facts 1 and 2, we assume, without loss of generality, that $d(0, C) = 1$ and $x = 0$. Let $H = \{x \in X : x^*(x) = 1\}$, $x^* \in S_{X^*}$, be a hyperplane separating C and B_X. Let $x_n \in P_C(0, \frac{1}{n})$ for all n. Use Lemma 8 and Theorem 15 and follow the steps of the proof of (b) $\Rightarrow$ (c).

(e) $\Rightarrow$ (f): Let $C \in CC(X)$ and $x \in X$. If $y_n \in P_C(x, \frac{1}{n})$ for every n, then by (e), (y_n) is Cauchy, hence it converges to an element of $P_C(x)$. This shows that C is proximinal and (e) further implies that diam$(P_C(x)) = 0$. Therefore, C is Chebyshev. The approximative compactness follows from Proposition 1 and Theorem 15.

(f) $\Rightarrow$ (b): Let $x^* \in S_{X^*}$ and $C = \{x \in X : x^*(x) = 1\}$. Suppose $x_n \in S\left(X, x^*, \frac{1}{n}\right)$ for every $n \in \mathbb{N}$. Then there exist sequences (y_n) and (δ_n) such that

$$\delta_n \to 0, \quad y_n \in P_C(0, \delta_n) \text{ and } \|x_n - y_n\| \to 0.$$

Since (y_n) is a minimizing sequence for 0 in C, by (f), every subsequence of (y_n) has a convergent subsequence converging to $P_C(0)$ which is a singleton. Therefore (x_n) converges to $P_C(0) = S(X, x^*, 0)$. Hence (b) follows from Theorem 15. □

5.5 Characterizations of Radon-Riesz Property

We now generalize Theorem 35 by relaxing the condition "strict convexity".

Theorem 36 *The following statements are equivalent.*

(a) *X is reflexive and has the Radon-Riesz property.*
(b) $S(X, x^*, 0)$ *compact and* $S(X, x^*, \frac{1}{n}) \xrightarrow{H} S(X, x^*, 0)$ *for all* $x^* \in S_{X^*}$.

(c) $P_H(0)$ *is compact and* $P_H(0, \frac{1}{n}) \xrightarrow{H} P_H(0)$ *for all* $H \in CH_1(X)$.
(d) $P_C(x)$ *compact and* $P_C(x, \frac{1}{n}) \xrightarrow{H} P_C(x)$ *for all* $C \in CC(X)$ *and* $x \in X$.
(e) *Every* $C \in CC(X)$ *is approximatively compact.*

Proof The equivalence (a) $\Leftrightarrow$ (b) is a part of Theorem 24.

(b) $\Rightarrow$ (c): Let $H = \{x \in X : x^*(x) = 1\}, x^* \in S_{X^*}$ and $x_n \in P_H\left(0, \frac{1}{n}\right)$ for all n. Then, by Lemma 8, there exist sequences (y_n) and (δ_n) such that $\delta_n \to 0, y_n \in S(X^*, x^*, \delta_n)$ and $\|x_n - y_n\| \to 0$. By (b) and Theorem 10, (y_n) has a convergent subsequence and hence (x_n) has a convergent subsequence converging to an element in $P_H(0)$. By Theorem 10, (c) follows.

(c) $\Rightarrow$ (d): Let $C \in CC(X)$ and $x \in C$. By Facts 1 and 2 we can assume that $x = 0$ and $d(0, C) = 1$. Let $x_n \in P_C(0, \frac{1}{n})$ for all n. Use Lemma 8 and repeat the proof of (b) $\Rightarrow$ (c).

(d) $\Rightarrow$ (e): This follows from Proposition 1.

(e) $\Rightarrow$ (b): This is a consequence of Lemma 8 and Theorem 10. The proof is similar to the proof of (b) $\Rightarrow$ (c). □

5.6 Characterizations of Strong Subdifferentiability of the Dual Norm

Theorem 36 leads to the following theorem if we drop the compactness condition.

Theorem 37 *Consider the following statements.*

(a) *The norm of* X^* *is ssd on* S_{X^*}.
(b) $S\left(X, x^*, \frac{1}{n}\right) \xrightarrow{H} S(X, x^*, 0)$ *for all* $x^* \in S_{X^*}$.
(c) $P_H\left(0, \frac{1}{n}\right) \xrightarrow{H} P_H(0)$ *for all* $H \in CH_1(X)$.
(d) $P_H\left(x, \frac{1}{n}\right) \xrightarrow{H} P_H(x)$ *for all* $H \in CH(X)$ *and* $x \in X$.
(e) $P_C\left(x, \frac{1}{n}\right) \xrightarrow{H} P_C(x)$ *for all* $C \in CC(X)$ *and* $x \in X$.
(f) *Every* $C \in CC(X)$ *is strongly proximinal.*

Then (a) $\Leftrightarrow$ (b) $\Leftrightarrow$ (c) $\Leftrightarrow$ (d) $\Leftarrow$ (e) $\Leftrightarrow$ (f).

Proof The equivalence (a) $\Rightarrow$ (b) is a part of Theorem 26. For the proofs of the implications (b) $\Leftrightarrow$ (c) $\Leftrightarrow$ (d), use Lemma 8 and Theorem 9 and follow the same lines of proof of (b) $\Rightarrow$ (c) of Theorem 36.
(e) $\Leftrightarrow$ (f): This is stated in Proposition 1.

(e) $\Rightarrow$ (d): This is obvious. □

Remark 6 If we compare Theorem 37 with Theorems 34, 35 and 36, it is expected that all the statements in the previous result should be equivalent. Interestingly the implication (d) $\Rightarrow$ (e), in Theorem 37, is not true. Surprisingly, (e) forces the set $P_C(x)$ to be compact for every $C \in CC(X)$ and $x \in X$, which will be proved below.

We need the following result of Osman [30].

Lemma 10 *For $x^* \in S_X$ and $c > 0$, let $H = \{x : x^*(x) = c\}$. Suppose that (x_n) is a sequence in X such that $x_n \to x_0$ weakly for some x_0. Let $x^*(x_n) > c$ for all n. Then $(\overline{co}\{x_n : n \in \mathbb{N}\}) \cap H \neq \emptyset$ if and only if $x_0 \in H$, and in this case*

$$(\overline{co}\{x_n : n \in \mathbb{N}\}) \cap H = \{x_0\}.$$

Theorem 38 *The following statements are equivalent.*

(a) *X is reflexive and has the Radon-Riesz property.*
(b) *The norm of X^* is ssd on S_{X^*} and $S(X, x^*, 0)$ is compact for all $x^* \in S_{X^*}$.*
(c) *Every $C \in CC(X)$ is approximatively compact.*
(d) *Every $C \in CC(X)$ is strongly proximinal.*

Proof The equivalence (a) $\Leftrightarrow$ (b) is stated in Theorem 27 and (a) $\Leftrightarrow$ (c) is stated in Theorem 36.

(c) $\Rightarrow$ (d): This is obvious.

(d) $\Rightarrow$ (b): By Theorem 37, (d) implies that the norm of X^* is ssd on S_{X^*}. It remains to show that $S(X, x^*, 0)$ is compact for all $x^* \in S_{X^*}$.

Note that, by Corollary 4, (d) implies that X is reflexive. Suppose $x^* \in S_{X^*}$. Then $S(X, x^*, 0)$ is nonempty and weakly compact. Let (x_n) be a sequence in $S(X, x^*, 0)$. Define $y_n = (1 + \frac{3}{n})x_n$ for every n. Then $x^*(y_n) > 1$ for every n and $d(y_n, S(X, x^*, 0)) \to 0$. Since (y_n) is bounded, it has a weakly convergent subsequence. Let us denote the subsequence as (y_n) for simplicity and let $y_n \to y_0$ weakly for some $y_0 \in S(X, x^*, 0)$. By Lemma 10,

$$(\overline{co}\{y_n : n \in \mathbb{N}\}) \cap H = \{y_0\}$$

where $H = \{x \in X : x^*(x) = 1\}$. Since $y_0 \in S_X$ and

$$(\overline{co}\{y_n : n \in \mathbb{N}\}) \cap S_X \subseteq (\overline{co}\{y_n : n \in \mathbb{N}\}) \cap H,$$

we have $\overline{co}\{y_n : n \in \mathbb{N}\}) \cap S_X = \{y_0\}$.

Take $C = \overline{co}\{y_n : n \in \mathbb{N}\}$. Since $\|y_n\| \to 1$, $d(0, C) = 1$ and $y_n \in P_C(0, \delta_n)$ for some $\delta_n \to 0$. By the strong proximinality of C, we have $d(y_n, P_C(0)) \to 0$. Since

$$P_C(0) = (\overline{co}\{y_n : n \in \mathbb{N}\}) \cap S_X,$$

we have $y_n \to y_0 \in S(X, x^*, 0)$. This implies that (x_n) has a convergent subsequence which converges to y_0 and the proof is complete. □

The implication (d) ⇒ (b) in the previous theorem is proved in [12].

5.7 Notes and Remarks

The property strong proximinality is relatively recent compared to the property approximative compactness. The properties associated with approximative compactness can be seen in [29, 34, 36]. Strong proximinality of certain subspaces of some classical Banach spaces are studied in the literature [10, 11, 20, 26].

We observed that the statement (d) in Theorem 37 is equivalent to the ssd of the norm of X^* and (f) forces the faces $S(X, x^*, 0)$ to be compact for all $x^* \in S_{X^*}$ (see Theorem 38). The natural question is the following: Is there a condition which is stronger than (d) and weaker than (f) and is equivalent to (a) of Theorem 37? The question can also be posed as follows: Can we find a class $\mathscr{F}$ of subsets of X such that $CH(X) \subset \mathscr{F} \subset CC(X)$ and the norm of X^* is ssd on S_{X^*} if and only if every $C \in \mathscr{F}$ is strongly proximinal?

In this section, we studied V^+ and H^+ convergence of $(P_C(x, \frac{1}{n}))$ and associated the convergence with the corresponding convergence of slices. Convergence of $(P_C(x, \frac{1}{n}))$ with respect to other convergence notions can still be explored.

References

1. Asplund, E., Rockafellor, R.T.: Gradients of convex functions. Trans. Amer. Math. Soc. **139**, 443–467 (1969)
2. Banaś, J., Goebel, K.: Measures of Noncompactness in Banach Spaces. Lecture Notes in Pure and Applied Mathematics, vol. 60. Marcel Dekker Inc, New York (1980)
3. Banaś, J., Sadarangani, K.: Compactness conditions and strong subdifferentiability of a norm in geometry of banach spaces. Nonlinear Anal. **49**, 623–629 (2002)
4. Beer, C.: Topologies on Closed and Closed Convex Sets. Kluwer Academic Publishers, London (1993)
5. Bishop, E., Phelps, R.R.: A proof that every banach space is subreflexive. Bull. Amer. Math. Soc. **67**, 97–98 (1961)
6. Bollabas, B.: An extension to the theorem of bishop and phelps. Bull. London Math. Soc. **2**, 181–182 (1970)
7. Chakraborty, A.K.: Some contributions to set-valued and convex analysis. Ph.D. Thesis, IIT, Kanpur (2006)
8. Chakraborty, A.K., Shunmugaraj, P., Zălinescu, C.: Continuity properties for the subdifferential and ε-subdifferential of a convex function and its conjugate. J. Convex Anal. **14**, 479–514 (2007)
9. Dolecki, S., Rolewicz, S.: Metric characterizations of upper semicontinuity. J. Math. Anal. Appl. **69**, 146–152 (1979)
10. Dutta, S., Narayana, D.: Strongly proximinal subspaces in Banach spaces. Contemp. Math. Amer. Math. Soc. **435**, 143–152 (2007)

11. Dutta, S., Narayana, D.: Strongly proximinal subspaces of finite codimension in $c(k)$. Colloquium Math. **109**, 119–128 (2007)
12. Dutta, S., Shunmugaraj, P.: Strong proximinality of closed convex sets. J. Approx. Theor. **163**, 547–553 (2011)
13. Fabian, M., Habala, P., Hajek, P., Montesinos, V., Pelant, J., Zizler, V.: Functional Analysis and Infinite Dimensional Geometry. Springer, New York (2001)
14. Fan, K., Glicksberg, I.: Some geometric properties of the sphere in a normed linear space. Duke Math. J. **25**, 553–568 (1958)
15. Franchetti, C., Payá, R.: Banach spaces with strongly subdifferentiable norm. Boll. Un. Mat. Ital. B **7**, 45–70 (1993)
16. Giles, J.R.: Convex Analysis with Application in the Differentiation of Convex Functions. Pitman Advanced Publishing Program, Boston (1982)
17. Giles, J.R., Gregory, D.A., Sims, B.: Geometrical implications of upper semi-continuity of the duality mapping on a banach space. Pacific J. Math. **79**, 99–109 (1978)
18. Giles, J.R., Sims, B., Yorke, A.C.: On the drop and weak drop properties for banach space. Bull. Austral. Math. Soc. **41**, 503–507 (1990)
19. Giles, J.R., Sciffer, S.: On weak hadamard differentiability of convex functions on banach spaces. Bull. Austral. Math. Soc. **54**, 155–166 (1996)
20. Godefroy, G., Indumathi, V.: Strong proximinality and polyhedral spaces. Rev. Mat. Complut. **14**, 105–125 (2001)
21. Godefroy, G., Indumathi, V.: Norm-to-weak upper semi-continuity of the duality and preduality mappings. Set-Valued Anal. **10**, 317–330 (2002)
22. Godefroy, G., Indumathi, V., Lust-Piquard, F.: Strong subdifferentiability of convex functionals and proximinality. J. Approx. Theor. **116**, 397–415 (2002)
23. Gregory, D.A.: Upper semicontinuity of sub-differential mappings. Canad. Math. Bull. **23**, 11–19 (1980)
24. Holmes, R.B.: Geometric Functional Analysis and Its Applications. Springer, New York (1975)
25. Hu, Z., Lin, B.-L.: Smoothness and the asymptotic-norming properties of banach spaces. Bull. Austral. Math. Soc. **45**, 285–296 (1992)
26. Indumathi, V.: Proximinal and strongly proximinal subspaces of finite co-dimension. Report, Pondicherry University, Pondicherry (2010)
27. James, R.C.: Characterizations of reflexivity. Studia Math. **23**, 205–216 (1964)
28. Luenberger, D.G.: Optimization by Vector Space Methods. Wiley, New York (1969)
29. Megginson, R.E.: An Introduction to Banach Space Theory. Springer, New York (1998)
30. Osman, E.V.: On the continuity of metric projections in banach spaces. Math. USSR Sb. **9**, 171–182 (1969)
31. Sonntag, Y., Zălinescu, C.: Set convergence : an attempt of classifications. Trans. Amer. Math. Soc. **340**, 199–226 (1993)
32. Šmulian, V.L.: On some geometrical properties of the sphere in a space of the type (B). Mat. Sb. (N.S.) **6**, 77–94 (1939)
33. Šmulian, V.L.: Sur la dérivabilité de la norme dans l'espace de Banach. C. R. (Doklady) Acad. Sci. URSS (N.S.) **27**, 643–648 (1940)
34. Vlasov, L.P.: Approximative properties of sets in normed linear spaces (Russian). Uspehi Mat. Nauk. (Russian Math. Surveys **28**, 1–66 (1973)) **28**, 3–66 (1973)
35. Zălinescu, C.: Convex Analysis in General Vector Spaces. World Scientific, Singapore (2002)
36. Zhang, Z., Shi, Z.: Convexities and approximateive compactness and continuity of metric projection in banach spaces. J. Approx. Theory **161**, 802–812 (2009)

Measures of Noncompactness and Well-Posed Minimization Problems

Józef Banaś

Abstract This chapter presents facts concerning the theory of well-posed minimization problems. We recall some classical results obtained in the framework of the theory but focus mainly on the detailed presentation of the application of the theory of measures of noncompactness to investigations of the well-posedness of minimization problem.

Keywords Measure of noncompactness · Kuratowski measure of noncompactness · Hausdorff measure of noncompactness · Minimization problems · Well-posedness · Well-posedness in the sense of Tikhonov · Well-posedness in the sense of Levitin and Polyak · Nearly uniform convex spaces

1 Introduction

Numerous problems of control theory and optimization theory are connected with finding of a sequence minimizing some functional related to an investigated problem. It turns out that in practice we are often not able to find an exact solution of a considered minimum (or maximum) problem. Nevertheless, sometimes it is possible to construct a sequence (the so-called minimizing or maximizing sequence) that is convergent to the solution of an investigated minimum or maximum problem. Such a situation is very desirable since it creates the possibility to obtain an approximate solution of an investigated minimum (maximum) problem.

Such an approach generates, in a natural way, the problem associated with the structure of possible minimizing (or maximizing) sequences.

J. Banaś (✉)
Department of Mathematics, Rzeszów University of Technology, al. Powstańców Warszawy 8, 35-959 Rzeszów, Poland
e-mail: jbanas@prz.edu.pl

Q. H. Ansari (ed.), *Nonlinear Analysis*, Trends in Mathematics,
DOI: 10.1007/978-81-322-1883-8_4, © Springer India 2014

Indeed, we can raise the following questions: Does the minimizing sequence converge? In the case of a positive answer we can ask about the accuracy of the approximation of a point realizing minimum of a considered functional with help of a constructed minimizing sequence.

The next question arising naturally in the situation when a minimizing sequence is not convergent, is connected to the structure of the set containing all accumulation points of obtained minimizing sequence. Thus, we can ask if the set of all accumulation points of an arbitrary minimizing (maximizing) sequence is compact or, in some sense, is a set being "almost" compact etc. Obviously, the above approach creates further questions related to the posedness of a considered minimum problem.

Before further relevant discussion of the above raised problem let us establish that in what follows we will always consider the minimum problem only. Obviously, the considerations of the maximum problem are similar and can be therefore conducted in the same way as the minimum problem.

To explain our approach let us take into account the well-known and, simultaneously, the fundamental problem considered in several branches of mathematics; the problem of the existence and uniqueness of a solution of the equation having the form

$$x = F(x), \tag{1}$$

where F is a given operator defined on a metric space X and with values in the same space X.

The above formulated problem may be studied as the problem of the minimum of the functional $J : X \to \mathbb{R}$ ($\mathbb{R}$ denotes the set real numbers) defined in the following way:

$$J(x) = d(F(x), x), \tag{2}$$

where d is a metric given in the assumed metric space X. It is clear that the values of the functional J defined by (2) are located in the positive real half-axis $\mathbb{R}_+ = [0, \infty)$. This implies that there exists infimum of the functional J in the metric space X. On the other hand, Eq. (1) has a solution if and only if the functional J has minimum equal to zero. Moreover, Eq. (1) has exactly one solution if and only if the functional J attains its minimum exactly at one point belonging to the metric space X. Apart from this every minimizing sequence of the functional J defined by (2) is a sequence approximating a fixed point of the operator F related to Eq. (1). The behavior of the minimizing sequence of the functional J is reflected by the behavior of the sequence approximating solutions of Eq. (1).

This connection suggests the possibility of the use of tools of fixed point theory [17, 18] in the theory of optimization (cf. [2]).

In this chapter we follow the ideas presented above. More precisely, we show that the tools associated with the technique of measures of noncompactness (cf. [12] and references therein) can be used in the study of well-posed minimum problems.

2 The Minimization Problem and Its Well-Posedness in the Classical Sense of Tikhonov

In this section we recall the fundamental classical concepts and results related to the minimization problem of a functional J acting from a nonempty subset D of a metric space X ($X = (X, d)$) into the set $\mathbb{R}$ of real numbers.

Following A. N. Tikhonov [33] we state that the minimization problem for the functional J is *well posed* if and only if every minimizing sequence of the functional J is convergent in the metric space X.

To express the above concept more precisely, let us recall that the sequence $\{x_n\}$, $\{x_n\} \subset D$, is called *minimizing* for the functional J if

$$\lim_{n\to\infty} J(x_n) = \inf\{J(x) : \ x \in D\}. \tag{3}$$

It is clear that if J is bounded from below then the sequence $\{J(x_n)\}$ is also bounded from below and the infimum $\inf\{J(x) : \ x \in D\}$ is a real number. In such a case we obtain a refinement of the above definition which is discussed below.

Let us regard some simple consequences of the above accepted definition of the well-posedness of minimization problem in the sense of Tikhonov.

First of all observe that if $\{x_n\}$ is a minimizing sequence of the functional J then its limit must be the cluster point of the set D.

Further, it is also worthwhile noticing that the requirement that the minimizing sequence $\{x_n\}$ of the functional J is convergent implies that $\{x_n\}$ has always a unique limit, i.e., every minimizing sequence of the functional J tends to the same limit, say x, being the cluster point of the set D.

Consequently, if the minimization problem for the functional J is well-posed in the sense of Tikhonov, then J has exactly one minimum in the set D.

It turns out that there are some natural problems considered, for example, in functional analysis, when the corresponding minimization problem is well-posed.

We provide the following well-known example:

Example 1 Let D be a nonempty subset of a metric space X with a metric d. Fix an element $y \in X$ and consider the functional J_y defined on D by the formula

$$J_y(x) = d(x, y). \tag{4}$$

Obviously such a functional has always a real infimum and

$$\inf_{x\in D} J_y(x) = \text{dist}(y, D). \tag{5}$$

On the other hand it is also well known that the mentioned infimum need not be unique and not every minimizing sequence is convergent.

To guarantee that infimum of the functional J_y on the set D is attained we impose some additional requirements [20].

Thus, assume that our space under consideration is a reflexive Banach space $(E, ||\cdot||)$. Further, let D be a nonempty, closed, and convex subset of E and, as before, let y be an arbitrarily fixed point in E.

It is well known that the functional J_y defined by (4), which can be now written in the form

$$J_y(x) = ||x - y||, \tag{6}$$

attains its infimum on the set D.

To prove this fact denote by $d_D(y)$ the distance $\text{dist}(y, D)$ i.e., $d_D(y) = \text{dist}(y, D)$. Next, consider the closed ball $B(y, d_D(y) + r)$ centered at y and with radius $d_D(y) + r$, where $r > 0$ is an arbitrary number.

Now, consider the family $\{D_r\}_{r>0}$ of the sets D_r defined as the intersection of the ball $B(y, d_D(y) + r)$ and the set D:

$$D_r = B(y, d_D(y) + r) \cap D. \tag{7}$$

It is clear that the set D_r is nonempty, closed, and convex. In view of reflexivity of the space E this implies that D_r is weakly compact (and, of course, weakly closed) [15]. Thus, $\{D_r\}_{r>0}$ is a centered family of weakly compact sets (i.e., compact with respect to the weak topology in the Banach space E). This yields that the intersection of this family

$$D_0 = \bigcap_{r>0} D_r \tag{8}$$

is nonempty, convex, and weakly compact. Obviously, the set D_0 is contained in the sphere $S(y, d_D(y))$ centered at the point y and with radius $d_D(y)$, i.e.,

$$S(y, d_D(y)) = \{x \in E : \ ||x - y|| = d_D(y)\}. \tag{9}$$

Finally, observe that this allows us to conclude that the functional J_y attains its minimum on the nonempty set D_0.

Now, let us observe that if we additionally assume that the space E is *strictly convex* (this means that spheres in E do not contain nontrivial segments [17]) then the functional J_y defined by (6) attains its minimum at exactly one point of the set D.

Let us recall that such a situation is realized if we assume, among others, that the Banach space E is uniformly convex (for definition of uniformly convex Banach spaces and their properties we refer to [2, 17]). Thus, in uniformly convex Banach spaces the functional J_y defined by (6) attains its minimum at exactly one point.

From the above presented reasoning we infer that every minimizing sequence $\{x_n\}$ of the functional J_y (in the case when E is reflexive and strictly convex) is convergent to one point at which the functional J_y attains its minimum.

To what extent we can generalize the above assertion is an interesting question, i.e., which assumptions should be imposed on a Banach space in order to guarantee that the set M_y consisting of points at which the considered functional J_y attains its minimum is not a "big" set. For example, the set M_y is compact in the strong topology of the Banach space E.

In the next section we will discuss the above problem.

3 Some Generalizations of the Concept of Well-Posed Minimization Problem and Their Consequences

In this section we discuss some generalizations of the concept of the well-posed minimization problem in the sense of Tikhonov, which was presented in the previous section.

Similarly as in the previous section, for the sake of generality, we place our considerations in a metric space X (with a metric d).

The following definition by Levitin and Polyak [26] generalizes the concept of the *well-posed minimization problem in the sense of Tikhonov*.

Definition 1 Let $J : X \to \mathbb{R}$ be a given functional. We state that the problem of minimization of the functional J is *well-posed in the sense of Levitin and Polyak* if each minimizing sequence of the functional J is compact.

Obviously, if the minimization problem for the functional J is well-posed in the sense of Tikhonov then that problem is also well-posed in the sense of Levitin and Polyak. Below we provide an example showing, among others, that the converse implication is not always true.

To this end we provide first the definition of the concept of nearly strictly convex Banach space, which will be needed further on. This definition is taken from [8].

Definition 2 A Banach space E is said to be *nearly strictly convex* (NSC, in short) if its unit sphere $S_E = \{x \in E : ||x|| = 1\}$ does not contain noncompact convex sets.

In order to formulate the concept of NSC Banach space in another, equivalent way, denote by E^* the dual space of E. Let S^* denote the unit sphere in E^*. Then, a Banach space E is NSC if and only if for every functional $x^* \in S^*$ the set $\{x \in S_E : x^*(x) = 1\}$ is nonempty and compact [7].

Observe that every strictly convex Banach space is NSC but the converse implication is not always true [7].

Next, we discuss an example being the extension of the previously considered Example 1.

Example 2 Similarly as in Example 1 take a nonempty, closed, and convex subset D of a Banach space E. We will assume additionally that the space E is reflexive and

NSC. Next, fix arbitrarily a point $y \in E$ and denote by J_y the functional defined by (6). Obviously, the same reasoning as that conducted in Example 1 leads to the conclusion that the set D_0 consisting of all points belonging to D at which the functional J_y attains its minimum, is nonempty, convex, and weakly compact. To describe the set D_0 more precisely, recall the notation introduced previously in Example 1, i.e.,

$$d_D(y) = \text{dist}(y, D). \tag{10}$$

Observe that the set D_0 is contained in the sphere $S(y, d_D(y))$. Thus, in view of the imposed assumption that E is NSC, we conclude that the set D_0 is compact.

On the other hand consider a minimizing sequence $\{x_n\}$ for the functional J_y, $\{x_n\} \subset D$. This means that

$$\lim_{n\to\infty} ||y - x_n|| = \text{dist}(y, D). \tag{11}$$

Without loss of generality we can assume that the sequence $\{||y - x_n||\}$ is nonincreasing, i.e., $||y - x_{n+1}|| \leq ||y - x_n||$ for $n = 1, 2, \ldots$. Next, for arbitrarily fixed natural number n consider the ball $B\left(y, ||y - x_n|| + \frac{1}{n}\right)$.

Further, consider the set D_n defined as follows:

$$D_n = D \cap B\left(y, ||y - x_n|| + \frac{1}{n}\right), \quad \text{for } n = 1, 2, \ldots. \tag{12}$$

Obviously the set D_n is nonempty, closed, and convex. Apart from this we see that the set D_n is weakly compact in view of reflexivity of the space E. Moreover, $x_n \in D$ for any $n = 1, 2, \ldots$.

Now, observe that the set $D_0 = \bigcap\limits_{n=1}^{\infty} D_n$ is nonempty, closed, and convex. It is clear that D_0 is a subset of the sphere $S(y, d_D(y))$.

On the other hand the set D_0 contains all accumulation points of the sequence $\{x_n\}$. Since the space E is assumed to be NSC, this allows us to deduce that the set A of those accumulation points is compact.

This means that the minimization problem for the functional J_y on the set D is well-posed in the sense of Levitin and Polyak.

In order to recall a generalization of the well-posed minimization problem given by Furi and Vignoli [16] we need to give the concept of the so-called Kuratowski measure of noncompactness.

To this end assume that X is a given complete metric space with a metric d. Denote by $\mathfrak{M}_X$ the family of all nonempty and bounded subsets of the space X and by $\mathfrak{N}_X$ its subfamily consisting of all relatively compact sets.

Now, for $A \in \mathfrak{M}_X$, denote by $\alpha(A)$ the nonnegative number defined in the following way:

$$\alpha(A) = \inf\{\varepsilon > 0 : \text{ A can be covered by a finite number of sets of diameters smaller than } \varepsilon\}. \tag{13}$$

Equivalently, we have

$$\alpha(A) = \inf\Big\{\varepsilon > 0 : \ A \subset \bigcup_{i=1}^{n} A_i,\ A_i \subset X,\ \operatorname{diam} A_i < \varepsilon\ (i = 1, 2, \ldots, n), n \text{ is an arbitrary natural number}\Big\}, \tag{14}$$

where the symbol $\operatorname{diam} B$ denotes the diameter of the set B ($B \subset X$). The quantity $\alpha(A)$ is called the *Kuratowski measure of noncompactness* of the set A and was introduced by Kuratowski [24].

Observe that the function $\alpha : \mathfrak{M}_X \to \mathbb{R}_+ = [0, \infty)$ and has the following properties being an immediate consequence of the definition:

$$\alpha(A) = 0 \Leftrightarrow A \in \mathfrak{N}_X, \tag{15}$$

$$A \subset B \Rightarrow \alpha(A) \leq \alpha(B), \tag{16}$$

$$\alpha(\overline{A}) = \alpha(A), \tag{17}$$

where the symbol $\overline{A}$ denotes the closure of the set A.

Moreover, for any set $A \in \mathfrak{M}_X$ the following inequality holds:

$$\alpha(A) \leq \operatorname{diam} A. \tag{18}$$

Indeed, the above inequality is an immediate consequence of the definition of the Kuratowski measure of noncompactness α.

The most important and useful property of the Kuratowski measure of noncompactness α is contained in the below formulated theorem.

Theorem 1 *Let $\{A_n\}$ be a sequence of nonempty, bounded, and closed subsets of the space X such that $A_n \supset A_{n+1}$ for $n = 1, 2, \ldots$ and $\lim_{n\to\infty} \alpha(A_n) = 0$. Then the intersection set A_∞ of the sequence $\{A_n\}$, i.e., the set*

$$A_\infty = \bigcap_{n=1}^{\infty} A_n, \tag{19}$$

is nonempty and compact.

The above theorem was proved by Kuratowski and it creates the generalization of the well-known Cantor intersection theorem.

Note that in the case when we consider the Kuratowski measure of noncompactness α in a Banach space E it has some additional useful properties related to the algebraic structure of the space E. The presentation of these properties are given in detail in the next chapter.

Now, we formulate the definition of the concept of well-posedness of minimization problem in the sense of Furi and Vignoli.

Thus, analogously as earlier, assume that D is a nonempty and closed subset of a complete metric space X with a metric d. Let $J : D \to \mathbb{R}$ be a given functional which is lower semicontinuous and lower bounded on the set D. Hence we infer that the functional J has a real minimum m_J on the set D, i.e., there exists a real number m_J such that

$$m_J = \inf_{x \in D} J(x). \tag{20}$$

Next, for an arbitrary given number $\varepsilon > 0$ denote:

$$D_\varepsilon = \{x \in D : \ J(x) \leq m_J + \varepsilon\}. \tag{21}$$

Note that in view of lower semicontinuity of the functional J we conclude that the set D_ε is closed.

Further observe that if $0 < \varepsilon_1 < \varepsilon_2$ then $D_{\varepsilon_1} \subset D_{\varepsilon_2}$.

In what follows we always assume that there exists a number $\varepsilon_0 > 0$ such that the set D_{ε_0} is bounded. Thus, for every ε such that $0 < \varepsilon \leq \varepsilon_0$ the set D_ε is bounded.

It is worth noting that if the set D is bounded then set D_ε is bounded for every $\varepsilon > 0$.

Definition 3 We say that the minimization problem for the functional J is *well-posed on the set D in the sense of Furi and Vignoli* if

$$\lim_{\varepsilon \to 0} \alpha(D_\varepsilon) = 0. \tag{22}$$

If the minimization problem is well-posed in the sense of Furi and Vignoli then it is well-posed in the sense of Levitin and Polyak.

To prove this fact note first that the set D_0 defined as the intersection of the family $\{D_\varepsilon\}_{\varepsilon > 0}$, i.e.,

$$D_0 = \bigcap_{\varepsilon > 0} D_\varepsilon \tag{23}$$

is nonempty, closed, and compact. The assertion concerning the compactness of the set D_0 is a consequence of Theorem 1.

On the other hand, the set D_0 contains of all accumulation points of an arbitrary minimizing sequence of the functional J on the set D. This means that the set of all accumulation points of all minimizing sequences of functional J is compact and the minimization problem is well-posed in the sense of Levitin and Polyak.

It is worth mentioning that the relations among various types of well-posed minimization problems were discussed in several papers (cf. [13, 14, 21, 27, 29, 34], for example). Obviously, in these papers well-posed minimization problems were mostly discussed for functionals under some additional constraints.

In the example given below we discuss the well-posedness for the "distance" functional considered in Examples 1 and 2.

Example 3 Assume that $(E, ||\cdot||)$ is a Banach space with the zero element θ. Denote by B_E the unit ball, i.e., $B_E = B(\theta, 1)$ and let S_E stand for the unit sphere in E. Similarly as before, let E^* denote the dual space of E. Denote by $S^* = S_{E^*}$. Further assume that α is the Kuratowski measure of noncompactness in the space E.

We will say that the space E is *nearly uniformly convex* (NUC, in short) [22] if for any $\varepsilon > 0$ there exists $\delta > 0$ such that whenever a closed convex subset X of the ball B_E has $\text{dist}(\theta, X) \geq 1 - \delta$ then $\alpha(X) \leq \varepsilon$.

It is well known that every NUC space E is reflexive and has some additional properties [4, 5, 7, 18, 22, 28, 30–32].

There is also another approach to the concept of NUC space. Namely, for an arbitrary $\varepsilon \in (0, 1]$ define the quantity $\beta_E(\varepsilon)$ in the following way [5]:

$$\beta_E(\varepsilon) = \sup\{\alpha(X) : \ X \subset B_E, \ X \text{ is convex}, \ \text{dist}(\theta, X) \geq 1 - \varepsilon\}. \tag{24}$$

Function $\beta_E : [0, 1] \to [0, 2]$ is a kind of a *modulus of near convexity* (cf. [5], for details). It is easily seen that E is NUC if and only if $\lim_{\varepsilon \to 0} \beta_E(\varepsilon) = 0$.

Moreover, we need also some other characterization of NUC spaces. To this end, for a fixed functional $f \in S^*$ denote by $F(f, \varepsilon)$ the slice of the unit ball B_E defined as follows:

$$F(f, \varepsilon) = \{x \in E_E : \ f(x) \geq 1 - \varepsilon\}. \tag{25}$$

Then we have the following result [5]:

Theorem 2 *A space E is NUC if and only if*

$$\lim_{\varepsilon \to 0} \alpha(F(f, \varepsilon)) = 0 \tag{26}$$

uniformly with respect to $f \in S^*$.

Particularly, from the above theorem we deduce that if a Banach space E is NUC then for an arbitrarily fixed functional f belonging to the sphere S^* we have that $\lim_{\varepsilon \to 0} \alpha(F(f, \varepsilon)) = 0$.

Additionally note that every NUC space is NSC [4] but the converse implication is not always true [32].

In what follows, extending the considerations conducted in Examples 1 and 2, assume that $(E, ||\cdot||)$ is an NUC Banach space and D is a nonempty, closed, and convex subset of the space E. Fix an element $y \in E$ and consider distance functional J_y defined on the set D by formula (6), i.e.,

$$J_y(x) = ||x - y||. \tag{27}$$

We show that the minimization problem for the functional J_y on the set D is well-posed in the sense of Furi and Vignoli.

For the proof observe that the functional J_y attains its minimum m_J on the set D and

$$m_J = \text{dist}(y, D). \tag{28}$$

This is an immediate consequence of the above-mentioned reflexivity of the space E and the facts established in Example 1.

Further, let us fix arbitrarily a number $\varepsilon > 0$ and consider the set

$$\begin{aligned} D_\varepsilon &= \{x \in D : \ J_y(x) \leq m_J + \varepsilon\} = \{x \in D : \ ||x - y|| \leq m_J + \varepsilon\} \\ &= D \cap B(y, m_J + \varepsilon). \end{aligned} \tag{29}$$

Since the set D_ε is closed and convex and there exists a point $z \in D_\varepsilon$ such that $||z - y|| = \text{dist}(y, D)$, then from some well-known facts (cf. [15, p. 452]) we infer that there exists a functional $f \in E^*$ supporting the set D_ε at the point z (in fact, f is the functional tangent to D_ε at z). Multiplying the functional f by a suitable positive number we see that the set D_ε may be considered as a subset of the slice

$$F(f, \varepsilon) = \{x \in B(y, m_J + \varepsilon) : \ f(x) \geq d_y + \varepsilon\}. \tag{30}$$

Since the space E is assumed to be NUC we deduce from Theorem 2 that $\lim_{\varepsilon \to 0} \alpha(F(f, \varepsilon)) = 0$. This implies that $\lim_{\varepsilon \to 0} \alpha(D_\varepsilon) = 0$. This means that the minimization problem for the functional J_y is well-posed in the sense of Furi and Vignoli.

4 Measures of Noncompactness

In this section we present basic facts concerning measures of noncompactness. We focus here on the axiomatic approach to this concept contained in the monograph [9]. Such an approach is sufficiently general although there are somc more general definitions of the concept of a measure of noncompatness (cf. [1]). Nevertheless, our definition admits several natural realizations. Apart from this measures of noncompactness satisfying our axiomatics have useful properties and are handy in numerous applications (cf. [9, 12] and references therein). Moreover, based on the mentioned axiomatic definition we are able to construct measures of noncompactness in those

Banach spaces in which we do not know necessary and sufficient conditions for relative compactness of sets.

The considerations of this section will be mainly conducted in Banach spaces although some of them have also sense in the setting of metric spaces.

Thus, let us assume that $(E, ||\cdot||)$ is a given Banach space. For subsets X, Y of E and for a number $c \in \mathbb{R}$ denote by $X + Y$, cX the usual algebraic operations on sets. By the symbol Conv$X$ we denote the closed convex hull of X, while the symbol conv$X$ stands for the convex hull of X.

For completeness of our considerations recall that the ball centered at x and with radius r is denoted by $B(x, r)$. We write B_E to denote the unit ball $B(\theta, 1)$. Moreover, the symbol $\overline{X}$ stands for the closure of the set X.

If X is a given nonempty subset of E then the symbol $B(X, r)$ denotes the ball centered at the set X and with radius r, i.e.,

$$B(X, r) = \bigcup_{x \in X} B(x, r). \tag{31}$$

For an arbitrary family $\mathscr{P}$ of some subsets of E we denote by $\mathscr{P}^c$ the family of all closed sets belonging to $\mathscr{P}$.

Further, similarly as in the preceding section, denote by $\mathfrak{M}_E$ the family of all nonempty and bounded subsets of the space E and by $\mathfrak{N}_E$ its subfamily consisting of all relatively compact sets.

If $X, Y \in \mathfrak{M}_E$ then by $h(X, Y)$ we denote the so-called *nonsymmetric Hausdorff distance* between sets X and Y, defined as follows:

$$h(X, Y) = \inf\{r > 0 : \ X \subset B(Y, r)\}. \tag{32}$$

Finally, we put

$$H(X, Y) = \max\{h(X, Y), \ h(Y, X)\}. \tag{33}$$

The quantity $H(X, Y)$ is called the *Hausdorff distance* between sets X and Y. This distance generates the pseudometric on the family $\mathfrak{M}_E$ and it is a complete metric on the family $\mathfrak{M}_E^c$ [25].

If $\mathscr{Z}$ is a nonempty subfamily of the family $\mathfrak{M}_E$ then for an arbitrary $X \in \mathfrak{M}_E$ we denote by $H(X, \mathscr{Z})$ the distance of X to $\mathscr{Z}$ with respect to the Hausdorff distance H, i.e.,

$$H(X, \mathscr{Z}) = \inf\{H(X, Z) : \ Z \in \mathscr{Z}\}. \tag{34}$$

We accept the following definition of a measure of noncompactness [9]:

Definition 4 The function $\mu : \mathfrak{M}_E \to \mathbb{R}_+ = [0, \infty)$ is said to be a *measure of noncompactness* in the space E if it satisfies the following conditions:

(i) The family $\ker\mu = \{X \in \mathfrak{M}_E : \mu(X) = 0\}$; is nonempty and $\ker\mu \subset \mathfrak{M}_E$;
(ii) $X \subset Y \Rightarrow \mu(X) \le \mu(Y)$;
(iii) $\mu(\overline{X}) = \mu(X)$;
(iv) $\mu(\mathrm{Conv}X) = \mu(X)$;
(v) $\mu(\lambda X + (1-\lambda)Y) \le \lambda\mu(X) + (1-\lambda)\mu(Y)$ for $\lambda \in [0,1]$;
(vi) If $\{X_n\}$ is a sequence of closed sets from $\mathfrak{M}_E$ such that $X_{n+1} \subset X_n$ for $n = 1, 2, \ldots$ and if $\lim_{n\to\infty} \mu(X_n) = 0$ then the intersection set $X_\infty = \bigcap_{n=1}^{\infty} X_n$ is nonempty.

The family $\ker\mu$ described in axiom (i) is called the *kernel* of the measure of noncompactness μ.

Observe that in axiom (vi), from the inclusion $X_\infty \subset X_n$ which is valid for $n = 1, 2, \ldots$, we infer that $\mu(X_\infty) \le \mu(X_n)$ which implies that $\mu(X_\infty) = 0$ i.e., the set X_∞ is a member of the family $\ker\mu$.

Let us point out also that we frequently use an equivalent approach to the concept of a measure of noncompactness. In that approach the role of the kernel of the measure of noncompactness is exposed.

Definition 5 A nonempty family $\mathscr{P} \subset \mathfrak{N}_E$ is called the *kernel (of a measure of noncompactness)* provided the following conditions are satisfied:

(i) $X \in \mathscr{P} \Rightarrow \overline{X} \in \mathscr{P}$;
(ii) $X \in \mathscr{P}, Y \subset X, Y \neq \emptyset \Rightarrow Y \in \mathscr{P}$;
(iii) $X \in \mathscr{P} \Rightarrow \mathrm{Conv}X \in \mathscr{P}$;
(iv) $X, Y \in \mathscr{P} \Rightarrow \lambda X + (1-\lambda)Y \in \mathscr{P}$ for $\lambda \in [0,1]$;
(v) $\mathscr{P}^c$ is closed in $\mathfrak{M}_E^c$ with respect to Hausdorff metric.

Observe that the family $\mathfrak{N}_E$ may serve as an example of the kernel of a measure of noncompactness. Indeed, it is easily seen that $\mathfrak{N}_E$ satisfies all conditions of Definition 5. Another example of the kernel of a measure of noncompactness may serve the family $\mathfrak{N}_E^0$ consisting of all singletons belonging to E.

In what follows we provide the definition of a measure of noncompactness related to the concept of the kernel given in Definition 5.

Definition 6 The function $\mu : \mathfrak{M}_E \to \mathbb{R}_+$ is said to be a *measure of noncompactness with the kernel* $\mathscr{P}$ ($\ker\mu = \mathscr{P}$) if it satisfies the following conditious:

(i) $\mu(X) = 0 \Leftrightarrow X \in \mathscr{P}$;
(ii) $X \subset Y \Rightarrow \mu(X) \le \mu(Y)$;
(iii) $\mu(\overline{X}) = \mu(X)$;
(iv) $\mu(\mathrm{Conv}X) = \mu(X)$;
(v) $\mu(\lambda X + (1-\lambda)Y) \le \lambda\mu(X) + (1-\lambda)\mu(Y)$ for $\lambda \in [0,1]$;
(vi) If $X_n \in \mathfrak{M}_E$, $X_n = \overline{X}_n$ and $X_{n+1} \subset X_n$ for $n = 1, 2, \ldots$ and if $\lim_{n\to\infty} \mu(X_n) = 0$ then the set $X_\infty = \bigcap_{n=1}^{\infty} X_n$ is nonempty.

Let us pay attention to the situation when we consider a complete metric space X instead of a Banach space E. Then the above given Definitions 4, 5, and 6 can be adopted to such a situation. To this end we have to delete axioms connected with algebraic structure of the family $\mathfrak{M}_E$. For example, in this case the definition of the kernel runs as follows:

Definition 7 Let X be a complete matric space. A nonempty family $\mathscr{P} \subset \mathfrak{N}_X$ is referred to as the kernel (of a measure of noncompactness) if it satisfies the following conditions:

(i) $X \in \mathscr{P} \Rightarrow \overline{X} \in \mathscr{P}$;
(ii) $X \in \mathscr{P}, Y \subset X,\ Y \neq \emptyset \Rightarrow Y \in \mathscr{P}$;
(iii) $\mathscr{P}^c$ is closed in $\mathfrak{M}_X^c$ with respect to the topology generated by Hausdorff's metric.

In a similar way we can formulate the definition of a measure of noncompactness in the space X. Indeed, the function $\mu : \mathfrak{M}_X \to \mathbb{R}_+$ is a measure of noncompactness provided it satisfies conditions (i), (ii), (iii), and (vi) of Definition 4 or the same conditions of Definitions 6.

Now, note the fact that Kuratowski's measure of noncompactness α discussed in Sect. 3 is the measure of noncompactness in the sense of Definition 4 (or Definition 6). Indeed, from (15) we infer that the function α satisfies condition (i) of Definition 4 and $\ker \mu = \mathfrak{N}_E$. Moreover, from (16), (17) and Theorem 1 we deduce that α satisfies conditions (ii), (iii), and (vi) of Definition 4 (or Definition 6). Thus, α is the measure of noncompactness in any metric space. It can be also shown [9] that the function α satisfies also axioms (iv) and (vi) of Definition 4 or 6 when we consider it in a Banach space E.

Below we provide further examples of measures of noncompactness in the sense of Definitions 4 and 6.

Example 4 Assume that E is a Banach space (although we can also treat the case of a metric space X). For $X \in \mathfrak{M}_E$ let us consider the quantity $\chi(X)$ defined in the following way (cf. [9, 19]):

$$\chi(X) = \inf\{\varepsilon > 0 : \ X \text{ can be covered by a finite number of balls of radii } \varepsilon\}. \tag{35}$$

The function χ is called the *Hausdorff measure of noncompactness.*

It is not hard to show that the function χ satisfies the conditions of Definition 4. Indeed, the equivalence

$$\chi(X) = 0 \Leftrightarrow X \text{ is relatively compact} \tag{36}$$

is an easy consequence of the famous Hausdorff theorem. Thus condition (i) of Definition 4 is satisfied with $\ker \chi = \mathfrak{N}_E$. For the proof of the remaining conditions we refer to [9].

It is worth mentioning that the Hausdorff measure of noncompactness can be expressed in terms of the Hausdorff distance H. In fact, we can show that for an arbitrary set $A \in \mathfrak{M}_E$ the following equality holds:

$$\chi(A) = H(A, \mathfrak{N}_E). \tag{37}$$

It turns out that this equality is also true if we replace the Banach space E by a complete metric space X. The details of the proof can be found in [9].

Example 5 Now, according to Definitions 5 and 6 let us take as the family $\mathscr{P}$ described by Definition 5 the family $\mathfrak{N}_E^0$ consisting of all singletons in E. It is easily seen that the family $\mathfrak{N}_E^0$ is the kernel (of a measure of noncompactness) in the sense of Definition 5.

Further, for $X \in \mathfrak{M}_E$, let us put

$$\mu(X) = \operatorname{diam} X, \tag{38}$$

i.e., $\mu(X)$ is equal to the diameter of the set X.

It is easy to verify that the function μ defined by formula (38) is the measure of noncompactness in the sense of Definition 6 (or Definition 4). For example, axiom (vi) of Definition 6 coincides with the classical Cantor intersection theorem.

Note the fact that formula (38) defines the measure of noncompactness in an arbitrary complete metric space.

In what follows we see that every kernel generates at least one measure of noncompactness. Indeed, for the sake of generality assume that X is a given metric space. Let $\mathscr{P} \subset \mathfrak{N}_X$ be an arbitrary kernel (of a measure of noncompactness) in the sense of Definition 7. For $A \in \mathfrak{M}_X$ let us put

$$\mu(A) = H(A, \mathscr{P}), \tag{39}$$

where $H(A, \mathscr{P})$ denotes the Hausdorff distance of the set A to the family $\mathscr{P}$.

It can be shown that formula (39) defines a measure of noncompactness in the metric space X such that $\ker \mu = \mathscr{P}$. This theorem was given in [9] but the detailed proof can be found in [10].

It is also worth mentioning that formula (39) defines also the measure of noncompactness in the sense of Definitions 6 and 4, in the setting of a Banach space.

It turns out that the most convenient and useful in applications seems to be the Hausdorff measure of noncompactness. It is, among others, a consequence of equality (37). But another, very important reason is a result of the fact that in some Banach spaces we are able to express the Hausdorff measure χ with the help of a convenient formula associated with the structure of a Banach space under considerations. We illustrate this assertion by a few examples.

Example 6 Let $C[a, b]$ denote the classical Banach space consisting of all real functions defined and continuous on the interval $[a, b]$. We consider $C[a, b]$ furnished with the standard maximum norm, i.e.,

$$||x|| = \max\{|x(t)| : t \in [a, b]\}. \tag{40}$$

Keeping in mind the Arzéla-Ascoli criterion for compactness in $C[a, b]$ we can express the Hausdorff measure of noncompactness in the below described manner.

Namely, for $x \in C[a, b]$ denote by $\omega(x, \varepsilon)$ the modulus of continuity of the function x:

$$\omega(x, \varepsilon) = \sup\{|x(t) - x(s)| : t, s \in [a, b], \ |t - s| \leq \varepsilon\}, \quad \text{for } \varepsilon > 0. \tag{41}$$

Next, for an arbitrary set $X \in \mathfrak{M}_{C[a,b]}$ let us put:

$$\omega(X, \varepsilon) = \sup\{\omega(x, \varepsilon) : x \in X\}, \tag{42}$$

$$\omega_0(X) = \lim_{\varepsilon \to 0} \omega(X, \varepsilon). \tag{43}$$

It can be shown [9, 19] that for $X \in \mathfrak{M}_{C[a,b]}$ the following equality holds:

$$\chi(X) = \frac{1}{2}\omega_0(X). \tag{44}$$

This equality is very useful in applications.

Example 7 Let c_0 denote the space of all real sequences $x = \{x_n\}$ converging to zero and endowed with the maximum norm, i.e.,

$$||x|| = ||\{x_n\}|| = \max\{|x_n| : n = 1, 2, \ldots\}. \tag{45}$$

To describe the formula expressing the Hausdorff measure χ in the space c_0 fix arbitrarily a set $X \in \mathfrak{M}_{c_0}$. Then, it can be shown that the following equality holds (cf. [9]):

$$\chi(X) = \lim_{n \to \infty} \left\{ \sup_{x \in X} \{\max\{|x_k| : k \geq n\}\} \right\}. \tag{46}$$

In the space c consisting of real sequences converging to a proper limit and furnished with the supremum norm

$$||x|| = ||\{x_n\}|| = \sup\{|x_n| : n = 1, 2, \ldots\}, \tag{47}$$

the situation is more complicated and we only know estimates of the Hausdorff measure χ. Indeed, the function $\mu : \mathfrak{M}_c \to \mathbb{R}_+$ defined by the formula

$$\mu(X) = \lim_{n\to\infty} \left\{ \sup_{x\in X} \left\{ \sup \left\{ |x_k - \lim_{i\to\infty} x_i| : k \geq n \right\} \right\} \right\} \tag{48}$$

satisfies the following inequalities:

$$\frac{1}{2}\mu(X) \leq \chi(X) \leq \mu(X), \tag{49}$$

which were proved in [9].

On the other hand observe that the function μ defined by (48) may serve as an example of a measure of noncompactness in the space c with the kernel $\ker \mu = \mathfrak{N}_c$.

The formula expressing the Hausdorff measure of noncompactness is also known in the space l^p for $1 \leq p < \infty$ [9]. On the other hand in the classical Banach spaces $L^p(a, b)$ and l^∞ we only know some estimates of the Hausdorff measure of noncompactness with the help of formulas that define measures of noncompactness in those spaces. Refer to [9] for details.

Finally, note that there are some Banach spaces in which we do not know the criteria for compactness similar to those of Arzéla-Ascoli in the space $C[a, b]$ or Riesz and Kolmogorov in the space $L^p(a, b)$.

In Banach spaces of such type we do not know how to construct formulas for measures of noncompactness with kernels equal to the family of all nonempty and relatively compact sets.

In the below example we discuss a Banach space of such type.

Example 8 Denote by $BC(\mathbb{R}_+)$ the space consisting of all functions defined, continuous, and bounded on the real half-axis $\mathbb{R}_+$ and having real values. This space is furnished with the supremum norm, i.e., for $x \in BC(\mathbb{R}_+)$ we put

$$||x|| = \sup\{|x(t)| : t \in \mathbb{R}_+\}. \tag{50}$$

In the space $BC(\mathbb{R}_+)$ the Arzéla-Ascoli criteriion for relative compactness of sets fails to work. What is more, we do not even know a necessary and sufficient condition for the relative compactness of sets which is connected with the structure of this space. By these regards we can only define measures of noncompactness such that their kernels are essentially smaller than the family $\mathfrak{N}_{BC(\mathbb{R}_+)}$.

In order to construct the mentioned measures in the space $BC(\mathbb{R}_+)$ take an arbitrary set $X \in \mathfrak{M}_{BC(\mathbb{R}_+)}$ and choose a function $x \in X$. Next, fix numbers $\varepsilon > 0$, $T > 0$ and let us define the following quantities:

$$\omega^T(x, \varepsilon) = \sup\{|x(t) - x(s)| : t, s \in [0, T], |t - s| \leq \varepsilon\}, \tag{51}$$

$$\omega^T(X, \varepsilon) = \sup\{\omega^T(x, \varepsilon); x \in X\}, \tag{52}$$

$$\omega_0^T(X) = \lim_{\varepsilon\to 0} \omega^T(X, \varepsilon), \tag{53}$$

$$\omega_0(X) = \lim_{T\to\infty} \omega_0^T(X). \tag{54}$$

Further, let us define the set-functions $a(X)$, $b(X)$, $c(X)$ by putting

$$a(X) = \lim_{T\to\infty} \left\{ \sup_{x\in X} \{\sup\{|x(t)| : t \geq T\}\} \right\}, \tag{55}$$

$$b(X) = \lim_{T\to\infty} \left\{ \sup_{x\in X} \{\sup\{|x(t) - x(s)| : t, s \geq T\}\} \right\}, \tag{56}$$

$$c(X) = \limsup_{t\to\infty} \operatorname{diam} X(t), \tag{57}$$

where $X(t) = \{x(t) : x \in X\}$ and the symbol $\operatorname{diam} X(t)$ stands for the diameter of the set $X(t)$ i.e.,

$$\operatorname{diam} X(t) = \sup\{|x(t) - y(t)| : x, y \in X\}. \tag{58}$$

Finally, let us take the functions μ_a, μ_b, μ_c defined on the family $\mathfrak{M}_{BC(\mathbb{R}_+)}$ in the following way:

$$\mu_a(X) = \omega_0(X) + a(X), \tag{59}$$

$$\mu_b(X) = \omega_0(X) + b(X), \tag{60}$$

$$\mu_c(X) = \omega_0(X) + c(X). \tag{61}$$

It can be shown [3] (cf. also [6]) that these functions are measures of noncompactness in the space $BC(\mathbb{R}_+)$ with kernels essentially smaller than the family $\mathfrak{N}_{BC(\mathbb{R}_+)}$. Note also that the kernel $\ker \mu_a$ consists of all bounded sets X such that functions from X are locally equicontinuous on $\mathbb{R}_+$ and tend to zero at infinity with the same rate. Similarly, the kernel $\ker \mu_b$ contains bounded sets X such that functions from X are locally equicontinuous on $\mathbb{R}_+$ and tend to limits at infinity with the same rate (i.e., uniformly with respect to the set X).

Finally, the kernel $\ker \mu_c$ contains all bounded subsets X of the space $BC(\mathbb{R}_+)$ such that functions from X are locally equicontinuous on $\mathbb{R}_+$ and the thickness of the bundle formed by graphs of functions from X tends to zero at infinity. As we mentioned above, we have that $\ker \mu_y \subset \mathfrak{N}_{BC(\mathbb{R}_+)}$ but $\ker \mu_y \neq \mathfrak{N}_{BC(\mathbb{R}_+)}$ for $y \in \{a, b, c\}$.

5 Generalized Definition of the Well-Posed Minimum Problem

The final section of this chapter is devoted to present some generalization of the concept of well-posed minimization problem in the sense of Furi and Vignoli. Our generalization contains, as special cases, all definitions of the well-posedness of minimization problem discussed previously in this chapter.

Let us assume that X is a given complete metric space and let D be a nonempty and closed subset of the space X. Further, let $J : D \to \mathbb{R}$ be a lower semicontinuous and lower bounded functional on the set D. Similarly as in Sect. 3 denote by m_J the minimum of J on the set D, i.e., $m_J = \inf_{x \in D} J(x)$. Next, for $\varepsilon > 0$ let us put

$$D_\varepsilon = \{x \in D : \ J(x) \leq m_J + \varepsilon\}. \tag{62}$$

In the sequel we always assume that there exists $\varepsilon_0 > 0$ such that the set D_{ε_0} is bounded (cf. Sect. 3).

Now assume that the family $\mathscr{P} \subset \mathfrak{N}_X$ is a kernel (of a measure of noncompactness) in the sense of Definition 7.

Definition 8 We say that the minimization problem for the functional J is well-posed on the set D with respect to the kernel $\mathscr{P}$ if there exists a measure of noncompactness μ in the space X with $\ker \mu = \mathscr{P}$ and such that

$$\lim_{\varepsilon \to 0} \mu(D_\varepsilon) = 0. \tag{63}$$

From the above definition it follows that the set $D_0 = \{x \in D : \ J(x) = m_J\}$ consisting of all points belonging to D at which the functional J attains its minimum, is compact, and belongs to the kernel $\mathscr{P}$.

Indeed, similarly as in Sect. 3 observe that $D_0 = \bigcap_{\varepsilon > 0} D_\varepsilon$. Hence, in view of axiom (vi) from Definition 6 we infer that the set D_0 is a member of the family $\mathscr{P}$ (cf. also [9]). Moreover, keeping in mind the assumption requiring the lower semicontinuity of the functional J we obtain that D_0 is closed. Thus the set D_0 is compact.

As a consequence of Definition 8 we derive a few properties of the well-posed minimization problem with respect to a kernel. These properties are formulated in the theorems given below.

Theorem 3 *If the minimization problem for the functional J on the set D is well-posed with respect to the kernels $\mathscr{P}_1$ and $\mathscr{P}_2$, then this problem is well-posed with respect to the kernel $\mathscr{P} = \mathscr{P}_1 \cap \mathscr{P}_2$.*

Proof Let D_0 be the set defined above, i.e., $D_0 = \{x \in D : \ J(x) = m_J\}$. Then, from the properties of the set D_0 established above we deduce that $D_0 \in \mathscr{P}_1$ and $D_0 \in \mathscr{P}_2$. Hence $D_0 \in \mathscr{P}_1 \cap \mathscr{P}_2$ which implies that the family $\mathscr{P} = \mathscr{P}_1 \cap \mathscr{P}_2$ is nonempty. It is easy to check that the family $\mathscr{P}$ satisfies the conditions of Definition 7 which means that it is a kernel (of a measure of noncompactness).

In view of imposed assumptions there exist measures of noncompactness μ_1, μ_2 defined in X with $\ker \mu_1 = \mathscr{P}_1$, $\ker \mu_2 = \mathscr{P}_2$ and such that

$$\lim_{\varepsilon \to 0} \mu_1(D_\varepsilon) = \lim_{\varepsilon \to 0} \mu_2(D_\varepsilon) = 0. \tag{64}$$

Further, let us define the measure of noncompactness μ in the space X by putting $\mu = \mu_1 + \mu_2$. It is easy to check that the kernel of the measure μ is equal to $\mathscr{P}$ i.e., $\ker \mu = \mathscr{P}$. Moreover, from equalities (64) we get

$$\lim_{\varepsilon \to 0} \mu(D_\varepsilon) = \lim_{\varepsilon \to 0} [\mu_1(D_\varepsilon) + \mu_2(D_\varepsilon)] = 0 \tag{65}$$

which means that the minimization problem for the functional J on the set D is well-posed with respect to the kernel $\mathscr{P}$. The proof is complete. □

Theorem 4 *Assume that the minimization problem for the functional J on the set D is well-posed with respect to the kernel $\mathscr{P}_1$. If $\mathscr{P}_2$ is such a kernel that $\mathscr{P}_1 \subset \mathscr{P}_2$ then the minimization problem for the functional J is well-posed with respect to the kernel $\mathscr{P}_2$.*

Proof In view of our assumptions there exists a measure of noncompactness μ_1 with $\ker \mu_1 = \mathscr{P}_1$ such that the minimization problem for the functional J is well-posed with respect to $\mathscr{P}_1$ i.e.,

$$\lim_{\varepsilon \to 0} \mu_1(D_\varepsilon) = 0. \tag{66}$$

Next, let us define the measure of noncompactness μ_2 in the space X by putting

$$\mu_2(A) = \mu_1(A) H(A, \mathscr{P}_2), \tag{67}$$

where $H(A, \mathscr{P}_2)$ denotes the distance of the set A to the family $\mathscr{P}_2$ with respect to the Hausdorff distance. It is easy to verify that μ_2 is a measure of noncompactness in the space X such that $\ker \mu_2 = \mathscr{P}_2$. Moreover, from (66) and (67) we obtain

$$\lim_{\varepsilon \to 0} \mu_2(D_\varepsilon) = \lim_{\varepsilon \to 0} \mu_1(D_\varepsilon) H(D_\varepsilon, \mathscr{P}_2) = 0, \tag{68}$$

and the proof is complete. □

Remark 1 Observe that if the minimization problem for a functional J on the set D is well-posed with respect to the kernel $\mathscr{P}$ then it does not guarantee that for every measure of noncompactness μ with $\ker \mu = \mathscr{P}$ equality (63) is satisfied.

Indeed, if μ_1 is a measure of noncompactness with $\ker \mu_1 = \mathscr{P}$ and such that (63) is satisfied, then not every measure of noncompactness μ_2 with $\ker \mu_2 = \mathscr{P}$ has to satisfy condition (63) even if $\mathscr{P} = \mathfrak{N}_X$ (cf. [11])

In light of Remark 1 and Theorems 3 and 4 we can raise the problem of the existence of the minimal kernel $\mathscr{P}$ (with respect to the relation of inclusion) such that for $\mathscr{P}$ the minimization problem for the functional J is well-posed.

Below we present the solution of this problem.

Theorem 5 *Assume that $\{\mathscr{P}_\lambda\}_{\lambda\in\Lambda}$ is a family of kernels in X such that for any $\lambda \in \Lambda$ the minimization problem for the functional J on the set D is well-posed with respect to the kernel $\mathscr{P}_\lambda$, i.e., there exists a measure of noncompactness μ_λ with* $\ker \mu_\lambda = \mathscr{P}_\lambda$ *such that* (63) *holds with $\mu = \mu_\lambda$. Then, the minimization problem for the functional J is well-posed on the set D with respect to the kernel $\mathscr{P}$ defined as intersection of all kernels $\mathscr{P}_\lambda$, i.e.,*

$$\mathscr{P} = \bigcap_{\lambda\in\Lambda} \mathscr{P}_\lambda. \tag{69}$$

Proof Arguing in the same way as in the proof of Theorem 3 we see that the family $\mathscr{P}$ is nonempty. On the other hand, in the standard way we can show that $\mathscr{P}$ is the kernel in the sense of Definition 7.

Now, let us consider the function $\mu : \mathfrak{M}_X \to \mathbb{R}_+$ defined by the formula

$$\mu(A) = \sup\{\mu_\lambda(A) : \lambda \in \Lambda\}, \tag{70}$$

for an arbitrary set $A \in \mathfrak{M}_X$. Obviously we have that $\ker \mu = \mathscr{P}$. Moreover, it is easily seen that the function μ is a measure of noncompatness in the metric space X which means that μ satisfies conditions (i), (ii), (iii), and (vi) of Definition 6. This completes the proof. □

Note that Theorem 5 describes the best possible kernel with respect to which the minimization problem is well-posed for the functional J on the set D.

Obviously, Theorem 5 is a generalization of Theorem 3.

Now, we provide a few theorems on the connection of the well-posedness of minimization problem for the functional J on the set D with some properties of minimizing sequences.

Theorem 6 *If the minimization problem for the functional J on the set D is well-posed with respect to the kernel $\mathscr{P}$ then the set of all accumulation points of each minimizing sequence $\{x_n\}$ for the functional J on the set D is a member of the family $\mathscr{P}$.*

Proof Let $\{x_n\}$ be a minimizing sequence for the functional J on the set D, i.e.,

$$\lim_{n\to\infty} J(x_n) = m_J, \tag{71}$$

where the number m_J was defined previously. Then, the set all accumulation points of the sequence $\{x_n\}$ can be represented in the form

$$\tilde{A} = \bigcap_{n=1}^{\infty} \overline{A}_n, \tag{72}$$

where

$$A_n = \{x_i : i \geq n\}. \tag{73}$$

Next, let μ be a measure of noncompactness in X chosen according to Definition 8. Then, in view of (66) we infer that

$$\lim_{n\to\infty} \mu(\overline{A}_n) = 0. \tag{74}$$

But this yields that $\tilde{A} \in \mathscr{P}$ and the proof is complete. □

Theorem 7 *If each minimizing sequence for the functional J on the set D has at least one accumulation point and $D_0 \in \mathscr{P}$, then the minimization problem for the functional J on the set D is well posed with respect to the kernel $\mathscr{P}$.*

Proof Let us define the measure of noncompactness on the space X by putting $\mu(A) = H(A, \mathscr{P})$ for an arbitrary set $A \in \mathfrak{M}_X$ (cf. (39)). To prove our assertion it is sufficient to show that

$$\lim_{n\to\infty} \mu(D_{1/n}) = 0. \tag{75}$$

Suppose contrarily, i.e., there exists a constant $\gamma > 0$ such that

$$\lim_{n\to\infty} H(D_{1/n}, \mathscr{P}) = \gamma. \tag{76}$$

Observe that the above limit does exist since the sequence of sets $\{D_{1/n}\}$ is decreasing. Thus, in virtue of (76) we get

$$H(D_{1/n}, D_0) \geq H(D_{1/n}, \mathscr{P}) \geq \gamma. \tag{77}$$

Based on the above inequality we can find a sequence $\{x_n\}$ such that

$$x_n \in D_{1/n}, \quad \text{for } n = 1, 2, \ldots, \tag{78}$$

and

$$\text{dist}(x_n, D_0) \geq \frac{\gamma}{2}. \tag{79}$$

From (78) we infer that $\{x_n\}$ is a minimizing sequence for the functional J on the set D. On the other hand inequality (79) allows us to deduce that $\{x_n\}$ has no accumulation points. The obtained contradiction completes the proof. □

Now we formulate a theorem characterizing functional for which the minimization problem is well posed in the generalized sense accepted in Definition 8.

Theorem 8 *The minimization problem for the functional J on the set D is well posed with respect to the kernel $\mathscr{P}$ if and only if there exists a measure of noncompactness μ with* $\ker \mu = \mathscr{P}$ *such that for any $A \in \mathfrak{M}_X$ the following inequality holds:*

$$m_J + \mu(A) \leq \sup J(A). \tag{80}$$

Proof First, let us assume that the minimization problem for the functional J is well posed on the set D with respect to the kernel $\mathscr{P}$. Then there exists a measure of noncompactness $\overline{\mu}$ with $\ker \overline{\mu} = \mathscr{P}$ and such that

$$\lim_{\varepsilon \to 0} \overline{\mu}(D_\varepsilon) = 0. \tag{81}$$

Now, let us consider the function φ acting from the set $\overline{\mathbb{R}_+} = [0, \infty]$ into itself and defined in the following way:

$$\varphi(r) = \begin{cases} 0, & \text{if } r = 0, \\ \overline{\mu}(D_r), & \text{if } \operatorname{diam} D_r < \infty, \\ \infty, & \text{if } \operatorname{diam} D_r = \infty, \\ \infty, & \text{if } r = \infty. \end{cases} \tag{82}$$

It is easily seen that φ is nondecreasing. Moreover, φ is continuous at the point $r = 0$.

Further, denote by $\overline{r}$ the constant

$$\overline{r} = \sup\{r > 0 : \ \operatorname{diam} D_r < \infty\}. \tag{83}$$

In what follows we define the measure of noncompactness μ on the space X with $\ker \mu = \mathscr{P}$ by putting

$$\mu(A) = \min\left\{1, \overline{r}, \inf \varphi^{-1}([\overline{\mu}(A), \infty])\right\}. \tag{84}$$

To finish this part of the proof it is sufficient to show that inequality (80) is satisfied. To this end, let us fix a set $A \in \mathfrak{M}_X$. If $\sup J(A) = \infty$ then inequality (80) does hold. So, assume that $\sup J(A) < \infty$. Further, choose the number $r_1 = \sup J(A) - m_J$. Let $\overline{r} \in [1, \infty)$. If $r_1 \geq \overline{r}$, we have

$$m_J + \mu(A) \leq m_J + \overline{r} \leq m_J + r_1 = \sup J(A). \tag{85}$$

If $r_1 < \overline{r}$ then $\operatorname{diam} D_{r_1} < \infty$ and we get

$$A \subset D_{r_1} = \{x \in D : \ J(x) \leq \sup J(A)\}. \tag{86}$$

Hence, we obtain

$$m_J + \mu(A) \leq m_J + \mu(D_{r_1}) \leq m_J + r_1 = \sup J(A). \tag{87}$$

Similarly, we show the validity of inequality (80) in the case $\overline{r} < 1$.

Conversely, suppose that inequality (80) is satisfied with a measure of noncompactness μ. Fix $\varepsilon \in (0, \varepsilon_0)$, where ε_0 is such a number that $\operatorname{diam} D_{\varepsilon_0} < \infty$. Then we have

$$m_J + \mu(D_\varepsilon) \leq \sup J(D_\varepsilon) \leq m_J + \varepsilon. \tag{88}$$

This implies that $\mu(D_\varepsilon) \leq \varepsilon$. Consequently we derive that

$$\lim_{\varepsilon \to 0} \mu(D_\varepsilon) = 0 \tag{89}$$

and the proof is complete. □

In what follows we provide a few examples illustrating our considerations and obtained results.

Example 9 Let X be a complete metric space and let J be a lower semicontinuous and lower bounded functional defined on a nonempty and closed subset D of the space X. Next, let $\mathfrak{N}_X^0$ denote the kernel consisting of all singletons in X (cf. Example 5). Then we can assert that the minimization problem for the functional J on the set D is well-posed with respect to the kernel $\mathfrak{N}_X^0$ if and only if it is well-posed in the sense of Tikhonov.

Indeed, it suffices to put $\mu(A) = \operatorname{diam} A$ for $A \in \mathfrak{M}_X$.

If we take the kernel $\mathscr{P} = \mathfrak{N}_X$, then the minimization problem for the functional J on the set D is well-posed with respect to $\mathfrak{N}_X$ if and only if it is well-posed in the sense of Furi and Vignoli. In fact, it is sufficient to put $\mu(A) = \alpha(A)$ for $A \in \mathfrak{N}_X$, where α denotes the Kuratowski measure of noncompactness (cf. Sect. 3).

Example 10 Assume that $\mathscr{P}$ is an arbitrarily fixed kernel (of a measure of noncompactness) in a Banach space E. Further, let μ be a measure of noncompactness in the space E with the kernel $\ker \mu = \mathscr{P}$, which is defined by the formula

$$\mu(A) = H(A, \mathscr{P}), \quad \text{for } A \in \mathfrak{M}_E. \tag{90}$$

Next, assume that D is a nonempty, bounded, closed, and convex subset of the Banach space E and $T : D \to D$ is a continuous operator such that

$$\mu(T(A)) \leq k\mu(A), \tag{91}$$

for any nonempty subset A of the set D, where k is a constant from the interval $[0, 1)$.

Then the minimization problem for the functional $J : D \to D$, defined by the formula

$$J(x) = ||Tx - x||, \tag{92}$$

is well-posed with respect to the kernel $\mathscr{P}$.

Indeed, we have

$$m_J = \inf\{J(x) : x \in D\} = 0, \tag{93}$$

since the operator T has a fixed point in the set D [9]. Moreover, we have

$$\begin{aligned}\mu(A) &= H(A, \mathscr{P}) \leq H(A, T(A)) + H(T(A), \mathscr{P}) \\ &= H(A, T(A)) + \mu(T(A)) \leq H(A, T(A)) + k\mu(A) \\ &\leq \sup J(A) + k\mu(A).\end{aligned} \tag{94}$$

Hence, we obtain

$$(1 - k)\mu(A) \leq \sup J(A) \tag{95}$$

which means that inequality (80) is satisfied for the measure of noncompactness $(1 - k)\mu$.

Example 11 Similarly to the preceding example assume that E is a Banach space. Further, let D be a nonempty, closed, and convex subset of the space E. Consider a functional $J : D \to D$ which is quasiconvex in the set D, i.e., for all $x_1, x_2 \in D$ and for each $\alpha \in [0, 1]$ the following inequality is satisfied:

$$J(\alpha x_1 + (1 - \alpha)x_2) \leq \max\{J(x_1), J(x_2)\}. \tag{96}$$

Next, let us assume that there exists a measure of noncompactness in the Banach space E (cf. Definition 4) with the kernel $\ker \mu$ and such that

$$m_J + \mu(A) \leq \sup J(A), \tag{97}$$

for any convex set $A \in \mathfrak{M}_E$. Then the minimization problem for the functional J on the set D is well-posed with respect to $\ker \mu$.

To prove, this fact let us take an arbitrary set $A \in \mathfrak{M}_E$. Then, in view of (97) we obtain

$$m_J + \mu(A) = m_J + \mu(\text{conv}A) \leq \sup J(\text{conv}A). \tag{98}$$

Since the functional J is assumed to be quasiconvex, we have

$$\sup J(\operatorname{conv} A) \leq \sup J(A). \tag{99}$$

Joining inequalities (98) and (99) we deduce that inequality (80) is satisfied. This completes the proof of our assertion.

Finally, let us mention that considerations presented in this section are partly based on [23].

References

1. Akhmerov, R.R., Kamenski, M.I., Potapov, A.S., Rodkina, A.E., Sadovskii, B.N.: Measures of Noncompactness and Condensing Operators. Birkháuser, Basel (1992)
2. Ansari, Q.H. (ed.): Topics in Nonlinear Analysis and Optimization. World Education, Delhi (2012)
3. Banaś, J.: Measures of noncompactness in the space of continuous tempered functions. Demonstratio Math. **14**, 127–133 (1981)
4. Banaś, J.: On drop property and nearly uniformly smooth Banach spaces. Nonlinear Anal. Theory Meth. Appl. **14**, 927–933 (1990)
5. Banaś, J.: Compactness conditions in the geometric theory of Banach spaces. Nonlinear Anal. Theory Meth. Appl. **16**, 669–682 (1991)
6. Banaś, J.: Measures of noncompactness in the study of soultions of nonlinear differential and integral equations. Cent. Eur. J. Math. **10**, 2003–2011 (2012)
7. Banaś, J., Frączek, K.: Conditions involving compactness in geometry of Banach spaces. Nonlinear Anal. Theory Meth. Appl. **20**, 1217–1230 (1993)
8. Banaś, J., Frączek, K.: Locally nearly uniformly smooth Banach spaces. Collect. Math. **44**, 13–22 (1993)
9. Banaś, J., Goebel, K.: Measures of Noncompactness in Banach spaces. Lecture Notes in Pure and Applied Mathematics, vol. 60. Marcel Dekker, New York (1980)
10. Banaś, J., Martinon, A.: Some properties of the Hausdorff distance in metric spaces. Bull. Austral. Math. Soc. **42**, 511–516 (1990)
11. Banaś, J., Martinon, A.: Measures of noncompactness in Banach sequence spaces. Math. Slovaca **42**, 497–503 (1992)
12. Banaś, J., Sadarangani, K.: Compactness conditions in the study of functional, differential and integral equations. Abstr. Appl. Anal. **2013**, Article ID 819315 (2013)
13. Bednarczuk, E., Penot, J.P.: Metrically well-set minimization problems. Appl. Math. Optim. **26**, 273–285 (1992)
14. Dontchev, A.L., Zolezzi, T.: Well-Posed Optimization Problems. Lecture Notes in Mathematics, vol. 1543. Springer, Berlin (1993)
15. Dunford, N., Schwartz, J.T.: Linear Operators I. International Publications, Leyden (1963)
16. Furi, M., Vignoli, A.: About well-posed optimization problems for functionals in metric spaces. J. Optim. Theory Appl. **5**, 225–229 (1970)
17. Goebel, K., Kirk, W.A.: Topics in Metric Fixed Point Theory. Cambridge University Press, Cambridge (1990)
18. Goebel, K., Sękowski, T.: The modulus of noncomapct convexity. Ann. Univ. Mariae Curie-Skłodowska Sect. A **38**, 41–48 (1984)
19. Goldenštein, L.S., Markus, A.S.: On a measure of noncompactness of bounded sets and linear operators. In: Studies in Algebra and Mathematical Analysis, Kishinev pp. 45–54 (1965)
20. Granas, A., Dugundji, J.: Fixed Point Theory. Springer, New York (2003)

21. Hu, R., Fang, Y., Huang, N., Wong, M.: Well-posedness of systems of equilibrium problems. Taiwan. J. Math. **14**, 2435–2446 (2010)
22. Huff, R.: Banach spaces which are nearly uniformly convex. Rocky Mount. J. Math. **10**, 743–749 (1980)
23. Knap, Z., Banaś, J.: Characterization of the well-posed minimum problem (in Polish). Tow. Nauk. w Rzeszowie Met. Numer. **6**, 51–62 (1980)
24. Kuratowski, K.: Sur les espaces complets. Fund. Math. **15**, 301–309 (1930)
25. Kuratowski, K.: Topology. Academic Press, New York (1968)
26. Levitin, E.S., Polyak, B.T.: Convergence of minimizing sequences in conditional extremum problem. Soviet Math. Dokl. **7**, 764–767 (1966)
27. Long, X.J., Huang, N.J., Teo, K.L.: Levitin-Polyak well-posedness for equilibrium problems with functional constrains. J. Inequal. Appl. **2008**, Article ID 657329 (2008)
28. Montesinos, V.: Drop property equals reflexivity. Studia Math. **87**, 93–100 (1987)
29. Revalski, J.P.: Hadamard and strong well-posedness for convex programs. SIAM J. Optim. **7**, 519–526 (1997)
30. Rolewicz, S.: On drop property. Studia Math. **85**, 27–35 (1987)
31. Rolewicz, S.: On δ-uniform convexity and drop property. Studia Math. **87**, 181–191 (1987)
32. Sękowski, T., Stachura, A.: Noncompact smoothness and noncompact convexity. Atti. Sem. Mat. Fis. Univ. Modena **36**, 329–338 (1988)
33. Tikhonov, A.N.: On the stability of the functional optimization problem. USSR J. Comput. Math. Math. Phys. **6**, 631–634 (1966)
34. Zolezzi, T.: Extended well-posedness of optimization problem. J. Optim. Theory Appl. **91**, 257–266 (1996)

Well-Posedness, Regularization, and Viscosity Solutions of Minimization Problems

D. V. Pai

Abstract This chapter is divided into two parts. The first part surveys some classical notions for well-posedness of minimization problems. The main aim here is to synthesize some known results in approximation theory for best approximants, restricted Chebyshev centers, and prox points from the perspective of well-posedness of these problems. The second part reviews Tikhonov regularization of ill-posed problems. This leads us to revisit the so-called viscosity methods for minimization problems using the modern approach of variational convergence. Lastly, some of these results are particularized to convex minimization problems, and also to ill-posed inverse problems.

Keywords Well-posedness · Best approximants · Well-posedness of restricted Chebyshev centers · Best simultaneous approximation · Prox pairs · Well-posedness of prox-pairs · Tikhonov regularization · Viscosity solutions · Epi-convergence · Hierarchial minimization

1 Introduction and Preliminaries

Well-posedness, regularization, and viscosity methods are topics of continuing interest in the literature on variational analysis and optimization (cf., e.g., [3–5, 12, 15, 24, 26, 43, 47, 48]). In the present chapter, we will attempt to highlight the following scheme of development:

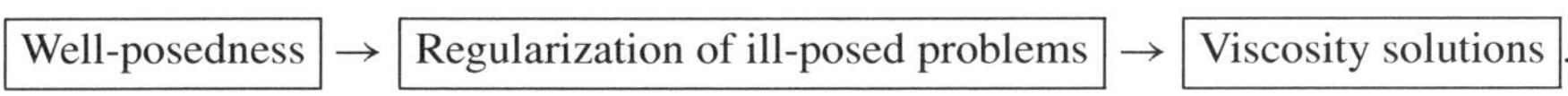

D. V. Pai (✉)
Indian Institute of Technology Gandhinagar, VGEC Campus, Visat-Gandhinagar Highway, Chandkheda 382424, Ahmedabad, India
e-mail: dvp@iitgn.ac.in

Q. H. Ansari (ed.), *Nonlinear Analysis*, Trends in Mathematics,
DOI: 10.1007/978-81-322-1883-8_5,

In Sect. 2, we review some well-posedness notions for minimization problems starting from the classical notion due to Tikhonov [47]. This leads us to Sect. 3 where our main goal is to synthesize some classical as well as newer results in approximation theory from the angle of well-posedness of the underlying minimization problems. In Sect. 4 we begin by considering some more standard results on Tikhonov well-posedness as well as a classical example motivating *Tikhonov regularization*. The main result here (Theorem 14) apparently originated in [27]. Its present refinement could be traced to [15].

Section 5 is devoted to viscosity solutions. Viscosity methods have been used for a longtime in diverse problems arising in variational analysis and optimization (cf., e.g., [4, 14, 16, 26, 48]). A central feature of these methods is to come up, as a limit of solutions of a sequence of approximating problems, a particular solution of the underlying problem, the so-called *viscosity solution* which enjoys interesting properties. Although the first abstract formulation for studying viscosity approximations was given in [47], it is really the elegant article due to Attouch [4], which provided an efficient abstract framework for exploring the viscosity solutions using the modern tools of variational convergence for sequences of functions and operators. Here, our main aim is to revisit some of these results in [4] with a view to provide a greater flexibility to these results.

1.1 Preliminaries

In the sequel, X will be a convergence space endowed with convergence of nets (or sequences) denoted by $\to$, satisfying the "Kuratowski" axioms [22, pp. 83–84]. When X is a topological space, the convergence of nets (or sequences) will be understood as the one induced by the given topology. For the most part X will be a normed linear space over $\mathbb{K}$ (either $\mathbb{R}$ or $\mathbb{C}$). Its normed dual will be denoted by X^* and w (resp. w^*) will denote the weak topology (resp. weak* topology). S (resp. S^*) will denote the unit sphere (norm one elements) of X (resp. X^*). The closed unit ball of X (resp. X^*) is denoted by U (resp. U^*). The open (resp. closed) ball of center x and radius r will be denoted by $B(x, r)$ (resp. $B[x, r]$). We distinguish the following classes of normed spaces:

- $(Rf) :=$ the reflexive Banach spaces
- $(R) :=$ the rotund (strictly convex) normed spaces
- $(A) :=$ the normed spaces for which the norm satisfies the Kadec property: w convergence of a sequence in S entails its norm convergence
- $(UR) :=$ the uniformly convex Banach spaces

Following [50], we denote the class of spaces $(Rf) \cap (A)$ by (CD) and the class of spaces $(CD) \cap (R) = (Rf) \cap (R) \cap (A)$ by (D). It is well known [19, pp. 147–149] that the class (D) coincides with the class of Banach spaces whose dual norms are Fréchet differentiable except at the origin.

We also distinguish the following classes of subsets of X:

- $CL(X) :=$ the nonempty closed subsets of X
- $CLB(X) :=$ the nonempty closed and bounded subsets of X
- $CLC(X) :=$ the nonempty closed and convex subsets of X
- $CLBC(X) :=$ the nonempty closed, bounded, and convex subsets of X
- $K(X) :=$ the nonempty compact subsets of X
- $KC(X) :=$ the nonempty compact and convex subsets of X
- $WCL(X) :=$ the nonempty w-closed subsets of X
- $WK(X) :=$ the nonempty w-compact subsets of X
- $WKC(X) :=$ the nonempty weakly compact and convex subsets of X

Recall that in case X is a normed space, the *Hausdorff distance* H between sets A, B in $CL(X)$ is defined by

$$H(A, B) := \inf\{\alpha : A \subset B + \alpha U \text{ and } B \subset A + \alpha U\}.$$

Equivalently, $H(A, B) = \max\{e(A, B), e(B, A)\}$, where $e(A, B) := \sup\{d(a, B) : a \in A\}$ denotes the *Hausdorff excess* of A over B. Hausdorff distance so defined yields an infinite-valued metric on $CL(X)$, which is complete when X is complete [21]. We denote the topology of the Hausdorff distance by τ_H. Since $CLB(X), CLBC(X), K(X), KC(X)$ are closed subsets of $\langle CL(X), \tau_H\rangle$, Hausdorff distance restricted to these classes yields a metric on them which is complete if X is complete [21, p. 45].

During the last 40 years or so, Mosco convergence [32] of convex sets—a much weaker notion of convergence than Hausdorff metric convergence—has been a convergence notion of choice in convex analysis and approximation theory, especially in the framework of reflexive Banach spaces. Specifically, a sequence $\{C_n\}$ in $CLC(X)$, is said to be *Mosco convergent* to an element C in $CLC(X)$, written $C_n \overset{M}{\to} C$, provided that: (M_1) for each $x \in C$ there exists a sequence $\{x_n\}$ convergent to x such that eventually $x_n \in C_n$, and (M_2) whenever $\{n(i)\}$ is an increasing sequence of positive integers and $x_{n(i)} \in C_{n(i)}$ for each i, then the w-convergence of $\{x_{n(i)}\}$ to x in X implies $x \in C$. In the framework of reflexive Banach spaces, Mosco convergence on the hyperspace $CLC(X)$ has been shown to be a fundamental notion for the convergence of metric projections and distance functions [3, 16, 44, 49] as well as for the convergence of restricted Chebyshev centers and restricted Chebyshev radii [10, 39, 40, 42]. This convergence has been observed to be stable with respect to duality [8, 33]. In [7], a Vietoris-type topology on $CLC(X)$ compatible with Mosco convergence was introduced. This topology was considered for the larger class $WCL(X)$ in [11]. This topology now called the *Mosco-Beer topology* τ_{MB} is generated by all sets of the form $V^- := \{A \in WCL(X) : A \cap V \neq \emptyset\}$, where V runs over norm open subsets of X, and all sets of the form $(K^c)^+ := \{A \in WCL(X) : A \subset K^c\} = \{A \in WCL(X) : A \cap K = \emptyset\}$, where K runs over $WK(X)$. It is the weakest topology on $WCL(X)$ for which the *gap functional* $A \to d(A, K)$, where $d(A, K) := \inf\{\|a - k\| : a \in A, k \in K\}$, is continuous [11, Theorem 3.1]. Thus, identifying each $A \in WCL(X)$ with the distance functional $d(., A)$ as an element of $C(X)$, where $d(x, A) := \inf\{\|x - a\| : a \in A\}$, this topology is the topology of uniform convergence of the distance functionals $d(., A)$, $A \in WCL(X)$ on the members of the class $WK(X)$. Clearly, this topology is weaker than the Hausdorff distance topology τ_H on $WCL(X)$, which can be identified with the topology of uniform convergence of the distance functionals $d(., A)$ on X, and it is stronger than the Wijsman topology [9, Chap. 2] on $WCL(X)$ which can be identified with the topology of pointwise convergence of the distance functionals $d(., A)$ as A varies over $WCL(X)$.

Let us recall that a function f in $X \to (-\infty, \infty]$ is called *proper*, if it is finite somewhere. Given a function $f : X \to \mathbb{R}$ and $\alpha \in \mathbb{R}$, we denote by $\text{slev}(f; \alpha)$ (resp. $\text{lev}\ (f; \alpha)$) the sublevel set $\{x \in X : f(x) \leq \alpha\}$ (resp. the level set $\{x \in X : f(x) = \alpha\}$) of f at height α. The function f is said to be *inf-bounded* (resp. *w-inf-compact*) if $\text{slev}(f; \alpha)$ is bounded (resp. w-compact) for each $\alpha \in \mathbb{R}$. Let us denote by $\sum(X)$ (resp. $\Lambda(X)$) the class of all real functions on X which are continuous and w-inf compact (resp. convex, continuous, and inf-bounded). Clearly, if $X \in (Rf)$, then $\Lambda(X) \subset \sum(X)$.

The following weak topology result for $WCL(X)$ was already noted in [42] for $CLC(X)$.

Theorem 1 *Suppose $X \in (Rf)$. Then τ_{MB} is the weakest topology on $WCL(X)$ for which the function $C \to \inf I(C) := v_C(I)$ of $\langle WCL(X), \tau_M \rangle$ into $\mathbb{R}$ is continuous for each $I \in \sum(X)$.*

2 A Review of Some Well-Posedness Notions for Minimization Problems

Given a nonempty subset V of a convergence space X and a function $f : E \to (-\infty, \infty]$ which is a proper extended real-valued function, let us consider well-posedness of the following abstract minimization problem:

$$\min f(v), \quad \text{for all } v \in V,$$

which we denote by (V, f). Let $v_V(f) := \inf\{f(v) : v \in V\}$ denote the *optimal value function*. We assume f to be lower bounded on V, i.e., $v_V(f) > -\infty$, and let $\arg\min_V(f)$ denote the (possibly void) set $\{v \in V : f(v) = v_V(f)\}$ of optimal solutions of problem (V, f). For $\epsilon \geq 0$, let us also denote by $\epsilon\text{-}\arg\min_V(f)$ the nonempty set $\{v \in V : f(v) \leq v_V(f)+\epsilon\}$ of ϵ-approximate minimizers of f. Recall (cf., e.g., [15, p. 1]) that problem (V, f) is said to be

(a) *Tikhonov well-posed* if f has a unique global minimizer on V toward which every *minimizing sequence* (i.e., a sequence $\{v_n\} \subset V$, such that $f(v_n) \to v_V(f)$) converges. Put differently, there exists a point $v_0 \in V$ such that $\arg\min_V(f) = \{v_0\}$, and whenever a sequence $\{v_n\} \subset V$ is such that $f(v_n) \to f(v_0)$, one has $v_n \to v_0$;

(b) *generalized well-posed* (abbreviated g.w.p) if $\arg\min_V(f)$ is nonempty and every minimizing sequence for (V, f) has a subsequence convergent to an element of $\arg\min_V(f)$.

In case $V \in WCL(X)$, where X is a normed linear space, the problem (V, f) is said to be *w-T.w.p.* (resp. *w*-g.w.p.), if it is Tikhonov well-posed (resp. generalized well-posed) for w-convergence of sequences and simply T.w.p. (resp. g.w.p.) if it is Tikhonov well-posed (resp. generalized well-posed) for strong convergence of sequences.

Proposition 1 *Let $V \subset X$, a convergence space (resp. $V \in WCL(X)$, X a normed space). Then problem (V, f) is T.w.p. (resp. w-T.w.p.) if and only if* $\arg\min_V(f)$ *is a singleton and (V, f) is g.w.p. (resp. w-g.w.p).*

The concept of Tikhonov well-posedness has been extended to minimization problems admitting nonunique optimal solutions. For our purpose here, the most appropriate well-posedness notion for such problems is the one introduced in Bednarczuk and Penot [6] (cf. also [15, p. 26]):

In case X is a metric space and $V \subset X$, problem (V, f) is called *metrically well-set* (or *M-well set*) if $\arg\min_V(f) \neq \emptyset$ and for every minimizing sequence $\{v_n\}$, one has

$$\mathrm{dist}(v_n, \arg\min_V(f)) \to 0 \text{ as } n \to \infty.$$

(Here $\mathrm{dist}(x, S)$ denotes the distance of x from the set S.) Equivalently, it is easily seen that problem (V, f) is M-well set if and only if $\arg\min_V(f) \neq \emptyset$ and the multifunction

$$\epsilon \rightrightarrows \epsilon - \arg\min_V(f)$$

is upper Hausdorff semicontinuous (uHsc) at $\epsilon = 0$. We mention that in [15, p. 46], problem (V, f) is also called *stable* in this case.

Tikhonov well-posedness as well as M-well setness of problem (V, f) are conveniently characterized in terms of the notion of a firm function (or a forcing function). A function $c : T \to [0, \infty)$ is called a *firm function* or a *forcing function* if

$$0 \in T \subset [0, \infty), \quad c(0) = 0 \text{ and } t_n \in T, \quad c(t_n) \quad \to \quad 0 \Rightarrow t_n \to 0.$$

It is well known (cf., e.g., [15, p. 6]) that problem (V, f) is Tikhonov well-posed if and only if there exists a firm function c and a point $v_0 \in V$ such that

$$f(v) \geq f(v_0) + c[d(v, v_0)], \quad \text{for all } v \in V.$$

Likewise, it is well known (cf. [6]) that if f is a proper lower semicontinuous function then problem (V, f) is M-well set if and only if $\arg\min_V(f) \neq \emptyset$ and f is firmly conditioned, i.e., there exists a firm function c on $\mathbb{R}^+ := \{x \in \mathbb{R} : x \geq 0\}$ such that

$$f(v) \geq v_V(f) + c(\mathrm{dist}(v, \arg\min_V(f))), \quad \text{for all } v \in V.$$

3 Well-Posedness in Approximation Theory

We will be mainly concerned below with well-posedness of minimization problems involving best approximants, restricted Chebyshev centers, and prox-points.

3.1 Well-Posedness of Best Approximants

Let X be a normed linear space over $\mathbb{K}$ (either $\mathbb{R}$ or $\mathbb{C}$), $V \in CL(X)$ and $x \in X$. The problem of finding a best approximant v_0 to x in $V : \|x - v_0\| = d(x, V) = \inf_{v \in V} \|x - v\|$, is the problem (V, I_x), where $I_x(v) = \|x - v\|$. Recall that the set V is called

(i) *Chebyshev* if each $x \in X$ has a unique best approximant in V;
(ii) *almost Chebyshev* if each x in a dense and G_δ subset X_0 of X admits a unique best approximant in V;
(iii) *approximatively compact* (resp. *approximatively w-compact*) if each minimizing sequence has a subsequence convergent (resp. w-convergent) to an element of V.

Here, the multifunction $x \rightrightarrows P_V(x)$ of X to V, where $P_V(x) = \arg\min_V(I_x)$ is called the *metric projection* of X onto V.

Remark 1 It follows from Proposition 1 that

(a) the best approximation problems $(V, I_x), x \in X$ are all T.w.p. (resp. w-T.w.p) if and only if the set V is Chebyshev and approximatively compact (resp. approximatively w-compact).
(b) If X is a Hilbert space and $V \in CLC(X)$, then the best approximation problems $(V, I_x), x \in X$ are all T.w.p. This result also holds for a uniformly convex Banach space. More generally, the following result is well known [50].
(c) A Banach space X is in the class $(D) = (Rf) \cap (R) \cap (A)$ if and only if each member of $CLC(X)$ is Chebyshev and approximatively compact. Hence, it follows from the first remark that a Banach space X is in the class (D) if and only if for each $V \in CLC(X)$, each problem $(V, I_x), x \in X$ is T.w.p.

Let us recall the following definitions from [13].

Definition 1 A subset V of a Banach space X is called *boundedly relatively w-compact* if $V \cap B[0, r]$ has a w-compact closure for each $r > 0$.

Definition 2 Given $V \in CL(X)$ where X is a Banach space, let $\Omega(V)$ denote the following set as defined in [13].

$$\Omega(V) = \{x \in X \backslash V : \exists x^* \in S^* \text{ such that } \forall \epsilon > 0, \exists \delta > 0 \text{ so that}$$
$$\inf\{\langle x^*, x - v\rangle : v \in V \cap B(x, d_V(x)) > (1 - \epsilon)d_V(x)\}.$$

Let us recall here the following theorem in [13].

Theorem 2 *If X is a Banach space and $V \in CL(X)$ is a bounded relatively w-compact set then $\Omega(V)$ is a dense G_δ subset of $X \backslash V$.*

Let us be given a set C in $CLC(X)$ and $V \in CL(X)$. We need the following definitions from [2].

Definition 3 The set C is said to be *rotund (strictly convex)* w.r.t. V, written *V-rotund* if

$$x, y \in C, \quad x - y \in V - V, \quad \|x\| = \|y\| = \left\| \frac{x+y}{2} \right\| \quad \Rightarrow \quad x = y.$$

Definition 4 The set C is said to be *sequentially Kadec w.r.t. V*, written *V-Kadec* if $\{x_n\} \subset C$, $x_0 \in C$, $x_n - x_0 \in V - V$, $w - \lim x_n = x_0$, and

$$\lim \|x_n\| = \|x_0\| \quad \Rightarrow \quad \|x_n - x_0\| \to 0.$$

The proof of the next proposition follows on the same lines as in the proof of [13, Theorem 6.1].

Proposition 2 *Let X be a Banach space, $C \in CLC(X)$ and $V \in CL(X)$. If C is V-rotund then V is semi-Chebyshev w.r.t. $C \cap \Omega(V)$.*

The proof of the next proposition follows on similar lines as in the proof of [13, Corollary 8].

Proposition 3 *Let X be a Banach space, $V \in CL(X)$ be boundedly relatively w-compact, and $C \in CLC(X)$ be V-Kadec. Then problem (V, I_x) is g.w.p. for each $x \in C \cap \Omega(V)$.*

The next proposition is due to [29].

Proposition 4 *If $V \in CL(X)$ is boundedly relatively w-compact and $C \in CL(X)$ then $C \cap \Omega(V)$ is a dense G_δ-subset of $C \backslash V$.*

The last two propositions in conjunction with Theorem 11 (Lau-Konjagin) in [13] yields:

Theorem 3 *For a Banach space X, the following statements are equivalent.*

(a) *$X \in (CD) = (Rf) \cap (A)$.*
(b) *For each $V \in CL(X)$, the family of problems (V, I_x), $x \in X \backslash V$ is generically g.w.p.*

The proof of the next theorem due to [29] follows easily from Propositions 2, 3, and 4.

Theorem 4 *Let X be a Banach space and $V \in CL(X)$ be boundedly relatively w-compact. If $C \in CLC(X)$ is V-rotund and V-Kadec then problem (V, I_x) is T.w.p. for each $x \in C \cap \Omega(V)$. Thus, the family of problems (V, I_x), $x \in C \backslash V$ is generically T.w.p.*

3.2 Well-Posedness of Restricted Chebyshev Centers (Best Simultaneous Approximation)

Let X be a normed linear space, $V \in CL(X)$ and $F \in CLB(X)$. Let

$$r(F; x) := \sup\{\|x - y\| : y \in F\}.$$

The function $x \to r(F; x)$ is a proper continuous and w-l.s.c. convex function on X. Its *sublevel set at height* α is the set

$$\text{slev}(r(F; .), \alpha) = \{x \in X : r(F; x) \leq \alpha\} = \bigcap_{y \in F} B[y, \alpha].$$

For simplicity, we denote this set by $\text{slev}_F(\alpha)$. Let $I_F : V \to \mathbb{R}$ denote the function $I_F(v) = r(F; v)$, and let

$$\text{rad}_V(F) := \inf I_F(V)$$

and

$$\text{Cent}_V(F) := \arg\min_V(I_F).$$

The number $\text{rad}_V(F)$ is called the *Chebyshev radius* of F in V and in case $\text{Cent}_V(F) \neq \emptyset$, a typical element $v_0 \in \text{Cent}_V(F)$ is called a *restricted (Chebyshev) center* or a *best simultaneous approximant* of F in V.

Let $\mathscr{F} \subset CLB(X)$. Then a set $V \in CL(X)$ is called *cent-compact* for $\mathscr{F}$ (resp. *w-cent-compact for* $\mathscr{F}$ in case X is a Banach space) if each minimizing sequence for problem (V, I_F) has a subsequence convergent (resp. w-convergent) in V. Clearly V is cent-compact (resp. w-cent-compact) if and only if (V, I_F) is g.w.p. (resp. w-g.w.p.). The terminology was employed in [10] in terms of minimizing nets rather than sequences. We note, however, that since for a subset of a Banach space X, w-compactness is equivalent to its w-sequential compactness by Eberlain-Smulyan theorem, this stronger requirement is really not necessary. Let us denote by $\text{remote}_V(X)$ the family of all sets in $CLB(X)$ which are "remotal," w.r.t. V, i.e., possessing farthest points for points of V. For the proofs of the next lemma and the following proposition, we refer the reader to [37] (See also Theorems 5 and 9 in Sect. 5.4, Chap. viii of [31]) .

Lemma 1 *If* $X \in (Rf) \cap (A)$ *and* $V \in CLC(X)$, *then* V *is cent-compact for* $\text{remote}_V(X)$.

Proposition 5 (a) *If* $X \in (D)$, $V \in CLC(X)$ *and* $F \in \text{remote}_V(X)$, *then problem* (V, I_F) *is T.w.p.*

(b) *If* $X \in (UR)$, $V \in CLC(X)$ *and* $F \in CLB(X)$, *then problem* (V, I_F) *is T.w.p.*

By [7, Theorem 4.3], when X is reflexive and separable, $CLC(X)$ equipped with the Mosco-Beer topology τ_{MB} is a Polish space (second countable and completely metrizable). Since $\langle CLC(X), \tau_{MB}\rangle$ is a Baire space, it is of interest to consider the following generic theorem for Tikhonov well-posedness of restricted centers.

Theorem 5 *Let X in $(Rf) \cap (A)$ be separable. Let $K(X)$ be equipped with the topology τ_H, let, $CLC(X)$ be equipped with the topology τ_{MB}, and let the set*

$$\Omega = \{(F, V) \in (K(X) \times CLC(X)) : \mathrm{Cent}_X(F) \cap V = \emptyset\}$$

be equipped with the relative topology. Then there exists a dense G_δ subset Ω_0 of Ω such that for each (F, V) in Ω_0, the problem (V, I_F) is T.w.p.

Proof We observe that $\langle K(X), \tau_H\rangle$ is complete. Also note that if $X \in (Rf) \cap (A)$ and $V \in CLC(X)$, then by Lemma 1, V is cent-compact for $K(X)$. The desired result now follows from [10, Theorem 4.3] in conjunction with Proposition 1. □

For exploring generic well-posedness of restricted centers, a somewhat different approach was followed recently in [29] using the following embedding theorem due to Radstrom [41].

Theorem 6 *Given a Banach space X, there exists a Banach space $(E, \|\cdot\|)$ such that $(KC(X), H)$ is embedded as a convex cone in E in such a manner that:*

(i) *The embedding is isometric: $H(A, B) = \|A - B\|, \forall A, B \in KC(X)$;*
(ii) *X is a linear subspace of E;*
(iii) *Addition and multiplication by nonnegative scalars in E induce the corresponding operations in KC(X).*

Furthermore, if X is reflexive, then the above statements also hold for $CLBC(X)$.

The following lemma and the next theorem are given in [29].

Lemma 2 *Let X be a Banach space, $V \in CL(X)$, and let E be as given in the preceding theorem. If $X \in (R) \cap (A)$, then $KC(X)$ is both V-rotund and V-Kadec.*

In view of Theorem 5, Lemma 2, and Theorem 6, one obtains the next theorem.

Theorem 7 *Let $X \in (R) \cap (A)$, and $V \in CL(X)$ be bounded relatively w-compact. Then there exists a dense G_δ subset Σ of $KC(X)\backslash V$ such that problem (V, I_F) is T.w.p. for each $F \in \Sigma$. Put differently, the family of problems (V, I_F), $F \in KC(X)$ is generically T.w.p.*

Observing that if $X \in (D)$ and $V \in CL(X)$, then V is boundedly relatively w-compact, the preceding theorem in conjunction with Theorem 6.6 in [13] yields:

Theorem 8 *For a Banach space X, the following statements are equivalent.*

(a) *$X \in (D) = (Rf) \cap (R) \cap (A)$.*
(b) *For each $V \in CL(X)$, the family of problems (V, I_F), $F \in KC(X)$ is generically T.w.p.*

3.3 Well-Posedness of the Prox Pairs

Let X be a normed linear space over $\mathbb{K}$. Given A, B in $CL(X)$, a pair (b, a) in $B \times A$ is called a *prox pair* of the pair (B, A) of sets if

$$\|b - a\| = d(B, A) := \inf\{\|b - a\| : b \in B, a \in A\}.$$

We denote the (possibly void) set of all prox pairs of (B, A) by Prox (B, A). Note that $(b, a) \in$ Prox (B, A) if and only if $a - b$ is a best approximant to 0 in $A - B$. Prox pairs of pairs of convex sets are studied in [34] in relation to mutually nearest points giving rise to a characterization of smooth normed linear spaces. More general results of this type are given in [35] for multioptima and Nash equilibrium points of convex functionals defined on (finite) products of locally convex spaces. In [11, Theorem 4.3], a generic uniqueness result is given for prox points. Here, we will revisit this theorem as a generic Tikhonov well-posedness result.

It is easily seen that Prox $(B, A) \neq \emptyset$ whenever X is in (Rf) and (B, A) is in $WKC(X) \times CLC(X)$. In what follows, we consider the multifunction

$$\text{Prox} : WKC(X) \times CLC(X) \rightrightarrows X \times X.$$

As observed in [11], if $WKC(X)$ is equipped with the topology $\mathscr{T}_H$ and $CLC(X)$ is equipped with τ_{MB}, then the product space $WKC(X) \times CLC(X)$ is completely metrizable whenever X is reflexive and separable. The same thing can be said about its subspace $KC(X) \times CLC(X)$, since τ_H restricted to $KC(X)$ is complete. It is therefore meaningful to ask generic well-posedness questions about the multifunction Prox defined on $KC(X) \times CLC(X)$.

Let $B \in KC(X)$ and $A \in CLC(X)$. We equip $B \times A$ with the convergence structure: a sequence (b_n, a_n) in $B \times A$ converges to (b, a) in $B \times A$ if and only if $b_n \to b$ and $a_n \to a$. Let $I : B \times A \to \mathbb{R}$ be defined by: $I(b, a) = \|b - a\|$, $(b, a) \in B \times A$. We need the next lemma, whose proof is left to the reader.

Lemma 3 *Let $X \in (Rf) \cap (A)$. If $(B, A) \in KC(X) \times CLC(X)$, then problem $(B \times A, I)$ is g.w.p.*

In conjunction with [11], the above lemma yields:

Theorem 9 *Let $X \in (Rf) \cap (A)$ be separable. Suppose $KC(X)$ is equipped with τ_H and $CLC(X)$ is equipped with τ_{MB}. Then there exists a dense G_δ subset Ω_0 of*

$$\Omega := \{(B, A) \in KC(X) \times CLC(X) : d(B, A) > 0\}$$

such that for each $(B, A) \in \Omega_0$, problem $(B \times A, I)$ is T.w.p.

3.4 Strong Uniqueness of Best Simultaneous Approximation

In the classical Chebyshev theory (cf., e.g., [31, 45]) as well as in the more recent theory of best approximants in normed linear spaces, there has been a lot of interest in studying "strong unicity" of best approximants: An element $v_0 \in V$, a finite dimensional linear subspace of a normed linear space X, is called a *strongly unique best approximant* (SUBA) to x in V if there exists a constant $\lambda = \lambda(x)$, $0 < \lambda < 1$, such that

$$\|x - v\| \geq \|x - v_0\| + \lambda\|v - v_0\|, \quad \text{for all } v \in V.$$

Put differently, the strong uniqueness of a best approximant $v_0 \in V$ to x is precisely the Tikhonov well-posedness of problem (V, I_x) where $I_x(v) := \|x - v\|, v \in V$, with the associated firm function being linear: $c(t) = \lambda t,\ t \in T$. The problem (V, I_x) is also said to be *linearly conditioned* in this case.

Given a finite dimensional subspace V of a normed linear space X and $x \in X$, let us denote by $P_V(x)$ the (nonempty) set $\{v_0 \in V : \|x - v_0\| = \text{dist}(x, V)\}$ of *best approximants* to x in V. In this case the multifunction $X : x \rightrightarrows P_V(x)$ of X into V is called the *metric projection multifunction* is said to be *Chebyshev* if $P_V(x) \neq \emptyset$, for each $x \in X$. In case V is not Chebyshev, Li [30] introduced the following definition: The metric projection multifunction $P_V : X \rightrightarrows V$ is said to be *Hausdorff strongly uniquely* at $x \in X$ if there exists a constant $\lambda_V(x) > 0$, such that $\|x - v\| \geq \text{dist}(x, V) + \lambda_V(x)\text{dist}(v, P_V(x))$, for all $v \in V$. Note that Hausdorff strong uniqueness of the multifunction P_V at x is precisely M-well setness of the problem (V, I_x) with the associated firm function c_x being linear: $c_x(t) = \lambda_V(x)t$. In this case problem (V, I_x) is also said to be *linearly conditioned.*

Consider the problem of approximating simultaneously a *data set* in a given space by a single element of an approximating family. Such a problem arises naturally in many practical situations (cf., e.g., [18, 19, 25]). One way to treat this is to cover the given dataset (assumed to be bounded) by a ball of minimal radius among those centered at the points of the approximating family. The problem of *best simultaneous approximation* in this sense coincides with problem (V, I_F), where V, a finite dimensional subspace of a normed linear space X, is the approximating family, and F, a nonempty bounded subset of X, is the dataset. The objective function in this problem is $I_F : V \to \mathbb{R}$, which measures "worstness" of an element $v \in V$ as a representer of F, defined by

$$I_F(v) = r(F; v), \quad \text{where } r(F; v) := \sup\{\|f - v\| : f \in F\}.$$

The optimal value function $v_V(I_F)$ in this case is denoted by $\text{rad}_V(F)$. Thus the "intrinsic error" in the problem of approximating simultaneously all the elements $f \in F$ by the elements of V is the number $\text{rad}_V(F) := \inf\{r(F; v) : v \in V\}$, called the *Chebyshev radius* of F in V. It is the minimal radius of a ball (if one such exists) centered at a point in V and covering F. The centers of all such balls are precisely the elements of the set $\arg\min_V(I_F)$ which in this case will be denoted by $\text{Cent}_V(F)$.

A typical element of the set

$$\mathrm{Cent}_V(F) := \{v_0 \in V : r(F; v_0) = r_V(F)\}$$

is called a *best simultaneous approximant* or a *restricted center* of F in V. When the bounded sets F are allowed to range over a certain family $\mathscr{F}$ of nonempty closed and bounded subsets of X, the multifunctions $\mathrm{Cent}_V : \mathscr{F} \rightrightarrows V$, with values $\mathrm{Cent}_V(F)$, $F \in \mathscr{F}$, is called the *restricted center multifunction.* Note that in case F is a singleton $\{x\}, x \in X, r_V(F)$ is the distance of x from V, denoted by $\mathrm{dist}(x, V)$, and $\mathrm{Cent}_V(F)$ is precisely the set $P_V(x)$ of all best approximants to x in V.

Let $F \in \mathscr{F}$. Analogously, as in the case of a SUBA, an element $v_0 \in V$ is called a *strongly unique best simultaneous approximant* (*SUBSA*) to F in V if there exists a constant $\lambda = \lambda_V(F) > 0$ such that

$$r(F; v) \geq r(F; v_0) + \lambda\|v - v_0\|, \quad \text{for all } v \in V.$$

Likewise, in case $\mathrm{Cent}_V(F)$ is not a singleton, the set F is said to admit *Hausdorff strongly unique best simultaneous approximant* (*H-SUBSA*) in V if there exists a constant $\lambda = \lambda_V(F) > 0$ such that for all $v \in V$,

$$r(F; v) \geq r_V(F) + \lambda \mathrm{dist}(v, C_V(F)).$$

Clearly, F admits a SUBSA (resp. a H-SUBSA) in V if and only if problem (V, I_F) is Tikhonov well-posed (resp. M-well set) and linearly conditioned. The triplet $(X, V, \mathscr{F})$ is said to satisfy *property SUBSA* (resp. *property H-SUBSA*) if F admits SUBSA (resp. H-SUBSA) in V for every $F \in \mathscr{F}$. Let us recall that $\mathscr{C}_0(T)$ consists of all continuous functions $f : T \to \mathbb{K}$ vanishing at infinity, i.e., a continuous function f is in $\mathscr{C}_0(T)$ if and only if, for every $\epsilon > 0$, the set $\{t \in T : \|f(t)\| \geq \epsilon\}$ is compact. The space $\mathscr{C}_0(T)$ is endowed with the norm:

$$\|f\| := \max\{|f(t)| : t \in T\}, \quad f \in \mathscr{C}_0(T).$$

Let us now take $X = \mathscr{C}_0(T)$ and V a finite dimensional subspace of X. Recall that V is called a *Haar subspace* or that it satisfies the *Haar condition* if for each $v \in V\setminus\{0\}$, card $Z(v) \leq \dim V - 1$. Here, we use the notation card(A) to denote the cardinality of A and $Z(v)$ to denote the set of all zeros of v. Let

$$\Omega_V(X) := \{F \in CLB(X) : r_X(F) < r_V(F)\}.$$

Although uniqueness of best simultaneous approximants was studied previously in many articles (cf., e.g., [1, 23, 25]), strong uniqueness was not treated in these articles. See, however, [28, 36, 38]. Triplets $(X, V, \mathscr{F})$ satisfying SUBSA and other related properties were investigated in [36].

For finite dimensional subspaces V of $\mathscr{C}_0(T)$, the following extension of Haar condition is due to Li [30].

Definition 5 V is said to satisfy *property (Li)* if for every $v \in V\backslash\{0\}$,

$$\text{card } bdZ(v) \leq \dim\left\{p \in V : p|_{\text{int}Z(v)} = 0\right\} - 1.$$

Note that if T is connected, then property (Li) coincides with the Haar condition. Li [30] has shown that this property (Li) of V is equivalent to Hausdorff Lipschitz continuity of the metric projection multifunction $P_V : X \rightrightarrows V$. This result was extended in [20] to the restricted center multifunction as follows.

Theorem 10 [20] *For a finite dimensional subspace V of* $\mathscr{C}_0(T)$ *the following statements are equivalent.*

(i) *The multifunction* $C_V : K_V(X) \rightrightarrows V$ *is lsc.*
(ii) V *satisfies property (Li).*

We also recall here the following theorem which was established in [20]. This theorem extends to restricted center multifunction a similar result due to Li [30] for metric projection multifunction.

Theorem 11 [20] *Let* V *be a finite dimensional subspace of* $\mathscr{C}_0(T)$. *If* V *satisfies property (Li) then the triplet* $(\mathscr{C}_0(T), V, K_V(X))$ *satisfies property H-SUBSA.*

4 Tikhonov Regularization

Let us begin by recalling the following theorem which lists some classical sufficient conditions for Tikhonov well-posedness of problem (X, f).

Theorem 12 *Under any one of the following conditions, problem* (X, f) *is T.w.p.*

(i) X *is sequentially compact,* f *is proper and sequentially lower semicontinuous,* $\arg\min_X(f)$ *is a singleton.*
(ii) $X = \mathbb{R}^n$, $f : X \to \mathbb{R}$ *is strictly convex, and coercive:*

$$f(x) \to +\infty \quad \textit{as } \|x\| \to \infty.$$

(iii) E *is a reflexive Banach space,* $X \subset E$ *is a nonempty closed convex set,* $f : X \to \mathbb{R} \cup \{+\infty\}$ *is proper, strictly convex, l.s.c., and coercive.*
(iv) $X = \mathbb{R}^k$, $f : X \to \mathbb{R}$ *is convex and l.s.c.,* $\arg\min_X(f)$ *is a singleton.*

Proof We will only prove Tikhonov well-posedness of (X, f) under the last set of conditions:

$$X = \mathbb{R}^k,\ f : X \to \mathbb{R} \text{ is convex and l.s.c., } \arg\min_X(f) \text{ is a singleton.}$$

Indeed, by replacing f by $x \to f(x + x_0) - f(x_0)$, one may assume, without loss of generality, that $f(0) = 0 < f(x), x \neq 0$. Let $< x_n >$ be a minimizing sequence. We claim that $< x_n >$ is bounded. Indeed, if we assume the contrary that $\|x_n\| \to \infty$ for some subsequence, then by convexity of f,

$$0 \leq f\left(\frac{x_n}{\|x_n\|}\right) \leq \frac{1}{\|x_n\|} f(x_n) \to 0.$$

Again, for a subsequence, $\frac{x_n}{\|x_n\|} \to y$ with $\|y\| = 1$. However, the lower semicontinuity of f gives $0 < f(y) \leq \liminf f\left(\frac{x_n}{\|x_n\|}\right) = 0$, which is a contradiction. The proofs of the remaining parts are left to the reader. □

In what follows, we need the following definition and the next theorem.

Definition 6 Let K be a nonempty convex subset of a normed space. Recall that a function $f : K \to \mathbb{R}$ is said to be *uniformly quasi-convex* if there exists a forcing function $c : [0, +\infty) \to [0, +\infty)$ such that

$$f(\alpha x + (1-\alpha)y) \leq \max\{f(x), f(y)\} - c(\|x - y\|), \quad \forall x, y \in K \text{ and } \alpha \in (0, 1). \tag{1}$$

Theorem 13 [15] *Let K be a nonempty closed and convex subset of a Banach space X,and $f : K \to \mathbb{R}$ be lower semicontinuous, bounded below, and uniformly quasi-convex. Then problem (K, f) is T.w.p.*

Proof Let $< x_n >$ be a minimizing sequence for (K, f). Then

$$v_K(f) \leq f\left(\frac{x_n + x_m}{2}\right) \leq \max\{f(x_n), f(x_m)\} - c(\|x_n - x_m\|).$$

Let $\epsilon > 0$ be given. Pick $N \in \mathbb{N}$ such that

$$f(x_n) < v_K(f) + \epsilon,\ f(x_m) < v_K(f) + \epsilon, \quad \forall n, m \geq N.$$

Hence, $v_K(f) < v_K(f) + \epsilon - c(\|x_n - x_m\|) \Rightarrow c(\|x_n - x_m\|) < \epsilon$. This implies $<x_n>$ is Cauchy, and if $x_n \to x_0$, by lower semicontinuity of f, $x_0 \in \arg\min_K(f)$. Uniform quasi-convexity of f entails $\arg\min_K(f) = \{x_0\}$. □

To motivate the idea of Tikhonov regularization, let us begin with the following interesting example.

Example 1 [15] Let $X := U(L^2(0, 1))$ be equipped with the strong convergence. Given $u \in X$, let x_u denote the unique absolutely continuous solution of the IVP:

$$\dot{x} = u, \text{ a.e. in } (0, 1), \quad x(0) = 0.$$

Let

$$f(u) = \int_0^1 x_u^2(t)\mathrm{d}t, \quad g(u) = \int_0^1 u^2(t)\mathrm{d}t. \tag{2}$$

Then (X, f) is not T.w.p.; but $(X, f + \epsilon g)$ is T.w.p. for every $\epsilon > 0$. This follows from the fact that $u_n(t) = \frac{\sin nt}{\|\sin nt\|}$ is a minimizing sequence for (X, f), which does not converge to 0 in X.

Let us consider problem (K, f) where $K \subset X$, a Banach space and $f : X \to \mathbb{R}$ are such that (K, f) is *Tikhonov ill-posed*. Our aim here is to explore a strongly convergent minimizing sequence for (K, f) by approximately solving *appropriate perturbations* of (K, f) by adding to f a small *regularizing term*. This procedure originally due to Tikhonov [47] is robust since only approximate knowledge of f is all that is required.

Fix up sequences $\alpha_n > 0, \epsilon_n \geq 0$ such that $\alpha_n \to 0$ and $\epsilon_n \to 0$. For regularizing (K, f), we add to f a small nonnegative *uniformly convex* term $\alpha_n g$ defined on the whole of X. So, g satisfies:

$$g(\alpha x + (1-\alpha)y) \leq \alpha g(x) + (1-\alpha)g(y) - c(\|x - y\|) \tag{3}$$

for all $x, y \in X, \alpha \in (0, 1)$ and for some forcing function $c : [0, +\infty) \to [0, +\infty)$.

The following theorem is an extension of Theorem 5 in Levitin-Polyak [27] apparently due to Dontchev and Zolezzi [15].

Theorem 14 *Let X be a Banach space, $f : X \to \mathbb{R}$ be w-sequentially l.s.c., $K \subset X$ be nonempty w-compact, and $g : X \to [0, +\infty)$ be l.s.c. and uniformly convex. Let $\alpha_n > 0$, $\epsilon_n \geq 0$ be given sequences of numbers such that $\alpha_n \to 0$ and $\epsilon_n \to 0$. Then the following conclusions hold:*

(a) *If f and K are both convex, then problem $(K, f + \alpha g)$ is T.w.p. for every $\alpha > 0$.*
(b) *If $u_n \in \epsilon_n - \arg\min_K(f + \alpha_n g)$, $n \in \mathbb{N}$, then $< u_n >$ is a minimizing sequence for $(K, f) : f(u_n) \to v_K(f)$.*
Also, if $\frac{\epsilon_n}{\alpha_n} \to 0$, then we have:
(c) $\emptyset \neq \limsup_n[\epsilon_n - \arg\min_K(f + \alpha_n g)] \subset \arg\min_{\arg\min_K(f)}(g)$.
Furthermore, if K and f are both convex, then we have:
(d) *$\arg\min_{\arg\min_K(f)}(g)$ is a singleton and denoting this set by $\{\tilde{u}\}$, we have $u_n \to \tilde{u}$ if $u_n \in \epsilon_n - \arg\min_K(f + \alpha_n g)$, $n \in \mathbb{N}$.*

Proof (a) Note that since f is w-l.s.c. and K is w-compact, $\arg\min_K(f) \neq \emptyset$. Also, since f is convex and g is uniformly convex, $f + \alpha g$ is uniformly convex. By Theorem 13, $(K, f + \alpha g)$ is T.w.p.

(b) Let $v \in \arg\min_K(f)$ and $u_n \in \arg\min_K(f + \alpha_n g)$, $n \in \mathbb{N}$. Then

$$\begin{aligned} f(u_n) + \alpha_n g(u_n) &\leq v_K(f + \alpha_n g) + \epsilon_n \\ &\leq f(v) + \alpha_n g(v) + \epsilon_n \\ &\leq f(u_n) + \alpha_n g(v) + \epsilon_n. \end{aligned}$$

This gives

$$g(u_n) \leq g(v) + \frac{\epsilon_n}{\alpha_n}. \tag{4}$$

Also, $f(v) \leq f(u_n) \leq f(v) + \alpha_n[g(v) - g(u_n)] + \epsilon_n \leq f(v) + \alpha_n g(v) + \epsilon_n$. This implies $\lim_n f(u_n) = f(v) = v_K(f)$, i.e., $< u_n >$ is a minimizing sequence for problem (K, f). Next, let $\frac{\epsilon_n}{\alpha_n} \to 0$.

(c) By w-compactness of K, we may assume by passing to a subsequence that $u_n \xrightarrow{w} \tilde{u} \in K$. Note that since g is convex and l.s.c., it is w-l.s.c. Therefore, by (4),

$$g(\tilde{u}) \leq \liminf_n g(u_n) \leq g(v).$$

This shows that $\tilde{u} \in \arg\min_{\arg\min_K(f)}(g)$. We claim that $u_n \xrightarrow{\|\cdot\|} \tilde{u}$. Indeed, by uniform convexity of g, $c(\|u_n - \tilde{u}\|) \leq \frac{1}{2}g(u_n) + \frac{1}{2}g(\tilde{u}) - g\left(\frac{u_n+\tilde{u}}{2}\right)$. This implies

$$c\left(\|u_n - \tilde{u}\|\right) \leq \frac{1}{2}\left[g(v) + \frac{\epsilon_n}{\alpha_n}\right] + \frac{1}{2}g(\tilde{u}) - g\left(\frac{u_n + \tilde{u}}{2}\right)$$

This implies

$$\limsup_n c(\|u_n - \tilde{u}\|) \leq g(\tilde{u}) - \liminf_n g\left(\frac{u_n + \tilde{u}}{2}\right) \leq 0,$$

since g is w-l.s.c. This implies $u_n \xrightarrow{\|\cdot\|} \tilde{u}$, which proves (c).

(d) The assumptions K and f are convex $\Rightarrow \arg\min_K(f)$ is convex. This implies $\arg\min_{\arg\min_K(f)}(g)$ is a singleton, since g is uniformly convex. □

5 Viscosity Solutions

Viscosity methods provide a very effective approach for tackling many global minimization problems arising in variational analysis and optimization (cf., for example, [4, 15, 16, 26, 46]). In various problems originating in the classical calculus of variations, viscosity method was also called *elliptic regularization*. It is convenient to begin with the following abstract framework which seems to have been first perfected in [4].

An Abstract Setting

- Let X be an arbitrary set to be equipped with a suitable topology τ.
- Let $f : X \to \mathbb{R} \cup \{+\infty\}$ be a given extended real-valued function whose definition may include some constraints.
- Consider the minimization problem

(P) $$\min\{f(x) : x \in X\}.$$

- Given $g : X \to \mathbb{R}^+ \cup \{+\infty\}$ called the *viscosity function* and a sequence $\epsilon_n \subset \mathbb{R}^+$ convergent to 0, consider the sequence of perturbed minimization problems:

(P$_n$) $$\min\{f(x) + \epsilon_n g(x) : x \in X\}.$$

- It is assumed that for each $n \in \mathbb{N}$, there exists a solution u_n of (P$_n$).

Our central goal here is to study the convergence of the sequence $<u_n>$ and to characterize its limit. To this end, we use the notion of variational convergence of functions called *epi-convergence* also called Γ*-convergence* given below.

5.1 Epi-convergence

Given a topological space (X, τ) and functions $< f, f_n : X \to \mathbb{R} \cup \{+\infty\}, n \in \mathbb{N} >$, the sequence $< f_n >$ is said to *epi-converge* to f, written $\tau - \text{epi} - \lim_{n\to\infty} f_n = f$, if for each $x \in X$, we have:

(i) There exists $< x_n >$ which is τ-convergent to x for which

$$\limsup_{n\to\infty} f_n(x_n) \leq f(x);$$

(ii) Whenever $< x_n >$ is τ-convergent to x, we have

$$f(x) \leq \liminf_{n\to\infty} f_n(x_n).$$

For convenience, we simply write $f_n \stackrel{\text{epi}}{\to} f$, whenever f_n epi-converges to f.

Remark 2 Recall that in case X is a first countable topological space and the sequence $<A_n>_{n\in\mathbb{N}} \subset CL(X)$, we define

$$\liminf_n A_n = \{x \in X : \exists <x_n> \subset <A_n>,\ \text{such that } x_n \to x\},$$

and

$$\limsup_n A_n = \{x \in X : \exists <x_{nk}> \subset <A_{nk}>,\ \text{such that } x_{nk} \to x\}.$$

The sequence $<A_n>$ is said to *converge* to A in $CL(X)$ in the Painlevé–Kuratowski sense, written, $A_n \stackrel{P-K}{\to} A$ if

$$A = \limsup_n A_n = \liminf_n A_n.$$

Given $<f, f_n : X \to \mathbb{R} \cup \{+\infty\}, n \in \mathbb{N}>$, a sequence of extended real-valued functions, we note that:

$$f_n \overset{\text{epi}}{\to} f \Leftrightarrow \text{epi}(f_n) \overset{P-K}{\to} \text{epi}(f).$$

We may also recall that the Painlevé-Kuratowski convergence of sequence of nonempty closed subsets of X (assumed first countable) is compatible with the Fell topology [17] τ_F of $CL(X)$ generated by the families V^-, $V \subset X$ open and $(K^c)^+, K$ nonempty and compact.

Remark 3 In case X is a reflexive Banach space and we use for τ the strong topology (the topology of the norm $\|.\|$) in (i) and the weak topology w in (ii), then the sequence $<f_n>$ is said to converge to f in the *Mosco sense* [32], written: $f_n \overset{M}{\to} f$.

Epi-convergence of Monotone Sequences

In general, there is no compatibility between *epi-convergence* and *pointwise convergence*. However, an important case frequently encountered in applications, where, these two notions coincide (up to *some closure operations*) is this case of monotone sequences:

(i) If

$$f_1 \le f_2 \le \cdots \le f_n \le \cdots,$$

then

$$\tau - \text{epi} - \lim_{n\to\infty} f_n = \sup_{n\in\mathbb{N}}(\text{lsc}_\tau f_n);$$

(ii) If

$$f_1 \ge f_2 \ge \cdots \ge f_n \ge \cdots,$$

then

$$\tau - \text{epi} - \lim_{n\to\infty} f_n = \text{lsc}_\tau(\inf_n f_n).$$

Here $\text{lsc}_\tau f$ denotes the lower semicontinuous regularization of f. It is partly due to this reason that the *monotone approximation schemes and viscosity methods* are popular tools in variational analysis and optimization.

The next theorem clarifies the epi-convergence approach.

Theorem 15 *Let us be given a sequence* $< \epsilon_n > \subset \mathbb{R}^+$ *such that* $\epsilon_n \to 0$ *and a sequence of minimization problems*

$$(\text{P}_\text{n}) \qquad \min\{f_n(x) : x \in X\}.$$

Assume that there exists a topology τ *on* X *such that:*

(i) *For every* $n \in \mathbb{N}$, *there exists an* ϵ_n*-approximate solution* u_n *to* (P_n), $u_n \in \epsilon_n - \arg\min_X(f_n)$, $n \in \mathbb{N}$, *such that the sequence* $< u_n : n \in \mathbb{N} >$ *is* τ*-relatively compact;*

(ii) $f = \tau - \text{epi} - \lim_{n\to\infty} f_n$.

Then

$$\lim_{n\to\infty} v_X(f_n) = v_X(f),$$

and every τ-cluster point $\hat{u}$ of $< u_n >$ minimizes f on X, i.e., $\hat{u} \in \arg\min_X(f)$.

Proof Since $u_n \in \epsilon_n - \arg\min_X(f_n)$, $n \in \mathbb{N}$, we have

$$f_n(u_n) \le v_X(f_n) + \epsilon_n.$$

By assumption (i) and property (ii) of epi-convergence, we may assume, without loss of generality, that $u_n \to \tilde{u}$ and that

$$v_X(f) \le f(\tilde{u}) \le \liminf_n f_n(u_n) \le \liminf_n v_X(f_n). \tag{5}$$

Also, by property (i) of epi-convergence, we have, for each $x \in X$, there is a sequence $<x_n> \subset X$ such that $f_n(x_n) \to f(x)$. Thus

$$\limsup_n v_X(f_n) \le \limsup_n f_n(x_n) \le f(x). \tag{6}$$

Thus, $\limsup_n v_X(f_n) \le v_X(f)$, and (5) completes the proof. □

The next theorem is a modified version of [4, Theorem 2.1].

Theorem 16 *Let $f : X \to \mathbb{R} \cup \{+\infty\}$ be a given function which is proper and bounded below. Consider the associated minimization problem:*

(P) $$\min\{f(x); x \in X\}.$$

Assume that the following conditions hold.

(1) *Let $<\epsilon_n>$, $<\alpha_n>$, $n \in \mathbb{N}$ be given sequences in $\mathbb{R}^+$, $\epsilon_n \neq 0$, such that $\epsilon_n \to 0$, $\alpha_n \to 0$, and letting $\beta_n := \frac{\alpha_n}{\epsilon_n}$, $\beta_n \to 0$.*
(2) *Let us be given a function $g : X \to \mathbb{R}^+$(called the viscosity function), and for each $n \in \mathbb{N}$,*
consider the perturbed minimization problem

(P_n) $$\min\{f(x) + \epsilon_n g(x) : x \in X\}.$$

(3) *Assume that there exists an α_n-approximate solution of (P_n) : $u_n \in \alpha_n - \arg\min_X(f_n)$, $n \in \mathbb{N}$, where $f_n := f + \epsilon_n g$ such that for some topology τ on X, we have:*

(i) *The sequence $<u_n>_{n\in\mathbb{N}}$ is τ-relatively compact;*
(ii) *The functions f and g are both τ-l.s.c.*

Then every τ-cluster point $\hat{u}$ of $<u_n>$ minimizes f on X and satisfies for all $v \in \arg\min_X(f)$,

(a) *the so-called viscosity selection criterion*

$$(VSC) \quad \hat{u} \in \arg\min_X(f),\ g(\hat{u}) \leq g(v) \Leftrightarrow \hat{u} \in \arg\min_{\arg\min_X(f)}(g).$$

Moreover, the sequence $<u_n>$ is a minimizing sequence for problem (P):

$$\lim_{n\to\infty} \frac{1}{\epsilon_n}(f(u_n) - v_X(f)) = 0, \tag{7}$$

(b) *and also*

$$\lim_{n\to\infty} g(u_n) = v_{\arg\min_X(f)}(g). \tag{8}$$

Proof Since g is a nonnegative finite-valued function, the sequence of functions $f_n = f + \epsilon_n g$ is pointwise convergent and monotonically decreases to f. This implies that

$$f_n \overset{\text{epi}}{\to} \text{lsc}_\tau f = f.$$

Since $u_n \in \alpha_n - \arg\min(f_n)$, and $\alpha_n \to 0$, by epi-convergence, if a subsequence $<u_{nk}>$ of $<u_n>$ converges to $\hat{u}$, then

$$\arg\min_X(f_n) \to \arg\min_X(f) \text{ and } \hat{u} \in \arg\min_X(f).$$

Next, we *rescale* the minimization problem (P_n). Let

$$h_n = \frac{1}{\epsilon_n}[f_n - v_X(f)] = \frac{1}{\epsilon_n}[f - v_X(f)] + g(x).$$

Note that

$$v_X(h_n) = \frac{1}{\epsilon_n}(v_X(f_n) - v_X(f)),$$

and

$$u_n \in \alpha_n - \arg\min_X(f_n) \quad \Rightarrow \quad f_n(u_n) \leq v_X(f_n) + \alpha_n.$$

This implies

$$h_n(u_n) = \frac{1}{\epsilon_n}(f_n(u_n) - v_X(f)) \leq v_X(h_n) + \beta_n,$$

which implies that $u_n \in \beta_n - \arg\min_X(h_n)$.

Next, note that since $f - v_X(f)$ is a nonnegative function, h_n monotonically increases to the function $h := g + \delta_{\arg\min_X(f)}$. Hence, $h_n \overset{epi}{\to} h$, and by variational properties of epi-convergence,

$$v_X(h_n) \to v_X(h), \;\; \hat{u} \in \arg\min_X(h) = \arg\min_{\arg\min_X(f)}(g).$$

Also note that, we have

$$h_n(u_n) = \frac{1}{\epsilon_n}(f_n(u_n) - v_X(f)) = \frac{1}{\epsilon_n}(f(u_n) - v_X(f)) + g(u_n).$$

Hence,

$$h_n(u_n) \le v_X(h_n) + \beta_n \le h_n(v) + \beta_n$$

This gives

$$h_n(u_n) = \left(\frac{1}{\epsilon_n}\right)(f(v) - v_X(f)) + g(v) + \beta_n, \quad \forall v \in X. \tag{9}$$

Thus, $g(u_n) \le g(v) + \beta_n, \forall v \in \arg\min_X(f)$ which implies that

$$\limsup_n g(u_n) \le g(v) \quad \Rightarrow \quad \limsup_n g(u_n) \le g(\hat{u}).$$

By τ-lower semicontinuity of g, we have

$$g(\hat{u}) \le \liminf_k g(u_{n(k)}) \le \limsup_k g(u_{n(k)}) \le g(\hat{u}).$$

This implies

$$\lim_k g(u_{n(k)}) = g(\hat{u}) = v_{\arg\min_X(f)}(g).$$

This being true for any extracted subsequence, $g(u_n)$ converges:

$$\lim_{n\to\infty} g(u_n) = v_{\arg\min_X(f)}(g). \tag{10}$$

Let us again go back to (9). Using (10), and taking $v = \hat{u}$ in (9), we obtain

$$\limsup_{n\to\infty} \frac{1}{\epsilon_n}(f(u_n) - v_X(f)) + g(\hat{u}) \le g(\hat{u}).$$

This implies $\lim_{n\to\infty} \dfrac{(f(u_n) - V_X(f))}{\epsilon_n} = 0$, and the proof is complete. □

Definition 7 Following [4, Definition 2.2], we call a solution $\hat{u}$ of the minimization problem (P) as in the last theorem, a *viscosity solution* corresponding to the viscosity function $g : X \to \mathbb{R}^+ \cup \{+\infty\}$ if $\hat{u}$ is a minimizer of g over $\arg\min_X(f)$.

Following [4], we consider below a natural extension of the preceding theorem where g takes on infinite values. Here, we need to assume that all the required information can be recovered just from the knowledge of f on the effective domain $\mathrm{dom}(g)$ of g.

Theorem 17 *Let the hypothesis of the previous theorem be all fulfilled except that we take the viscosity function* $g : X \to \mathbb{R}^+ \cup \{+\infty\}$ *to be extended real-valued, and that in place of*

(ii) f, g *are* τ*-l.s.c., we assume*
(ii)′ $lsc_\tau(f + \delta_{\mathrm{dom}(g)}) = \mathrm{lsc}_\tau(f)$.

Here δ_A *denotes the indicator function of A. Then every* τ*-cluster point* $\hat{u}$ *of* $< u_n >$ *minimizes the function* $lsc_\tau f$ *on* X *and satisfies*

$$\lim_{n\to\infty} f(u_n) = v_X(f).$$

Furthermore, assume that the following hypotheses hold:

(iii) g *is* τ*-l.s.c.*;
(iv) $\mathrm{lsc}_\tau(f + \epsilon_n g) = \mathrm{lsc}_\tau f + \epsilon_n g, \forall n \in \mathbb{N}$;
(v) $\mathrm{dom}(g) \cap \arg\min_X(\mathrm{lsc}_\tau f) \neq \emptyset$.

Then $\hat{u} \in \arg\min_X(\mathrm{lsc}_\tau f)$ *satisfies the following viscosity selection criterion*

$$\text{(VSC)} \qquad \hat{u} \in \arg \min_{\arg\min_X(\mathrm{lsc}_\tau f)} (g).$$

Proof We imitate the proof of the previous theorem. Indeed, note that the sequence $<f_n>$ is pointwise convergent and monotonically decreases to $f + \delta_{\mathrm{dom}(g)}$. Thus, by assumption (ii),

$$f_n \stackrel{\mathrm{epi}}{\to} \mathrm{lsc}_\tau(f + \delta_{\mathrm{dom}(g)}) = \mathrm{lsc}_\tau(f).$$

By variational properties of epi-convergence, $\hat{u} \in \arg\min_X(\mathrm{lsc}_\tau(f))$. As in Theorem 16, the rescaled minimization problem is

$$\min\{h_n(x) : x \in X\}, h_n = \frac{1}{\epsilon_n}(f_n - v_X(\mathrm{lsc}_\tau f)) = \frac{1}{\epsilon_n}[f - v_X(\mathrm{lsc}_\tau f)] + g.$$

By hypothesis (iv), this can be expressed as

$$h_n = \frac{1}{\epsilon_n}[\mathrm{lsc}_\tau f - v_X(\mathrm{lsc}_\tau f)] + g,$$

which monotonically increases to the function $h := g + \delta_{\arg\min_X(\mathrm{lsc}_\tau f)}$. Hence,

$$h_n \overset{\text{epi}}{\to} h = g + \delta_{\arg\min_X(\text{lsc}_\tau f)}.$$

Using variational properties of epi-convergence, we conclude that $\hat{u}$ minimizes g on $\arg\min_X(\text{lsc}_\tau f)$. The hypothesis (v) ensures that the function on the right-hand side is proper. □

5.2 Hierarchial Minimization

As before, given a function $f : X \to \mathbb{R} \cup \{+\infty\}$, consider the minimization problem

$$\text{(P)} \qquad \min\{f(x) : x \in X\}$$

which is approximated by the sequence

$$\text{(P}_\text{n}\text{)} \qquad \min\{f(x) + \epsilon_n g(x) + \epsilon_n^2 h(x) : x \in X\}$$

of minimization problems. Observe that we can write $f(x) + \epsilon_n g(x) + \epsilon_n^2 h(x) = f(x) + \epsilon_n g_n(x)$, where $g_n(x) = g(x) + \epsilon_n h(x)$. Clearly, it would be desirable if $g_n \overset{\text{epi}}{\to} g$. To this end, we have:

Lemma 4 (Attouch) [4] *Let* $\{\alpha_n : X \to \mathbb{R} \cup \{+\infty\} : n \in \mathbb{N}\}$ *and* $\{\beta_n : X \to \mathbb{R} \cup \{+\infty\} : n \in \mathbb{N}\}$ *be two sequences of functions that are converging both pointwise as well as* τ*-epi-convergence sense, respectively to some functions* g *and* h*. Then* $(\alpha_n + \beta_n) \overset{\text{epi}}{\to} (g + h)$.

The next theorem extends Theorem 16 to the situation as mentioned above, where g is replaced by a sequence g_n.

Theorem 18 *Let* $f : X \to \mathbb{R} \cup \{+\infty\}$ *be a given function which is proper and bounded below. Consider the associated minimization problem:*

$$\text{(P)} \qquad \min\{f(x); x \in X\}.$$

Assume that the following conditions hold.

(1) *Let* $<\epsilon_n>, <\alpha_n>, n \in \mathbb{N}$ *be given sequences in* $\mathbb{R}^+$, $\epsilon_n \neq 0$, *such that* $\epsilon_n \to 0$, $\alpha_n \to 0$, *and letting* $\beta_n := \frac{\alpha_n}{\epsilon_n}$, $\beta_n \to 0$.
(2) *Let us be given functions* $g, g_n : X \to \mathbb{R}^+, n \in \mathbb{N}$, *that are nonnegative and finite-valued.*
(3) *For each* $n \in \mathbb{N}$, *consider the perturbed minimization problem*

$$\text{(P}_\text{n}\text{)} \qquad \min\{f(x) + \epsilon_n g_n(x) : x \in X\}.$$

(4) *Assume that there exists an* α_n*-approximate solution of* (P_n) : $u_n \in \alpha_n - \arg\min_X(f_n), n \in \mathbb{N}$, *where* $f_n := f + \epsilon_n g_n$ *such that for some topology* τ *on* X, *we have:*

(i) *The sequence* $<u_n>_{n\in\mathbb{N}}$ *is* τ*-relatively compact;*
(ii) *The function* f *is* τ*-l.s.c.;*
(iii) *The sequence* $<g_n>$ *converges to* g *both pointwise as well as in* τ*-epi-convergence sense.*

Then every τ*-cluster point* $\hat{u}$ *of* $<u_n>$ *minimizes* f *on* X *and satisfies for all* $v \in \arg\min_X(f)$,

(a) *the so-called viscosity selection criterion*

$$\text{(VSC)} \qquad \hat{u} \in \arg\min_X(f),\ g(\hat{u}) \le g(v) \Leftrightarrow \hat{u} \in \arg\min_{\arg\min_X(f)}(g).$$

Moreover, the sequence $<u_n>$ *is a minimizing sequence for problem* (P):

$$\lim_{n\to\infty} \frac{1}{\epsilon_n}(f(u_n) - v_X(f)) = 0, \tag{11}$$

(b) *and also*

$$\lim_{n\to\infty} g_n(u_n) = v_{\arg\min_X(f)}(g). \tag{12}$$

Proof The proof is a straight forward imitation of the proof of Theorem 17. We need to apply the preceding lemma to the sequences $\{f + \epsilon_n g_n\}$ and $\{\frac{1}{\epsilon_n}(f - v_X(f)) + g_n : n \in \mathbb{N}\}$, noting that the sequence $\epsilon_n g_n \to 0$ both in the pointwise as well as in the τ-epi-convergence sense and that the sequence $\{\frac{1}{\epsilon_n}(f - v_X(f))\}$ converges both in the pointwise as well as in the τ-epi-convergence sense to $\delta_{\arg\min_X(f)}$. □

For the sake of completeness, we will merely state here the following theorem, whose proof is left to the reader.

Theorem 19 *Let* $f_0 : X \to \mathbb{R} \cup \{+\infty\}$ *be a given function which is proper and bounded below. Consider the associated minimization problem:*

$$\text{(P)} \qquad \min\{f_0(x); x \in X\}.$$

Let $<\epsilon_n>, <\alpha_n>, n \in \mathbb{N}$ *be given sequences in* $\mathbb{R}^+, \epsilon_n \neq 0$, *such that* $\epsilon_n \to 0, \alpha_n \to 0$, *and letting* $\beta_n := \frac{\alpha_n}{\epsilon_n}$, $\beta_n \to 0$. *Let* $f_1, f_2 : X \to \mathbb{R}^+$ *be given functions. Consider the sequence of approximate minimization problems*

$$(\mathrm{P_n}) \qquad \min\{f_0(x) + \epsilon_n f_1(x) + {\epsilon_n}^2 f_2(x) : x \in X\}$$

Let us denote $M_0 = \arg\min_X(f_0)$, $M_1 = \arg\min_{M_0}(f_1)$, $M_2 = \arg\min_{M_1}(f_2)$. *Assume that there exists an* α_n*-approximate solution of* (P_n) : $u_n \in \alpha_n -$

$\arg\min_X(f_n)$, $n \in \mathbb{N}$, *where* $f_n := f_0 + \epsilon_n f_1 + {\epsilon_n}^2 f_2$ *such that for some topology* τ *on* X, *we have:*

(i) *the sequence* $<u_n>$ *is* τ*-relatively compact:*
(ii) f_0, f_1, f_2 *are* τ*-l.s.c.*

Then every τ*-cluster point* $\hat{u}$ *of the sequence* $< u_n >$ *belongs to* M_1, $\lim_n f_0(u_n) = v_X(f_0)$, *and* $\lim_n f_1(u_n) = v_{M_0}(f_1)$. *In addition, if we assume*

$$\liminf_{n\to\infty} \frac{1}{\epsilon_n}(f_1(u_n) - v_{M_0}(f_1)) \geq 0,$$

then we have $\hat{u} \in M_2$.

6 Convex Minimization and the Viscosity Approach

In what follows X will be either a reflexive Banach space or the normed dual E^* of a separable normed linear space E. We will denote the weak (resp. weak*) topology of X by w (resp.w^*). Recall that a function $g : X \to \mathbb{R}$ is called *coercive* if $\lim_{\|x\|\to+\infty} g(x) = +\infty$. The next result is a modified version of [4, Theorem 5.1].

Theorem 20 *Let* X *be a reflexive Banach space (resp. the dual* E^* *of a separable normed space* E*). Consider the minimization problem:*

$$\text{(P)} \qquad \min\{f(x) : x \in X\}$$

where $f : X \to \mathbb{R} \cup \{+\infty\}$ *is a proper, convex, l.s.c. (resp.* w^**-l.s.c.) function which is bounded below. Let* $<\alpha_n> \subset \mathbb{R}^+$ *and* $<\epsilon_n> \subset \mathbb{R}^+\backslash\{0\}$ *be sequences such that* $\alpha_n \to 0, \epsilon_n \to 0$ *and* $\beta_n = \frac{\alpha_n}{\epsilon_n} \to 0$. *Let* $g : X \to \mathbb{R}^+$ *be convex, l.s.c.(resp.* w^**-l.s.c.) and coercive. For each* $n \in \mathbb{N}$, *consider the perturbed minimization problem:*

$$(\mathrm{P_n}) \qquad \min\{f_n(x) : x \in X\}, \qquad f_n := f + \epsilon_n g,$$

and consider a sequence $<u_n>$ *of* α_n*-approximate solutions:* $u_n \in \alpha_n-\arg\min_X(f_n)$ *of* (P_n). *We have:*

- *The sequence* $<u_n>$ *is bounded if and only if* $\arg\min_X(f) \neq \emptyset$.
  In that case, every w*-(resp.* w^**-) cluster point* $\hat{u}$ *of the sequence* $<u_n>$ *minimizes* f *on* X *and satisfies the viscosity selection criterion:*
-

$$\text{(VSC)} \qquad \hat{u} \in \arg\min_{\arg\min_X(f)}(g).$$

Moreover, the sequence $<u_n>$ *is a minimizing sequence of problem* (P)*:*

$$\lim_{n\to\infty} \frac{1}{\epsilon_n}(f(u_n) - v_X(f)) = 0,$$

and also,

$$\lim_{n\to\infty} g(u_n) = v_{\arg\min_X(f)}(g).$$

Proof The conclusion follows easily from Theorem 16 by taking the topology τ to be w (resp.w^*). If $<u_n>$ is bounded, its w-(resp. w^*-) cluster point $\hat{u} \in \arg\min_X(f)$.

Conversely, if $\arg\min_X(f) \neq \emptyset$, then as in the proof of Theorem 16,

$$g(u_n) \leq g(v) + \beta_n, \forall v \in \arg\min_X(f).$$

From the coercivity of g we conclude that $<u_n>$ is bounded. □

Let X be a Hilbert space and take $g(x) = \frac{1}{2}\|x\|^2$. The *Fenchel-Moreau conjugate* of g is:

$$g^* = \sup\{\langle x, y\rangle - \frac{1}{2}\|x\|^2 : x \in X\} = \frac{1}{2}\|y\|^2, y \in X.$$

This gives:

$$y \in \partial g(u_n) \quad \Leftrightarrow \quad g(u_n) + g^*(y) = \langle u_n, y\rangle.$$

(Here $\partial g(u_n)$ denotes the *subdifferential* of g at u_n.) The last step gives

$$\frac{1}{2}\|u_n\|^2 + \frac{1}{2}\|y\|^2 = \langle u_n, y\rangle \quad \Leftrightarrow \quad y = u_n.$$

Let $f : X \to \mathbb{R} \cup \{+\infty\}$ be convex, l.s.c., proper function, and let $f_n := f + \epsilon_n g$. Then

$$z \in \partial f_n(u_n) \quad \Leftrightarrow \quad 0 \in \partial(f_n - \langle ., z\rangle)(u_n).$$

Since the function $f_n - \langle ., z\rangle = (f - \langle ., z\rangle) + \epsilon_n g$ is strictly convex, $\arg\min_X(f_n - \langle ., z\rangle)$ is a singleton. Given a sequence $<\epsilon_n> \subset \mathbb{R}^+$ such that $\epsilon_n \to 0$, observing that

$$\partial f_n(u_n) = \partial f(u_n) + \epsilon_n u_n, n \in \mathbb{N},$$

the previous theorem leads us to the next corollary.

Corollary 1 [4] *Let X be a Hilbert space and $f : X \to \mathbb{R} \cup \{+\infty\}$ be a proper convex function which is l.s.c. Let us be given a sequence $<\epsilon_n>_{n\in\mathbb{N}} \subset \mathbb{R}^+$ such that $\epsilon_n \to 0$. Fix $z \in X$. Then for each $n \in \mathbb{N}$, there exists a unique solution u_n of the equation:*

$$z \in \partial f(u_n) + \epsilon_n u_n. \tag{13}$$

The sequence $<u_n>$ remains bounded if and only if the set $(\partial f)^{-1}z$ is nonvoid. In that case

$$\lim_{n\to\infty} u_n = \mathrm{proj}_{(\partial f)^{-1}z}(0) = \hat{u},$$

where $\hat{u}$ is the unique element of the minimum norm of $(\partial f)^{-1}z$.

Proof Let us first note that $(\partial f)^{-1}z = \arg\min_X(f - \langle ., z\rangle)$. The previous theorem yields (for a subsequence) $u_n \overset{w}{\to} \hat{u}$ and $g(u_n) \to g(\hat{u})$, i.e., $\|u_n\| \to \|\hat{u}\|$. Since norm in a Hilbert space is a Kadec norm, we conclude that $u_n \to \hat{u}$ in the norm. This completes the proof. □

6.1 A Revisit to Tikhonov Regularization

Under the same hypothesis as in Theorem 20: X is a reflexive Banach space (resp. the dual E^* of a separable normed space E), $f : X \to \mathbb{R} \cup \{+\infty\}$ a proper convex, l.s.c. (resp. w^*-l.s.c.) function which is bounded below. Assume problem $(P)(= (X, f))$ is *Tikhonov ill-posed*. Let us be given a function $g : X \to \mathbb{R}^+$ which is uniformly convex:

$$g(\alpha x + (1-\alpha)y) \le \alpha g(x) + (1-\alpha)g(y) - c(\|x-y\|), \quad \forall x, y \in X, \alpha \in (0,1)$$

for some *forcing function* $c : [0, +\infty) \to [0, +\infty)$ and a sequence $<\epsilon_n> \subset \mathbb{R}^+$ such that $\epsilon_n \to 0$.

Consider, following the standard Tikhonov regularization, the sequence of perturbed problems

$$(\mathrm{P_n}) \qquad \min\{f(x) + \epsilon_n g(x) : x \in X\}, \quad n \in \mathbb{N}.$$

In addition, assume that g is l.s.c. (resp. w^*-l.s.c.), bounded below and coercive. Since the function $f_n := f + \epsilon_n g$ is strictly convex, l.s.c., bounded below and coercive, problem (P_n) is Tikhonov well-posed for every $n \in \mathbb{N}$. Let u_n denote the unique solution of (P_n) for each $n \in \mathbb{N}$. Theorem 20 reveals that:

(i) The sequence $<u_n>$ is bounded $\Leftrightarrow \arg\min_X(f) \ne \emptyset$;
(ii) The sequence $<u_n>$ is a minimizing sequence of problem (P);
(iii) If $\arg\min_X(f) \ne \emptyset$, then $\arg\min_{\arg\min_K(f)}(g)$ is a singleton and denoting this set $\{\hat{u}\}$ we have

$$u_n \to \hat{u}.$$

This covers the standard Tikhonov regularization in which case X is a Hilbert space and $g(x) = \frac{1}{2}\|x\|^2$.

6.2 Regularization of Ill-Posed Operator Equations

Let X and Y be Hilbert spaces and $T : X \to Y$ be a bounded linear operator. Given $y \in Y$, consider the operator equation

$$Tx = y, \tag{14}$$

The Eq. (14) is called *well-posed*, if for every $y \in Y$, there exists a unique solution $x \in X$ which depends continuously on the *data* y; otherwise, it is called *ill-posed*.

Let us recall that if T is a compact operator of infinite rank, then its range $\mathscr{R}(T)$ is not closed. Hence, a solution of (14) may not exist for every $y \in Y$. Even if unique solution exists for some $y \in Y$, it need not depend continuously on the data y. (Recall the result: *If $T : X \to Y$ is an injective compact operator, then $T^{-1} : \mathscr{R}(T) \to X$ is bounded if and only if T is of finite rank.*) In this case one looks for a *least-residual norm (LRN)* solution which is the least squares solution of the convex quadratic minimization problem:

$$\text{(P)} \qquad \min\{\|Tx - y\|^2 : x \in X\}.$$

The *Euler equation* characterizing a solution of (P) is

$$T^*T(x) = T^*(y),$$

whose solution exists if and only if the following *compatibility condition* holds:

$$y \in \mathscr{R}(T) + \mathscr{R}(T)^{\perp}.$$

Tikhonov regularization overcomes the lack of stability of problem (P). Given a sequence $<\epsilon_n> \subset \mathbb{R}^+$ such that $\epsilon_n \to 0$, the perturbed problem

$$\text{(P}_\text{n}\text{)} \qquad \min\{\|Tx - y\|^2 + \epsilon_n \|x\|^2 : x \in X\}$$

is T.w.p. for each $n \in \mathbb{N}$, with a unique solution u_n. The sequence $<u_n>$ remains bounded in X if and only if problem (P) has a solution, in which case this sequence norm converges to an element $\hat{u}$ in X, which is the unique element of minimum norm in $\arg\min_X(f)$:

$$\|\hat{u}\| \le \|v\|, \quad \forall v \in \arg\min_X(f),$$

Here, $f(x) = \|Tx - y\|^2, x \in X$.

References

1. Amir, D.: Best simultaneous approximation (Chebyshev Centers). In: ISNM 72, pp. 19–35. Birkhauser-Verlag, Basel (1984)
2. Amir, D., Ziegler, Z.: Relative Chebyshev centers in normed linear spaces. J. Approx. Theory **29**, 235–252 (1980)
3. Attouch, H.: Variational Convergence for Functions and Operators. Pitman, Boston (1984)
4. Attouch, H.: Viscosity solutions of minimization problems. SIAM J. Optim. **6**, 769–806 (1996)
5. Attouch, H., Buttazzo, G., Michaille, G.: Variational Analysis in Sobolev and BV Spaces. MPS-SIAM Series on Optimization. SIAM, Philadelphia (2006)
6. Bednarczuk, E., Penot, J.P.: Metrically well-set minimization problems. Appl. Math. Optim. **26**, 273–285 (1992)
7. Beer, G.: On Mosco convergence of convex sets. Bull. Austral. Math. Soc. **38**, 239–253 (1988)
8. Beer, G.: On the Young-Fenchel transform for convex functions. Proc. Amer. Math. Soc. **104**, 1115–1123 (1988)
9. Beer, G.: Topologies on Closed and Closed Convex Sets. Kluwer Academic Publishers, Netherlands (1993)
10. Beer, G., Pai, D.V.: On convergence of convex sets and relative Chebyshev centers. J. Approx. Theory **62**, 147–179 (1990)
11. Beer, G., Pai, D.V.: The prox map. J. Math. Anal. Appl. **156**, 428–443 (1991)
12. Bensoussan, A., Kenneth, P.: Sur l'analogie entre les méthods de régularisation et de pénalisation. Rev. Inform. Recherche Opeér. **13**, 13–26 (1969)
13. Borwein, J.M., Fitzpatrick, S.: Existence of nearest points in Banach spaces. Can. J. Math. **41**, 702–720 (1989)
14. Crandall, M.G., Lions, P.L.: Viscosity solutions of Hamilton-Jacobi equations. J. Amer. Math. Soc. **277**, 1–42 (1983)
15. Dontchev, A.L., Zolezzi, T.: Well-Posed Optimization Problems, Lecture Notes in Mathematics, No. 1543. Springer, Berlin (1993)
16. Ekeland, I., Temam, R.: Analyse Convexe et Problèmes Variationnels. Dunod/Gauthiers-Villars, Paris (1974)
17. Fell, J.: A Hausdorff topology for the closed subsets of a locally compact non-Hausdorff space. Proc. Amer. Math. Soc. **13**, 472–476 (1962)
18. Golomb, M.: On the uniformly best approximation of functions given by incomplete data. M.R.C. Technical Summary Report #121, University of Wisconsin, Madison (1959)
19. Holmes, R.B.: A Course on Optimization and Best Approximation. Lecture Notes No.257. Springer, New York (1972)
20. Indira, K., Pai, D.: Hausdorff strong uniqueness in simultaneous approximation, Part I. In: Chui, C.K., Neamtu, M., Schumaker, L.L. (eds.) Approximation Theory XI : Gatlinburg 2004, pp. 101–118, Nashboro Press, Brentwood, TN, USA (2005)
21. Klein, E., Thompson, A.: Theory of Correspondences. Wiley, Toronto (1984)
22. Kuratowski, K.: Topologie, vol. 1. Panstowowe Wydawnictwo Naukowe, Warszawa (1958)
23. Lambert, J.M., Milman, D.D.: Restricted Chebyshev centers of bounded sets in arbitrary Banach spaces. J. Approx. Theory **26**, 71–78 (1979)
24. Laurent, P.J., Pai, D.V.: On simultaneous approximation. Numer. Funct. Anal. Optimiz. **19**, 1045–1064 (1998)
25. Laurent, P.J., Tuan, P.D.: Global approximation of a compact set by elements of a convex set in a normed space. Numer. Math. **15**, 137–150 (1970)
26. Lemaire, B.: Régularisation et pénalisation en optimisation convexes. Sémi. Anal. Convexe Montpellier 1, expose 17 (1971)
27. Levitin, E.S., Polyak, B.T.: Convergence of minimizing sequences in conditional extremum problems. Soviet Math. Dokl. **7**, 764–767 (1966)
28. Li, C.: Strong uniqueness of restricted Chebyshev centers with respect to RS-set in a Banach space. J. Approx. Theory **135**, 35–53 (2005)

29. Li, C., Lopez, G.: On generic well-posedness of restricted Chebyshev center problems in Banach spaces. Acta Math. Sinica **22**, 741–750 (2006)
30. Li, W.: Strong uniqueness and Lipschitz continuity of metric projections : a generalization of the classical Haar theory. J. Approx. Theory **56**, 164–184 (1989)
31. Mhaskar, H.N., Pai, D.V.: Fundamentals of Approximation Theory. Narosa Publishing House, New Delhi, India and CRC Press, Boca Raton, Florida, USA (2000)
32. Mosco, U.: Convergence of convex sets and solutions of variational inequalities. Adv. Math. **3**, 510–585 (1969)
33. Mosco, U.: On the continuity of the Young-Fenchel transform. J. Math. Anal. Appl. **35**, 518–535 (1971)
34. Pai, D.V.: A characterization of smooth normed linear spaces. J. Approx. Theory **17**, 315–320 (1976)
35. Pai, D.V.: Multioptimum of a convex functional. J. Approx. Theory **19**, 83–99 (1977)
36. Pai, D.V.: Strong uniqueness of best simultaneous approximation. J. Indian Math. Soc. **67**, 201–215 (2000)
37. Pai, D.V.: On well-posedness of some problems in approximation. J. Indian Math. Soc. **70**, 1–16 (2003)
38. Pai, D.V.: Strong unicity of restricted p-centers. Numer. Funct. Anal. Optimiz. **29**, 638–659 (2008)
39. Pai, D.V., Deshpande, B.M.: Topologies related to the prox map and the restricted center map. Numer. Funct. Anal. and Optimiz. **16**, 1211–1231 (1995)
40. Pai, D.V., Deshpande, B.M.: On continuity of the prox map and the restricted center map. In: Chui, C.K., Schumaker, L.L. (eds.) Approximation Theory VIII, Vol. 1: Approximation and Interpolation. World Scientific, New Jersey, 451–458 (1995)
41. Radstrom, H.: An embedding theorem for spaces of convex sets. Proc. Amer. Math. Soc. **3**, 165–169 (1952)
42. Shunmugaraj, P., Pai, D.V.: On approximate minima of convex functional and lower semicontinuity of metric projections. J. Approx. Theory **64**, 25–37 (1991)
43. Shunmugaraj, P., Pai, D.V.: On stability of approximate solutions of minimization problems. Numer. Funct. Anal. Optimiz. **12**, 599–610 (1991)
44. Sonntag, Y.: Convergence au sens de Mosco; théorie et applications a l'approximation des solutions d' inéquations. These d'Etat, Université de Provence, Marseille (1982)
45. Steckin, S.B.: Approximative properties of subsets of Banach spaces. Rev. Roumain Math. Pures Appl. **8**, 5–18 (1963)
46. Temam, R.: Problèmes Mathématiques en Plasticité. Gauthiers-Villars, Paris (1983)
47. Tihkonov, A.N.: Solution of incorrectly formulated problems and the regularization method. Soviet Math. Dokl. **4**, 1035–1038 (1963)
48. Tikhonov, A.N., Arsénine, V.: Méthods de résolution de problèmes mal posé, MIR (1976)
49. Tsukada, M.: Convergence of best approximations in a smooth Banach space. J. Approx. Theory **40**, 301–309 (1984)
50. Vlasov, L.P.: Approximative properties of sets in normed linear spaces. Russian Math. Surveys **28**, 1–66 (1973)

Best Approximation in Nonlinear Functional Analysis

S. P. Singh and M. R. Singh

Abstract An introduction to best approximation theory and fixed point theory are presented. Several known fixed point theorems are given. Ky Fan's best approximation is studied in detail. The study of approximating sequences followed by convergence of the sequence of iterative process is studied. An introduction to variational inequalities is also presented.

Keywords Best approximation theory · Fixed point theory · Ky Fan's best approximation · Iterative process · Variational inequalities · Hartman-Stampacchia theorem

1 Introduction

In this chapter the material organized is as follows. We have introduction, then best approximation theory, followed by fixed point theory. We cover Ky Fan's best approximation and after that the study of approximating sequences followed by convergence of the sequence of iterative process. In the end, a list of references is given.

The nonlinear analysis covers areas like fixed point theory, best approximation, variational inequality, complementarity problems, and nonlinear problems arising in economics, engineering, and physical sciences.

Multivalued analog of the present theory is not presented in this chapter. Though this topic is very useful in optimization theory, game theory, and mathematical eco-

S. P. Singh
Department of Mathematics and Statistics, Memorial University,
St. John's, NL A1C 5S7, Canada
e-mail: ssingh@mun.ca

M. R. Singh (✉)
Department of Physics and Astronomy, The University of Western Ontario,
London, ON N6A 3K7, Canada
e-mail: msingh@uwo.ca

Q. H. Ansari (ed.), *Nonlinear Analysis*, Trends in Mathematics,
DOI: 10.1007/978-81-322-1883-8_6,

nomics, Ky Fan's best approximation theory has important applications in fixed point theory, approximation theory, minimax theory, variational inequalities, and complementarity problems. The fixed point theory is a very useful tool in the study of nonlinear problems of mathematics and engineering. Recently, this theory has been applied in biology, chemistry, economics, game theory, optimization theory, and physics. Fixed point theory is mainly used in the study of existence of solutions for nonlinear problems arising in physical and biological sciences and engineering. It plays a very important role in the existence theory of differential equation, integral equations, functional equation, partial differential equations, eigen-value problems, and two-point boundary value problems. The study of variational inequality has become a very powerful tool for solving a wide variety of problems arising in physical sciences including engineering. The other applications are in the area of fluid dynamics, transportation and economic equilibrium problems, free boundary value problems, elasticity problems, and hydrodynamics. The well-known result of variational inequality, due to Hartman and Stampacchia [29], in finite dimensional case, is stated below:

Let C be a compact convex subset of $\mathbb{R}^n$ and $f : C \to \mathbb{R}^n$ a continuous function. Then there exists a point $x \in C$ such that

$$\langle f(x), y - x \rangle \geq 0, \quad \text{for all } y \in C. \tag{1}$$

The Complementarity Problem (CP) provides a unified model for problems arising in game theory, engineering, and mathematical economics.

Let $f : \mathbb{R}^n \to \mathbb{R}^n$ be a continuous function. Then the (CP) [62] is to find a solution of the system

$$y = f(x), \quad x \geq 0, \quad y \geq 0 \quad \text{and} \quad \langle x, y \rangle = 0. \tag{2}$$

In this system x and y are nonnegative vectors in $\mathbb{R}^n$, so either x and y are orthogonal $\langle x, y \rangle = 0$, or the component wise product of x and y is the zero vector. The (CP) requires to find a nonnegative vector whose image is also nonnegative and such that the two vectors are orthogonal.

The equilibrium problem (EP) has applications in optimization theory, fixed point theory, and other related areas. It is stated below.

Let X be a real topological vector space, C a closed convex subset of X and $f : C \times C \to \mathbb{R}$ such that $f(x, x) = 0$ for all $x \in C$. Then the *equilibrium problem* is to find an $x_0 \in X$ such that

$$x_0 \in C, \quad f(x_0, y) \geq 0, \quad \text{for all } y \in C. \tag{3}$$

For details see Blum and Oettli [3, 8]. It is worth to mention that the optimization problems, variational inequalities, minimax problems, complementarity problems, and fixed point problems are particular cases of an equilibrium problem.

In Sect. 5, on convergence of approximating sequences with applications, we give the following: If a sequence of contraction maps $\{f_n\}$ with a sequence of fixed points

$\{x_n\}$ is given and the sequence $\{f_n\}$ converges to a function f, then we discuss the convergence of $\{x_n\}$ to a fixed point of f.

In case the limit function f is a nonexpansive map, then it is also given in detail.

In the last section, detailed discussion on the convergence of the sequence of the iterative process is given. The tools available in fixed point theory on the convergence of iterative process can be applied to the study of variational inequalities and approximation theory.

For example, if C_1 and C_2 are closed convex sets in Hilbert space H and P_1, P_2 are projection operators, then a fixed point of composition of $P_1 \circ P_2$ is a point of C_1 nearest C_2 [18].

If C_1 and C_2 are closed convex sets in Hilbert space H and g the composition $P_1 \circ P_2$ of their projection operators. Let $x \in C_1$ be arbitrary. Then convergence of $\{g^n x\}$ to a fixed point of g is guaranteed if one set is finite dimensional and the distance between the sets is attained [18].

The following result is in variational inequality [42].

Let C be a closed convex subset of Hilbert space H and $f : C \to H$ a continuous function such that $I - rf$ is a contraction function (r is a constant). Then there exists a unique solution $u \in C$ of

$$\langle fu, v - u\rangle \geq 0, \quad \text{for all } v \in C, \tag{4}$$

and $u = \lim_{n\to\infty} u_n$, where $u_{n+1} = P(I - rf)u_n$, $u_0 \in C$. P is the proximity map on C.

2 Theory of Best Approximation

Let C be a subset of a Banach space X and let $x \in X$, $x \notin C$. We define the distance between x and C by $d(x, C) = \inf\{\|x - z\| : z \in C\}$. The problem of *best approximation* is to find an element $y \in C$ such that

$$\|y - x\| = d(x, C).$$

The element $y \in C$ is said to be a *nearest point* or a *closest point* or an *element of best approximation* to $x \in X$.

In other words, an element $y \in C$ is called an *element of best approximation* to x, if

$$\|x - y\| = d(x, C) = \inf\{\|x - z\| : z \in C\}.$$

Let Px denote the set of all points in C closest to x, that is,

$$Px = \{y \in C : \|x - y\| = d(x, C)\}.$$

If Px is nonempty for each $x \in X$, then the set C is said to be *proximinal*. If Px is singleton for each $x \in X$, then C is said to be *Chebyshev set*. Thus, if C is a Chebyshev set, then P is a single-valued map.

The mapping $P : X \to 2^C$ is a multivalued map and it is called the *metric projection* or the *best approximation operator* [17, 18, 62, 65].

We have the following properties of a proximinal set C:

(i) A proximinal set is always closed.
(ii) A compact set is always a proximinal set.
(iii) A closed convex subset of a Hilbert space is a proximinal set.
(iv) A closed convex subset of a reflexive Banach space is a proximinal set.

Remark 1 (a) If C is a compact set of a Banach space X, then the problem of best approximation is considered as an optimization problem. In this case, the projection map P is a continuous function and attains its minimum on set C.
(b) If X is a finite dimensional normed linear space and C is a closed, bounded subset of X, then, for each point $x \in X$, there is a unique nearest point in C.

The geometry of spaces plays a key role in the theory of projection operators.

Definition 1 A normed linear space X is said to be *strictly convex* if $\|x\| \le r$ and $\|y\| \le r$ imply that $\|x + y\| < 2r$, unless $x = y$. In other words, if $\|x + y\| = \|x\| + \|y\|$ for all $x, y \in X$, then $x = ry$, for $r > 0$.

Definition 2 A Banach space is called *uniformly convex* if for all $\varepsilon > 0$, there is a $\delta > 0$ such that

$$\|x\| = \|y\| = 1 \text{ and } \|x - y\| \ge \varepsilon \quad \Rightarrow \quad \|(x + y)/2\| \le 1 - \delta.$$

The strict convexity is a sufficient condition for the uniqueness of the best approximation. Indeed, let y and z be two distinct elements nearest to x. Then $\|x - y\| = \|x - z\| = d(x, C) = d$. Since $\|x - y\|$ and $\|x - z\|$ are distinct, strict convexity implies that $\|x - y + x - z\| < 2d$, that is, $\|x - (z + y)/2\| < d$, a contradiction to the fact that d is the infimum. Hence $y = z$ and the nearest point are unique.

In case X is not strictly convex, then the nearest point need not be unique. The following example illustrates the fact. In case the norm on a finite dimensional normed linear space X is not the Euclidean, then the nearest point need not be unique. For example, let $X = \mathbb{R}^2$ with norm $\|(x, y)\| = \max(\|x\|, \|y\|)$. Let C be the set of all point $(a, 0)$ with $x = (2, 1)$. Then $d\,((2, 1), (a, 0)) = \max\,(\|2 - a\|, \|1 - 0\|)$, which takes a minimum value 1 for all $(a, 0)$ such that $1 < a < 3$.

This example suggests that the uniqueness question may have a different answer if the norm is replaced by an equivalent norm. If we take Euclidean norm then there is a unique nearest point.

Remark 2 A Hilbert space is the uniformly convex Banach space and so strictly convex.

Definition 3 Let C be a nonempty set of a normed linear space X. A sequence $\{x_n\}$ in C is called a *minimizing sequence* for $x \in X$, if $\{\|x - x_n\|\}$ converges to $d(x, C)$.

If a minimizing sequence converges weakly to $y \in C$, then y is closest to x.

Definition 4 [56] A set C in a normed linear space X is called an *approximatively compact* if each minimizing sequence has a convergent subsequence.

If C is an approximatively compact subset of a normed linear space X, then

(i) C is a set of existence, that is, each point $x \notin C$ has a nearest point in C,
(ii) C is closed.
(iii) C is P-compact.

If C is an approximatively compact subset of X and Px is a singleton for some $x \in X$, then every minimizing sequence for x converges to Px.

A compact set is approximatively compact, but not conversely. A closed ball of a uniformly convex Banach space is approximatively compact, but it is not compact.

A closed convex subset C of an uniformly convex Banach space X is approximatively compact. If C is an approximatively compact set in a Banach space X, then for each $x \in X$, $Px = \{y \in C : \|x - y\| = d(x, C)\}$ is nonempty and the map $P : X \to 2^C$ is upper semicontinuous.

Let C be a nonempty, approximatively compact subset of a Banach space X and $P : X \to 2^C$ be a metric projection of X onto C. Then, for $x \in X$, we have the following:

(i) $P(x)$ is nonempty,
(ii) $P(x)$ is compact,
(iii) $P(x)$ is convex if C is convex,
(iv) $P(A)$ is compact for every compact set A of C.

The following result holds in Hilbert space.

Theorem 1 *If C is a nonempty closed convex subset of a Hilbert space H, then each $x \in H$ has a unique nearest point in C.*

Proof Let $d = \inf\{\|x - z\| : z \in C\}$. Then we take a sequence of points $\{y_n\}$ in C such that $\{\|y_n - x\|\}$ converges to d. Now, by using the parallelogram law, we show that $\{y_n\}$ is a Cauchy sequence. For $n, m \in \mathbb{N}$, we have

$$2\|y_n - x\|^2 + 2\|y_m - x\|^2 = \|y_n - y_m\|^2 + \|y_n + y_m - 2x\|^2,$$

it follows that

$$\|y_n - y_m\|^2 = 2\|y_n - x\|^2 + 2\|y_m - x\|^2 - 4\|(y_n + y_m)/2 - x\|^2.$$

Since C is a convex set, $(y_n + y_m)/2 \in C$. Then $d \leq \|(y_n + y_m)/2 - x\|$ and hence

$$\|y_n - y_m\|^2 \leq 2\|y_n - x\|^2 + 2\|y_m - x\|^2 - 4d^2 \to 0, \text{ as } n \text{ and } m \to \infty.$$

Hence, $\{y_n\}$ is a Cauchy sequence. Since C is a closed subset of Hilbert space H, therefore $\{y_n\}$ converges to $y \in C$. Hence, $\|y_n - x\| \to d = \|y - x\|$. It is easy to show that y is a unique nearest point to $x \in H$. □

Let C be a closed convex subset of a Hilbert space H. Then, for each $x \in H$, there is a unique closest point $y \in C$, called its *projection* Px. Thus,

$$\|x - y\| = \|x - Px\| = d(x, C),$$

and

$$\|x - Px\| \leq \|x - z\| \text{ for all } z \in C.$$

Here P is called the *proximity map* of H onto C.

The proximity map P has the following properties:

(i) $Px = x$ for all $x \in C$;
(ii) $\langle x - Px, Px - Py \rangle \geq 0$ for all x and $y \in H$;
(iii) P is nonexpansive, i.e, $\|Px - Py\| \leq \|x - y\|$ for all $x, y \in H$;
(iv) $\|x - Px\|^2 + \|Px - y\|^2 \leq \|x - y\|^2$ for all $y \in C$.

Remark 3 If H is a Hilbert space and $f : H \to H$ is a nonexpansive map, then $I - f$ satisfies $\langle (I - f)x - (I - f)y, x - y \rangle \geq 0$ for all $x, y \in H$, that is, $I - f$ is monotone, where I is the identity map on H.

To show this, let $g = I - f$ and $x, y \in H$. Then $g : H \to H$. Since f is nonexpansive, we have $\|fx - fy\| \leq \|x - y\|$, and

$$\|fx - fy\|^2 \leq \|x - y\|^2 + \|gx - gy\|^2.$$

Note

$$\begin{aligned}\|(I - g)x - (I - g)y\|^2 &= \|(x - y) - (gx - gy)\|^2 \\ &= \|x - y\|^2 + \|gx - gy\|^2 - 2\langle gx - gy, x - y \rangle \\ &\leq \|x - y\|^2 + \|gx - gy\|^2,\end{aligned}$$

only if $\langle gx - gy, x - y \rangle \geq 0$. Therefore, $g = I - f$ is monotone.

Theorem 2 *Let C be a nonempty convex subset of a Hilbert space H. Then a point $x \in H$ $(x \notin C)$ has a closest element $y \in C$ if and only if $\langle x - y, y - z \rangle \geq 0$ for all $z \in C$.*

Proof Let $x \in H$ with $x \notin C$. Assume that $y \in C$ is nearest to x and take any $z \in C$. Since C is convex, $\alpha z + (1 - \alpha)y \in C$ for all $\alpha \in [0, 1]$. Note

$$0 \leq \|x - (\alpha z + (1 - \alpha)y)\|^2 - \|x - y\|^2 = \alpha^2 \|y - z\|^2 + 2\alpha \langle x - y, y - z \rangle.$$

If $\langle x - y, y - z \rangle < 0$, then for small α we find that the right hand side is < 0, a contradiction since left hand side is positive. Hence, $\langle x - y, y - z \rangle \geq 0$ for all $z \in C$.

On the other hand, if $\langle x - y, y - z\rangle \geq 0$, then

$$\|x - z\|^2 - \|x - y\|^2 = \|y - z\|^2 + 2\langle x - y, y - z\rangle \geq 0 \text{ for all } z \in C.$$

Consequently, $\|x - y\|^2 \leq \|x - z\|^2$, and hence $\|x - y\| = \inf\{\|x - z\| : z \in C\}$. This means that $x \in H$ has a closest element $y \in C$. □

We briefly introduce the variational inequalities [4, 10, 16, 35, 62].

Let $f : [a, b] \to \mathbb{R}$ be a differentiable function with continuous derivative. Then one tries to solve min fx for $x \in [a, b]$. Let $x_0 \in [a, b]$ such that

$$fx_0 = \min\{fx : x \in [a, b]\}.$$

Then x_0 is called the solution of the minimization problem. It is easy to see that $f'x_0 = 0$ for all $x \in (a, b)$, $f'x_0 \geq 0$ if $x_0 = a$, and $f'x_0 \leq 0$ if $x_0 = b$. Thus,

$$f'x_0(x - x_0) \geq 0, \quad \text{for all } x \in [a, b].$$

Inequality of this type is termed as the *variational inequality* (VI). The variational inequalities are useful in the study of calculus of variations and in general, in the study of optimization problems like minimize f on a given domain and thus we need to solve $f'x = 0$.

In general, $f'x = 0$ may not hold. For example, if we have $f : [2, 4] \to \mathbb{R}$ given by $fx = x$, then $f(2) = 2$ is the minimum but $f'2 = 1 \neq 0$. However, the inequality of the type $\langle f'x, y - x\rangle \geq 0$ holds for all $y \in [2, 4]$. If we replace $f'x$ by a continuous function say gx, then $\langle gx, y - x\rangle \geq 0$ for all y, is called the variational inequality. The theory of variational inequality has many applications in nonlinear functional analysis like fixed point theory and best approximation. There are also applications to the study of mathematical economics, engineering, and applied mathematics.

The study of variational inequality has become a very powerful tool for solving a wide range of problems arising in physical sciences including engineering. The other applications to the area of fluid dynamics, transportation, and economic equilibrium problems, free boundary value problems, and hydrodynamics are well known.

3 Fixed Point Theory

The fixed point theory is a very useful tool in the study of nonlinear problems of mathematics, engineering, and other physical sciences. Recently fixed point theorems have been applied in biology, chemistry, economics, game theory, optimization theory, and physics. This theory is useful in the study of existence of solutions of nonlinear problems arising in physical, biological, and social sciences and in problems of mathematical economics, optimization theory, and game theory. The main application is to find the solution of the nonlinear problems arising in mathematics, physics, social sciences, economics, engineering, and other sciences. The applications in the existence theory of differential equations, integral equations, partial differential equations, functional equations, eigen-value problems, periodic solutions to the Navier-Stokes equations, and two-point boundary value problems are well known.

Definition 5 Let X be a nonempty set. If $f : X \to X$ is a function such that $fy = y$ for $y \in X$, then y is said to be a *fixed point* of f.

Fixed point theory is useful in determining zeros of polynomial equations. A polynomial equation $Px = 0$ can be written as $fx - x = Px = 0$. For example, if $x^2-5x+4 = 0$, where $Px = x^2-5x+4$. We can write $fx-x = Px = x^2-5x+4$, so $x = (x^2 + 4)/5 = fx$. Here f has two fixed points, $f1 = 1$, and $f4 = 4$. Consequently, $Px = 0$ has two zeros $x = 1$ and $x = 4$.

The well-known Brouwer's fixed point theorem [11] is given below:

Theorem 3 (Brouwer's Fixed Point Theorem) (see [1, 10, 22, 27, 44, 62, 65]). *If $f : B \to B$ is a continuous function where B is a closed unit ball in $\mathbb{R}^n$, then f has a fixed point.*

Brouwer fixed point theorem simply guarantees the existence of a solution, but gives no information about the uniqueness and determination of the solution. For example, if $f : [0, 1] \to [0, 1]$ is given by $fx = x^2, x \in [0, 1]$, then $f0 = 0$ and $f1 = 1$, that is, f has two fixed points.

Several mathematicians have given different proofs of this theorem. Effective methods have been developed to approximate the fixed points.

Brouwer fixed point theorem is not true in infinite dimensional spaces. For example, if B is a closed unit ball in an infinite dimensional Hilbert space and $f : B \to B$ is a continuous function, then f does not have a fixed point [33].

The first fixed point theorem in an infinite dimensional Banach space was given by Schauder in 1930 stated below.

Theorem 4 (Schauder Fixed Point Theorem) [54] *If B is a compact, convex subset of a Banach space X and $f : B \to B$ is a continuous function, then f has a fixed point.*

The Schauder fixed point theorem has applications in approximation theory, game theory, and other areas like engineering, economics, and optimization theory. It is natural to prove the theorem by relaxing the condition of compactness. Schauder proved the following theorem.

Theorem 5 (Schauder Fixed Point Theorem) [54] *If B is a closed, bounded convex subset of a Banach space X and $f : B \to B$ is a continuous map such that $f(B)$ is compact, then f has a fixed point.*

A very useful result in fixed point theory is known as the Banach contraction principle [7] (see [1–3, 25, 27, 62] for details). The Banach contraction principle is important as a source of existence and uniqueness theorems in different branches of sciences.

This theorem provides an illustration of the unifying power of functional analytic methods and usefulness of fixed point theory in analysis. The important feature of the Banach contraction principle is that it gives the existence, uniqueness, and the sequence of the successive approximation converges to a solution of the problem. The important aspect of the result is that existence, uniqueness, and determination all are answered by the Banach contraction principle.

Definition 6 A map $f : X \to X$ is said to be a *contraction map* on a metric space X if there exists a constant $k \in (0, 1)$ such that

$$d(fx, fy) \leq kd(x, y), \quad \text{for all } x, y \in X.$$

Every contraction map is a continuous map, but a continuous map need not be a contraction map. For example, $fx = x^2$, $x \in [0, 1]$ is not a contraction map.

Theorem 6 (Banach Contraction Principle) *If X is a complete metric space and $f : X \to X$ is a contraction map, then f has a unique fixed point, or $fx = x$ has a unique solution.*

Proof It is easy to see the uniqueness. If $fx = x$ and $fy = y$ with $x \neq y$, then we get

$$d(x, y) = d(fx, fy) \leq kd(x, y),$$

a contradiction. Hence fixed point of f is unique.

We now define $x_{n+1} = fx_n$ for $n = 0, 1, 2, 3, \ldots$, starting with any $x_0 \in X$. Here $x_1 = fx_0$ and $x_2 = fx_1$ and so on. Note

$$\begin{aligned} d(x_2, x_1) &= d(fx_1, fx_0) \leq kd(x_1, x_0), \\ d(x_3, x_2) &= d(fx_2, fx_1) \\ &\leq kd(x_2, x_1) \\ &\leq k^2 d(x_1, x_0). \end{aligned}$$

Inductively, we have

$$d(x_{n+1}, x_n) \leq k^n (x_1, x_0).$$

We now show that $\{x_n\}$ is a Cauchy sequence. For $m > n$, we have

$$\begin{aligned} d(x_n, x_m) &\leq d(x_n, x_{n+1}) + d(x_{n+1}, x_{n+2}) + \cdots + d(x_{m-1}, x_m) \\ &\leq k^n d(x_1, x_0)[1 + k + k^2 + \cdots + k^{m-n-1}] \\ &\leq k^n d(x_1, x_0)[1 + k + k^2 + \cdots + k^{m-n-1} + \cdots] \\ &\leq k^n d(x_1, x_0)[1/(1-k)] \to 0 \text{ as } n \to \infty, \end{aligned} \tag{5}$$

since $k < 1$. It follows that $\{x_n\}$ is a Cauchy sequence. Since X is a complete metric space, $x_n \to y$ for some $y \in X$. From (5), we have $d(x_{n+1}, x_n) \to 0$ as $n \to \infty$ and hence $\lim_{n\to\infty} x_{n+1} = y$. Since f is continuous,

$$y = \lim_{n\to\infty} x_{n+1} = \lim_{n\to\infty} f x_n = f y. \qquad \square$$

Definition 7 If X is a metric space and $f : X \to X$ is a mapping such that $d(fx, fy) < d(x, y)$ for all $x, y \in X$ with $x \neq y$, then f is said to be a *contractive map*.

A contractive map need not have a fixed point in a complete metric space. For example, if we take $f : \mathbb{R} \to \mathbb{R}$ defined by $fx = x + \pi/2 - \arctan x$, then f is a contractive map but does not have a fixed point. Here $\arctan x < \pi/2$ for all x. In case a contractive map f has a fixed point, then it is always unique. For example, if $x = fx$ and $y = fy$, and $x \neq y$ then $d(x, y) = d(fx, fy) < d(x, y)$, a contradiction. So, $x = y$.

Theorem 7 *If X is a compact metric space and $f : X \to X$ is a contractive map, then f has a unique fixed point.*

Proof Consider $gx = d(x, fx)$. Then g is continuous and has a minimum. Let gx_0 be a minimum. Then $fx_0 = x_0$. If it is not, then, by taking $d(fx_0, ffx_0) < d(x_0, fx_0) = gx_0$ for some x_0 and we get a contradiction to the fact that gx_0 is a minimum. Hence, $fx_0 = x_0$. $\square$

Definition 8 Let X be a metric space and $f : X \to X$ a mapping. Then f is said to be a *nonexpansive* if $d(fx, fy) \leq d(x, y)$ for all $x, y \in X$.

If $f : \mathbb{R} \to \mathbb{R}''$ given by $fx = x + p$, a translation map for some $p \neq 0$, then f is a nonexpansive map, but f has no fixed point.

Theorem 8 [58, 62] *If C is a compact, convex nonempty subset of Banach space X, and $f : C \to C$, is a nonexpansive map, then f has a fixed point.*

Proof Let $0 \in C$. Define a sequence of maps $f_{r_i} = r_i f$, where $0 < r_i < 1, r_i \to 1$ as $i \to \infty$. Then each f_{r_i} is a contraction map, and by the Banach contraction principle, each f_{r_i} has a unique fixed point say x_{r_i}, that is, $f_{r_i} x_{r_i} = x_{r_i}$. Now

$$\|x_{r_i} - f x_{r_i}\| = \|f_{r_i} x_{r_i} - f x_{r_i}\| = \|r_i f x_{r_i} - f x_{r_i}\| = (1 - r_i)\|f x_{r_i}\|.$$

Since C is compact and f is a continuous map so $f(C)$ is compact. By taking limit we have $\|x_{r_i} - f x_{r_i}\| \to 0$ as $i \to \infty$. Let $\{x_{r_i}\}$ converge to $y \in C$, since C is compact so each sequence in C has a convergent subsequence. Then

$$\|| y - f_y || \leq || y - x_{r_i} || + || x_{r_i} - f x_{r_i} || + || f x_{r_i} - f_y ||,$$

which gives $y = fy$. Since $\{x_{r_i}\}$ converges to y, $\{f x_{r_i}\}$ converges to fy and $\|x_{r_i} - f x_{r_i}\| \to 0$. Hence f has a fixed point. $\square$

Remark 4 A nonexpansive map, unlike contraction map, need not have a fixed point and if it has a fixed point, then it may not be unique. For example, if $f : \mathbb{R} \to \mathbb{R}$ given by $fx = x + 10$, then f does not have a fixed point, and if identity function $I : \mathbb{R} \to \mathbb{R}$, is taken then each point of I is a fixed point.

The fixed point theorem for nonexpansive maps, given by Browder [12], Gohde [26] and Kirk [36] independently, is stated below. Here it is given in a Hilbert Space.

Theorem 9 [12, 26, 36] *If B is a closed, convex bounded subset of a Hilbert space H and $f : B \to B$ is a nonexpansive map, then f has a fixed point.*

Browder [13] proved the following results.

Theorem 10 *Suppose H is a real Hilbert space and $f : H \to H$ is a nonexpansive map. Let there be a closed convex bounded subset C of H and $f : C \to C$. Then $F(f)$ is nonempty closed and convex. If $y \in H$, then there is a unique nearest point $x \in F(f)$ and is given by*

$$\langle y - x, x - z\rangle \geq 0, \quad \textit{for all } z \in F(f).$$

Proof It is easy to see that $F(f)$, the set of fixed points of f, is closed set. Let x and y be fixed points in F. If $z = (1 - t)x + ty$, a linear combination of x and y, $t \in [0, 1]$, then $\|fz - x\| + \|fz - y\| \leq \|z - x\| + \|z - y\| = \|x - y\|$. Since H is strictly convex so $fz = z$ and the set $F(f)$ is convex set. Thus, $F(f)$ is closed and convex set. Existence and uniqueness of the closest point $x \in F = F(f)$ to $w \in H$ follows from Theorem 1 and Theorem 2, since $F(f)$ is closed and convex subset of H. □

Theorem 9 has been generalized by many researchers. The demiclosedness property is very useful in the theory of nonexpansive mappings.

Definition 9 Let H be a Hilbert space and C a closed convex subset H. If $f : C \to C$, then f is said to be *demiclosed* when $x_n \to y$ weakly and $fx_n \to z$ strongly, imply that $z = fy$.

Theorem 11 [13] *Let H be a Hilbert space and $f : H \to H$ a nonexpansive map. If $\{x_n\}$ is a sequence in H converging weakly to x and $\{x_n - fx_n\} \to 0$ converging strongly, then x is a fixed point of f.*

Proof Let $g = I - f$. Then g is monotone, that is, $\langle gx - gy, x - y\rangle \geq 0$ for all $x, y \in H$. It is given that $gx_n \to 0$, converges strongly. We claim that $gx = 0$, to get $x = fx$. Let $u \in H$. Then $0 \leq \langle gu - gx_n, u - x_n\rangle \to \langle gu, u - x\rangle$, since $x_n - fx_n = gx_n \to 0$. Set $u_t = x + tz$, where $z \in H$ is arbitrary, $t > 0$. Now we have $t\langle gu_t, z\rangle \geq 0$. If we cancel t, then we get $\langle gu_t, z\rangle \geq 0$. If $t \to 0^+$, then $\langle gx, z\rangle \geq 0$. Since this is true for all $z \in H$, so $gx = 0$, that is, $(I - f)x = 0$, and $x = fx$. □

Theorem 11 provides demiclosedness principle for the class of nonexpansive mappings. Here proof is based on theory of monotone operators. Another proof of Theorem 11 can be found in Chap. 7 based on Opial condition.

We now give existence theorem for nonself, nonexpansive mappings, which was is proved in [1].

Theorem 12 [1] *Let C be a closed bounded convex subset of a uniformly convex Banach space X, and $0 \in C$. Let $f : C \to X$ be a nonexpansive map. Then either*

(i) *f has a fixed point, or*
(ii) *there exists an $r \in (0, 1)$ such that $u = rfu$, for $u \in \partial C$ (boundary of C).*

One can see also [5] for maps when domain and range are different.

In the following result, C is not necessarily a bounded set but $f(C)$ is bounded.

Theorem 13 *Let C be a closed convex subset of a uniformly convex Banach space X with $0 \in C$. Let $f : C \to X$ be a nonexpansive map with $f(C)$ bounded. Then one of the following holds:*

(i) *f has a fixed point or*
(ii) *there is a $k \in (0, 1)$ such that $u = kfu$ for $u \in \partial C$ (boundary of C).*

The study of iterated contraction was initiated by Rheinboldt [50] in 1969. The concept of iterated contraction proves to be very useful in the study of certain iterative process and has wide applicability (for further work see [59]).

Definition 10 Let X be a metric space. If $f : X \to X$ is map such that $d(fx, ffx) \leq kd(x, fx)$, for all $x \in X$, $0 \leq k < 1$, then f is said to be an *iterated contraction map*.

In case $d(fx, ffx) < d(x, fx)$, $x \neq fx$, then f is an iterated contractive map.

Remark 5 A contraction map is continuous and is an iterated contraction.

For example, if $y = fx$, then $d(fx, ffx) \leq kd(x, fx)$ is satisfied.

However, converse is not true. for example, if $f : [-1/2, 1/2] \to [-1/2, 1/2]$ is given by $fx = x^2$, then f is an iterated contraction but not a contraction map.

If $f : \mathbb{R} \to \mathbb{R}$, is a mapping defined by $fx = 0$ for $x \in [0, 1/2)$ and $fx = 1$ for $x \in [1/2, 1]$, then f is not continuous at $x = 1/2$, and f is an iterated contraction.

An iterated contraction map may have more than one fixed point. For example, the iterated contraction function $fx = x^2$ on $[0, 1]$ has $f0 = 0$ and $f1 = 1$, two fixed points.

A discontinuous function need not always be an iterated contraction. Let $X = [0, 1]$. Define $f : [0, 1] \to [0, 1]$ by $fx = 1/2$, $x \in [0, 1/2)$, and $fx = 0$, $x \in [1/2, 1]$. Then f is discontinuous and is not an iterated contraction. If we take $x = 1/4$, then $d(fx, ffx) \leq kd(x, fx)$, $0 \leq k < 1$, is not satisfied and hence f is not an iterated contraction.

If $f : [0, 1] \to [0, 1]$ is defined by $fx = 0$ for $x \in [0, 1/2)$, and $fx = \frac{1}{2}$ for $x \in [1/2, 1]$, then f is discontinuous at $x = 1/2$ and f is an iterated contraction map and has a fixed point $f(1/2) = 1/2$.

Theorem 14 *If $f : X \to X$ is an iterated contraction map, and X is a complete metric space, then the sequence of iterates $\{x_n\}$ converges to $y \in X$. In case f is continuous at y, then $y = fy$, that is, f has a fixed point.*

Proof Let $x_{n+1} = fx_n$, $n = 1, 2, \ldots$ for $x_1 \in X$. It is easy to show that $\{x_n\}$ is a Cauchy sequence, since f is an iterated contraction. The Cauchy sequence $\{x_n\}$ converges to $y \in X$, since X is a complete metric space. Moreover, if f is continuous at y, then $\{x_n\}$ converges to fy. It then follows that $y = fy$. □

Remark 6 A continuous iterated contraction map on a complete metric space has a unique fixed point.

We give the following example to show that if f is an iterated contraction that is not continuous, then f may not have a fixed point.

Let $f : \mathbb{R} \to \mathbb{R}$, be a mapping defined by $fx = x/5 + 1/5$ for $x \leq 0$, and $fx = x/5$ for $x > 0$. Then f does not have a fixed point. Here f is discontinuous at $x = 0$.

Continuity of an iterated contraction map is sufficient but not necessary to have fixed point. For example, if $f : [0, 1] \to [0, 1]$ is defined by $fx = 0$ on $[0, \frac{1}{2})$, and $fx = 2/3$, for $x \in [1/2, 1]$. Then f is an iterated contraction and $f0 = 0$, $f(2/3) = 2/3$, but f is not continuous.

The following is a fixed point theorem for iterated contractive map.

Theorem 15 *If $f : X \to X$ is a continuous iterated contractive map and the sequence of iterates $\{x_n\}$ defined by $x_{n+1} = fx_n$, $n = 1, 2, \ldots$ for $x_1 \in X$, has a subsequence converging to $y \in X$, then f has a fixed point.*

Proof Note $x_{n+1} = fx_n$, $n = 1, 2, \ldots$. Then the sequence $\{d(x_{n+1}, x_n)\}$ is a non-increasing sequence. It is bounded below by 0, and therefore has a limit. Since the subsequence converges to y and f is continuous on X, so $f(x_{n_i})$ converges to fy and $f(f(x_{n_i})$ converges to $f(fy)$. Thus,

$$d(fy, y) = \lim_{n\to\infty} d(x_{n_i}, x_{n_i+1}) = \lim_{n\to\infty} d(x_{n_i+1}, x_{n_i+2}) = d(ffy, fy).$$

If $y \neq fy$, then $d(ffy, fy) < d(fy, y)$, since f is an iterated contractive map. Consequently, $d(fy, y) = d(ffy, fy) < d(fy, y)$, a contradiction and $fy = y$. □

Theorem 16 *If $f : C \to C$ is a continuous iterated contraction , where C is closed subset of a metric space X, with $f(C)$ is compact, then f has a fixed point.*

Remark 7 It is shown that the sequence $\{x_n\}$ has a convergent subsequence. By using iterated contraction and continuity of f we get that f has a fixed point as in Theorem 16.

Definition 11 Let X be a metric space and $f : X \to X$. Then f is said to be an *iterated nonexpansive map* if $d(fx, ffx) \leq d(x, fx)$ for all $x \in X$.

The following is a fixed point theorem for the iterated nonexpansive map.

Theorem 17 *Let X be a metric space and $f : X \to X$ an iterated nonexpansive map satisfying the following :*

(i) *if $x \neq fx$, then $d(ffx, fx) < d(fx, x)$,*
(ii) *for some $x_1 \in X$, the sequence of iterates $x_{n+1} = fx_n$ has a convergent subsequence converging to y say and f is continuous at y.*

Then f has a fixed point.

Proof It is easy to show that the sequence $\{d(x_{n+1}, x_n)\}$ is a nonincreasing sequence of positive reals bounded below by 0. The sequence has a limit. Hence

$$d(fy, y) = \lim_{n \to \infty} d(x_{n_i}, x_{n_i+1}) = \lim_{n \to \infty} d(x_{n_i+1}, x_{n_i+2}) = d(ffy, fy).$$

This is a contradiction to (i). Therefore, f has a fixed point, that is, $fy = y$. □

Noticing that if C is a compact subset of a metric space X and $f : C \to C$ is a continuous iterated nonexpansive map satisfying condition (i) of the above theorem, then f has a fixed point.

Remark 8 If C is compact, then condition (ii) of Theorem 17 is satisfied, as f is continuous iterated nonexpansive map, and hence the result.

The result given below is due to Cheney and Goldstein [18].

Theorem 18 *Let f be a map of a metric space X into itself such that*

(i) *f is a nonexpansive map on X, that is, $d(fx, fy) \leq d(x, y)$ for all $x, y \in X$,*
(ii) *if $x \neq fx$, then $d(fx, ffx) < d(x, fx)$, and*
(iii) *the sequence $x_{n+1} = f(x_n)$ has a convergent subsequence converging to y say.*

Then the sequence $\{x_n\}$ converges to a fixed point of f.

Kannan [34] considered the following map.

Definition 12 Let X be a metric space. Let $f : X \to X$ satisfy

$$d(fx, fy) \leq k[d(x, fx) + d(y, fy)]$$

for all $x, y \in X$, where $0 \leq k < 1/2$. Then f is said to be a *Kannan map*.

A Kannan map is an iterated contraction map. For example, if $y = fx$, then we get $d(fx, ffx) \leq k[d(x, fx) + d(fx, ffx)]$. This gives $d(fx, ffx) \leq k/(1 - k)d(x, fx)$, where $0 \leq k/(1 - k) < 1$, that is, f is an iterated contraction.

The following result due to Kannan [34] is valid for iterated contraction map.

Theorem 19 *Let X be a metric space. Let $f : X \to X$ satisfy $d(fx, fy) \leq k[d(x, fx) + d(y, fy)]$ for all $x, y \in X$, $0 \leq k < 1/2$, f continuous on X and let the sequence of iterates $\{x_n\}$ have a subsequence $\{x_{n_i}\}$ converging to y. Then f has a fixed point.*

Several researchers have used fixed point theory to prove results in approximation theory.

For example, let $f : X \to X$, where X is a normed linear space, C a closed convex subset of X, $f : C \to C$, and $fy = y$, $y \notin C$. If the set D of best $C-$approximation to y is nonempty closed convex and $f : D \to D$ is nonexpansive map with $f(D)$ compact, then f has a fixed point which is best approximation to y [58]. (see also [57, 62]). In this direction, Hicks and Humphries have shown that if $f(\partial C) \subseteq C$, then $f : D \to D$ [30].

4 Ky Fan's Best Approximation

The well-known best approximation theorem of Ky Fan has been of great importance in nonlinear analysis, approximation theory, minimax theory, game theory, fixed point theory, and variational inequalities.

Let C be a nonempty subset of a normed linear space X and $f : C \to X$. We seek a point $x \in C$ which is a best approximation for fx, that is, seek an $x \in C$ such that

$$\|x - fx\| = d(fx, C) = \inf\{\|fx - y\| : y \in C\}. \tag{6}$$

Several researchers have contributed to this field [16, 48, 49, 61].

Remark 9 We know that y is a solution of (6) if and only if y is a fixed point of $P \circ f$ (composition of P and f), where P is the metric projection onto C.

We state the Ky Fan's best approximation theorem below.

Theorem 20 (Fan's Best Approximation Theorem) [22] *Let C be a nonempty compact convex subset of a normed linear space X and $f : C \to X$ a continuous function. Then there exists a $y \in C$ such that $\|y - fy\| = d(fy, C)$.*

Reich [49] proved the following where compactness of C has been relaxed.

Theorem 21 [49] *Let C be a closed convex and nonempty subset of Banach space X such that the metric projection on C is upper semicontinuous. If $f : C \to X$ is continuous and $f(C)$ is relatively compact, then there exists a $y \in C$ such that $\|y - fy\| = d(fy, C)$.*

Proof Let P be the metric projection on C. Define $F(x) = (P \circ f)(x)$ for each $x \in C$. Then F is upper semicontinuous and $F(x)$ is nonempty, compact convex subset of C for $x \in C$. Since $f(C)$ is relatively compact, so $F(C)$ is also relatively compact because the image of a compact set under an upper semicontinuous map with compact point images is compact. The result follows from Himmelberg's theorem. □

The Himmelberg theorem is stated below [31].

If C is a nonempty convex subset of a locally convex Hausdorff topological vector space E and $F : C \to C$ is an upper semicontinuous multifunction with nonempty closed, convex values with $F(C)$ is contained in a compact set of C, then F has a fixed point.

The following result given in $\mathbb{R}^n$ is known as the Fan's best approximation theorem [22, 23]:

Theorem 22 (Fan's Best Approximation Theorem) *Let C be a closed bounded convex subset of* $\mathbb{R}^n$ *and* $f : C \to \mathbb{R}^n$ *a continuous function. Then there is a* $y \in C$ *such that*

$$\|y - fy\| = d(fy, C), \tag{7}$$

where $d(x, C) = \inf\{\|x - y\| \textit{ for all } y \in C\}$, $x \in \mathbb{R}^n$, $x \notin C$.

Proof We note that y is a solution of (7) if and only if y is a fixed point of $P \circ f$, where P is the metric projection from $\mathbb{R}^n$ onto C. Then P is a continuous function. Thus, $P \circ f : C \to C$ is a continuous function and has a fixed point in C by Brouwer fixed point theorem. On the other hand, if $P \circ f$ has a fixed point say, $(P \circ f)y = y$, $y \in C$, then $\|y - fy\| = d(fy, C)$. □

Theorem 22 has application in fixed point theory [1, 12, 30, 57, 61]. The following is the Brouwer fixed point theorem.

Theorem 23 *Let* $f : C \to C$ *be a continuous function, where C is a closed bounded convex subset of* $\mathbb{R}^n$. *Then* f *has a fixed point.*

In this case $d(fy, C) = 0$ and therefore, $fy = y$. In case $f : C \to \mathbb{R}^n$ is a continuous function and C is a closed bounded convex subset of $\mathbb{R}^n$, then f has a fixed point provided additional condition $f(\partial C) \subset C$ is satisfied.

Theorem 22 has application in approximation theory. For example, a closed convex bounded subset of $\mathbb{R}^n$ is a set of existence, that is, for each $x \in \mathbb{R}^n$, $x \notin C$, there is a $y \in C$ such that $\|y - x\| = d(x, C)$. We define $f : C \to \mathbb{R}^n$ by $fy = x$ for all $y \in C$, then $\|y - fy\| = d(fy, C)$, that is, $\|y - x\| = d(x, C)$.

Remark 10 $\|y - x\| = d(x, C)$ if and only if $\langle x - y, y - z\rangle \geq 0$ for all $z \in C$.

Theorem 22 is also applicable in deriving results of variational inequalities [29].

Theorem 24 *If C is a closed bounded and convex subset of* $\mathbb{R}^n$, *and* $f : C \to \mathbb{R}^n$ *is a continuous function, then there is a* $y \in C$ *such that* $\langle fy, x - y\rangle \geq 0$ *for all* $x \in C$.

Proof Let $g = I - f$. Then $g : C \to \mathbb{R}^n$, is a continuous function and by Theorem 22, we have $\|y - gy\| = d(gy, C)$. Thus $\|y - gy\| \leq \|gy - x\|$ for all $x \in C$, that is, $\langle gy - y, y - x\rangle \geq 0$. Hence, $\langle fy, x - y\rangle \geq 0$ for all $x \in C$. □

Theorem 25 [61, 62] *Let C be a closed convex subset of a Hilbert space H and $f : C \to H$ a nonexpansive map with $f(C)$ bounded. Then there is a $y \in C$ such that $\|y - fy\| = d(fy, C)$.*

It is easy to see that $P \circ f$ has a fixed point $y \in C$ and the result follows [13], where $P : H \to C$ is a metric projection, and is a nonexpansive map.

Remark 11 If in Theorem 25, $f : C \to C$, then f has a fixed point.

The following theorem was proved independently by Browder [12], Gohde [26] and Kirk [36] in the Banach space setting. Here it is derived as a corollary from Theorem 25 in Hilbert space.

Theorem 26 *If C is a closed bounded convex subset of a Hilbert space H and $f : C \to C$ is a nonexpansive map, then f has a fixed point.*

In case we take $B_r = \{x \in H : \|x\| \leq r\}$, a ball of radius r and center 0, in place of C in Theorem 21, then the following result holds [1]:

(A) Let $f : B_r \to H$ be a nonexpansive map. Then there is a $y \in B_r$ such that $\|y - fy\| = d(fy, B_r)$. If $f : B_r \to B_r$, then f has a fixed point.
(B) If in (A) we have one of the following additional boundary conditions, then f has a fixed point:

For $y \in \partial B_r$,

(i) $\|fy\| \leq \|y\|$
(ii) $\langle y, fy\rangle \leq \|y\|^2$
(iii) $\|fy\| \leq \|y - fy\|$
(iv) If $fy = ky$, then $k \leq 1$
(v) $\|fy\|^2 \leq \|y\|^2 + \|y - fy\|^2$.

For example, if $\langle y, fy\rangle \leq \|y\|^2$, then $\|fy\| \leq \|y\| = r$, implies that $fy \in B_r$. Hence, f has a fixed point.

Hartman and Stampacchia [29] proved the following interesting result of variational inequalities, that is applicable in mathematical, physical and economic problems.

Theorem 27 (Hartman–Stampacchia Theorem) *Let C be a compact convex subset of $\mathbb{R}^n$ and $f : C \to \mathbb{R}^n$ a continuous function. Then the following problem has a solution:*

(VIP) *find $y \in C$ such that $\langle fy, x - y\rangle \geq 0$, for all $x \in C$.*

The study of variational inequality is said to find a solution of the variational inequality problem (VIP).

Remark 12 The variational inequality problem (VIP) has a solution if and only if $P(I - f) : C \to C$ has a fixed point [16, 60].

The following is given in Hilbert space.

Theorem 28 *Let C be a closed bounded convex subset of H and $f : C \to H$ a monotone, continuous map. Then there is a $y \in C$ such that $\langle fy, x - y\rangle \geq 0$, for all $x \in C$.*

Recall that $f : C \to H$ is monotone if $\langle fx - fy, x - y\rangle \geq 0$ for all $x, y \in C$. An application of Theorem 28 is to prove the following [12].

Theorem 29 *Let C be a closed convex bounded subset of a Hilbert space H and $f : C \to C$ a nonexpansive map. Then f has a fixed point.*

Proof Since f is a nonexpansive map, it is continuous. Consider $g : C \to H$, where $g = I - f$. Then g is a continuous map. It is easy to see that g is monotone, that is, $\langle gx - gy, x - y\rangle \geq 0$. Since g is continuous and monotone, therefore, by Theorem 28, there is a $y \in C$ such that $\langle gy, x - y\rangle \geq 0$ for all $x \in C$, that is,

$$\langle y - fy, x - y\rangle \geq 0, \quad \text{for all } x \in C.$$

Since $f : C \to C$, so by taking $x = fy$ we get that

$$\langle y - fy, fy - y\rangle \geq 0,$$

that is,

$$\langle y - fy, y - fy\rangle \leq 0.$$

But $\|y - fy\|^2 \geq 0$, therefore, $y = fy$. □

The following theorem is an application of Theorem 28 in approximation theory [29, 58].

Theorem 30 *Let C be a closed convex bounded subset of a Hilbert space H. Then for each $y \notin C$, there is an $x \in C$ such that $\|x - y\| = d(y, C)$.*

The following result for a closed ball B of radius r and center the origin, in a Banach space X is given by Lin [38]. The definition and properties of a densifying map are discussed in Sect. 6.

Theorem 31 *Let B be a ball of radius r and center 0 in a Banach space X and $f : B \to X$ a continuous densifying map. Then there is an $x \in B$ such that $\|x - fx\| = d(fx, B)$.*

Proof Let $R : X \to B$ be a retraction map, defined by $Rx = x$ if $\|x\| \leq r$, and $Rx = rx/\|x\|$ if $\|x\| \geq r$. Then R is a continuous and 1-set contraction map [23, 43]. Let $gx = Rfx$. Then g is a continuous densifying map and $g : B \to B$. Hence $gx = x$ for some $x \in B$. It follows that $\|x - fx\| = d(fx, B)$. □

If $f : B \to \mathbb{R}^n$, then f is said to be a nonself map. Most of the fixed point theorems have been given for self maps like $f : B \to B$. Rothe gave the following fixed point theorem for nonself maps ([6, 58]).

Theorem 32 *Let B be the closed unit ball of a Banach space X with $0 \in B$, and let $f : B \to X$ be continuous. Further, assume that $cl(f(B))$ is compact and $f(\partial B) \subseteq B$, Then f has a fixed point.*

The following result due to Petryshyn [45] is derived from Theorem 31.

Theorem 33 *Let $f : B \to X$ be a continuous densifying map, where B is a closed ball of radius r and center at the origin in a Banach space X. Then f has a fixed point provided that one of the following conditions is satisfied for $x \in \partial B$:*

(i) *If $fx = \lambda x$, then $\lambda \leq 1$, (Leray Schauder condition).*
(ii) *$f(\partial B) \subseteq B$, (Rothe condition)*
(iii) *$\|fx\|^2 \leq \|fx - x\|^2 + \|x\|^2$ (Altman condition) [62].*

Some further work has been also given by Prolla [47], Carbone [14], Sehgal and Singh [55, 56] and Takahashi [63]. Approximatively compact set has been taken in [56].

5 Convergence of the Approximating Sequences

In this section, we give results dealing with convergence of approximating sequences.

The following is given by Bonsall [9].

Theorem 34 [9, Theorem 1.2] *Let X be a complete metric space and $\{f_n\}$ a sequence of contraction mappings of X into itself with the same Lipschitz constant $k < 1$ such that $f_n u_n = u_n$, $n \in \mathbb{N}$. Let $\lim_{n\to\infty} f_n x = fx$ for every $x \in X$, where f is a contraction map with Lipschitz constant $k < 1$ with fixed point u. Then $\lim_{n\to\infty} u_n = u$.*

Remark 13 In Theorem 34, it is not necessary to assume that f is a contraction map. As a matter of fact, it follows that f is a contraction map with the same Lipschitz constant $k < 1$.

The following result is given in general setting and extends Theorem 34.

Theorem 35 *Let X be a complete metric space and $\{f_n\}$ a sequence of mappings with fixed points u_n. Suppose that $\lim_{n\to\infty} f_n x = fx$ for every $x \in X$ such that f is a contraction map with Lipschitz constant $k < 1$. Then f has a unique fixed point $u \in X$, and $\lim_{n\to\infty} u_n = u$.*

Proof It is easy to derive that f has a unique fixed point u. Since $\lim_{n\to\infty} f_n x = fx$ for every $x \in X$, therefore, for given $\varepsilon > 0$, there exists $n_0 \in \mathbb{N}$ such that for $n \geq n_0$, we have $d(f_n u, fu) \leq (1-k)\varepsilon$. Now, for $n \geq n_0$,

$$d(u, u_n) = d(fu, f_n u_n) \leq d(fu, f_n u) + d(f_n u, f_n u_n) \leq (1-k)\varepsilon + kd(u, u_n).$$

It follows that $d(u, u_n) \leq \varepsilon$ for $n \geq n_0$. Therefore, $\lim_{n\to\infty} u_n = u$. □

In case we do not have a k as a constant, then the result need not be valid. For example, if we have a sequence of contraction maps converging to a nonexpansive map f say, then the sequence of fixed points of f_n need not converge to a fixed point of nonexpansive map f. There are examples of a nonexpansive mapping without fixed points, like translation map. Theorem 35 is valid if the sequence $\{k_n\}$ of the contraction constants is a decreasing sequence. In this case, we can replace k_n by k_1 since $k_n < k_1$ for all n.

The following simple example illustrates the facts.

Example 1 For $n \in \mathbb{N}$, let $f_n : \mathbb{R} \to \mathbb{R}$ be given by

$$f_n x = 1 - \frac{x}{n+1}, \quad \text{for all } x \in \mathbb{R}.$$

Then f_n is a contraction map with contraction constant $k_n = 1/(n+1)$. Here $k_1 = \frac{1}{2}$ so one could take $\frac{1}{2}$ as a contraction constant for all the maps. The unique fixed point for each f_n is $x_n = (n+1)/(n+2)$. Also, $\lim_{n\to\infty} f_n x = fx = 1$ for every x and $\lim_{n\to\infty} x_n = \lim_{n\to\infty} (n+1)/(n+2) = 1$. Here $f1 = 1$ is a unique fixed point of f.

In case $\{k_n\}$ is an increasing sequence, then the result is false. For example, let $f_n : \mathbb{R} \to \mathbb{R}$ be given by $f_n x = p + n/(n+1)x$ for all $x \in \mathbb{R}$, where $p \in \mathbb{R}$. Then $k_n = n/(n+1) \to 1$. The sequence of fixed points is given by $f_n x_n = x_n = (n+1)p$. Now the sequence $\{f_n\}$ of contraction maps converges to f, where $fx = p + x$ for every $x \in \mathbb{R}$ and f is a translation map has no fixed points. In this example, $\lim_{n\to\infty} x_n$ does not exist.

In case of uniform convergence, we have the following

Theorem 36 [40, Theorem 1] *Let X be a metric space and $\{f_n\}$ a sequence of mappings of X into itself with fixed points x_n and $f : X \to X$ a contraction map with fixed point x. If the sequence $\{f_n\}$ converges uniformly to f, then the sequence of fixed points x_n of f_n converges to $x = fx$.*

Proof Since f_n converges uniformly to f, therefore, for each $\varepsilon > 0$, there exists an $n_0 \in \mathbb{N}$ with

$$d(f_n x, fx) < (1-k)\varepsilon \quad \text{for all } x \in X, n \geq n_0,$$

where $0 < k < 1$ is a contraction constant of f. Note

$$d(x_n, x) = d(f_n x_n, fx) \leq d(f_n x_n, fx_n) + d(fx_n, fx)$$
$$\leq d(f_n x_n, fx_n) + kd(x_n, x),$$

which implies that

$$(1 - k)d(x_n, x) \leq d(f_n x_n, fx_n).$$

Hence, we derive that $d(x_n, x) < \varepsilon$ since f_n converges uniformly to f. Hence $\{x_n\}$ converges to $x = fx$. □

We give the following [59, 64].

Theorem 37 *Let X be a metric space and $\{f_n\}$ a sequence of mappings of X into itself with fixed points u_n. Let $\{f_n\}$ converge uniformly to f, where f is a contraction mapping with contraction constant $k < 1$. Let $f(X)$ be compact. Then the sequence $\{u_n\}$ converges to u, a unique fixed point of f.*

Proof Since f is a contraction map and $f(X)$ is compact, so by Schauder fixed point theorem f has a unique fixed point say $fu = u$. To prove this, let $B = f(X)$. Then B is compact. Define a sequence $\{x_n\}$ by Picard method:

$$x_{n+1} = fx_n, n = 0, 1, 2, \cdots.$$

The sequence $\{x_n\}$ has a convergent subsequence since B is compact. Let $\{x_{n_i}\}$ be a subsequence of $\{x_n\}$, and let $\{x_{n_i}\}$ converge to z. Since f is continuous, so fx_{n_i} converges to fz. Now

$$d(z, fz) \leq d(z, x_{n_i+1}) + d(x_{n_i+1}, fz) = d(z, x_{n_i+1}) + d(fx_{n_i}, fz)$$
$$\leq d(z, x_{i+1}) + kd(x_{n_i}, z) \to 0 \text{ as } i \to \infty.$$

Thus, $z = fz$, that is, f has a fixed point. Since f is a contraction map so f has a unique fixed point.

Now, since $\{f_n\}$ converges uniformly to f, so for given $\varepsilon > 0$, there exists $n_0 \in \mathbb{N}$ such that $n \geq n_0$ implies that $d(f_n x, fx) < (1 - k)\varepsilon$ for all $x \in X$. Hence

$$d(u_n, u) = d(f_n u_n, fu) \leq d(f_n u_n, fu_n) + d(fu_n, fu) \leq (1 - k)\varepsilon + kd(u_n, u).$$

Thus, $d(u_n, u) < \varepsilon$, that is, $\lim_{n\to\infty} u_n = u$. □

We give the following result due to Nadler [40].

Theorem 38 *Let X be a metric space and $\{f_n\}$ a sequence of contraction mappings of X into itself with fixed point u_n. Suppose that the sequence $\{f_n\}$ converges pointwise to f, where $f : X \to X$ is contraction mapping with a unique fixed point x. If there exists a sequence $\{u_{n_i}\}$ of $\{u_n\}$ such that $\{u_{n_i}\}$ converges to u, then $u = x$.*

Now we give results dealing with contractive approximation of a sequence of contraction mappings to a nonexpansive map.

Let C be a nonempty closed convex subset of a Banach space X, and let $f : C \to C$ be a nonexpansive mapping. Let $\{r_n\}$ be a sequence in $(0, 1)$. For any $x_0 \in C$ and $n \in \mathbb{N}$, define a mapping f_{r_n} by

$$f_{r_n}x = r_n fx + (1 - r_n)x_0, x \in C.$$

Then f_{r_n} maps C into itself and is a contraction map with Lipschitz constant r_n. By Banach contraction principle each f_{r_n} has a unique fixed point say x_{r_n}. Thus,

$$x_{r_n} = f_{r_n}x_{r_n} = r_n fx_{r_n} + (1 - r_n)x_0.$$

It is easy to see that if $\{r_n\}$ converges to 1, then the sequence $\{f_{r_n}\}$ of contraction mappings converges pointwise to the nonexpansive mapping f.

As the function f is a contractive approximation of f_{r_n}, so does the fixed point of f is the limit of the sequence of fixed points x_{r_n} (Browder [13]).

There are nonexpansive mappings without a fixed point, therefore, in general the convergence of $x_{r_n} = f_{r_n}x_{r_n}$ to a fixed point of f is not guaranteed. However, when $F(f)$, is nonempty, then under suitable restrictions the answer is positive.

Before to give an affirmative answer, we need the following:

Theorem 39 *Let C be a closed convex subset of a Hilbert space H and $f : C \to H$ a nonexpansive map with $f(C)$ bounded. Then there is a $y \in C$ such that $\|y - fy\| = d(fy, C)$.*

Browder [13] proved the following result using the monotonicity of $I - f$, where f is a nonexpansive mapping.

Theorem 40 *Let C be a closed convex bounded subset of a Hilbert space H and $f : C \to C$ a nonexpansive map. Define $f_{r_n}x = r_n fx + (1 - r_n)u$, where $u \in C$, $0 < r_n < 1$ and $r_n \to 1$. Let $f_{r_n}x_{r_n} = x_{r_n}$. Then the sequence $\{x_{r_n}\}$ converges to y, where y is a fixed point of f closest to u.*

Remark 14 If C is a closed convex subset of H and $f : C \to C$ is a nonexpansive map, then $f(C)$ is weakly compact.

Proof We know that f has a fixed point and the set of fixed points is closed and convex. Therefore, there is a unique nearest point to u say $x^* = fx^* \in F(f)$, that is,

$$\langle u - x^*, x^* - z\rangle \geq 0, \quad \text{for all } z \in F(f).$$

In order to show that $\{x_{r_n}\}$ converges strongly to x^*, it suffices to show that there is a subsequence $\{x_{r_{n_i}}\}$ with $r_{n_i} \to 1$ converges strongly to x^*. Let $v_i = x_{r_{n_i}}$. Since all the x_{r_n} lie in weakly compact set $f(C)$, we may assume that $\{v_i\}$ converges weakly to v for some $v \in H$. We now show that $v = fv$. Note

$$\|v_i - fv_i\| = \|r_{n_i} fv_i + (1 - r_{n_i})u - fv_i\| = (1 - r_{n_i})\|u - fv_i\|.$$

Since $\{fv_i\}$ is bounded, so $v_i - fv_i \to 0$ as $i \to \infty$. Set $g = I - f$. Then g is monotone. Now,

$$0 \le \langle gu - gv_i, u - v_i \rangle,$$

which implies that $\langle gu, u - v \rangle \ge 0$, for any $u \in C$. Set $u_t = v + tz$ for z arbitrary in C and $0 < t < 1$. So $\langle gu_t, u_t - v \rangle \ge 0$. Since $tz = u_t - v$, therefore $t\langle gu_t, z \rangle \ge 0$, that is, $\langle gu_t, z \rangle \ge 0$ for all z. Taking limit as $t \to 0$, we get $\langle gv, z \rangle \ge 0$. Putting $z = 0$, we have $gv = 0$, that is, $v = fv$. Since v lies in $f(C)$ as $f(C)$ is weakly closed, and $v = fv \in F(f)$.

Finally, we prove that $\{v_i\}$ converges strongly to x^*. Observe that

$$(1 - r_{n_i})v_i + r_{n_i}(v_i - fv_i) = (1 - r_{n_i})u, \quad \text{since } gv_i = 0 \tag{8}$$

and since $x^* = fx^*$, we get

$$(1 - r_{n_i})x^* + r_{n_i}(x^* - fx^*) = (1 - r_{n_i})x^*. \tag{9}$$

Taking (8) and (9) and inner product with $v_i - x^*$ we get

$$(1 - r_{n_i})\langle v_i - x^*, v_i - x^* \rangle + r_{n_i}\langle gv_i - gx^*, v_i - x^* \rangle = (1 - r_{n_i})\langle u - x^*, v_i - x^* \rangle.$$

Hence,

$$\|v_i - x^*\|^2 \le \langle u - x^*, v_i - x^* \rangle$$

since g is monotone. Note

$$\langle u - x^*, v_i - v + v - x^* \rangle = \langle u - x^*, x_i - v \rangle + \langle u - x^*, v - x^* \rangle.$$

On the right hand side x_i converges weakly to v and

$$\langle u - x^*, v - x^* \rangle \le 0,$$

since v lies in F, and x^* is the closest point to u in $F(f)$, therefore, $\{v_i\}$ converges strongly to x^*. □

The following result is a general one and derives Theorem 40 as a corollary.

Theorem 41 [61] *Let C be a closed convex subset of a Hilbert space H and $f : C \to H$ a nonexpansive map with $f(C)$ bounded and $f(\partial C) \subset C$. Define $f_{r_n}x = r_n fx + (1 - r_n)x_0$, where $0 < r_n < 1$ and $r_n \to 1$. Let $f_{r_n}x_{r_n} = x_{r_n}$. Then the sequence of fixed points $\{x_{r_n}\}$ of $f_{r_n}\{x_{r_n}\}$ converges to y where y is a fixed point of f closest to x_0.*

Proof First, we note that $F(f)$ is nonempty [12]. Here $F(f)$ is a closed and convex set [12], so has a unique nearest point from any point $\notin F(f)$. Thus, x_0 has a unique nearest point in $F(f)$ say $y = fy$. For the sake of convenience, we take $x_0 = 0$ and we write r for r_n. Note

$$\|x_r/r - y\|^2 = \|fx_r - y\|^2 = \|fx_r - fy\|^2 \le \|x_r - y\|^2.$$

On simplification we get that $\|x_r\|^2 \le \langle x_r, y\rangle$. Hence, $\|x_r\| \le \|y\|$, and $\{x_r\}$ is a bounded sequence. The sequence $\{x_r\}$ has a subsequence $\{x_k\}$ converging weakly to x. Further,

$$\|x_k - fx_k\| = \|r_k fx_k - fx_k\| = (1 - r_k)\|fx_k\| \to 0$$

as $k \to \infty$. Then $\{x_k - fx_k\}$ converges strongly to zero.

Since $I - f$ is demiclosed, therefore we get that $(I - f)x = 0$. Thus x is a fixed point of f. Since $\|x_n\| \le \|y\|$ so $\|x\| \le \|y\|$. But y is closest to $x = 0$ and the nearest point is unique, therefore $x = y$. Hence $\{x_k\}$ converges weakly to y. Again,

$$\|y\|^2 \ge \|x_k\|^2 = \|x_k - y + y\|^2 = \|x_k - y\|^2 + \|y\|^2 + 2\langle x_k - y, y\rangle.$$

The last part of right hand side goes to 0 as $k \to \infty$. Therefore,$\|x_k - y\| \to 0$, and the subsequence $\{x_k\}$ converges strongly to y. Since $\{x_k\}$ is any subsequence of $\{x_{r_n}\}$, therefore the sequence $\{x_{r_n}\}$ converges strongly to y. □

Halpern [28] gave the following.

Theorem 42 *Let C be a closed convex bounded subset of a Hilbert space H and $f : C \to C$ is a nonexpansive mapping. Let $\{t_n\}$ be a sequence in $(0, 1)$, and define a sequence $\{x_n\}$ by*

$$x_{n+1} = t_n z + (1 - t_n)fx_n, n \ge 0.$$

Then the sequence $\{x_n\}$ converges strongly to y, a fixed point of f, closest to z if $\lim_{n\to\infty} t_n = 0$, $\sum_{n=0}^{\infty} t_n = \infty$, and either $\sum_{n=0}^{\infty} \|t_{n+1} - t_n\| < \infty$ or $\lim_{n\to\infty} t_n/t_{n+1} = 1$ (for example, if $t_n = 1/(n + 1)$).

If $0 \in C$, then both results are good to find $y \in F(f)$, as $f_t = (1 - t)f$ is a mapping from C into itself.

Remark 15 If C is a closed convex subset of a Hilbert space H and $f : C \to C$ is a nonexpansive mapping, and for $x \in C$, the sequence $\{f^n x\}$ is bounded, then $F(f)$ is nonempty.

Recently, the following result is given by Cui and Liu [19].

Theorem 43 *Let H be a real Hilbert space, C a nonempty closed convex subset of H and $f : C \to C$ a nonexpansive mapping with $F(f)$ nonempty. Let $\{t_n\}$ be a sequence in $(0, 1)$ such that $\lim_{n\to\infty} t_n = 0$, $\sum_{n=0}^{\infty} t_n = \infty$, and either $\sum_{n=0}^{\infty} \|t_{n+1} - t_n\| < \infty$ or $\lim_{n\to\infty} t_n/t_{n+1} = 1$. Then the sequence the sequence $\{x_n\}$ defined by $x_{n+1} = P((1 - t_n)fx_n)$, $n \ge 0$ converges strongly to $y = fy$ closest to 0.*

6 Convergence of the Sequence of Iterates

The method of successive approximation is useful in determining solutions of integral, differential, algebraic, and nonlinear operator equations. This technique is used in approximation theory and variational inequalities as well. The method of successive approximations is quite useful in approximating fixed points, in finding solution of nonlinear equations [9, 42, 64]. The sequence of successive approximation for a contraction map $f : X \to X$, given by $x_{n+1} = fx_n$, $n = 0, 1, \ldots$ converges to a unique fixed point of f.

For example, if $x^2 = 2$, then writing $fx = \frac{1}{2}(x + 2/x)$, f is a contraction map and $x_{n+1} = fx_n$, $n = 0, 1, \ldots$ converges to $\sqrt{2}$.

On the other hand, a sequence $\{x_n\}$ of iterates for nonexpansive map need not converge. For example, if $f : \mathbb{R} \to \mathbb{R}$ defined by $fx = -x$, then $x_{n+1} = fx_n$ need not converge if $x_0 \neq 0$. Krasnoselskii [37] considered the iterative sequence given below:

Let $f_{\frac{1}{2}}x = \frac{1}{2}x + \frac{1}{2}fx$. Then $f_{\frac{1}{2}}$ is a nonexpansive map. Also, $F(f) = F(f_{\frac{1}{2}})$. It is easy to see that if $f_{\frac{1}{2}}x = x$, then $fx = x$ and conversely.

The sequence of iterates for the function $f_{\frac{1}{2}}$ under suitable restrictions converges to a fixed point of f. Schaefer [53] considered $f_r x = rfx + (1-r)x$, $0 < r < 1$. In this case $F(f) = F(f_r)$ and f_r is a nonexpansive map.

For example, if $fx = x$, then we show that $f_r x = x$. Since $f_r x = rfx + (1-r)x$, therefore $f_r x = rx + (1-r)x = x$. Further, if $f_r x = x$, then $fx = x$.

Definition 13 Let X be a Banach space and C a subset of X. The map $f : X \to X$ with $F(f) \neq \emptyset$ is said to be *quasi-nonexpansive* if

$$\|fx - p\| \leq \|x - p\|$$

for all $x \in C$ and $p \in F(f)$.

The quasi-nonexpansive map need not be continuous. A quasi-nonexpansive map need not be nonexpansive. However, a nonexpansive map with at least one fixed point is a quasi-nonexpansive. [2, 20, 21, 60]. The following example illustrates the facts.

Example 2 Let $f : [0, 1] \to [0, 1]$ be given by $fx = 0$ for $0 \leq x < 2/3$ and $fx = 2/3$ for $2/3 < x \leq 1$. Note $f0 = 0$ and f is a quasi-nonexpansive map, though f is discontinuous. However, f is not a nonexpansive map. Indeed, for $x = 2/3$ and $y = 1$, we have

$$|fx - fy| = |2/3 - 0| = 2/3 > |x - y| = |2/3 - 1| = 1/3.$$

Mann [39] introduced the sequence of iterative process as follows:

If $x_{n+1} = (1 - c_n)x_n + c_n fx_n$, $0 < c_n < 1$, $\lim\limits_{n\to\infty} c_n = 0$ and $\sum_{n=1}^{\infty} c_n$ diverges, then the sequence $\{x_n\}$ converges to a fixed point of f under suitable conditions on domain and space. In this case $F(f) = F(f_{c_n})$, where $f_{c_n}x$ stands for $(1 - c_n)x + c_n fx$.

Let $fx = x$, then $f_{c_n}x = c_n x + (1 - c_n)x = x$. If $f_{c_n}x = x$, then $x = c_n fx + (1 - c_n)x$, so $c_n fx = c_n x$ and hence, $fx = x$.

It is shown that f_{c_n} is quasi-nonexpansive, that is, if $p \in F(f_{c_n})$, then

$$\|f_{c_n}x - p\| = \|f_{c_n}x - f_{c_n}p\| \leq \|x - p\|, \quad \text{for all } x \in C.$$

Note

$$\|c_n fx + (1 - c_n)x - p\| = \|c_n fx - c_n p + (1 - c_n)x - p\| \leq c_n\|fx - p\| + (1 - c_n)\|x - p\|.$$

We know that $\|fx - p\| \leq \|x - p\|$, we have

$$\|f_{c_n}x - p\| \leq c_n\|x - p\| + (1 - c_n)\|x - p\| = \|x - p\|.$$

Therefore, f_{c_n} is a quasi-nonexpansive map.

Ishikawa [32] considered the following.

$$x_{n+1} = t_n f(r_n fx_n + (1 - r_n)x_n) + (1 - t_n)x_n,$$

$0 < r_n < 1$ and $0 < t_n < 1$ with suitable restrictions on r_n and t_n. In each case, it is shown that $x_n - fx_n \to 0$ as $n \to \infty$. If $f(C)$ is compact, then the sequence $\{x_n\}$ converges to a fixed point of f.

The following result due to Krasnoselskii [37] deals with the convergence of the iterative sequence of nonexpansive mappings..

Theorem 44 [37] *Let X be a uniformly convex Banach space and C a closed convex bounded subset of X. If $f : C \to C$ is a nonexpansive map with closure $(f(C))$ compact, then the sequence of iteration of the map defined by*

$$x_{n+1} = \frac{1}{2}x_n + \frac{1}{2}fx_n, \quad n \geq 1$$

converges to a fixed point of f.

Theorem 45 [20] *Let C be a closed subset of a Banach space X and $f : C \to C$ a quasi-nonexpansive map with $F(f)$, nonempty. For $x_1 \in C$, let $\{x_n\}$ be a sequence of Mann iteration defined by*

$$x_{n+1} = c_n fx_n + (1 - c_n)x_n, \quad n \geq 1.$$

Then $\{x_n\}$ converges to a fixed point of f, provided that $\lim_{n\to\infty} d(x_n, F) = 0$.

The following result due to Naimpally and Singh [41] is given in general setting.

Theorem 46 *Let X be a normed linear space and C a nonempty closed convex subset of X. Let $f : C \to C$ be a mapping satisfying*

$$\|fx - fy\| \leq k \max\{\|x-y\|, \|x-fx\|, \|y-fy\|, \|x-fy\|+\|y-fx\|\}, \text{ for all } x, y \in C,$$

where $0 \leq k < 1$ and let $\{x_n\}$ be a sequence defined by

$$\begin{cases} y_n = (1-\beta_n)x_n + \beta_n f x_n, \\ x_{n+1} = (1-\alpha_n)x_n + \alpha_n f y_n, n \geq 0, \end{cases}$$

$x_0 \in C$, where $0 \leq \alpha_n, \beta_n \leq 1$. If $\{\alpha_n\}$ is bounded away from 0 and if $\{x_n\}$ converges to p, then p is a fixed point of f.

For the proof and details see [41].

The following result is worth mentioning.

Theorem 47 *Let C be a closed convex subset of a Banach space X and $f : C \to X$ a function with $F(f)$ nonempty and satisfying the following:*

$$\|fx - p\| \leq \|x - p\|, \quad \text{for all } x \in C \text{ and } p \in F(f).$$

Define $x_{n+1} = c_n f x_n + (1 - c_n)x_n$, $n \geq 0$ for $x_0 \in C$, where $c_n \in [a, b]$ with $0 < a < b < 1$, and $\lim_{n\to\infty} d(x_n, F(f)) = 0$. Then $\{x_n\}$ converges to a fixed point of f.

Proof Let us denote $x_{n+1} = f_{c_n} x_n = c_n f x_n + (1 - c_n)x_n$. Then $F(f) = F(f_{c_n})$. The remaining proof follows on the lines as given in [20]. □

The contraction, contractive, and nonexpansive maps have been further extended to densifying, and 1- set contraction maps. In 1969 several interesting results of fixed points were given for densifying maps.

We now give results dealing with densifying mappings. First, we need a few preliminaries ([24, 43, 52]).

Definition 14 Let C be a bounded subset of a metric space X. Define the measure of noncompactness

$$\alpha(C) = \inf\{\varepsilon > 0 : C \text{ has a finite covering of subsets of diameter } \leq \varepsilon\}.$$

The following properties of α are well known.

Let A be a bounded subset of a metric space X. Then

(i) $\alpha(A) \leq \delta(A)$, $\delta(A)$ is the diameter of A.
(ii) $\alpha(\text{closure of } A) = \alpha(A)$.
(iii) If $A \subseteq B$, then $\alpha(A) \leq \alpha(B)$.

(iv) $\alpha(A \cup B) = \max\{\alpha(A), \alpha(B)\}$.
(v) $\alpha(A) = 0$ if and only if A is precompact.

A continuous mapping $f : X \to X$ is called a *densifying map* if for any bounded set A with $\alpha(A) > 0$, we have $\alpha f(A) < \alpha(A)$. If $\alpha f(A) \leq k\alpha(A)$, $0 < k < 1$, then f is a k-set contraction.

In case $\alpha f(A) \leq \alpha(A)$, then f is said to be 1-*set contraction*. A nonexpansive map is an example of 1-set contraction.

A contraction map is densifying and so is a compact mapping.

There are results in fixed point theory dealing with combination of two maps- say $f + g$, where f is a contraction map and g is a compact map.

Remark 16 If f and g both are continuous functions, then $f + g$ is also a continuous map and the fixed point theorem for continuous map is applicable for $f + g$. However, if f is a contraction map, then Banach contraction Principle is applied and if g is a compact map, then Schauder fixed point theorem is applicable. If f is densifying and g is densifying, then $f + g$ is also densifying.

The following is a well-known result [24, 43, 52].

Theorem 48 *Let $f : C \to C$ be a densifying map, where C is closed convex bounded subset of a Banach space X. Then f has at least one fixed point in C.*

For future work Nussbaum [43] is worth mentioning.

The following result is for densifying mappings [16, 24, 52].

Theorem 49 *Let C be a closed convex subset of a Banach space and $f : C \to C$ a nonexpansive densifying map such that $f(C)$ is bounded. Then the sequence of iterates $x_{n+1} = fx_n$ converges strongly to a fixed point of f in C.*

Proof Define f_r by $f_r x = rfx + (1 - r)u$ for some $u \in C$, $0 < r < 1$. Then we get $x_r = f_r x_r$. By using the Banach contraction principle, it is easy to see that $x_r - fx_r = 0$ as $r \to 1$. Let $A = \cup\{x_n : n = 0, 1, 2, \ldots\}$. Then $f(A) = \cup\{fx_n : n = 0, 1, 2, 3\}$ and $\alpha(f(A)) = \alpha(A)$, and hence $\alpha(A) = 0$, since f is densifying. Thus, A is relatively compact and $\{x_n\}$ has a convergent subsequence converging to y. Denote the subsequence of $\{x_n\}$ by $\{x_i\}$. Let $\{x_i\}$ converge to y. Now using the fact that f is continuous, we get that fx_i converges to fy. Since $x_i - fx_i \to 0$, we get that $y - fy = 0$. Hence, $\{x_i\}$ converges to $y = fy$. Since $\{x_i\}$ is any subsequence of $\{x_n\}$, therefore, $\{x_n\}$ converges to y a fixed point of f. □

Theorem 50 *Let C be a closed convex subset of a Banach space X and $f : C \to C$ a nonexpansive map. If the following conditions are satisfied, $F(f) \neq \phi$ and $\lim_{n\to\infty} d(x_n, F(f)) = 0$, then the sequence of iterates $x_{n+1} = r_n fx_n + (1 - r_n)x_n \in C$ for $x_1 \in C$, where $r_n \in [a, b]$ with $0 < a < b < 1$, converges to a fixed point of f.*

Proof Since $F(f) \neq \phi$, therefore for $p \in F(f)$, we have

$$\|x_{n+1} - p\| \leq \|x_n - p\|.$$

Since $\lim_{n\to\infty} d(x_n, F(f)) = 0$, therefore for given $\varepsilon > 0$, there exists a positive integer N such that $d(x_n, F) < \varepsilon/2$ for all $n \geq N$. Hence for large n, m we get for $p \in F(f)$,

$$\|x_n - x_m\| \leq 2\|x_N - p\| < \varepsilon.$$

Thus, $\{x_n\}$ is a Cauchy sequence and converges in C since C is a closed subset of a Banach space X and is therefore complete. Let $z \in C$ such that $\lim_{n\to\infty} x_n = z$. From $\lim_{n\to\infty} d(x_n, F) = 0$ we get that $z \in \overline{F(f)}$. Since for any $p \in F(f)$, $\|z - fz\| \leq 2\|z - p\|$, therefore $\|z - fz\| \leq 2d(z, F) = 0$. Hence, the sequence of iterates $\{x_n\}$ converges to a fixed point of f (see [20] for details). □

The following is given by Petryshyn [45, 46].

Theorem 51 *Let X be a strictly convex Banach space, C a closed convex bounded subset of X, and $f : C \to C$ a densifying nonexpansive mapping. Let $f_r x = rx + (1-r)fx$ for real r, $0 < r < 1$. Then the sequence $x_{n+1} = rx_n + (1-r)fx_n$, $n = 0, 1, \dots$ converges to a fixed point of f in C.*

Remark 17 In case $f : C \to C$ is a nonexpansive densifying map and C is a closed bounded convex subset of a Banach space X, then the sequence of iterates $x_{n+1} = c_n fx_n + (1 - c_n)x_n$, where $\{c_n\} \in (0, 1)$, $c_n \leq k < 1$ and $\sum_{n=1}^{\infty} c_n = \infty$ has the property that $\lim_{n\to\infty} \|x_{n+1} - x_n\| = 0$ [21, 41, 51]. Hence, if the range of f is compact, then the sequence converges to a fixed point of f.

The following result deals with the convergence of the sequence of iterates [42] in variational inequalities.

Theorem 52 *Let C be a closed convex subset of a Hilbert space H and $f : C \to H$ a continuous function such that $I - \lambda f$ is a contraction function. Then there is a unique solution $u \in C$ of $\langle fu, v - u\rangle \geq 0$ for all $v \in C$ and $u = \lim_{n\to\infty} u_n$, where $u_{n+1} = P(I - \lambda f)u_n$, $u_0 \in C$. (Here P is a proximity map onto C).*

It can be easily seen that if f is of Lipschitz class and strongly monotone, then $I - \lambda f$ is contraction for suitable $\lambda > 0$. For example, if $f : C \to H$ satisfies $\|fx - fy\| \leq k\|x - y\|$ for all $x, y \in C$ and for some $k > 0$, and f is strongly monotone, that is,

$$\alpha\|x - y\|^2 \leq \langle fx - fy, x - y\rangle, \quad \text{for all } x, y \in C \text{ and for some } k > 0,$$

then $I - \lambda f$ is a contraction map for $0 < \lambda < 2\alpha/k^2$.

It is seen as follows: For $x, y \in C$, we have

$$\begin{aligned}\|(I-\lambda f)x-(I-\lambda f)y\|^2 &= \|x-y\|^2 - 2\lambda\langle fx-fy, x-y\rangle + \lambda^2\|fx-fy\|^2\\ &= (1-2\lambda\alpha+k^2\lambda^2)\|x-y\|^2.\end{aligned}$$

Thus, $I-\lambda f$ is a contraction map if $1-2\lambda\alpha+k^2\lambda^2<1$, that is, $\lambda<2\alpha/k^2$.

Bae [6] has discussed results related to the generalized nonexpansive maps defined below.

Definition 15 Let X be a metric space. If $f: X \to X$ satisfies $d(fx, fy) \le ad(x, y) + b\{d(x, fx) + d(y, fy)\} + c\{d(x, fy) + d(y, fx)\}$ for all $x, y \in X$, where $a > 0, b > 0, c > 0, a + 2b + 2c \le 1$, then f is said to be a *generalized nonexpansive mapping*.

If $a + 2b + 2c < 1$ and X is a complete metric space, then f has a unique fixed point and the sequence of iterates converges to the fixed point of f. The following is also due to Bae [6].

Theorem 53 *Let (X, d) be a compact metric space and $f : X \to X$ a map satisfying*

$$d(fx, fy) \le ad(x, y) + bd(x, fy) + cd(y, fx) \text{ for all } x, y \in X, \tag{B}$$

where $a \ge 0$, $c > 0$, and $a + 2c = 1$. Then f has a fixed point and the sequence of iterates $x_{n+1} = fx_n, n = 0, 1, 2, \dots$ for any $x_0 \in X$ converges to a fixed point of f.

Definition 16 Let (X, d) be a metric space. A map $f : X \to X$ is called *asymptotically regular* if, for any $x \in X$, $d(f^{n+1}x, f^n x) \to 0$ as $n \to \infty$.

We give the following when X need not be compact.

Theorem 54 *Let X be a metric space and $f : X \to X$ satisfy Condition (B). Further, if X is bounded and the sequence of iterates $x_{n+1} = fx_n, \quad n = 0, 1, 2, \cdots$ has a convergent subsequence, then f has a fixed point and the sequence of iterates converges to the fixed point of f.*

Proof Let $x_0 \in X$ and $x_{n+1} = fx_n$ be a sequence of iterates. Then there is a subsequence $\{x_{n_i}\}$ of $\{x_n\}$ converging to $y \in X$. Let us write x_i for x_{n_i} for the sake of convenience. Then, by (B), we have

$$\begin{aligned}d(fy, x_{n_i}) = d(fy, x_i) &\le ad(y, x_{i-1}) + cd(y, x_i) + cd(fy, x_{i-1})\\ &= ad(y, x_i) + ad(x_i, x_{i-1}) + cd(y, x_i) + cd(fy, x_i) + cd(x_i, x_{i-1}).\end{aligned}$$

Hence

$$d(fy, x_{n_i}) \le (a+c)/(1-c)[d(y, x_i) + d(x_i, x_{i-1})] = d(y, x_i) + d(x_i, x_{i-1}).$$

Letting limit as $i \to \infty$, we get

$$d(fy, y) \leq 0,$$

since f is asymptotically regular [6, 62], therefore $d(x_i, x_{i-1}) \to 0$ as $i \to \infty$. Hence $fy = y$. Again, by using (B) we have

$$d(x_{n+1}, y) \leq ad(x_n, y) + cd(x_n, y) + cd(x_{n+1}, y),$$

so $d(x_{n+1}, y) \leq (a + c)/(1 - c)d(x_n, y) = d(x_n, y)$. Thus, $\{d(x_n, y)\}$ a bounded decreasing sequence and hence $\{x_n\}$ converges to y. □

Theorem 55 *Let C be a closed convex bounded subset of a Banach space X and $f : C \to C$ a densifying map satisfying*

$$\|fx - fy\| \leq a\|x - y\| + b\{\|x - fx\| + \|y - fy\|\} + c\{\|x - fy\| + \|y - fx\|\} \quad (10)$$

for all $x, y \in C$ with $a \geq 0$,$b \geq 0$, $c \geq 0$, and $a + 2b + 2c \leq 1$. Then the sequence of iterates $x_{n+1} = fx_n$ converges to a fixed point of f.

Proof Since C is a closed convex bounded subset of a Banach space and f is a densifying map, so $F(f)$ is nonempty [24]. Let $p = fp$. Then, for $x \in C$, we have

$$\|fx - p\| = \|fx - fp\| \leq a\|x - p\| + b\|x - fx\| + c\{\|x - p\| + \|p - fx\|\}.$$

By (10), we have

$$(1 - c)\|fx - p\| \leq (a + c)\|x - p\| + b\|p - fx\|.$$

Note

$$\|x - fx\| = \|x - p + p - fx\| \leq \|x - p\| + \|p - fx\|,$$

it follows that

$$(1 - c)\|fx - p\| \leq (a + c)\|x - p\| + b\|x - p\| + b\|p - fx\|,$$

and hence, $(1-b-c)\|fx-p\| \leq (a+b+c)\|x-p\|$. Therefore, $\|fx-p\| \leq \|x-p\|$. Thus. the sequence $\{\|x_n - p\|\}$ is a nonincreasing sequence and it is bounded below by 0. Hence it converges. Now, in order to show that the sequence $\{x_n\}$ converges strongly to a fixed point in $F(f)$, it is enough to show that the sequence $\{x_n\}$ has a subsequence converging to a point in $F(f)$.

For any $x_0 \in C$, the sequence ($\{x_n\}$ such that $n = 0, 1, 2 \ldots) = A_0$ is bounded and is transformed to ($\{x_n\}$ such that $n = 1, 2, \ldots) = A_1$. Hence $\alpha(A_0) = \alpha(A_1)$ and therefore $\alpha(A_0) = 0$ since f is densifying. Thus, $\{x_n\}$ has a convergent subsequence. Let $z = \lim_{n\to\infty} x_{n_i}$. Then $z \in F(f)$ and the sequence $\{x_n\}$ converges to z. □

Acknowledgments The first author wishes to thank the coordinator of the DST-CIMS, Banaras Hindu University, Varanasi, India for his very warm hospitality and providing the excellent facilities. He is also grateful to Professor H.H. Khan, Aligarh Muslim University, President of the Indian Mathematical Society for his kind invitation.

References

1. Agarwal, R.P., Meehan, M., O'Regan, D.: Fixed Point Theory and Applications. Cambridge University Press, Cambridge (2001)
2. Agarwal, R.P., O'Regan, D., Sahu, D.R.: Fixed Point Theory for Lipschitzian-type Mappings with Applications. Springer, New York (2009)
3. Ansari, Q.H.: Metric Spaces—Including Fixed Point Theory and Set-Valued Maps. Narosa Publishing House, New Delhi (2010) (Also Published by Alpha Science International Ltd., Oxford, U.K.) (2010)
4. Ansari, Q.H., Lalitha, C.S., Mehta, M.: Generalized Convexity, Nonsmooth Variational Inequalities, and Nonsmooth Optimization. CRC Press, Taylor & Francis Group, Boca Raton (2014)
5. Assad, N.A., Kirk, W.A.: Fixed point theorems for set-valued mappings of contractrive type. Pacific J. Math. **43**, 553–562 (1972)
6. Bae, J.S.: Studies on generalized nonexpansive maps. Ph.D. Thesis. Seoul National University, Seoul (1983)
7. Banach, S.: Sur les operations dans les ensembles abstrits et leur applications aux equations integrals. Fund. Math. **3**, 133–181 (1922)
8. Blum, E., Oettli, W.: From optimization and variational inequalities to equilibrium problems. Math. Student **63**, 123–145 (1994)
9. Bonsall, F.F.: Lecture on Some Fixed Point Theorems of Functional Analysis. Tata Institute of Fundamental Research, Bombay, India (1962)
10. Border, K.C.: Fixed Point Theorems with Applications to Economics and Game Theory. Cambridge University Press, Cambridge (1985)
11. Brouwer, L.E.J.: Uber Abbildung von Mannigfaltigkeiten. Math. Ann. **71**, 97–115 (1912)
12. Browder, F.E.: Fixed point theorems for noncompact mappings in Hilbert spaces. Proc. Nat. Acad. Sci. USA **53**, 1272–1276 (1965)
13. Browder, F.E.: Convergence of approximants to fixed points of nonexpansive nonlinear mappings in Banach spaces. Arch Rat. Mech Anal. **74**, 82–90 (1969)
14. Carbone, A.: A note on a Theorem of Prolla. Indian J. Pure Appl. Math. **25**, 257–260 (1991)
15. Carbone, A.: An extension of a best approximation theorem. Intern. J. Math. Math. Sci. **18**, 257–260 (1996)
16. Carbone, A., Singh, S.P.: On best approximation theorem in nonlinear analysis. Rendi. del Circolo Mat. Di Palermo **82**, 259–277 (2010)
17. Cheney, E.W.: Introduction to Approximation Theory. Chelsea Publishing Co., New York (1982)
18. Cheney, E.W., Goldstein, A.A.: Proximity maps for convex sets. Proc. Amer. Math. Soc. **10**, 448–450 (1959)
19. Cui, Y.-L., Liu, X.: Notes on Browder's and Halpern's methods for nonexpansive mappings. Fixed Point Theory **10**, 89–98 (2009)
20. Das, K.M., Singh, S.P., Watson, B.: A note on Mann iteration for quasi-nonexpansive mappings. Nonlinear Anal. **6**, 675–676 (1981)
21. Dotson Jr, W.G.: Fixed points of quasinonexpansive mappings. Bull. Austral. Math Soc. **13**, 167–170 (1972)
22. Fan, K.: A generalization of Tychonoff's fixed point theorem. Math. Ann. **142**, 305–310 (1961)
23. Fan, K.: Extensions of two fixed point theorems of F.E. Browder. Math Z. **112**, 234–240 (1969)

24. Furi, M., Vignoli, A.: Fixed point theorems in complete metric spaces. Bull. Unione Mat. Italiana **2**, 505–509 (1969)
25. Goebel, K., Kirk, W.A.: Topics in Metric Fixed Point Theory. Cambridge University Press, Cambridge, UK (1990)
26. Gohde, D.: Zum prinzip der kontraktiven abbildung. Math. Nachr. **30**, 251–258 (1965)
27. Granas, A.: Fixed Point Theory. Springer, New York (2003)
28. Halpern, B.: Fixed points of nonexpansive mappings. Bull. Amer. Math. Soc. **73**, 957–961 (1967)
29. Hartman, P., Stampacchia, G.: On some nonlinear elliptic differential equations. Acta Math. **115**, 271–310 (1966)
30. Hicks, T.L., Humphries, M.D.: A note on fixed point theorems. J. Approx. Theory **34**, 221–225 (1982)
31. Himmelberg, C.J.: Fixed points of compact multifunctions. J. Math. Anal. Appl. **38**, 205–207 (1972)
32. Ishikawa, S.: Fixed points by new iteration method. Proc. Amer. Math. Soc. **44**, 147–150 (1974)
33. Kakutani, S.: A generalization of Brouwer fixed point theorem. Duke Math. J. **8**, 457–459 (1941)
34. Kannan, R.: Some results on fixed points. Amer. Math. Monthly **76**, 405–408 (1969)
35. Kindrelehrer, D., Stampacchia, G.: An Introduction to Variational Inequalities and Their Applications. Academic Press, New York (1980)
36. Kirk, W.A.: A fixed point theorem for mappings which do not increase distances. Amer. Math. Monthly **72**, 1004–1006 (1965)
37. Krasnoselskii, M.A.: Two remarks on the method of successive approximations. Uspehi Math. Nauk. **10**, 123–127 (1955)
38. Lin, T.C.: A note on a theorem of Ky Fan. Can. Math. Bull. **22**, 513–515 (1979)
39. Mann, W.R.: Mean value methods in iteration. Proc. Amer. Math. Soc. **4**, 506–510 (1953)
40. Nadler Jr, S.B.: Sequence of contractions and fixed points. Pacific J. Math. **27**, 579–585 (1968)
41. Naimpally, S.A., Singh, K.L.: Extensions of some fixed point theorems of Rhoades. J. Math. Anal. Appl. **96**, 437–446 (1983)
42. Noor, M.A.: An iterative algorithm for variational inequalities. J. Math. Anal. Appl. **158**, 448–455 (1991)
43. Nussbaum, R.D.: Some fixed point theorems. Bull. Amer. Math. Soc. **77**, 360–65 (1971)
44. Park, S.: Ninety years of the Brouwer fixed point theorem. Vietnam J. Math. **27**, 187–222 (1999)
45. Petryshyn, W.V.: Structure of fixed point sets of the k-set contractions. Arch. Rat. Mech. Anal. **40**, 312–328 (1971)
46. Petryshyn, W.V., Williamson Jr, T.E.: Strong and weak convergence of sequence of successive approximations of quasi nonexpansive mappings. J. Math. Anal. Appl. **43**, 459–497 (1973)
47. Prolla, J.B.: Fixed point theorems for set-valued mappings and existence of best approximants. Numer. Funct. Anal. Optimiz. **5**, 449–455 (1982–1983)
48. Rao, G.S., Mariadoss, S.A.: Applications of fixed point theorems to best approx. Bulg. Math. Publ. **9**, 243–248 (1983)
49. Reich, S.: Approximate selections, best approximations, fixed points and invariant sets. J. Math. Anal. Appl. **62**, 104–113 (1978)
50. Rheinboldt, W.C.: A unified convergence theory for a class of iterative process. SIAM J. Math. Anal. **5**, 42–63 (1968)
51. Rhoades, B.E.: Comments on two fixed point iteration methods. J. Math. Anal. Appl. **56**, 741–750 (1976)
52. Sadovski, B.N.: A fixed point principle. Funct. Anal. Appl. **1**, 151–153 (1967)
53. Schaefer, H.: Ueber die methode sukzessive approximationen. Jahre Deutsch Math. Verein **59**, 131–140 (1957)
54. Schauder, J.: Der fixpunktsatz in funktionalraumen. Studia Math. **2**, 171–180 (1930)
55. Sehgal, V.M., Singh, S.P.: A variant of a fixed point theorem of Ky Fan. Indian J. Math. **25**, 171–174 (1983)

56. Sehgal, V.M., Singh, S.P.: A theorem on best approximation. Numer. Funct. Anal. Optimiz. **10**, 631–634 (1989)
57. Singh, K.L.: Applications of fixed point theory to approximation theory. In: S. P.Singh (ed.) Proceedings on Approximation Theory and Applications, pp. 198–213. Pitman Publishing Co., London (1985)
58. Singh, S.P.: Applications of fixed point theorem to approximation theory. J. Approx. Theory **25**, 88–89 (1979)
59. Singh, S.P., Singh, M.: Iterated contraction maps and fixed points. Bull. Allahabad Math. Soc. **21**, 393–400 (2008)
60. Singh, S.P., Watson, B.: Proximity maps and points. J. Approx. Theory **28**, 72–76 (1983)
61. Singh, S.P., Watson, B.: On approximating fixed points. In: F.E.Browder (ed.) Proceeding and Symposium in Pure Mathematics. Amer. Math. Soc. **45**, 393–395 (1986)
62. Singh, S.P., Watson, B., Srivastava, P.: Fixed Point Theory and Best Approximation: The KKM map principle. Kluwer Academic Publishers, Dordrecht (1997)
63. Takahashi, W.: Recent results in fixed point theory. Southeast Asian Bull. Math. **4**, 59–85 (1980)
64. Watson, B.: Convergence of sequences of mappings and fixed points. Varahmihir J. Math. Sci. **6**, 105–110 (2006)
65. Zeidler, E.: Nonlinear Functional Analysis and Applications I. Springer, New York (1985)

Hierarchical Minimization Problems and Applications

D. R. Sahu and Qamrul Hasan Ansari

Abstract In this chapter, several iterative methods for solving fixed point problems, variational inequalities, and zeros of monotone operators are presented. A generalized mixed equilibrium problem is considered. The hierarchical minimization problem over the set of intersection of fixed points of a mapping and the set of solutions of a generalized mixed equilibrium problem is considered. A new unified hybrid steepest descent-like iterative algorithm for finding a common solution of a generalized mixed equilibrium problem and a common fixed point problem of uncountable family of nonexpansive mappings is presented and analyzed.

Keywords Fixed point problems · Variational inequalities · Monotone operators · Generalized mixed equilibrium problems · Hierarchical minimization problem · Hybrid steepest descent-like iterative method · Nonexpansive mappings · Resolvent operators · Demiclosed principle · Projection gradient method

1 Introduction and Formulations

It is well known that the standard smooth convex optimization problem [47], given a convex, Fŕechet differentiable function $g : H \to \mathbb{R}$ and a closed convex subset C of a real Hilbert space H, find a point $x^* \in C$ such that

D. R. Sahu
Department of Mathematics, Banaras Hindu University, Varanasi, India
e-mail: drsahudr@gmail.com

Q. H. Ansari (✉)
Department of Mathematics, Aligarh Muslim University, Aligarh, India
e-mail: qhansari@gmail.com

Q. H. Ansari
Department of Mathematics and Statistics, King Fahd University of Petroleum and Minerals, Dhahran, Saudi Arabia

Q. H. Ansari (ed.), *Nonlinear Analysis*, Trends in Mathematics, DOI: 10.1007/978-81-322-1883-8_7,

$$g(x^*) = \min\{x \in C : g(x)\},$$

can be formulated equivalently as the *variational inequality problem* VIP(∇g, H) over C (see [18]):

$$\langle \nabla g x^*, y - x^* \rangle \geq 0, \quad \text{for all } y \in C,$$

where $\nabla g : H \to H$ is the gradient of g.

Convex minimization problems have a great impact and influence in the development of almost all branches of pure and applied sciences, see for example [3, 11, 13, 29]. Iterative methods for nonexpansive mappings have recently been applied to solve convex minimization problems.

Let H be a real Hilbert space. Let $A : H \to H$ be a bounded linear strongly positive operator with coefficient $\bar{\gamma} > 0$ (that is, there exists a constant $\bar{\gamma} > 0$ such that $\langle Ax, x \rangle \geq \bar{\gamma}\|x\|^2$ for all $x \in H$), $f : H \to H$ be a contraction, $\gamma > 0$ be a constant and h be a potential function for γf (that is, $h'(x) = \gamma f(x)$ for all $x \in H$). We consider the following *general convex minimization problem* over the set D:

$$\min_{x \in D} \frac{1}{2}\langle Ax, x \rangle - h(x), \tag{1}$$

where D is a nonempty closed convex subset of H.

A minimization problem defined over the set of solutions of another minimization problem or over the set of solutions of an equilibrium problem is called a *bilevel programming problem*. For further details on bilevel programming problems, we refer to [20, 24] and the references therein. If the minimization problem (1) is defined over the constrained set D which is solution set of another problem, then problem (1) is called the *hierarchical minimization problem* (in short, HMP) over the solution set D.

A typical problem to minimize a quadratic function over the set of the fixed points of a nonexpansive mapping $T : H \to H$ is the following:

$$\min_{x \in \mathrm{Fix}(T)} \frac{1}{2}\langle Ax, x \rangle - \langle x, b \rangle, \tag{2}$$

where b is a given point in H and $\mathrm{Fix}(T)$ denotes the set of all fixed points of T. The minimization problem (2) is a particular case of the general convex minimization problem (1).

On the other hand, in a variety of scenarios, the constrained set D can be expressed as the fixed point set of a nonexpansive mapping. A fixed point problem over the set of solutions of another fixed point problem is called a *hierarchical fixed point problem*. Since problem (2) is a minimization problem defined over the set of fixed points of an operator T, we call it *hierarchical minimization problem* (in short HMP). Xu [45] proposed an iterative method for finding the approximate solution of HMP (2) and studied the strong converge of the sequence generated by the proposed method to a unique solution of the HMP (2). Marino and Xu [21] introduced a general explicit iterative scheme by the viscosity approximation method for the minimization problem (1), where D is the set of fixed points of a nonexpansive mapping T and studied the strong convergence of the explicit iterative scheme. The HMP (1) over suitable solution set D has been extensively studied in recent years, see for example [16, 17, 27, 28, 35], and the references therein).

The hierarchical minimization problem over some solution set D (for example, the set of fixed point and generalized mixed equilibrium problem) has a very important role for the study of applied and computational mathematics.

In many applications, the family $\mathscr{T} = \{T(t) : 0 < t < \infty\}$ of nonexpansive self-mappings on C is not necessarily a semigroup, namely, the family of resolvents of maximal monotone operators and equilibrium problems.

In view of broad applicability of arbitrary family $\mathscr{T} = \{T(t) : 0 < t < \infty\}$ of nonexpansive self-mappings, in this survey chapter, we propose an iterative algorithm for computing solutions of the hierarchical VIP defined over the set $D = \text{Fix}(\mathscr{T}) \cap \Omega[\text{GMEP}(\Phi, \Psi, \varphi)]$, where $\text{Fix}(\mathscr{T})$ is the set of common fixed points of a family $\mathscr{T} = \{T(t) : 0 \leq t < \infty\}$ of nonexpansive self-mappings on C, and $\Omega[\text{GMEP}(\Phi, \Psi, \varphi)]$ is the set of solutions of generalized mixed equilibrium problem (see GMEP (14)). Then we establish the strong convergence of an iterative algorithm to the unique solution of the hierarchical minimization problem (1) over the set $D = \text{Fix}(\mathscr{T}) \cap \Omega[\text{GMEP}(\Phi, \Psi, \varphi)]$. In view of general theory of nonexpansive mappings, we also derive several known convergence theorems for zeros of monotone operators and fixed points of semigroups that have appeared in Hilbert space setting.

2 Preliminaries and Notations

Let H be a real Hilbert space whose inner product and norm are denoted by $\langle \cdot, \cdot \rangle$ and $\| \cdot \|$, respectively. For all $x, y \in H$, we have

$$\|x + y\|^2 \leq \|x\|^2 + 2\langle y, x + y\rangle. \tag{3}$$

Let C be a nonempty subset of H. A mapping $T : C \to H$ is said to be

(a) *η-strongly monotone* if there exists a positive real number η such that

$$\langle Tx - Ty, x - y\rangle \geq \eta||x - y||^2, \quad \text{for all } x, y \in C;$$

(b) *ν-inverse strongly monotone* if there exists a positive real number ν such that

$$\langle Tx - Ty, x - y\rangle \geq \nu||Tx - Ty||^2, \quad \text{for all } x, y \in C.$$

A subset C of H is called a *retract* of H if there exists a continuous mapping P from H onto C such that $Px = x$ for all x in C. We call such P a *retraction* of X

onto C. It follows that if a mapping P is a retraction, then $Py = y$ for all y in the range of P.

A retraction P is said to be *sunny* if $P(Px + t(x - Px)) = Px$ for each x in H and $t \geq 0$. If a sunny retraction P is also nonexpansive, then C is said to be a *sunny nonexpansive retract* of H.

Let C be a nonempty subset of H and let $x \in H$. An element $y_0 \in C$ is said to be a *best approximation* to x if $\|x - y_0\| = d(x, C)$, where $d(x, C) = \inf_{y \in C} \|x - y\|$. The set of all best approximations from x to C is denoted by:

$$P_C(x) = \{y \in C : \|x - y\| = d(x, C)\}.$$

This defines a mapping P_C from H into 2^C and is called the *nearest point projection mapping* (*metric projection mapping*) onto C.

It is well known that if C is a nonempty closed convex subset of a real Hilbert space H, then the nearest point projection P_C from H onto C is the unique sunny nonexpansive retraction of H onto C. It is also known that $P_C x \in C$ and

$$\langle x - P_C x, P_C x - y \rangle \geq 0, \quad \text{for all } x \in H, \ y \in C. \tag{4}$$

A Banach space X is said to satisfy the *Opial condition* if for each sequence $\{x_n\}$ in X which converges weakly to a point $x \in X$, we have

$$\liminf_{n \to \infty} \|x_n - x\| < \liminf_{n \to \infty} \|x_n - y\|, \quad \text{for all } y \in X, \ y \neq x.$$

Note that "$\liminf_{n \to \infty}$" can be replaced by "$\limsup_{n \to \infty}$." It is well that every Hilbert space enjoys the Opial condition (see [1]).

The following lemma is a consequence of the Opial condition.

Lemma 1 (Demiclosed Principle) *Let C be a nonempty closed convex subset of a Hilbert space H and $T : C \to C$ a nonexpansive mapping. Then $I - T$ is demiclosed at zero, that is, if $\{x_n\}$ is a sequence in C converges weakly to x and $\{(I - T)x_n\}$ converges strongly to zero, then $(I - T)x = 0$.*

Proof Suppose that $\{x_n\}$ is a sequence in C such that it converges weakly to x and $\{(I - T)x_n\}$ converges strongly to zero. Suppose, for contradiction, that $x \neq Tx$. Then, by the Opial condition, we have

$$\begin{aligned}
\limsup_{n \to \infty} \|x_n - x\| &< \limsup_{n \to \infty} \|x_n - Tx\| \\
&\leq \limsup_{n \to \infty} (\|x_n - Tx_n\| + \|Tx_n - Tx\|) \\
&\leq \limsup_{n \to \infty} \|x_n - x\|,
\end{aligned}$$

a contradiction. This proves that $(I - T)x = 0$.

We present some lemmas which will be used in the sequel. Some of them are known and others are not hard to prove.

Lemma 2 *Let C be a nonempty closed convex subset of a Hilbert space H and $T : C \to C$ be a nonexpansive mapping. Then* $\mathrm{Fix}(T)$ *is closed and convex.*

Proof If $\mathrm{Fix}(T) = \emptyset$, then clearly $\mathrm{Fix}(T)$ is closed and convex. Next, we assume that $\mathrm{Fix}(T) \neq \emptyset$. Let $\{x_n\}$ be a sequence in C such that $x_n \to x \in H$ as $n \to \infty$. Then $Tx_n = x_n$ for all $n \in \mathbb{N}$ and $\lim_{n\to\infty} \|x_n - x\| = 0$. Now,

$$\begin{aligned}\|Tx - x\| &\leq \|Tx - Tx_n\| + \|Tx_n - x\| \\ &\leq \|x - x_n\| + \|x_n - x\| \to 0 \text{ as } n \to \infty.\end{aligned}$$

This implies that $x \in C$, which shows that C is closed.

Now, let $x, y \in \mathrm{Fix}(T)$ and $\alpha \in [0, 1]$. Then for $z = \alpha x + (1 - \alpha)y$, we have

$$\|x - Tz\| = \|Tx - Tz\| \leq \|x - z\| = (1 - \alpha)\|x - y\|$$

and

$$\|y - Tz\| = \|Ty - Tz\| \leq \|y - z\| = \alpha\|x - y\|.$$

Hence

$$\|x - y\| \leq \|x - Tz\| + \|Tz - y\| \leq \|x - z\| + \|y - z\| = \|x - y\|.$$

This implies that $\|x - Tz\| = \|x - z\|$ and $\|y - Tz\| = \|y - z\|$. Since H is strictly convex, we have $Tz = z$. This shows that $\mathrm{Fix}(T)$ is convex. □

Lemma 3 [21] *Let C be a nonempty closed convex subset of a Hilbert space H. Let $f : C \to H$ be an α-contraction mapping and A be a strongly positive linear bounded operator on H with the coefficient $\bar{\gamma} > 0$. Then, for $0 < \gamma < \bar{\gamma}/\alpha$,*

$$\langle (A - \gamma f)x - (A - \gamma f)y, x - y\rangle \geq (\bar{\gamma} - \gamma\alpha)\|x - y\|^2, \quad \textit{for all } x, y \in C.$$

That is, $A - \gamma f : C \to H$ is strongly monotone with coefficient $\bar{\gamma} - \alpha\gamma$.

Remark 1 Taking $\gamma = 1$ and $A = I$, the identity mapping, we have the following inequality:

$$\langle (I - f)x - (I - f)y, x - y\rangle \geq (1 - \alpha)\|x - y\|^2, \quad \text{for all } x, y \in C.$$

Furthermore, if f is a nonexpansive mapping, then we have

$$\langle (I - f)x - (I - f)y, x - y\rangle \geq 0, \quad \text{for all } x, y \in C.$$

Lemma 4 [21] *Assume that A is a strongly positive linear bounded self-adjoint operator on a real Hilbert space H with coefficient $\bar{\gamma} > 0$ and $0 < \rho \leq \|A\|^{-1}$. Then $\|I - \rho A\| \leq 1 - \rho\bar{\gamma}$.*

Proof Recall that if T is a bonded linear self-adjoint operator on H, then

$$\|T\| = \sup\{|\langle Tx, x\rangle| : x \in H, \|x\| = 1\}.$$

Now, for $x \in H$ with $\|x\| = 1$, we see that

$$\langle (I - \rho A)x, x\rangle = 1 - \rho \langle Ax, x\rangle \geq 1 - \rho \|A\| \geq 0,$$

that is, $I - \rho A$ is positive. It follows that

$$\begin{aligned} \|I - \rho A\| &= \sup\{| \langle (I - \rho A)x, x\rangle | : x \in H, \|x\| = 1\} \\ &= \sup\{\langle (I - \rho A)x, x\rangle : x \in H, \|x\| = 1\} \\ &= \sup\{1 - \rho \langle Ax, x\rangle : x \in H, \|x\| = 1\} \\ &\leq 1 - \rho\overline{\gamma}. \end{aligned}$$

□

Lemma 5 *Let C be a nonempty subset of a real Hilbert space H and $\Psi : C \to H$ be an inverse strongly monotone mapping with coefficient $\nu > 0$. Let $0 < s < 2\nu$. Then $I - s\Psi$ is nonexpansive.*

Proof For $x, y \in C$, we have

$$\begin{aligned} \|(I - s\Psi)x - (I - s\Psi)y\|^2 &\leq \|(I - s\Psi)x - (I - s\Psi)y\|^2 \\ &= \|x - y\|^2 - 2s\langle \Psi x - \Psi y, x - y\rangle + s^2\|\Psi x - \Psi y\|^2 \\ &\leq \|x - y\|^2 - 2s\nu\|\Psi x - \Psi y\|^2 + s^2\|\Psi x - \Psi y\|^2 \\ &= \|x - y\|^2 - s(2\nu - s)\|\Psi x - \Psi y\|^2 \qquad (5) \\ &\leq \|x - y\|^2. \end{aligned}$$

□

Lemma 6 [44] *Let $\{\alpha_n\}$ be a sequence of nonnegative real numbers such that*

$$\alpha_{n+1} \leq (1 - \gamma_n)\alpha_n + \gamma_n\sigma_n, \quad \forall n \geq 0, \tag{6}$$

where $\{\gamma_n\}$ is a sequence in $(0, 1)$ and $\{\sigma_n\}$ is a sequence of real numbers such that

(i) $\sum_{n=0}^{\infty} \gamma_n = \infty$,

(ii) either $\limsup_{n\to\infty} \sigma_n \leq 0$ *or* $\sum_{n=0}^{\infty} |\gamma_n\sigma_n| < \infty$.

Then $\{\alpha_n\}_{n=0}^{\infty}$ converges to zero.

Proof For any $\varepsilon > 0$, let N be an integer big enough so that

$$\sigma_n < \varepsilon, \quad n \geq N.$$

Using (6) and by induction, we obtain, for $n > N$,

$$\alpha_{n+1} \leq \left(\prod_{k=N}^{n}(1-\gamma_k)\right)\alpha_N + \left((1-\prod_{k=N}^{n}(1-\gamma_k)\right)\varepsilon.$$

Then condition (ii) implies that $\limsup_{n\to\infty} \alpha_n \leq 2\varepsilon$. □

3 Variational Inequality Problems

The classical VIP is to find a point $x^* \in C$ such that

$$\langle \mathscr{F}x^*, y - x^*\rangle \geq 0, \quad \text{for all } y \in C,$$

where C is a nonempty closed convex subset of a real Hilbert space H and $\mathscr{F} : C \to H$ is a given operator. This problem is denoted by VIP $(\mathscr{F}, C)$.

It is important to note that the theory of variational inequalities has been playing an important role in the study of many diverse disciplines, for instance, partial differential equations, optimal control, optimization, mathematical programming, mechanics, finance, etc., see, for example, [18, 46, 47] and the references therein.

The relationship between VIP $(\mathscr{F}, C)$ and a fixed point problem can be made through the characterization (4) of the projection operator P_C as follows:

Theorem 1 *Let C be a nonempty closed convex subset of a real Hilbert space H and $\mathscr{F} : C \to H$ a given operator. Then $x^* \in C$ is the solution of variational inequality problem VIP$(\mathscr{F}, C)$ if and only if, for any $\mu > 0$, x^* is the fixed point of the mapping $P_C(I - \mu\mathscr{F}) : C \to C$, that is,*

$$x^* = P_C(I - \mu\mathscr{F})x^*.$$

Proof Suppose that $x^* \in C$ is a solution of *VIP*$(\mathscr{F}, C)$, that is,

$$\langle \mathscr{F}x^*, y - x^*\rangle \geq 0, \quad \text{for all } y \in C. \tag{7}$$

Let $\mu > 0$ such that $P_C(I - \mu\mathscr{F}) : C \to C$. Multiplying $-\mu$ in (7) and adding $\langle x^*, y - x^*\rangle$ both sides we get

$$\langle x^* - \mu\mathscr{F}x^*, y - x^*\rangle \leq \langle x^*, y - x^*\rangle, \quad \text{for all } y \in C.$$

From (4), we get $x^* = P_C(I - \mu\mathscr{F})x^*$.

Conversely, suppose that $\mu > 0$ and x^* is the fixed point of the mapping $P_C(I - \mu\mathscr{F}) : C \to C$. It is easy to see from (4) that $x^* \in C$ is a solution of *VIP*$(\mathscr{F}, C)$. □

This equivalence formulation is useful for existence, uniqueness, and computation of solutions of the varialtional ineqauality problem $VIP(\mathcal{F}, C)$. In particular, the Banach contraction principle guarantees that $VIP(\mathcal{F}, C)$ has unique solution x^* and the sequence of Picard iteration process converges strongly to x^*. In fact, we have

Theorem 2 (Projection Gradient Method) [36, 47] *Let C be a nonempty closed convex subset of a real Hilbert space H and $\mathcal{F} : C \to H$ a κ-Lipschitzian and η-strongly monotone. Let μ be a positive constant such that $\mu < 2\eta/\kappa^2$. Then*

(a) $P_C(I - \mu\mathcal{F}) : C \to C$ is contraction and there exists the unique solution $x^* \in C$ of varialtional ineqauality problem $VIP(\mathcal{F}, C)$.
(b) The sequence $\{x_n\}$ of Picard iteration process, given by,

$$x_{n+1} = P_C(I - \mu\mathcal{F})x_n, \quad \text{for all } n \in \mathbb{N} \tag{8}$$

converges strongly to x^*.

Proof Let $x, y \in C$. Then

$$\begin{aligned}\|P_C(I - \mu\mathcal{F})x - P_C(I - \mu\mathcal{F})y\|^2 &\le \|(I - \mu\mathcal{F})x - (I - \mu\mathcal{F})y\|^2 \\ &= \|x - y\|^2 - 2\mu\langle\mathcal{F}x - \mathcal{F}y, x - y\rangle + \mu^2\|\mathcal{F}x - \mathcal{F}y\|^2 \\ &\le (1 - 2\mu\eta + \mu^2\kappa^2)\|x - y\|^2.\end{aligned}$$

Thus, Theorem 2 follows from the Banach contraction principle. □

Motivated by nonexpansiveness of P_C in Theorem 2, Yamada [46] introduced the following hybrid steepest descent method for solving the $VIP(\mathcal{F}, \text{Fix}(T))$:

$$u_{n+1} = Tu_n - \lambda_n\mu\mathcal{F}Tu_n, \quad \text{for all } n \in \mathbb{N}, \tag{9}$$

where $0 < \mu < 2\eta/\kappa^2$, $T : H \to H$ is a nonexpansive mapping with $\text{Fix}(T) \neq \emptyset$ and $\mathcal{F} : H \to H$ is a κ-Lipschitzian and η-strongly monotone. Yamada proved that hybrid steepest descent method converges strongly to the unique solution $x^* \in \text{Fix}(T)$ of variational inequality problem $VIP(\mathcal{F}, \text{Fix}(T))$.

Note that, by letting $x_n = Tu_n$ ($\{u_n\}$ being the sequence given by (9)) we immediately obtain

$$x_{n+1} = T(x_n - \lambda_n\mu\mathcal{F}x_n), \quad \text{for all } n \in \mathbb{N}. \tag{10}$$

The hybrid steepest descent method (HSDM) has been extensively studied in recent years (see, e.g., [6, 31, 32, 45], and references therein).

If $\mathcal{F} : H \to H$ is monotone and hemicontinuous and if $C \subset H$ is nonempty, compact, and convex, the existence of a solution of the VIP for $\mathcal{F}$ over C is guaranteed.

4 Zeros of Monotone Operators

Let H be a real Hilbert space and let $\mathbb{A} \subset H \times H$ be an operator on H. The set $D(\mathbb{A})$ defined by

$$D(\mathbb{A}) = \{x \in H : \mathbb{A}x \neq \emptyset\}$$

is called the *domain* of $\mathbb{A}$, the set $R(\mathbb{A})$ defined by

$$R(\mathbb{A}) = \bigcup_{x \in X} \mathbb{A}x$$

is called the *range* of $\mathbb{A}$ and the set $G(\mathbb{A})$ defined by

$$G(\mathbb{A}) = \{(x, y) \in H \times H : x \in D(\mathbb{A}),\, y \in \mathbb{A}x\}$$

is called the *graph* of $\mathbb{A}$.

An operator $\mathbb{A} \subset H \times H$ with domain $D(\mathbb{A})$ is said to be *monotone* if for each $x_i \in D(\mathbb{A})$ and $y_i \in \mathbb{A}x_i$ $(i = 1, 2)$, we have

$$\langle x_1 - x_2, y_1 - y_2 \rangle \geq 0.$$

A monotone operator $\mathbb{A}$ is said to be *maximal monotone* if the graph $G(\mathbb{A})$ is not properly contained in the graph of any other monotone operator on H. If $\mathbb{A} : H \to 2^H$ is maximal monotone, then we can define, for each $\lambda > 0$, a nonexpansive single-valued mapping $J_\lambda^{\mathbb{A}} : H \to H$ by

$$J_\lambda^{\mathbb{A}} := (I + \lambda\mathbb{A})^{-1}.$$

It is called the *resolvent* of $\mathbb{A}$.

Let $\mathbb{A}^{-1}0 = \{x \in D(\mathbb{A}) : 0 \in \mathbb{A}x\}$. It is easy to see that $\mathbb{A}^{-1}0$ is closed and convex. It is well known, from [38], that if $\mathbb{A} \subset H \times H$ is a maximal monotone operator, then

$$\frac{1}{r}\left\| J_t x - J_r^{\mathbb{A}} J_t^{\mathbb{A}} x \right\| \leq \frac{1}{t}\left\| x - J_t^{\mathbb{A}} x \right\|, \quad \text{for all } x \in H \text{ and } r, t > 0. \tag{11}$$

It is well also known that [38] for each $\lambda, \mu > 0$ and $x \in H$,

$$\left\| J_\lambda^{\mathbb{A}} x - J_\mu^{\mathbb{A}} x \right\| \leq \frac{|\lambda - \mu|}{\lambda}\left\| x - J_\lambda^{\mathbb{A}} x \right\|. \tag{12}$$

One of the most interesting and important problems in the theory of maximal monotone operators is to find an efficient iterative algorithm to compute approximately zeroes of maximal monotone operators. One method for solving zeros of maximal monotone operators is *proximal point algorithm*. Let $\mathbb{A}$ be a maximal monotone operator in a Hilbert space H. The proximal point algorithm generates, for starting $x_1 \in H$, a sequence $\{x_n\}$ in H by

$$x_{n+1} = J^{\mathbb{A}}_{c_n} x_n, \quad \text{for all } n \in \mathbb{N}, \tag{13}$$

where $\{c_n\}$ is a regularization sequence in $(0, \infty)$. This iterative procedure is based on the fact that the proximal map $J^{\mathbb{A}}_{c_n}$ is single-valued and nonexpansive. Note that (13) is equivalent to

$$x_n \in x_{n+1} + c_n \mathbb{A} x_{n+1}, \quad \text{for all } n \in \mathbb{N}.$$

This algorithm was first introduced by Martinet [22]. If $\psi : H \to \mathbb{R} \cup \{\infty\}$ is a proper lower semicontinuous convex function, then the algorithm reduces to

$$x_{n+1} = \operatorname{argmin}_{y \in H} \left\{ \psi(y) + \frac{1}{2c_n} \|x_n - y\|^2 \right\}, \quad \text{for all } n \in \mathbb{N}.$$

Rockafellar [29] studied the proximal point algorithm in the framework of Hilbert space and he proved that if $\mathbb{A} \subset H \times H$ is a maximal monotone operator with $\mathbb{A}^{-1}0 \neq \emptyset$ and $\{x_n\}$ is a sequence in H defined by (13), where $\{c_n\}$ is a sequence in $(0, \infty)$ such that $\liminf_{n \to \infty} c_n > 0$, then $\{x_n\}$ converges weakly to an element of $\mathbb{A}^{-1}0$.

The proximal point algorithm has been improved and generalized by Ceng et al. [6], Lehdili and Moudafi [19], Sahu et al. [31, 32], Song and Yang [34], Takahashi [39], Tossings [42],and Xu [43] in different aspects.

5 Generalized Mixed Equilibrium Problems

Let C be a nonempty closed convex subset of a real Hilbert space H. Let $\Phi : C \times C \to \mathbb{R}$ be the equilibrium-bi function, i.e.,

$$(\Phi 0) \quad \Phi(u, u) = 0, \quad \text{for all } u \in C.$$

Assume the bifunction $\Phi : C \times C \to \mathbb{R}$ holds the following conditions :

(Φ1) $\Phi(x, x) = 0$ for all $x \in C$;
(Φ2) Φ is monotone, that is, $\Phi(x, y) + \Phi(y, x) \leq 0$ for all $x, y \in C$;
(Φ3) for each $x \in C$, $y \mapsto \Phi(x, y)$ is convex and lower semicontinuous;
(Φ4) for all $x, y, z \in C$, $\limsup_{t \downarrow 0} \Phi\big(tz + (1-t)x, y\big) \leq \Phi(x, y)$.

Let $\Psi : C \to H$ be a nonlinear mapping, $\varphi : C \to \mathbb{R}$ a convex function and $\Phi : C \times C \to \mathbb{R}$ a bifunction. Then, we consider the following *generalized mixed equilibrium problem* (in short, GMEP (Φ, Ψ, φ)) of finding $u \in C$ such that

$$\Phi(u, v) + \langle \Psi u, v - u \rangle + \varphi(v) - \varphi(u) \geq 0, \quad \text{for all } v \in C. \tag{14}$$

We denote by $\Omega[\text{GMEP}(\Phi, \Psi, \varphi)]$ the set of solutions of GMEP (14). The GMEP (14) is considered and studied in [7, 9].

Such types of nonlinear inequalities model some equilibrium problems drawn from operations research, as well as some unilateral boundary value problems stemming from mathematical physics. Several problems arising in optimization, such as fixed point problems, (Nash) economic equilibrium problems, complementarity problems, and generalized set-valued mixed variational inequalities, for instance, have the same mathematical formulation (see [5, 12, 15]), which may be formulated as GMEP (Φ, Ψ, φ).

Given a real-valued function $\varphi : H \to \mathbb{R}$ on a real Hilbert space H, we consider the following *generalized mixed equilibrium problem* (in short, GMEP (Φ, Ψ, φ)) of finding $u \in H$ such that

$$\Phi(u, v) + \langle \Psi u, v - u\rangle + \varphi(v) - \varphi(u) \geq 0, \quad \text{for all } v \in H. \tag{15}$$

We denote by $\Omega[\text{GMEP}(\Phi, \Psi, \varphi)]$ the set of solutions of GMEP (15).

Ceng et al. [7–9] studied and developed extragradient-like method and relaxed extragradient-like method for generalized mixed equilibrium problem (15). For different choices of Φ, Ψ and φ, we get different kinds of equilibrium problems and variational inequalities; see, for example, [2, 5, 9, 12, 14, 15, 18, 23, 25, 41] and the references therein. For instance, we have the following:

(1) If $\Psi \equiv 0$, then GMEP (14) reduces to the following mixed equilibrium problem (for short, MEP (Φ, φ)):

$$\text{find } u \in C \text{ such that } \Phi(u, v) + \varphi(v) - \varphi(u) \geq 0, \quad \text{for all } v \in C. \tag{16}$$

The computation of solutions of such problems is studied in Ceng and Yao [9].

(2) If $\varphi \equiv 0$, then GMEP (14) reduces to the following generalized equilibrium problem (for short, GEP (Φ, Ψ)):

$$\text{find } u \in C \text{ such that } \Phi(u, v) + \langle \Psi u, v - u\rangle \geq 0, \quad \text{for all } v \in C. \tag{17}$$

The problem (17) was studied by Moudafi [23] and Takahashi and Takahashi [41].

(3) If $\Psi \equiv 0$ and $\varphi \equiv 0$, then GMEP (14) reduces to the following equilibrium problem:

$$EP(\Phi) \quad \text{find } u \in C \text{ such that } \Phi(u, v) \geq 0, \quad \text{for all } v \in C.$$

We denote $\Omega[\text{EP}(\Phi)]$ for the set of solutions of EP (Φ).

(4) If $\Phi \equiv 0$ and $\varphi \equiv 0$, then GMEP (14) reduces to the classical variational inequality:

$$\text{find } u \in C \text{ such that } \langle \Psi(u), v - u\rangle \geq 0, \quad \text{for all } v \in C. \tag{18}$$

6 Hierarchical Minimization Problem Over Set of Fixed Point and Generalized Mixed Equilibrium Problem

In this section, we deal with existence and approximation of solutions of the hierarchical minimization problem (1) over $D = \mathrm{Fix}(\mathscr{T}) \cap \Omega[\mathrm{GMEP}(\Phi, \Psi, \varphi)]$, where $\mathrm{Fix}(\mathscr{T})$ is the set of common fixed points of a family $\mathscr{T} = \{T(t) : 0 \leq t < \infty\}$ of nonexpansive self-mappings on a nonempty closed and convex subset C of a real Hilbert space H, and $\Omega[\mathrm{GMEP}(\Phi, \Psi, \varphi)]$ is the set of solutions of GMEP (14).

The following *auxiliary mixed equilibrium problem* is an important tool for finding solution of MEP (16):

Let $r > 0$. For a given point $x \in H$,

$$\text{find } z \in C \text{ such that } \Phi(z, w)+\varphi(w)-\varphi(z)+\frac{1}{r}\langle w-z, z-x\rangle \geq 0, \quad \forall w \in C. \tag{19}$$

The existence of the solution of auxiliary mixed equilibrium problem (19) is guaranteed by the following result.

It is well known ([48]) that if C is a nonempty closed convex subset of a real Hilbert space H; $\Phi : C \times C \to \mathbb{R}$ is a bifunction satisfy the conditions (Φ1)-(Φ4); $\varphi : C \to \mathbb{R}$ is a proper lower semicontinuous convex function, $r > 0$ and $T_r^{(\Phi,\varphi)} : H \to C$ is a mapping defined by

$$T_r^{(\Phi,\varphi)}(x) := \left\{z \in C : \Phi(z, y) + \varphi(y) - \varphi(z) + \frac{1}{r}\langle y - z, z - x\rangle \geq 0, \quad \forall y \in C\right\}, x \in H,$$

then, the following assertions hold:

(a) $T_r^{(\Phi,\varphi)}(x) \neq \emptyset$ for each $x \in H$ and $T_r^{(\Phi,\varphi)}$ is single-valued;

(b) $T_r^{(\Phi,\varphi)}$ is firmly nonexpansive, that is,

$$\left\|T_r^{(\Phi,\varphi)}x - T_r^{(\Phi,\varphi)}y\right\|^2 \leq \left\langle T_r^{(\Phi,\varphi)}x - T_r^{(\Phi,\varphi)}y, x - y\right\rangle, \quad \text{for all } x, y \in H;$$

(c) $\mathrm{Fix}\left(T_r^{(\Phi,\varphi)}\right) = \Omega[MEP(\Phi, \varphi)]$;

(d) $\Omega[MEP(\Phi, \varphi)]$ is closed and convex.

Following [30], first we collect some properties of $T_r^{(\Phi,\varphi)}$ in the following propositions.

Proposition 1 *Let C be a nonempty closed convex subset of a real Hilbert space H. Let $\Psi : C \to H$ be an inverse strongly monotone mapping with coefficient $\nu > 0$. Let $0 < r < 2\nu$ and $T_r : H \to C$ be a firmly nonexpansive mapping. Assume that x^* is an element in C such that $x^* = T_r(I - r\Psi)x^*$. Then*

$$\left\|T_r(I - r\Psi)x - x^*\right\|^2 \leq \|x-x^*\|^2-\left\|x - T_r(I - r\Psi)x - r(\Psi x - \Psi x^*)\right\|^2, \quad \textit{for all } x \in C.$$

Proof Lemma 5 shows that $I - r\Psi$ is nonexpansive. For $x \in C$, we have

$$\begin{aligned}\|T_r(I - r\Psi)x - x^*\|^2 &= \|T_r(I - r\Psi)x - T_r(I - r\Psi)x^*\|^2 \\ &\le \langle T_r(I - r\Psi)x - x^*, (I - r\Psi)x - (I - r\Psi)x^*\rangle \\ &\le \frac{1}{2}[\|(I - r\Psi)x - (I - r\Psi)x^*\|^2 + \|T_r(I - r\Psi)x - x^*\|^2 \\ &\quad - \|T_r(I - r\Psi)x - x^* - ((I - r\Psi)x - (I - r\Psi)x^*)\|^2] \\ &\le \frac{1}{2}[\|x - x^*\|^2 + \|T_r(I - r\Psi)x - x^*\|^2 \\ &\quad - \|x - T_r(I - r\Psi)x - r(\Psi x - \Psi x^*)\|^2],\end{aligned}$$

which implies that

$$\|T_r(I - r\Psi)x - x^*\|^2 \le \|x - x^*\|^2 - \|x - T_r(I - r\Psi)x - r(\Psi x - \Psi x^*)\|^2. \square$$

Proposition 2 [30] *Let C be a nonempty closed convex subset of a real Hilbert space H. Let $\Phi : C \times C \to \mathbb{R}$ be a bifunction satisfying the conditions ($\Phi 1$)-($\Phi 4$) and let $\varphi : C \to \mathbb{R}$ be a proper lower semicontinuous convex function. For $r > 0$ and $x \in H$, define a mapping $T_r^{(\Phi,\varphi)} : H \to C$ by*

$$T_r^{(\Phi,\varphi)}(x) := \left\{z \in C : \Phi(z, y) + \varphi(y) - \varphi(z) + \frac{1}{r}\langle y - z, z - x\rangle \ge 0, \quad \text{for all } y \in C\right\}.$$

Let $\Psi : C \to H$ be an inverse strongly monotone mapping with coefficient $\nu > 0$. Let $\{r_n\}$ be sequence of real numbers such that $0 < a \le r_n \le b < 2\nu$ for all $n \in \mathbb{N}$. Let $\{x_n\}$ be a bounded sequence in C such that it converges weakly to $\bar{x} \in C$ and $\left\|x_n - T_{r_n}^{(\Phi,\varphi)}(I - r_n\Psi)x_n\right\| \to 0$ as $n \to \infty$. Then $\bar{x} \in \Omega[GMEP(\Phi, \Psi, \varphi)]$.

Proof Set $u_n := T_{r_n}^{(\Phi,\varphi)}(I - r_n\Psi)x_n$, we have

$$\Phi(u_n, v) + \langle \Psi x_n, v - u_n\rangle + \varphi(v) - \varphi(u_n) + \frac{1}{r_n}\langle u_n - x_n, v - u_n\rangle \ge 0, \text{ for all } v \in C. \tag{20}$$

From ($\Phi 2$), we have

$$\langle \Psi x_n, v - u_n\rangle + \varphi(v) - \varphi(u_n) + \frac{1}{r_n}\langle u_n - x_n, v - u_n\rangle \ge \Phi(v, u_n), \quad \text{for all } v \in C.$$

Let $v \in C$. Put $u_t = tv + (1 - t)\bar{x}$ for all $t \in (0, 1]$. Then $u_t \in C$, and from (20), we have

$$\langle \Psi x_n, u_t - u_n\rangle + \varphi(u_t) - \varphi(u_n) + \frac{1}{r_n}\langle u_n - x_n, u_t - u_n\rangle \ge \Phi(u_t, u_n).$$

By monotonicity of Ψ, we have

$$\begin{aligned}\langle \Psi u_t, u_t - u_n\rangle &\geq \langle \Psi u_t, u_t - u_n\rangle - \langle \Psi x_n, u_t - u_n\rangle + \Phi(u_t, u_n)\\ &\quad + \varphi(u_n) - \varphi(u_t) - \frac{1}{r_n}\langle u_n - x_n, u_t - u_n\rangle\\ &= \langle \Psi u_t - \Psi u_n, u_t - u_n\rangle + \langle \Psi u_n - \Psi x_n, u_t - u_n\rangle + \Phi(u_t, u_n)\\ &\quad + \varphi(u_n) - \varphi(u_t) - \frac{1}{r_n}\langle u_n - x_n, u_t - u_n\rangle\\ &\geq \langle \Psi u_n - \Psi x_n, u_t - u_n\rangle + \Phi(u_t, u_n)\\ &\quad + \varphi(u_n) - \varphi(u_t) - \frac{1}{r_n}\langle u_n - x_n, u_t - u_n\rangle.\end{aligned}$$

Since $\|x_n - u_n\| \to 0$ as $n \to \infty$, we obtain by Lipschitz continuity of Ψ that $\|\Psi x_n - \Psi u_n\| \to 0$ as $n \to \infty$. Moreover,

$$\frac{\|u_n - x_n\|}{r_n} \leq \frac{\|u_n - x_n\|}{a} \to 0 \text{ as } n \to \infty$$

By (Φ3), weak lower semicontinuity of φ and $u_n \rightharpoonup \bar{x}$, we have

$$\langle \Psi u_t, u_t - \bar{x}\rangle \geq \Phi(u_t, \bar{x}) + \varphi(\bar{x}) - \varphi(u_t).$$

From (Φ1) and (Φ3), we get

$$\begin{aligned}0 &= \Phi(u_t, u_t) + \varphi(u_t) - \varphi(u_t)\\ &\leq t\Phi(u_t, v) + (1-t)\Phi(u_t, \bar{x}) + t\varphi(v) + (1-t)\varphi(\bar{x}) - \varphi(u_t)\\ &= t\,[\Phi(u_t, v) + \varphi(v) - \varphi(u_t)] + (1-t)\,[\Phi(u_t, \bar{x}) + \varphi(\bar{x}) - \varphi(u_t)]\\ &\leq t\,[\Phi(u_t, v) + \varphi(v) - \varphi(u_t)] + (1-t)\langle \Psi u_t, u_t - \bar{x}\rangle\\ &= t\,[\Phi(u_t, v) + \varphi(v) - \varphi(u_t)] + (1-t)t\langle \Psi u_t, v - \bar{x}\rangle.\end{aligned}$$

It follows that

$$0 \leq \Phi(u_t, v) + \varphi(v) - \varphi(u_t) + (1-t)\langle \Psi u_t, v - \bar{x}\rangle.$$

Letting $t \to 0$, we have, for each $v \in C$,

$$0 \leq \Phi(\bar{x}, v) + \varphi(v) - \varphi(\bar{x}) + \langle \Psi\bar{x}, v - \bar{x}\rangle.$$

This implies that $\bar{x} \in \Omega[\mathrm{GMEP}(\Phi, \Psi, \varphi)]$. □

Proposition 3 [30] *Let C be a nonempty closed convex subset of a real Hilbert space H. Let $\Psi : C \to H$ be an inverse strongly monotone mapping with coefficient $\nu > 0$. Let $\Phi : C \times C \to \mathbb{R}$ be satisfy conditions (Φ1)–(Φ4) and $\varphi : C \to \mathbb{R}$ be* *a proper lower semicontinuous convex function. Then for* $r, s > 0$ *with* $r < 2\nu$ *and* $s < 2\nu$*, we have*

$$\left\| T_s^{(\Phi,\varphi)}(I - s\Psi)y - T_r^{(\Phi,\varphi)}(I - r\Psi)x \right\| \leq \|y - x\| + |1 - r/s|\,\|v - y\|, \quad \textit{for all } x, y \in C,$$

where $v := T_s^{(\Phi,\varphi)}(I - s\Psi)y$.

Proof Let $x, y \in C$. Set $u := T_r^{(\Phi,\varphi)}(I - r\Psi)x$ and $v := T_s^{(\Phi,\varphi)}(I - s\Psi)y$. Then, we have

$$\Phi(u, w) + \varphi(w) - \varphi(u) + \frac{1}{r}\langle w - u, u - (I - r\Psi)x\rangle \geq 0, \quad \text{for all } w \in C \tag{21}$$

and

$$\Phi(v, w) + \varphi(w) - \varphi(v) + \frac{1}{s}\langle w - v, v - (I - s\Psi)y\rangle \geq 0, \quad \text{for all } w \in C. \tag{22}$$

Put $w = v$ in (21) and $w = u$ in (22), we get

$$\Phi(u, v) + \varphi(v) - \varphi(u) + \frac{1}{r}\langle v - u, u - (I - r\Psi)x\rangle \geq 0 \tag{23}$$

and

$$\Phi(v, u) + \varphi(u) - \varphi(v) + \frac{1}{s}\langle u - v, v - (I - s\Psi)y\rangle \geq 0. \tag{24}$$

Adding inequalities (23) and (24), we obtain

$$0 \leq \left\langle v - u, \frac{1}{r}[u - (I - r\Psi)x] - \frac{1}{s}[v - (I - s\Psi)y]\right\rangle,$$

and hence,

$$\begin{aligned} 0 &\leq \left\langle v - u, [u - (I - r\Psi)x] - \frac{r}{s}[v - (I - s\Psi)y]\right\rangle \\ &= \left\langle v - u, u - v + \left(1 - \frac{r}{s}\right)(v - y) - (x - r\Psi x) + y - r\Psi y\right\rangle, \end{aligned}$$

which implies that

$$\|v - u\|^2 \leq \|v - u\|[|1 - r/s|\,\|v - y\| + \|x - r\Psi x - (y - r\Psi y)\|].$$

By Lemma 5, we obtain

$$\|v - u\| \leq \|y - x\| + |1 - r/s|\,\|v - y\|. \qquad \square$$

6.1 Existence and Uniqueness of Solutions of Hierarchical Variational Inequality Problem

First, we show that the solution of a certain hierarchical VIP over the set $D = \text{Fix}(\mathscr{T}) \cap \Omega[\text{GMEP}(\Phi, \Psi, \varphi)]$ is unique.

Proposition 4 *Let C be a nonempty closed convex subset of a real Hilbert space H. Let $\mathscr{T} = \{T(t) : t \in (0, \infty)\}$ be a family of nonexpansive self-mappings on C such that $Fix(\mathscr{T}) \neq \emptyset$. Assume that $A : H \to H$ is a strongly positive linear bounded self-adjoint operator with the coefficient $\bar{\gamma} > 0$, $f : C \to H$ is $\mathscr{L}$-contraction, $\Psi : C \to H$ is an inverse strongly monotone mapping with coefficient $\nu > 0$, $\varphi : C \to \mathbb{R}$ is a lower semicontinuous convex functional and $\Phi : C \times C \to \mathbb{R}$ is a bifunction satisfying conditions (Φ1)–(Φ4). Then, for $\gamma \in (0, \overline{\gamma}/\mathscr{L})$, the mapping $Q(I - A + \gamma f) : C \to C$ has a unique fixed point $x^* \in C$ and $x^* \in Fix(\mathscr{T}) \cap \Omega[GMEP(\Phi, \Psi, \varphi)]$, where $Q = P_{Fix(\mathscr{T}) \cap \Omega[GMEP(\Phi, \Psi, \varphi)]}$.*

Proof Let $Q = P_{\text{Fix}(\mathscr{T}) \cap \Omega[\text{GMEP}(\Phi, \Psi, \varphi)]}$ and $\gamma \in (0, \overline{\gamma}/\mathscr{L})$. First, we show that $Q(I - A + \gamma f)$ is a contraction mapping from C into itself. Let $x, y \in C$, From Lemma 4, we have

$$\begin{aligned}
\|Q(I - A + \gamma f)x &- Q(I - A + \gamma f)y\| \\
&\leq \|(I - A + \gamma f)x - (I - A + \gamma f)y\| \\
&\leq \|(I - A)x - (I - A)y\| + \gamma \|fx - fy\| \\
&\leq \|I - A\| \|x - y\| + \gamma \|fx - fy\| \\
&\leq (1 - \overline{\gamma}) \|x - y\| + \gamma \mathscr{L} \|x - y\| \\
&= (1 - (\overline{\gamma} - \gamma \mathscr{L})) \|x - y\|.
\end{aligned}$$

Since $Q(I - A + \gamma f)$ is a contraction, therefore, there exists a unique element $x^* \in C$ such that $x^* = Q(I - A + \gamma f)x^*$. Since Q is onto, it follows that $x^* \in \text{Fix}(\mathscr{T}) \cap \Omega[\text{GMEP}(\Phi, \Psi, \varphi)]$. □

6.2 Computation of Unique Solution of Hierarchical Optimization Problem

Let C be a nonempty closed convex subset of a real Hilbert space H and let $\mathscr{T} = \{T(t) : t \in (0, \infty)\}$ be a family of nonexpansive self-mappings on C with $\text{Fix}(\mathscr{T}) \neq \emptyset$. Let $\{\sigma_n : n \in \mathbb{N}\}$ be a sequence of nonexpansive self-mappings on C such that $\text{Fix}(\mathscr{T}) \subseteq \bigcap_{n \in \mathbb{N}} \text{Fix}(\sigma_n)$. Assume that $A : H \to H$ is a strongly positive linear bounded self-adjoint operator with the coefficient $\bar{\gamma} > 0$ such that $(I - \alpha A)(C) \subseteq C$ for each $\alpha \in (0, 1)$, $f : C \to H$ is $\mathscr{L}$-contraction, $\Psi : C \to H$ is an inverse strongly monotone mapping with coefficient $\nu > 0$, $\varphi : C \to \mathbb{R}$ is a lower semicontinuous convex functional and $\Phi : C \times C \to \mathbb{R}$ is a bifunction satisfying conditions (Φ1)–(Φ4).

Proposition 4 guarantees that, for $\gamma \in (0, \overline{\gamma}/\mathscr{L})$, the mapping

$$P_{\mathrm{Fix}(\mathscr{T})\cap\Omega[\mathrm{GMEP}(\Phi,\Psi,\varphi)]}(I - A + \gamma f) : C \to C$$

has a unique fixed point $x^* \in \mathrm{Fix}(\mathscr{T}) \cap \Omega[\mathrm{GMEP}(\Phi, \Psi, \varphi)] =: D$. It means that $x^* \in C$ is the unique solution of the following hierarchical VIP:

$$\text{find } z \in D \text{ such that } \langle(\gamma f - A)z, z - v\rangle \geq 0, \quad \text{for all } v \in D. \tag{25}$$

We remark that if $C = H$, then the hierarchical variational inequality problem (25) reduces to the hierarchical optimization problem (1) over the set $\mathrm{Fix}(\mathscr{T}) \cap \Omega[\mathrm{GMEP}(\Phi, \Psi, \varphi)]$.

To compute $x^* \in D = \mathrm{Fix}(\mathscr{T}) \cap \Omega[\mathrm{GMEP}(\Phi, \Psi, \varphi)]$, we propose the following iterative algorithm:

$$\begin{cases} x_1 \in C; \\ \Phi(u_n, v) + \langle \Psi x_n, v - u_n\rangle + \varphi(v) - \varphi(u_n) + \frac{1}{r_n}\langle u_n - x_n,\ v - u_n\rangle \geq 0, \quad \forall v \in C; \\ z_n = \beta_n x_n + (1 - \beta_n)\sigma_n(u_n); \\ x_{n+1} = \gamma\alpha_n f x_n + (I - \alpha_n A)z_n, \quad \text{for all } n \in \mathbb{N}, \end{cases} \tag{26}$$

where $\{\alpha_n\}$ and $\{\beta_n\}$ are sequences in $(0, 1]$ and $\{r_n\}$ is a sequence in $(0, \infty)$.

For convergence of algorithm (26), we use the method employed in [30]. We investigate the asymptotic behavior of the sequence $\{x_n\}$ generated, from an arbitrary point $x_1 \in C$, by algorithm (26) under the following conditions:

(C1) $\lim_{n\to\infty} \alpha_n = 0$ and $\sum_{n=1}^{\infty} \alpha_n = \infty$,
(C2) $0 < a \leq \beta_n \leq b < 1$ for all $n \in \mathbb{N}$ and $\lim_{n\to\infty} |\beta_n - \beta_{n+1}| = 0$,
(C3) $0 < \overline{r} \leq r_n \leq \overline{\overline{r}} < 2\nu$ for all $n \in \mathbb{N}$ and $\lim_{n\to\infty} |r_n - r_{n+1}| = 0$,
(C4) $\lim_{n\to\infty} \|\sigma_n(u_n) - \sigma_{n+1}(u_n)\| = 0$.

Some basic properties of Algorithm (26) are detailed below:

Lemma 7 *Let $\{x_n\}$ be a sequence of Algorithm* (26). *Then*

(a) $\{x_n\}$ is bounded, and
(b) $\lim_{n\to\infty} ||x_{n+1} - x_n|| = \lim_{n\to\infty} ||u_{n+1} - u_n|| = \lim_{n\to\infty} ||z_{n+1} - z_n|| = 0.$

Proof (a) Define $v_n := \sigma_n(u_n)$. Note that $u_n = T_{r_n}^{(\Phi,\varphi)}(I - r_n\Psi)x_n$ and $x^* = T_{r_n}^{(\Phi,\varphi)}(I - r_n\Psi)x^*$ for all $n \in \mathbb{N}$. Since $T_r^{(\Phi,\varphi)}$ is nonexpansive, it follows from Lemma 5 that $T_r^{(\Phi,\varphi)}(I - r\Psi)$ is also nonexpansive for each $r > 0$. Thus, $\|u_n - x^*\| \leq \|x_n - x^*\|$, and hence, $\|v_n - x^*\| \leq \|x_n - x^*\|$ for all $n \in \mathbb{N}$. Moreover,

$$\begin{aligned} \|z_n - x^*\| &\leq \beta_n\|x_n - x^*\| + (1 - \beta_n)\|\sigma_n(u_n) - x^*\| \\ &\leq \beta_n\|x_n - x^*\| + (1 - \beta_n)\|u_n - x^*\| \\ &\leq \|x_n - x^*\|, \quad \forall n \in \mathbb{N}. \end{aligned}$$

Noticing that $\lim_{n\to\infty} \alpha_n = 0$, we may assume, without loss of generality, that $\alpha_n < \|A\|^{-1}$ for all $n \in \mathbb{N}$. From (26), we have

$$\begin{aligned}
\|x_{n+1} - x^*\| &= \|\gamma\alpha_n f x_n - \alpha_n Ax^* + (I - \alpha_n A)(z_n - x^*)\| \\
&\le \alpha_n \|\gamma f x_n - Ax^*\| + (1 - \alpha_n \overline{\gamma})\|z_n - x^*\| \\
&\le \alpha_n (\|\gamma f x_n - \gamma f x^*\| + \|\gamma f x^* - Ax^*\|) + (1 - \overline{\gamma}\alpha_n)\|x_n - x^*\| \\
&\le [1 - \alpha_n(\overline{\gamma} - \gamma \mathscr{L})]\|x_n - x^*\| + \alpha_n \|\gamma f x^* - Ax^*\| \\
&\le \max\left\{\|x_n - x^*\|, \frac{\|\gamma f x^* - Ax^*\|}{\overline{\gamma} - \gamma\mathscr{L}}\right\} \\
&\le \max\left\{\|x_1 - x^*\|, \frac{\|\gamma f x^* - Ax^*\|}{\overline{\gamma} - \gamma\mathscr{L}}\right\}, \text{ for all } n \in \mathbb{N}.
\end{aligned} \tag{27}$$

Therefore, $\{x_n\}$ is bounded.

(b) Part (a) implies that iterates $\{x_n\}$ of Algorithm (26) is bounded. Therefore, $\{u_n\}$, $\{v_n\}$, and $\{z_n\}$ are bounded.
Taking $y = x_{n+1}$, $x = x_n$, $s = r_{n+1}$ and $r = r_n$ in Proposition 3, we have

$$\begin{aligned}
\|u_{n+1} - u_n\| &= \left\|T_{r_{n+1}}^{(\Phi,\varphi)}(x_{n+1} - r_{n+1}\Psi x_{n+1}) - T_{r_n}^{(\Phi,\varphi)}(x_n - r_n \Psi x_n)\right\| \\
&\le \|x_{n+1} - x_n\| + \left|1 - \frac{r_n}{r_{n+1}}\right| \|u_{n+1} - x_{n+1}\| \\
&\le \|x_{n+1} - x_n\| + \frac{|r_{n+1} - r_n|}{\overline{r}} \|u_{n+1} - x_{n+1}\|.
\end{aligned} \tag{28}$$

We rewrite x_{n+1} as

$$x_{n+1} = (1 - \lambda_n)x_n + \lambda_n w_n,$$

where

$$w_n = \frac{1}{\lambda_n}\left[\alpha_n \beta_n (I - A)x_n + (1 - \beta_n)(I - \alpha_n A)v_n + \alpha_n \gamma f(x_n)\right]. \tag{29}$$

and

$$\lambda_n = 1 - (1 - \alpha_n)\beta_n.$$

From the assumptions $\lim_{n\to\infty} \alpha_n = 0$ and $0 < a \le \beta_n \le b < 1$ for all $n \in \mathbb{N}$, we have

$$0 < \liminf_{n\to\infty} \lambda_n \le \limsup_{n\to\infty} \lambda_n < 1.$$

By (29), we have

$$
\begin{aligned}
&w_{n+1} - w_n \\
&= \frac{1}{\lambda_{n+1}} \left[\alpha_{n+1}\beta_{n+1}(I - A)x_{n+1} + (1 - \beta_{n+1})(I - \alpha_{n+1}A)v_{n+1} \quad + \alpha_{n+1}\gamma f x_{n+1}\right] \\
&- \frac{1}{\lambda_n} \left[\alpha_n \beta_n (I - A)x_n + (1 - \beta_n)(I - \alpha_n A)v_n + \alpha_n \gamma f x_n\right] \\
&= \frac{\alpha_{n+1}\beta_{n+1}}{\lambda_{n+1}}(I - A)x_{n+1} - \frac{\alpha_n \beta_n}{\lambda_n}(I - A)x_n \quad + \frac{1 - \beta_{n+1}}{\lambda_{n+1}}(v_{n+1} - v_n) \\
&+ \left(\frac{1 - \beta_{n+1}}{\lambda_{n+1}} - \frac{1 - \beta_n}{\lambda_n}\right)v_n - \frac{\alpha_{n+1}(1 - \beta_{n+1})}{\lambda_{n+1}}A(v_{n+1} - v_n) \\
&- (1 - \beta_{n+1})\left(\frac{\alpha_{n+1}}{\lambda_{n+1}} - \frac{\alpha_n}{\lambda_n}\right)Av_n + (\beta_{n+1} - \beta_n)\frac{\alpha_n}{\lambda_n}Av_n \\
&+ \frac{\alpha_{n+1}\gamma}{\lambda_{n+1}}\left[f x_{n+1} - f x_n\right] + \left(\frac{\alpha_{n+1}}{\lambda_{n+1}} - \frac{\alpha_n}{\lambda_n}\right)\gamma f x_n.
\end{aligned}
$$

Consequently,

$$
\begin{aligned}
&||w_{n+1} - w_n|| - ||x_{n+1} - x_n|| \\
&\leq \frac{\alpha_{n+1}\beta_{n+1}}{\lambda_{n+1}}\|(I - A)x_{n+1}\| + \frac{\alpha_n \beta_n}{\lambda_n}\|(I - A)x_n\| \\
&+ \frac{1 - \beta_{n+1}}{\lambda_{n+1}}(\|\sigma_{n+1}(u_{n+1}) - \sigma_{n+1}(u_n)\| + \|\sigma_{n+1}(u_n) - \sigma_n(u_n)\|) \\
&+ \left(\frac{1 - \beta_{n+1}}{\lambda_{n+1}} - \frac{1 - \beta_n}{\lambda_n}\right)\|v_n\| + \frac{\alpha_{n+1}(1 - \beta_{n+1})}{\lambda_{n+1}}\|A(v_{n+1} - v_n)\| \\
&+ \left[(1 - \beta_{n+1})\left|\frac{\alpha_{n+1}}{\lambda_{n+1}} - \frac{\alpha_n}{\lambda_n}\right| + \frac{\alpha_n|\beta_{n+1} - \beta_n|}{\lambda_n}\right]\|Av_n\| - ||x_{n+1} - x_n|| \\
&+ \frac{\alpha_{n+1}\gamma}{\lambda_{n+1}}\|f x_{n+1} - f x_n\| + \left|\frac{\alpha_{n+1}}{\lambda_{n+1}} - \frac{\alpha_n}{\lambda_n}\right|\gamma\|f x_n\|. \qquad (30)
\end{aligned}
$$

Note that $\frac{1-\beta_{n+1}}{\lambda_{n+1}} < 1$ for all $n \in \mathbb{N}$, $\frac{1-\beta_{n+1}}{\lambda_{n+1}} \to 1$ and $\alpha_n \to 0$. Using (28) and (30), we obtain

$$\limsup_{n\to\infty} (||w_{n+1} - w_n|| - ||x_{n+1} - x_n||) \leq 0,$$

and hence, by [37, Lemma 2.2], we deduce that $\lim_{n\to\infty} ||w_n - x_n|| = 0$. Therefore,

$$\lim_{n\to\infty} ||x_{n+1} - x_n|| = \lim_{n\to\infty} \lambda_n ||w_n - x_n|| = 0.$$

From (28), we immediately have $\lim_{n\to\infty}||u_{n+1}-u_n||=0$. Further, since $\{\sigma_n : n\in\mathbb{N}\}$ is a sequence of nonexpansive self-mappings, we have

$$\begin{aligned}
&||z_n - z_{n-1}||\\
&= \|\beta_n x_n + (1-\beta_n)\sigma_n(u_n) - \beta_{n-1}x_{n-1} - (1-\beta_{n-1})\sigma_{n-1}(u_{n-1})\|\\
&= \|\beta_n(x_n - x_{n-1}) + (\beta_n - \beta_{n-1})x_{n-1} + (1-\beta_n)\sigma_n(u_n) - \sigma_{n-1}(u_{n-1})\\
&\quad - (\beta_n - \beta_{n-1})\sigma_{n-1}(u_{n-1})\|\\
&\le \beta_n\|x_n - x_{n-1}\| + (1-\beta_n)\|\sigma_n(u_n) - \sigma_{n-1}(u_{n-1})\|\\
&\quad + |\beta_n - \beta_{n-1}|\,\|x_{n-1} - \sigma_{n-1}(u_{n-1})\|\\
&\le \beta_n\|x_n - x_{n-1}\| + (1-\beta_n)(\|\sigma_n(u_n) - \sigma_n(u_{n-1})\| + \|\sigma_n(u_{n-1}) - \sigma_{n-1}(u_{n-1})\|)\\
&\quad + |\beta_n - \beta_{n-1}|\,\|x_{n-1} - \sigma_{n-1}(u_{n-1})\|\\
&\le \beta_n\|x_n - x_{n-1}\| + (1-\beta_n)(\|u_n - u_{n-1}\| + \|\sigma_n(u_{n-1}) - \sigma_{n-1}(u_{n-1})\|)\\
&\quad + |\beta_n - \beta_{n-1}|\,\|x_{n-1} - \sigma_{n-1}(u_{n-1})\|.
\end{aligned}$$

Therefore,

$$\begin{aligned}
&||z_n - z_{n-1}||\\
&\le \beta_n\|x_n - x_{n-1}\| + (1-\beta_n)(\|x_n - x_{n-1}\| + \frac{|r_n - r_{n-1}|}{\bar{r}}\,\|u_n - x_n\|\\
&\quad + \|\sigma_n(u_{n-1}) - \sigma_{n-1}(u_{n-1})\|) + |\beta_n - \beta_{n-1}|\|x_{n-1} - \sigma_{n-1}(u_{n-1})\|\\
&\le \|x_n - x_{n-1}\| + (1-\beta_n)\left(\frac{|r_n - r_{n-1}|}{\bar{r}}\,\|u_n - x_n\| + \|\sigma_n(u_{n-1}) - \sigma_{n-1}(u_{n-1})\|\right)\\
&\quad + |\beta_n - \beta_{n-1}|\|x_{n-1} - \sigma_{n-1}(u_{n-1})\|.
\end{aligned}$$

By the conditions (C2)–(C4), we obtain that $\|z_{n+1} - z_n\| \to 0$ as $n\to\infty$. □

Now we prove strong convergence of the sequence generated by the proposed algorithm (26) to the unique solution of the hierarchical VIP (25) over the set $D = \text{Fix}(\mathscr{T}) \cap \Omega[\text{GMEP}(\Phi, \Psi, \varphi)]$.

Theorem 3 *Let C be a nonempty closed convex subset of a real Hilbert space H. Let $\mathscr{T} = \{T(t) : t\in(0,\infty)\}$ be a family of nonexpansive self-mappings on C with $Fix(\mathscr{T}) \neq \emptyset$, and $\Gamma = \{\sigma_{t_n} : n\in\mathbb{N}\}$ be a sequence of nonexpansive self-mappings on C such that $Fix(\mathscr{T}) \subseteq \bigcap_{n\in\mathbb{N}} Fix(\sigma_{t_n})$ and $Fix(\mathscr{T}) \cap \Omega[GMEP(\Phi,\Psi,\varphi)] \neq \emptyset$. For given $x_1 \in C$, let $\{x_n\}$ be a sequence in C generated by* (26), *where $\{\alpha_n\}$ and $\{\beta_n\}$ are two sequence in $(0, 1]$ and $\{r_n\}$ is a sequence in $(0,\infty)$ satisfying conditions (C1)–(C4). Assume that $0 < \gamma < \bar{\gamma}/\mathscr{L}$ and $\{\sigma_n(u_n)\}$ is an approximating fixed point sequence of the family $\mathscr{T}$. Then, the sequence $\{x_n\}$ converges strongly to $x^* \in Fix(\mathscr{T}) \cap \Omega[GMEP(\Phi,\Psi,\varphi)]$, which solves the hierarchical VIP* (25) *over the set $Fix(\mathscr{T}) \cap \Omega[GMEP(\Phi,\Psi,\varphi)]$.*

Proof First, we show that $\limsup_{n\to\infty}\langle(\gamma f - A)x^*, x_{n+1} - x^*\rangle \le 0$.

Set $v_n := \sigma_n(u_n)$ and $\Upsilon_n := \langle \gamma f(x_n) - Ax^*, x_{n+1} - x^*\rangle$. Note that $\{\Upsilon_n\}$ is bounded. Indeed,

$$\begin{aligned}|\Upsilon_n| &= |\langle \gamma f(x_n) - \gamma f(x^*) + \gamma f(x^*) - Ax^*, x_{n+1} - x^*\rangle| \\ &\le \left(\gamma\|f(x_n) - \gamma f(x^*)\| + \|\gamma f(x^*) - Ax^*\|\right)\|x_{n+1} - x^*\| \\ &\le \left(\gamma\mathscr{L}\|x_n - x^*\| + \|\gamma f(x^*) - Ax^*\|\right)\|x_{n+1} - x^*\| \text{ for all } n \in \mathbb{N}.\end{aligned}$$

Since $\{x_n\}$ is bounded and so is $\{\Upsilon_n\}$. Using (5), we have

$$\begin{aligned}\|u_n - x^*\|^2 &= \|T_{r_n}^{(\Phi,\varphi)}(I - r_n\Psi)x_n - T_{r_n}^{(\Phi,\varphi)}(I - r_n\Psi)x^*\|^2 \\ &\le \|(I - r_n\Psi)x_n - (I - r_n\Psi)x^*\|^2 \\ &\le \|x_n - x^*\|^2 - r_n(2\nu - r_n)\|\Psi x_n - \Psi x^*\|^2. \end{aligned} \tag{31}$$

From (3), we have

$$\|x_{n+1} - x^*\|^2 \le \|(I - \alpha_n A)(z_n - x^*)\|^2 + 2\alpha_n\langle \gamma f(x_n) - Ax^*, x_{n+1} - x^*\rangle.$$

So, from (31), we have

$$\begin{aligned}\|x_{n+1} - x^*\|^2 &\le (1 - \alpha_n\overline{\gamma})^2\|z_n - x^*\|^2 + 2\alpha_n\langle \gamma f(x_n) - Ax^*, x_{n+1} - x^*\rangle \\ &\le (1 + \alpha_n^2\overline{\gamma}^2)[\beta_n\|x_n - x^*\|^2 + (1 - \beta_n)\|\sigma_n(u_n) - x^*\|^2] + 2\alpha_n\Upsilon_n \\ &\le (1 + \alpha_n^2\overline{\gamma}^2)[\beta_n\|x_n - x^*\|^2 + (1 - \beta_n)\|u_n - x^*\|^2] + 2\alpha_n\Upsilon_n \\ &\le (1 + \alpha_n^2\overline{\gamma}^2)[\beta_n\|x_n - x^*\|^2 + (1 - \beta_n)(\|x_n - x^*\|^2 \\ &\quad - r_n(2\nu - r_n)\|\Psi x_n - \Psi x^*\|^2)] + 2\alpha_n\Upsilon_n \\ &\le (1 + \alpha_n^2\overline{\gamma}^2)\|x_n - x^*\|^2 \\ &\quad - r_n(2\nu - r_n)(1 - \beta_n)(1 + \alpha_n^2\overline{\gamma}^2)\|\Psi x_n - \Psi x^*\|^2 + 2\alpha_n\Upsilon_n. \end{aligned} \tag{32}$$

It follows that

$$\begin{aligned}&r_n(2\nu - r_n)(1 - \beta_n)(1 + \alpha_n^2\overline{\gamma}^2)\|\Psi x_n - \Psi x^*\|^2 \\ &\qquad \le \|x_n - x^*\|^2 - \|x_{n+1} - x^*\|^2 + \alpha_n^2\overline{\gamma}^2\|x_n - x^*\|^2 + 2\alpha_n\Upsilon_n \\ &\qquad \le \|x_{n+1} - x_n\|(\|x_n - x^*\| + \|x_{n+1} - x^*\|) + \alpha_n^2\overline{\gamma}^2\|x_n - x^*\|^2 \\ &\qquad\quad + 2\alpha_n\Upsilon_n.\end{aligned}$$

Since $\overline{r} \le r_n \le \overline{\overline{r}} < 2\nu$ and $\beta_n \le b$ for all $n \in \mathbb{N}$, $\lim_{n\to\infty}\alpha_n = 0$ and $\lim_{n\to\infty}\|x_{n+1} - x_n\| = 0$, we obtain that $\|\Psi x_n - \Psi x^*\| \to 0$ as $n \to \infty$. From Proposition 1, we get

$$\begin{aligned}\|u_n - x^*\|^2 &\le \|x_n - x^*\|^2 - \|x_n - u_n - r(\Psi x_n - \Psi x^*)\|^2 \\ &= \|x_n - x^*\|^2 - \|x_n - u_n\|^2 + 2r_n\langle x_n - u_n, \Psi x_n - \Psi x^*\rangle \\ &\quad - r_n^2\|\Psi x_n - \Psi x^*\|^2. \end{aligned} \tag{33}$$

From (32) and (33), we have

$$\begin{aligned}\|x_{n+1} - x^*\|^2 &\le (1 + \alpha_n^2\overline{\gamma}^2)[\beta_n\|x_n - x^*\|^2 + (1 - \beta_n)(\|x_n - x^*\|^2 - \|x_n - u_n\|^2 \\ &\quad + 2r_n\langle x_n - u_n, \Psi x_n - \Psi x^*\rangle - r_n^2\|\Psi x_n - \Psi x^*\|^2)] + 2\alpha_n\Upsilon_n \\ &\le (1 + \alpha_n^2\overline{\gamma}^2)\|x_n - x^*\|^2 + (1 + \alpha_n^2\overline{\gamma}^2)(1 - \beta_n)[2r_n\|x_n - u_n\|\|\Psi x_n - \Psi x^*\| \\ &\quad - \|x_n - u_n\|^2 - r_n^2\|\Psi x_n - \Psi x^*\|^2] + 2\alpha_n\Upsilon_n,\end{aligned}$$

and hence,

$$\begin{aligned}&(1 + \alpha_n^2\overline{\gamma}^2)(1 - b)\|x_n - u_n\|^2 \\ &\quad\le (1 + \alpha_n^2\overline{\gamma}^2)\|x_n - x^*\|^2 - \|x_n+1 - x^*\|^2 \\ &\qquad + (1 + \alpha_n^2\overline{\gamma}^2)(1 - \beta_n)[2r_n\|x_n - u_n\|\|\Psi x_n - \Psi x^*\| \\ &\qquad - r_n^2\|\Psi x_n - \Psi x^*\|^2] + 2\alpha_n\Upsilon_n \\ &\quad\le (1 + \alpha_n^2\overline{\gamma}^2)\|x_n - x^*\|^2 - \|x_{n+1} - x^*\|^2 \\ &\qquad + 2r_n(1 + \alpha_n^2\overline{\gamma}^2)(1 - \beta_n)\|x_n - u_n\|\|\Psi x_n - \Psi x^*\| + 2\alpha_n\Upsilon_n \\ &\quad\le \|x_{n+1} - x_n\|(\|x_n - x^*\| + \|x_{n+1} - x^*\|) + \alpha_n^2\overline{\gamma}^2\|x_n - x^*\|^2 \\ &\qquad + 2r_n(1 + \alpha_n^2\overline{\gamma}^2)\|x_n - u_n\|\|\Psi x_n - \Psi x^*\| + 2\alpha_n\Upsilon_n.\end{aligned}$$

Since $r_n \le \overline{\overline{r}} < 2\nu$ for all $n \in \mathbb{N}$, $\lim_{n\to\infty} \alpha_n = 0$ and $\lim_{n\to\infty} \|x_{n+1} - x_n\| = 0$, we have $\|x_n - u_n\| \to 0$ as $n \to \infty$. From the condition $\lim_{n\to\infty} \alpha_n = 0$, we have

$$\|x_{n+1} - z_n\| = \alpha_n\|\gamma f x_n - Az_n\| \to 0 \text{ as } n \to \infty,$$

which gives, $\lim_{n\to\infty} \|x_n - z_n\| = 0$. Observe that

$$\begin{aligned}\|x_n - v_n\| &\le \|x_n - z_n\| + \|z_n - v_n\| \\ &\le \|x_n - z_n\| + \beta_n\|x_n - v_n\|,\end{aligned}$$

and hence,

$$(1 - b)\|x_n - v_n\| \le (1 - \beta_n)\|x_n - v_n\| \le \|x_n - z_n\| \to 0 \text{ as } n \to \infty.$$

Define

$$D_0 := \left\{ w \in C : \|w - x^*\| \le \max \left\{ \|x_1 - x^*\|, \frac{\|\gamma f x^* - Ax^*\|}{\bar{\gamma} - \gamma \mathscr{L}} \right\} \right\}. \tag{34}$$

From (27) we see that D_0 is a nonempty closed convex bounded subset of C which is $T(t)$-invariant for each $t \in (0, \infty)$ and it contains $\{x_n\}, \{u_n\}, \{v_n\}, \{z_n\}$. Therefore, without loss of generality, we may assume that $\mathscr{T} = \{T(t) : t \in (0, \infty)\}$ is a family of nonexpansive self-mappings on D_0. Taking a suitable subsequence $\{x_{n_i}\}$ of $\{x_n\}$, we see that

$$\limsup_{n\to\infty} \langle (\gamma f - A)x^*, x_n - x^* \rangle = \lim_{i\to\infty} \langle (\gamma f - A)x^*, x_{n_i} - x^* \rangle. \tag{35}$$

Since the sequence $\{x_n\}$ is bounded in C, we may assume that $x_{n_i} \rightharpoonup \bar{x} \in C$. Note that $\{v_n\}$ is an approximating fixed point sequence of family $\mathscr{T}$, that is,

$$\lim_{n\to\infty} \|v_n - T(s)v_n\| = 0 \ \text{ for all } \ s \in (0, \infty). \tag{36}$$

Using (36) we obtain, from the demiclosedness principle, that $\bar{x} \in \text{Fix}(\mathscr{T})$. Since $\|x_{n_i} - u_{n_i}\| \to 0$ as $i \to \infty$, by Proposition 2, we obtain that $\bar{x} \in \Omega[\text{GMEP}(\Phi, \Psi, \varphi)]$. Thus, $\bar{x} \in \text{Fix}(\mathscr{T}) \cap \Omega[\text{GMEP}(\Phi, \Psi, \varphi)]$. Therefore, from (35), we have

$$\begin{aligned}
\limsup_{n\to\infty} \langle (\gamma f - A)x^*, x_{n+1} - x^* \rangle &= \lim_{i\to\infty} \langle (\gamma f - A)x^*, x_{n_i} - x^* \rangle \\
&= \langle (\gamma f - A)x^*, \overline{x} - x^* \rangle \le 0.
\end{aligned}$$

Observe that

$$\begin{aligned}
\|x_{n+1} - x^*\|^2 &= \|\alpha_n(\gamma f x_n - \gamma f x^* + \gamma f x^* - Ax^*) + (I - \alpha_n A)(z_n - x^*)\|^2 \\
&\le \|(I - \alpha_n A)(z_n - x^*) + \alpha_n(\gamma f x_n - \gamma f x^*\|^2 \\
&\quad + 2\alpha_n \langle \gamma f x^* - Ax^*, x_{n+1} - x^*) \rangle \\
&\le (\|(I - \alpha_n A)(z_n - x^*)\| + \gamma\alpha_n \|f x_n - \gamma f x^*\|)^2 + 2\alpha_n \Upsilon_n \\
&\le ((1 - \alpha_n \overline{\gamma})\|z_n - x^*\| + \gamma\alpha_n \|f x_n - f x^*\|)^2 + 2\alpha_n \Upsilon_n \\
&\le (1 - \alpha_n(\overline{\gamma} - \gamma \mathscr{L}))^2 \|x_n - x^*\|^2 + 2\alpha_n \Upsilon_n.
\end{aligned}$$

Therefore,

$$\|x_{n+1} - x^*\|^2 \le (1 - \alpha_n(\overline{\gamma} - \gamma \mathscr{L}))\|x_n - x^*\|^2 + 2\alpha_n \Upsilon_n.$$

Note $\sum_{n=1}^{\infty} \alpha_n = \infty$ and $\limsup_{n\to\infty} \Upsilon_n \le 0$. Therefore, we conclude from Lemma 6 that $\{x_n\}$ converges strongly to x^*. □

We now establish strong convergence of the sequence generated by the algorithm (37) to a solution of minimization problem (1) over the set $D = \mathrm{Fix}(\mathscr{T}) \cap \Omega[\mathrm{GMEP}(\Phi, \Psi, \varphi)]$.

Corollary 1 [30, Theorem 3.1] *Let* $\mathscr{T} = \{T(t) : t \in (0, \infty)\}$ *be a family of nonexpansive self-mappings on* H *with* $Fix(\mathscr{T}) \neq \emptyset$, *and* $\Gamma = \{\sigma_{t_n} : n \in \mathbb{N}\}$ *be a sequence of nonexpansive self-mappings on* H *such that* $Fix(\mathscr{T}) \subseteq \bigcap_{n\in\mathbb{N}} Fix(\sigma_{t_n})$ *and* $Fix(\mathscr{T}) \cap \Omega[GMEP(\Phi, \Psi, \varphi)] \neq \emptyset$. *For given* $x_1 \in H$, *let* $\{x_n\}$ *be a sequence in* H *generated by*

$$\begin{cases} x_1 \in C; \\ \Phi(u_n, v) + \langle \Psi x_n, v - u_n \rangle + \varphi(v) - \varphi(u_n) + \frac{1}{r_n}\langle u_n - x_n,\ v - u_n \rangle \geq 0, \quad \forall v \in C; \\ z_n = \beta_n x_n + (1 - \beta_n)\sigma_n(u_n); \\ x_{n+1} = \gamma \alpha_n f x_n + (I - \alpha_n A) z_n, \quad \textit{for all } n \in \mathbb{N}, \end{cases} \tag{37}$$

where $\{\alpha_n\}$ *and* $\{\beta_n\}$ *are two sequence in* $(0, 1]$ *and* $\{r_n\}$ *is a sequence in* $(0, \infty)$ *satisfying conditions (C1)–(C4). Assume that* $0 < \gamma < \bar{\gamma}/\mathscr{L}$ *and* $\{\sigma_n(u_n)\}$ *is an approximating fixed point sequence of the family* $\mathscr{T}$. *Then, the sequence* $\{x_n\}$ *converges strongly to* $x^* \in Fix(\mathscr{T}) \cap \Omega[GMEP(\Phi, \Psi, \varphi)]$, *which solves optimization problem (1) over the set* $Fix(\mathscr{T}) \cap \Omega[GMEP(\Phi, \Psi, \varphi)]$.

Corollary 1 is a far more general result than those in the existing literature of this nature. Therefore, it unifies a number of results and includes several convergence theorem of this nature. In particular, Kamraksa and Wangkeeree [17] considered the optimization problem (1) when $D = \mathrm{Fix}(\mathscr{T}) \cap \Omega[GEP(\Phi, \Psi)]$, where $\mathrm{Fix}(\mathscr{T})$ is the set of common fixed points of a nonexpansive semigroup $\mathscr{T} = \{T(t) : 0 \leq t < \infty\}$ and $\Omega[\mathrm{GEP}(\Phi, \Psi)]$ is the set of solutions of the *generalized equilibrium problem* (in short, GEP (Φ, Ψ)) [2].

7 Special Cases of Hierarchical Equilibrium Problems

7.1 When $\mathscr{T}$ is a Semigroup Nonexpansive Self-Mappings

Let C be a nonempty subset of a (real) Hilbert space H. We call a one parameter family $\mathscr{T} := \{T(t) : t \in [0, +\infty)\}$ of mappings from C into C a *strongly continuous semigroup of nonexpansive mappings* if

(S1) for each $t > 0$,

$$\|T(t)x - T(t)y\| \leq \|x - y\| \quad \text{for all } x, y \in C;$$

(S2) $T(0)x = x$ for all x in C;
(S3) $T(s + t) = T(s)T(t)$ for all s, t in $\mathbb{R}^+$;
(S4) for each x in C, the mapping $T(\cdot)x$ from $\mathbb{R}^+$ into C is continuous.

We need the following:

Lemma 8 [33] *Let C be a nonempty closed convex bounded subset of a Hilbert space H and $\mathscr{T} = \{T(t) : t \geq 0\}$ be a strongly continuous semigroup of nonexpansive mappings from C into itself. Let $\sigma_t(x) := \frac{1}{t}\int_0^t T(s)x ds$. Then*

$$\lim_{t\to\infty}\sup_{x\in C}\|\sigma_t(x) - T(h)\sigma_t(x)\| = 0, \quad \forall h > 0.$$

Proof Fix $h > 0$ and let $t > h$. Then, for $n \in \mathbb{N}$, there is a nonnegative integer i_n, such that $(\frac{t}{n})i_n \leq h \leq (\frac{t}{n})(i_n + 1)$. Thus, we have $\lim\limits_{n\to\infty}(\dfrac{t}{n})i_n = h$. Now, by an idea in [4], for $\{x_{i,n}\}_{i,n=1}^{\infty} \subseteq C$ and $y_n = \frac{1}{n}\sum_{i=1}^n x_{i,n} \in C$, we have

$$\|y_n - v\|^2 = \frac{1}{n}\sum_{i=1}^{n}\left\|x_{i,n} - v\right\|^2 - \frac{1}{n}\sum\nolimits_{i=1}^{n}\left\|x_{i,n} - y_n\right\|^2, \quad \textit{for each } v \in H.$$

Put $x_{i,n} = T((\frac{t}{n})i)x$, for $x \in C$ and $v = T((\frac{t}{n})i_n)y_n$. Then, we have

$$\begin{aligned}
\left\|y_n - T((\frac{t}{n})i_n)y_n\right\|^2 &= \frac{1}{n}\sum_{i=1}^{n}\left\|x_{i,n} - v\right\|^2 - \frac{1}{n}\sum_{i=1}^{n}\left\|x_{i,n} - y_n\right\|^2\\
&= \frac{1}{n}\sum_{i=1}^{i_n}\left\|x_{i,n} - v\right\|^2 + \frac{1}{n}\sum_{i=i_{n+1}}^{n}\left\|x_{i,n} - v\right\|^2 - \frac{1}{n}\sum_{i=1}^{n}\left\|x_{i,n} - y_n\right\|^2\\
&= \frac{1}{n}\sum_{i=1}^{i_n}\left\|x_{i,n} - v\right\|^2 + \frac{1}{n}\sum_{i=i_{n+1}}^{n}\left\|\left\{T\left(\frac{t}{n}\right)\right\}^{i} x - \left\{T\left(\frac{t}{n}\right)\right\}^{i_n} y_n\right\|^2\\
&\quad - \frac{1}{n}\sum_{i=1}^{n}\left\|x_{i,n} - y_n\right\|^2\\
&\leq \frac{1}{n}\sum_{i=1}^{i_n}\left\|x_{i,n} - v\right\|^2 + \frac{1}{n}\sum_{i=i_{n+1}}^{n}\left\|\left\{T\left(\frac{t}{n}\right)\right\}^{i-i_n} x - y_n\right\|^2\\
&\quad - \frac{1}{n}\sum_{i=1}^{n}\left\|x_{i,n} - y_n\right\|^2\\
&= \frac{1}{n}\sum_{i=1}^{i_n}\left\|x_{i,n} - v\right\|^2 + \frac{1}{n}\sum_{i=1}^{n-i_n}\left\|x_{i,n} - y_n\right\|^2 - \frac{1}{n}\sum_{i=1}^{n}\left\|x_{i,n} - y_n\right\|^2\\
&= \frac{1}{n}\sum_{i=1}^{i_n}\left\|x_{i,n} - v\right\|^2 - \frac{1}{n}\sum_{i=n-i_n+1}^{n}\left\|x_{i,n} - y_n\right\|^2\\
&\leq \frac{1}{n}\sum_{i=1}^{i_n}\left\|x_{i,n} - v\right\|^2 \leq \frac{i_n}{n}\operatorname{diam}(C).
\end{aligned}$$

For $\varepsilon > 0$, from $\frac{i_n}{n} \leq \frac{h}{t}$, there is a positive real number t_1 such that $0 < \frac{i_n}{n}\operatorname{diam}(C) \leq \frac{h}{t}\operatorname{diam}(C) < \varepsilon^2$, for each $t \geq t_1$. Hence, we have $\|y_n - T((\frac{t}{n})i_n)y_n\| < \varepsilon$, for each $t \geq t_1$, $n \in \mathbb{N}$, and $x \in C$. We also have that

$$\lim_{n\to\infty} y_n = \lim_{n\to\infty} \frac{1}{n}\sum_{i=1}^{n} T\left(\frac{t}{n}i\right)x = \lim_{n\to\infty} \frac{1}{t}\sum_{i=1}^{n} \frac{t}{n}T\left(\frac{t}{n}i\right)x = \frac{1}{t}\int_0^t T(s)xds = \sigma_t(x),$$

for each $x \in C$. Hence, for each $x \in C$, we have $\lim_{n\to\infty} \|y_n - \sigma_t(x)\| = 0$.

On the other hand, we have

$$\begin{aligned}
\|\sigma_t(x) - T(h)\sigma_t(x)\| &\leq \|\sigma_t(x) - y_n\| + \left\|y_n - T\left(\frac{t}{n}i_n\right)y_n\right\| \\
&\quad + \left\|T\left(\frac{t}{n}i_n\right)y_n - T\left(\frac{t}{n}i_n\right)\sigma_t(x)\right\| \\
&\quad + \left\|T\left(\frac{t}{n}i_n\right)\sigma_t(x) - T(h)\sigma_t(x)\right\| \\
&\leq 2\|y_n - \sigma_t(x)\| + \left\|y_n - T\left(\frac{t}{n}i_n\right)y_n\right\| \\
&\quad + \left\|T\left(\frac{t}{n}i_n\right)\sigma_t(x) - T(h)\sigma_t(x)\right\| \\
&< 2\|y_n - \sigma_t(x)\| + \left\|T\left(\frac{t}{n}i_n\right)\sigma_t(x) - T(h)\sigma_t(x)\right\| + \varepsilon,
\end{aligned}$$

for each $t \geq t_1$, $n \in \mathbb{N}$, and $x \in C$. From $\lim_{n\to\infty} \|y_n - \sigma_t(x)\| = 0$ and $\lim_{n\to\infty} (\frac{t}{n})i_n = h$, we have $\|\sigma_t(x) - T(h)\sigma_t(x)\| \leq \varepsilon$ for each $t \geq t_1$ and $x \in C$. Therefore,

$$\lim_{t\to\infty} \sup_{x\in C} \|\sigma_t(x) - T(h)\sigma_t(x)\| = 0$$

for each $h > 0$. □

We observe that Theorem 3 extends and improves [10, Theorem 4.1], [17, Theorem 4.1] and [27, Theorem 3.1] from semigroup of nonexpansive self-mappings to the general family of nonexpansive self-mappings. We now derive the following result as corollary:

Corollary 2 *Let C be a nonempty subset of a Hilbert space H and let $\mathscr{T} = \{T(t) : t \in [0, \infty)\}$ be a strongly continuous semigroup of nonexpansive self-mappings on C such that $Fix(\mathscr{T}) \cap \Omega[GMEP(\Phi, \Psi, \varphi)] \neq \emptyset$. Assume that $0 < \gamma < \bar{\gamma}/\mathscr{L}$ and that $\{t_n\}$ is a positive real divergent sequence such that $\lim_{n\to\infty} \frac{|t_n - t_{n-1}|}{t_n} = 0$. For given $x_1 \in C$, let $\{x_n\}$ be a sequence in C generated by the following algorithm:*

$$\begin{cases} x_1 \in C; \\ \Phi(u_n, v) + \langle \Psi x_n, v - u_n \rangle + \varphi(v) - \varphi(u_n) + \frac{1}{r_n}\langle u_n - x_n, \ v - u_n \rangle \geq 0, \quad \forall v \in C; \\ z_n = \beta_n x_n + (1 - \beta_n)\frac{1}{t_n}\int_0^{t_n} T(s)u_n \, ds; \\ x_{n+1} = \gamma \alpha_n f x_n + (I - \alpha_n A)z_n, \quad \forall n \in \mathbb{N}, \end{cases}$$

where $\{\alpha_n\}$ *and* $\{\beta_n\}$ *are two sequences in* $(0, 1]$ *and* $\{r_n\}$ *is a sequence in* $(0, \infty)$ *satisfying conditions (C1)~(C4). Then* $\{x_n\}$ *converges strongly to* $x^* \in Fix(\mathscr{T}) \cap \Omega[GMEP(\Phi, \Psi, \varphi)]$*, which solves optimization problem* (1) *over the set* $Fix(\mathscr{T}) \cap \Omega[GMEP(\Phi, \Psi, \varphi)]$.

Proof For each $n \in \mathbb{N}$, define $\sigma_n(x) = \frac{1}{t_n}\int_0^{t_n} T(s)x \, ds, x \in C$. Note that $\{\sigma_n(u_n)\}$ is in a bounded set D_0 defined by (34). Set $v_n := \sigma_n(u_n)$. As in the proof of Theorem 3, $\mathscr{T} = \{T(t) : 0 \leq t < \infty\}$ is a semigroup of nonexpansive self-mappings on D_0. It follows from Lemma 8 that $\{v_n\}$ is an approximating fixed point sequence of semigroup $\mathscr{T}$. Observe that

$$\begin{aligned} \|\sigma_n(u_{n-1}) - \sigma_{n-1}(u_{n-1})\| &= \left\| \left(\frac{1}{t_n} - \frac{1}{t_{n-1}}\right) \int_0^{t_{n-1}} [T(s)u_{n-1} - x^*] \, ds \right. \\ &\quad \left. + \frac{1}{t_n}\int_{t_{n-1}}^{t_n} [T(s)u_{n-1} - x^*] \, ds \right\| \\ &\leq \frac{2|t_n - t_{n-1}|}{t_n} \|u_{n-1} - x^*\| \\ &\leq \frac{2|t_n - t_{n-1}|}{t_n} K_1 \end{aligned}$$

for some constant $K_1 > 0$. From the condition $\lim_{n\to\infty} \frac{|t_n - t_{n-1}|}{t_n} = 0$, we conclude that the condition (C4) holds. Therefore, Corollary 2 follows from Theorem 3. □

Remark 2 (a) If $C = H$ and $\varphi \equiv 0$, then Corollary 2 reduces a main result of Kamraksa and Wangkeeree [17]. In Corollary 2, the conditions $\sum_{n=1}^{\infty} |\alpha_n - \alpha_{n+1}| < \infty$ and $\sum_{n=1}^{\infty} |r_n - r_{n+1}| < \infty$ are not used, therefore, Corollary 2 is a significant improvement on [10, Theorem 4.1].
(b) If $C = H$ and $f = I - F$, where $F : H \to H$ is δ-strongly monotone and λ-strictly pseudocontractive, then Corollary 2 reduces to [30, Corollary 3.2]

Remark 3 Compared with the results obtained in [8, 26, 27, 40, 41, 45], Theorem 3 deals with the problem of finding an element of $\text{Fix}(\mathscr{T}) \cap \Omega[\text{GMEP}(\Phi, \Psi, \varphi)]$ which involves the fixed point problem of uncountable nonexpansive mappings and the generalized mixed equilibrium problem.

7.2 When $\mathscr{T}$ is a Family of Resolvent Operators of Monotone Operators

From Corollary 1, we derive an interesting result, which is a combination of proximal point algorithm for zeros of maximal monotone operators and an iterative method for finding solutions of the generalized mixed equilibrium problems.

Corollary 3 [30] *Let H be a real Hilbert space. Let $\mathbb{A} \subseteq H \times H$ be a maximal monotone operator such that $\mathbb{A}^{-1}0 \cap \Omega[GMEP(G, \Psi)] \neq \emptyset$. Assume that $0 < \gamma < \bar{\gamma}/\mathscr{L}$ and $\{t_n\}$ is a divergent sequence of positive real numbers such that $\lim_{n\to\infty} \frac{|t_n - t_{n-1}|}{t_n} = 0$. For given $x_1 \in H$, let $\{x_n\}$ be a sequence in H generated by the following algorithm:*

$$\begin{cases} x_1 \in H; \\ \Phi(u_n, v) + \langle \Psi x_n, v - u_n\rangle + \varphi(v) - \varphi(u_n) + \frac{1}{r_n}\langle u_n - x_n,\ v - u_n\rangle \geq 0, \quad \forall v \in H; \\ z_n = \beta_n x_n + (1 - \beta_n) J_{t_n}^{\mathbb{A}} x_n\, ds; \\ x_{n+1} = \gamma \alpha_n f x_n + (I - \alpha_n A) z_n, \quad \forall n \in \mathbb{N}, \end{cases}$$

where $\{\alpha_n\}$ and $\{\beta_n\}$ are two sequences in $(0, 1]$ and $\{r_n\}$ is a sequence in $(0, \infty)$ satisfying conditions (C1)–(C4). Then $\{x_n\}$ converges strongly to $x^* \in \mathbb{A}^{-1}0 \cap \Omega[GMEP(\Phi, \Psi, \varphi)]$, which solves optimization problem (1) over the set $\mathbb{A}^{-1}0 \cap \Omega[GMEP(\Phi, \Psi, \varphi)]$.

Proof Set $\sigma_n := J_{t_n}^{\mathbb{A}}$. Then $\{\sigma_n\}$ is a sequence of firmly nonexpansive mappings from H into itself such that $\mathrm{Fix}(\sigma_n) = \mathbb{A}^{-1}0$ for every $n \in \mathbb{N}$. Set $v_n := \sigma_n(u_n)$. We show that $\{v_n\}$ is an approximating fixed point sequence of the family $\{J_t^{\mathbb{A}} : t > 0\}$ of resolvent operators of $\mathbb{A}$. As in the proof of Theorem 3, we can see that $\{v_n\}$ is bounded. Then, there exists a positive real number M such that $\|u_n - J_{t_n}^{\mathbb{A}} u_n\| \leq M$ for all $n \in \mathbb{N}$. For any fixed $r > 0$, by (11), we have

$$\begin{aligned} \left\| J_{t_n}^{\mathbb{A}} u_n - J_r^{\mathbb{A}} J_{t_n}^{\mathbb{A}} u_n \right\| &\leq \frac{r}{t_n} \left\| u_n - J_{t_n}^{\mathbb{A}} u_n \right\| \\ &\leq \frac{r}{t_n} M. \end{aligned}$$

Thus, in particular, we derive $\left\| v_n - J_r^{\mathbb{A}} v_n \right\| \to 0$ as $n \to \infty$ for all $r > 0$.

We now show that the condition (C4) holds. From (12), we have

$$\left\| J_{t_n}^{\mathbb{A}} u_{n-1} - J_{t_{n-1}}^{\mathbb{A}} u_{n-1} \right\| \leq \frac{|t_n - t_{n-1}|}{t_n} \left\| u_{n-1} - J_{t_n}^{\mathbb{A}} u_{n-1} \right\|.$$

Since $\lim_{n\to\infty} \frac{|t_n - t_{n-1}|}{t_n} = 0$, we conclude that the condition (C4) holds. □

Acknowledgments The work of second author was done during his visit to King Fahd University of Petroleum and Minerals (KFUPM), Dhahran, Saudi Arabia. He expresses his thanks to KFUPM for providing excellent research facilities.

References

1. Agarwal, R.P., O'Regan, D., Sahu, D.R.: Fixed Point Theory for Lipschitzian-type Mappings with Applications. Springer, New York (2009)
2. Ansari, Q.H., Wong, N.-C., Yao, J.-C.: The existence of nonlinear inequalities. Appl. Math. Lett. **12**, 89–92 (1999)
3. Bauschke, H.H., Borwein, J.M.: On projection algorithms for solving convex feasibility problems. SIAM Rev. **38**, 367–426 (1996)
4. Brézis, H., Browder, F.E.: Nonlinear ergodic theorems. Bull. Am. Math. Soc. **82**, 959–961 (1976)
5. Blum, E., Oettli, W.: From optimization and variational inequalities to equilibrium problems. Math. Student **63**, 123–145 (1994)
6. Ceng, L.C., Ansari, Q.H., Yao, J.C.: Mann-type steepest-descent and modified hybrid steepest-descent methods for variational inequalities in Banach spaces. Numer. Funct. Anal. Optim. **29**, 987–1033 (2008)
7. Ceng, L.C., Ansari, Q.H., Schaible, S.: Hybrid extragradient-like methods for generalized mixed equilibrium problems, systems of generalized equilibrium problems and optimization problems. J. Global Optim. **53**, 69–96 (2012)
8. Ceng, L.C., Ansari, Q.H., Wong, N.C., Yao, J.C.: An extragradient-like approximation method for variational inequalities and fixed point problems. Fixed Point Theory Appl. **2011**(1), 1–18 (2011)
9. Ceng, L.C., Yao, J.C.: A relaxed extragradient-like method for a generalized mixed equilibrium problem, a general system of generalized equilibria and a fixed point problem. Nonlinear Anal. **72**, 1922–1937 (2010)
10. Cianciaruso, F., Marino, G., Muglia, L.: Iterative methods for equilibrium and fixed point problems for nonexpansive semigroups in Hilbert spaces. J. Optim. Theory Appl. **146**, 491–509 (2010)
11. Combettes, P.L.: Hilbertian convex feasibility problem: convergence of projection method. Appl. Math. Optim. **35**, 311–330 (1997)
12. Combettes, P.L., Hirstoaga, S.A.: Equilibrium programming in Hilbert spaces. J. Nonlinear Convex Anal. **6**, 117–136 (2005)
13. Eckstein, J., Bertsekas, D.P.: On the douglas-rachford splitting method and the proximal point algorithm for maximal monotone operators. Math. Program. **55**, 293–318 (1992)
14. Gwinner, J.: Stability of monotone variational inequalities. In: Giannessi, F., Maugeri, A. (eds.). Variational Inequalities and Network Equilibrium Problems, pp. 123–142. Plenum Press, New York (1995)
15. Iusem, A.N., Sosa, W.: Iterative algorithms for equilibrium problems. Optimization **52**, 301–316 (2003)
16. Jaiboon, C., Kumam, P.: A general iterative method for solving equilibrium problems, variational inequality problems and fixed point problems of an infinite family of nonexpansive mappings. J. Appl. Math. Comput. **34**, 407–439 (2010)
17. Kamraksa, U., Wangkeeree, R.: Generalized equilibrium problems and fixed point problems for nonexpansive semigroups in Hilbert spaces. J. Global Optim. (2011) doi: 10.1007/s10898-011-9654-9
18. Kinderlehrer, D., Stampacchia, G.: An Introduction to Variational Inequalities and Their Applications. Academic Press, New York (1980)

19. Lehdili, N., Moudafi, A.: Combining the proximal algorithm and Tikhonov regularization. Optimization **37**, 239–252 (1996)
20. Luo, Z.Q., Pang, J.S., Ralph, D.: Mathematical Programs with Equilibrium Constraints. Cambridge University Press, New York (1996)
21. Marino, G., Xu, H.K.: A general iterative method for nonexpansive mappings in Hilbert spaces. J. Math. Anal. Appl. **318**, 43–52 (2006)
22. Martinet, B.: Regularisation dinequations variationnelles par approximations successives. Rev. FranMcaise Inf. Rech. Oper. **4**, 154–158 (1970)
23. Moudafi, A.: Weak convergence theorems for nonexpansive mappings and equilibrium problems. J. Nonlinear Convex Anal. **9**, 37–43 (2008)
24. Outrata, J., Kocvara, M., Zowe, J.: Nonsmooth Approach to Optimization Problems with Equilibrium Constraints. Kluwer, Dordrecht (1998)
25. Peng, J.W., Yao, J.C.: A new hybrid-extragradient method for generalized mixed euqilibrium problems, fixed point problems and variational inequality problems. Taiwanese J. Math. **12**, 1401–1432 (2008)
26. Plubtieng, S., Punpaeng, R.: Fixed point solutions of variational inequalities for nonexpansive semigroups in Hilbert spaces. Math. Comput. Model. **48**, 279–286 (2008)
27. Plubtieng, S., Punpaeng, R.: A general iterative method for equilibrium problems and fixed point problems in Hilbert spaces. J. Math. Anal. Appl. **336**, 455–469 (2007)
28. Qin, X., Shang, M., Su, Y.: A general iterative method for equilibrium problems and fixed point problems in Hilbert spaces. Nonlinear Anal. **69**, 3897–3909 (2008)
29. Rockafellar, R.T.: Monotone operators and the proximal point algorithm. SIAM J. Control Optim. **14**, 877–898 (1976)
30. Sahu, D.R., Ansari, Q.H., Yao, J.C.: A unified hybrid iterative method for hierarchical minimization problems. J. Comput. Appl. Math. **253**, 208–221 (2013)
31. Sahu, D.R., Wong, N.C., Yao, J.C.: A generalized hybrid steepest-descent method for variational inequalities in Banach spaces. Fixed Point Theory Appl. **2011**, Article ID 754702 (2011)
32. Sahu, D.R., Wong, N.C., Yao, J.C.: A unified hybrid iterative method for solving variational inequalities involving generalized pseudo-contractive mappings. SIAM J. Control Optim. **50**, 2335–2354 (2012)
33. Shimizu, T., Takahashi, W.: Strong convergence to common fixed points of families of nonexpansive mappings. J. Math. Anal. Appl. **211**, 71–83 (1997)
34. Song, Y., Yang, C.: A note on a paper "a regularization method for the proximal point algorithm". J. Global Optim. **43**, 171–174 (2009)
35. Su, Y., Shang, M., Qin, X.: A general iterative scheme for nonexpansive mappings and inverse-strongly monotone mappings. J. Appl. Math. Comput. **28**, 283–294 (2008)
36. Goldstein, A.A.: Convex programming in Hilbert space. Bull. Am. Math. Soc. **70**, 709–710 (1964)
37. Suzuki, T.: Strong convergence theorems for infinite families of nonexpansive mappings in general Banach spaces. Fixed Point Theory Appl. **2005**, 103–123 (2005)
38. Takahashi, W.: Nonlinear functional analysis. Fixed Point Theory and Its Applications. Yokohama Publishers, Yokohama (2000)
39. Takahashi, W.: Viscosity approximation methods for resolvents of accretive operators in Banach spaces. J. Fixed Point Theory Appl. **1**, 135–147 (2007)
40. Takahashi, S., Takahashi, W.: Viscosity approximation methods for equilibrium problems and fixed point problems in Hilbert spaces. J. Math. Anal. Appl. **331**, 506–515 (2007)
41. Takahashi, S., Takahashi, W.: Strong convergence theorem for a generalized equilibrium problem and a nonexpansive mapping in a Hilbert space. Nonlinear Anal. **69**, 1025–1033 (2008)
42. Tossings, P.: The perturbed proximal point algorithm and some of its applications. Appl. Math. Optim. **29**, 125–159 (1994)
43. Xu, H.K.: A regularization method for the proximal point algorithm. J. Global Optim. **36**, 115–125 (2006)
44. Xu, H.K.: Iterative algorithms for nonlinear operators. J. London Math. Soc. **66**, 240–256 (2002)

45. Xu, H.K.: An iterative approach to quadratic optimization. J. Optim. Theory Appl. **116**, 659–678 (2003)
46. Yamada, I.: The hybrid steepest descent method for the variational inequality over the intersection of fixed point sets of nonexpansive mappings. In: Butnariu, D., Censor, Y., Reich, S. (eds.) Inherently Parallel Algorithms in Feasibility and Optimization and Their Applications, pp. 473–504. Elsevier, Amsterdam (2001)
47. Zeidler, E.: Nonlinear Functional Analysis and its Applications, III: Variational Methods and Applications. Springer, New York (1985)
48. Zhang, S.S.: Generalized mixed equilibrium problem in Banach spaces. Appl. Math. Mech. **30**, 1105–1112 (2009)

Triple Hierarchical Variational Inequalities

Qamrul Hasan Ansari, Lu-Chuan Ceng and Himanshu Gupta

Abstract In this chapter, we give a survey on hierarchical variational inequality problems and triple hierarchical variational inequality problems. By combining hybrid steepest descent method, Mann's iteration method, and projection method, we present a hybrid iterative algorithm for computing a fixed point of a pseudo-contractive mapping and for finding a solution of a triple hierarchical variational inequality in the setting of real Hilbert space. We prove that the sequence generated by the proposed algorithm converges strongly to a fixed point which is also a solution of this triple hierarchical variational inequality problem. On the other hand, we also propose another hybrid iterative algorithm for solving a class of triple hierarchical variational inequality problems concerning a finite family of pseudo-contractive mappings in the setting of real Hilbert spaces. Under very appropriate conditions, we derive the strong convergence of the proposed algorithm to the unique solution of this class of problems.

Keywords Variational inequalities · Hierarchical variational inequalities · Triple hierarchical variational inequalities · Fixed point problems · Hybrid steepest descent method · Mann's iterative method · Projection method · Strong convergence results · Pseudo-contractive mappings

---

Q. H. Ansari (✉) · H. Gupta
Department of Mathematics, Aligarh Muslim University, Aligarh 202002, India
e-mail: qhansari@gmail.com

H. Gupta
e-mail: hmnshu08@gmail.com

L.-C. Ceng
Department of Mathematics, Shanghai Normal University, Shanghai 200234, China
e-mail: zenglc@hotmail.com

L.-C. Ceng
Scientific Computing Key Laboratory of Shanghai Universities, Shanghai, China

Q. H. Ansari (ed.), *Nonlinear Analysis*, Trends in Mathematics,
DOI: 10.1007/978-81-322-1883-8_8,

1 Introduction

The theory of variational inequalities has tremendous applications in many areas of science, social science, engineering, and management. It is a powerful unified methodology to study partial differential equations, optimization problems, optimal control problems, mathematical programming problems, financial management problems, industrial management problems, equilibrium problems from traffic network, spatial price equilibrium problems, oligopolistic market equilibrium problems, financial equilibrium problems, migration equilibrium problems, environmental network problems, knowledge network problems, and so on. It was started by the pioneer work of Fichera [22] and Stampacchia [69] in connection with Signorini's contact problem. Subsequently, Stampacchia gave out a series of classical papers [26, 45, 49, 70] on variational inequalities. In 1980, Dafermos [20] recognized that the traffic network equilibrium conditions as stated by Smith [68] is a structure of a variational inequality in the setting of finite dimensional spaces. Since then, the variational inequalities have been extended, generalized, and studied in the setting of finite/infinite dimensional spaces. For further details on variational inequalities, their generalizations and applications, we refer to [1–3, 21, 24, 25, 35, 40, 42, 43, 58, 63] and the references therein.

A constrained optimization problem in which the constrained set is a solution set of another optimization problem is called a bilevel programming problem. In the last two decades, such problems have been extensively studied in the literature because of their applications in mechanics, network design, etc. For further details on bilevel programming problems, we refer two monographs [46, 62]. If the first-level problem is a variational inequality problem and the second-level problem is a set of fixed points of a mapping, then the bilevel problem is called hierarchical variational inequality problem. In other words, variational inequality problem defined over the set of fixed points of a mapping is called a hierarchical variational inequality problem, also known as hierarchical fixed point problem. The signal recovery [18], beamforming [67] and power control [30] problems can be written in the form of a hierarchical variational inequality problem. For further details on hierarchical variational inequality problems and their applications, we refer to [8, 10, 17, 18, 30, 34, 47, 52, 56, 57, 67, 78, 82–84, 92] and the references therein. In the recent past, several iterative methods have been proposed and analyzed by several authors, see, for example [10, 17, 19, 28, 30, 34, 47, 48, 50, 52, 56, 57, 78, 80–83, 85, 92] and the references therein.

Recently, Iiduka [27, 29] considered a problem which has triple structure, that is, a variational inequality problem defined over the set of solutions of another variational inequality problem which is defined over the set of fixed points of a mapping. Because of the triple structure of the problem, it is called triple hierarchical variational inequality problem. So, a variational inequality problem defined over the set of solutions of a hierarchical variational inequality problem is called a triple hierarchical variational inequality problem (in short, THVIP). Iiduka [27, 29] proposed some iterative methods for computing the approximate solutions of THVIP. The strong convergence of the sequences generated by the proposed methods is also studied.

Some examples of triple hierarchical variational inequality problems are provided in [29]. Subsequently, Iiduka [31] translated the nonconcave utility bandwidth allocation problem with compoundable constraints into a triple hierarchical variational inequality problem. Then, he suggested some iterative method, so-called fixed point optimization algorithm, to find the solution of THVIP. The strong convergence of the iterative method is studied. Recently, several iterative methods for finding the solutions of THVIP have been proposed and analyzed; See, for example, [9, 11–13, 27, 29, 31, 36, 37, 41, 73, 89] and the references therein. Ceng et al. [11] combined the regularization method, hybrid steepest descent method, and the projection method to propose an implicit scheme that generates a net in an implicit way. They studied its convergence to a unique solution of THVIP. They also introduced an explicit scheme that generates a sequence via an iterative algorithm and proved that this sequence converges strongly to a unique solution of THVIP. Ceng et al. [12] considered a monotone variational inequality problem defined over the set of solutions of another variational inequality problem which is defined over the intersection of the fixed point sets of N nonexpansive mappings. They proposed two relaxed hybrid steepest descent algorithms with variable parameters for computing the approximate solutions of these two problems. The strong convergence of these two algorithms is also studied. Recently, Zeng et al. [89] presented strong convergence of relaxed hybrid steepest descent method under some mild conditions on parametric sequences. The THVIP is further investigated and generalized in [13, 36, 37, 73].

Yamada [80] considered a hierarchical variational inequality problem defined over the set of common fixed points of a finite family of nonexpansive mappings. Such problem is called hierarchical variational inequality problem for a family of nonexpansive mappings. An application of this kind of problem is given in [80]. Motivated by the work of Yamada [80], Ceng et al. [12, 89] considered the variational inequality problem with the variational inequality constraint which is defined over the intersection of the fixed point sets of a family of N nonexpansive mappings $T_i : H \to H$, where $N \geq 1$ an integer. Such problem is called triple hierarchical variational inequality problem for a family of nonexpansive mappings.

In this chapter, we give a survey on hierarchical variational inequality problems and triple hierarchical variational inequality problems. By combining hybrid steepest descent method, Mann's iteration method and projection method, we also present a hybrid iterative algorithm for computing a fixed point of a pseudo-contractive mapping and for finding a solution of a triple hierarchical variational inequality in the setting of real Hilbert space. Under very mild assumptions, we prove that the sequence generated by the proposed algorithm converges strongly to a fixed point which is also a solution of this triple hierarchical variational inequality problem. On the other hand, we also propose another hybrid iterative algorithm for solving a class of triple hierarchical variational inequality problems concerning a finite family of pseudo-contractive mappings in the setting of real Hilbert spaces. Under very appropriate conditions, we derive the strong convergence of the proposed algorithm to the unique solution of this class of problems. Our algorithms are quite general and very flexible and include some other iterative algorithms in the literature as special cases.

2 Preliminaries

Throughout the chapter, we use the following notations: "$\rightharpoonup$" and "$\to$" stand for the weak convergence and strong convergence, respectively. Moreover, we use the following notation: for a given sequence $\{x_n\} \subseteq H$, $\omega_w(x_n)$ denotes the weak ω-limit set of $\{x_n\}$, that is,

$$\omega_w(x_n) := \left\{x \in H : x_{n_j} \rightharpoonup x \text{ for some subsequence } \{n_j\} \text{ of } \{n\}\right\}.$$

The following lemma is an immediate consequence of an inner product.

Lemma 1 *Let H be a real Hilbert space. Then, for all $x, y \in H$,*

(a) $\|x - y\|^2 = \|x\|^2 - \|y\|^2 - 2\langle x - y, y\rangle$.
(b) $\|x + y\|^2 \leq \|x\|^2 + 2\langle y, x + y\rangle$.

Let C be a nonempty closed convex subset of a real Hilbert space H. For each point $x \in H$, there exists a unique nearest point in C, denoted by $P_C(x)$, such that

$$\|x - P_C(x)\| \leq \|x - y\|, \quad \text{for all } y \in C,$$

where P_C is called the *metric projection* of H onto C. It is known that P_C is a nonexpansive mapping.

The following lemmas are well known.

Lemma 2 *Let C be a nonempty closed convex subset of a real Hilbert space H. Given $x \in H$ and $z \in C$. Then, $z = P_C(x)$ if and only if*

$$\langle x - z, y - z\rangle \leq 0, \quad \textit{for all } y \in C.$$

Lemma 3 [53] *Let C be a nonempty closed convex subset of a real Hilbert space H. Let $\{x_n\}$ be a sequence in H, $u \in H$ and $q = P_C(u)$. If $\{x_n\}$ is such that $\omega_w(x_n) \subseteq C$ and*

$$\|x_n - u\| \leq \|u - q\|, \quad \textit{for all } n \geq 0.$$

Then, $x_n \to q$.

Definition 1 A mapping $T : H \to H$ is called

(a) *L-Lipschtz continuous* if there exists a constant $L > 0$ such that for all $x, y \in H$,

$$\|T(x) - T(y)\| \leq L\|x - y\|;$$

Further, if $L = 1$, then T is called *nonexpansive*; if $L \in (0, 1)$, then T is called *contraction.*

(b) *firmly nonexpansive* if $2T - I$ is nonexpansive, or equivalently,

$$\langle T(x) - T(y), x - y \rangle \geq \|T(x) - T(y)\|^2, \quad \text{for all } x, y \in H;$$

alternatively, T is firmly nonexpansive if and only if T can be expressed as

$$T = \frac{1}{2}(I + S),$$

where $S : H \to H$ is nonexpansive; projections are firmly nonexpansive;

(c) *pseudo-contractive* if

$$\langle T(x) - T(y), x - y \rangle \leq \|x - y\|^2, \quad \text{for all } x, y \in H; \tag{1}$$

(d) *ζ-strictly pseudo-contractive* if there exists a constant $\zeta \in [0, 1)$ such that

$$\|T(x) - T(y)\|^2 \leq \|x - y\|^2 + \zeta \|(I - T)(x) - (I - T)(y)\|^2, \quad \text{for all } x, y \in H. \tag{2}$$

The condition (1) is equivalent to the following condition.

$$\|Tx - Ty\|^2 \leq \|x - y\|^2 + \|(I - T)x - (I - T)y\|^2, \quad \text{for all } x, y \in H. \tag{3}$$

Indeed, observe that

$$\begin{aligned}
&\|Tx - Ty\|^2 \\
&= \|x - y - [(I - T)x - (I - T)y]\|^2 \\
&= \|x - y\|^2 - 2\langle x - y, (I - T)x - (I - T)y \rangle + \|(I - T)x - (I - T)y\|^2 \\
&= \|x - y\|^2 - 2[\|x - y\|^2 - \langle Tx - Ty, x - y \rangle] + \|(I - T)x - (I - T)y\|^2.
\end{aligned}$$

It is well known that the class of strictly pseudo-contractive mappings strictly includes the class of nonexpansive mappings.

Definition 2 A mapping $A : H \to H$ is said to be

(a) *monotone* if

$$\langle A(x) - A(y), x - y \rangle \geq 0, \quad \text{for all } x, y \in H; \tag{4}$$

Further, if strict inequality holds in (4) for all $x \neq y$, then A is called *strictly monotone*.

(b) *strongly monotone* or *β-strongly monotone* if there is a constant $\beta > 0$ such that

$$\langle A(x) - A(y), x - y \rangle \geq \beta \|x - y\|^2, \quad \text{for all } x, y \in H;$$

(c) *ν-inverse strongly monotone* (ν-ism) (also called *cocoercive*) if there exists constant $\nu > 0$ such that

$$\langle A(x) - A(y), x - y\rangle \geq \nu\|A(x) - A(y)\|^2, \quad \text{for all } x, y \in H.$$

Lemma 4 [32] *Let $A : H \to H$ be an α-inverse strongly monotone operator. For $\lambda \in [0, 2\alpha]$, define $S_\lambda : H \to H$ by $S_\lambda x := x - \lambda Ax$, for all $x \in H$. Then, S_λ is nonexpansive.*

Lemma 5 [80, Lemma 3.1] *Let $A : H \to H$ be β-strongly monotone and L-Lipschitz continuous and $\mu \in (0, 2\beta/L^2)$. For $\lambda \in [0, 1]$, define $T^\lambda : H \to H$ by*

$$T^\lambda(x) := x - \lambda\mu A(x), \quad \textit{for all } x \in H.$$

Then,

$$\|T^\lambda(x) - T^\lambda(y)\| \leq (1 - \lambda\tau)\|x - y\|, \quad \textit{for all } x, y \in H,$$

where, $\tau := 1 - \sqrt{1 - \mu(2\beta - \mu L^2)} \in (0, 1]$.

A Banach space X is said to satisfy *Opial's condition* if whenever $\{x_n\}$ is a sequence in X which converges weakly to x, then

$$\liminf_{n\to\infty} \|x_n - x\| < \liminf_{n\to\infty} \|x_n - y\|, \quad \text{for all } y \in X, \ y \neq x.$$

It is well known that every Hilbert space H satisfies Opial's condition (see, for example, [23]).

Definition 3 Let X be a Banach space and C be a closed convex subset X. A mapping $T : C \to C$ is said to be *demiclosed* when $x_n \rightharpoonup y$ (converges weakly) and $T(x_n) \to z$ (converges strongly), imply that $z = Ty$.

Lemma 6 [94] *Let X be a real reflexive Banach space which satisfies Opial's condition. Let C be a nonempty closed convex subset of X and $T : C \to C$ be a continuous pseudo-contractive mapping. Then, $I - T$ is demiclosed at zero.*

Lemma 7 [23, Demiclosed Principle] *Let $C \subset H$ be a nonempty, closed, and convex set and $T : C \to C$ be a nonexpansive mapping. If T has a fixed point, then $I - T$ is demiclosed, that is, whenever $\{x_n\}$ is a sequence in C and weakly converges to some $x \in C$ and the sequence $\{(I - T)x_n\}$ strongly converges to some y, it follows that $(I - T)x = y$, where I is the identity operator on H.*

Proposition 1 [23] *Let C be a nonempty closed convex subset of a real Hilbert space H and $T : H \to H$ be a nonexpansive mapping. Then,*

(a) $\text{Fix}(T)$ *is closed and convex;*
(b) $\text{Fix}(T)$ *is nonempty if C is bounded.*

3 Variational Inequalities

Throughout the section, unless otherwise specified, let H be a real Hilbert space whose inner product and norm are denoted by $\langle ., . \rangle$ and $\|\cdot\|$, respectively. Let C be a nonempty closed convex subset of H and $A : H \to H$ be a nonlinear operator. The *variational inequality problem* is to find $x^* \in C$ such that

$$\langle A(x^*), x - x^* \rangle \geq 0, \quad \text{for all } x \in C. \tag{5}$$

The solution set of the variational inequality problem defined on the set C and by the operator A is denoted by VIP(C, A). The inequality (5) is called a *variational inequality*. When the operator A is monotone, it is called *monotone variational inequality*. This kind of variational inequality was first introduced and studied by Browder [5]. However, a more general form of the variational inequality was introduced by Fichera [22] and Stampacchia [69] in connection with Signorini's contact problem. As we have pointed out in the first section that the theory of variational inequalities is well established and well recognized. Nowadays, people are using this theory as a tool to solve the problems from mechanics, partial differential equations, game theory, optimization, etc.

Geometrically speaking, a variational inequality problem is to find a vector x^* in a closed, convex set C of a Hilbert Space H such that the vector $A(x^*)$ forms a nonobtuse angle with every vector of the form $x - x^*$, for all $x \in C$.

For further details on variational inequalities and their applications, we refer to [1–3, 5, 20–22, 24–26, 35, 40, 42–45, 54, 58, 63, 69, 70] and the references therein.

During the last three decades, the theory of variational inequalities has been studied in different directions, namely, applications, existence results and solution methods. A large number of solution methods has been proposed. Among all the solution methods, the simplest iterative procedure for variational inequality problem is the *projection gradient method*, whose iterative scheme is

$$x_{n+1} = P_C(x_n - \mu A(x_n)), \quad n = 0, 1, 2, \ldots,$$

where $P_C : H \to C$ is a projection operator from H onto C and μ is a positive real number.

When A is strongly monotone and Lipschitz continuous, the projection gradient method with any initial choice $x_0 \in C$ and $\mu > 0$, generates a sequence $\{x_n\}$ that converges strongly to a unique solution of the variational inequality problem (5).

Proposition 2 [29, Proposition 2.3] *Let C be a nonempty closed convex subset of a real Hilbert space H and $A : H \to H$ be α-inverse strongly monotone. Then, for $\lambda \in [0, 2\alpha]$, the mapping $S_\lambda : H \to H$, defined by*

$$S_\lambda(x) := P_C(x - \lambda A(x)), \quad \textit{for all } x \in H,$$

is nonexpansive and $\text{Fix}(S_\lambda) = \text{VIP}(C, A)$.

Definition 4 Let C be a nonempty convex subset of a real Hilbert space H. A mapping $A : C \to H$ is said to be *hemicontinuous* if for any fixed $x, y \in C$, the mapping $\lambda \mapsto A(x + \lambda(y - x))$ defined on $[0, 1]$ is continuous, that is, if A is continuous along the line segments in C.

The following proposition is well known and the proof can be found in [1, 40, 42, 43, 63].

Proposition 3 *Let $C \subseteq H$ be a nonempty closed and convex subset of real Hilbert space H and $A : C \to H$ be a mapping.*

(a) *If A is monotone and hemicontinuous, then VIP(C, A) is equivalent to*

$$MVIP(C, A) := \{x^* : \langle Ay, y - x^* \rangle \geq 0, \text{ for all } y \in C\}.$$

(b) *VIP$(C, A) \neq \emptyset$, when C is bounded and A is monotone and hemicontinuous.*
(c) *VIP$(C, A) = \mathrm{Fix}(P_C(I - \lambda A))$, for all $\lambda > 0$, where I is identity map on H.*
(d) *VIP(C, A) consists only one point, if A is strongly monotone and Lipschitz continuous.*
(e) *VIP$(C, \nabla f) = \arg\min_{x \in C} f(x) := \{x^* \in C : f(x^*) = \min_{x \in C} f(x)\}$, where $f : C \to \mathbb{R}$ is a convex and Fréchet differentiable function and ∇f is the gradient of f.*

4 Hierarchical Variational Inequalities

A variational inequality problem defined over a set of fixed points of a mapping on a Hilbert space H, is called *hierarchical variational inequality problem*.

Let H be a real Hilbert space, $T : H \to H$ be a nonexpansive mapping, and $A : H \to H$ be a mapping. We assume that the set $\mathrm{Fix}(T)$ of all fixed points of T is nonempty.

The *hierarchical variational inequality problem* (in short, HVIP) is to find $x^* \in \mathrm{Fix}(T)$ such that

$$\langle A(x^*), y - x^* \rangle \geq 0, \quad \text{for all } y \in \mathrm{Fix}(T), \tag{6}$$

that is, to find

$$x^* \in \mathrm{VIP}(\mathrm{Fix}(T), A) := \{x^* \in \mathrm{Fix}(T) : \langle A(x^*), y - x^* \rangle \geq 0, \text{ for all } y \in \mathrm{Fix}(T)\}.$$

Several problems, namely, image recovery problem in 3D electron microscopy [65], quadratic signal recovery problem [18], robust wideband beamforming problem [67], power control problem in code-division multiple-access (CDMA) data network [30], etc can be modeled as a HVIP. Sezan [65] formulated the image recovery problem in the form of a hierarchical minimization problem, that is, a minimization problem defined over the set of fixed points of a mapping. She / He used projection method to study this problem. Combettes [18] proposed a block-iterative parallel decomposition method to solve quadratic signal recovery problem. Slavakis and Yamada [67] used hybrid steepest descent method (HSDM) to design robust smart antennas. Iiduka [30] formulated a power control problem for a CDMA data network in the form of a hierarchical variational inequality problem. He proposed an iterative algorithm to solve this problem. Because of the applications of HVIP, it has been extensively studied during the last decade by several researchers; See, for example [8, 10, 17, 19, 28, 30, 34, 47, 48, 50, 52, 56, 57, 67, 78, 80–85, 92] and the references therein.

Yamada [80] proposed a hybrid steepest descent method for solving variational inequality problem which is defined over the set of fixed points of a nonexpansive mapping. He also studied the solution method for HVIP defined by means of the set of common fixed points of N nonexpansive mappings. He presented the following result.

Lemma 8 [80, Lemma 3.1] *Let $T : H \to H$ be a nonexpansive mapping with* $\mathrm{Fix}(T) \neq \emptyset$. *Suppose that $A : H \to H$ is L-Lipschitzian and β-strongly monotone over $T(H)$. By using arbitrary fixed $\mu \in \left(0, \frac{2\beta}{L^2}\right)$, define $T^\lambda : H \to H$ by*

$$T^\lambda(x) := T(x) - \lambda\mu A(T(x)), \quad \textit{for all } \lambda \in [0, 1]. \tag{7}$$

Then,

(a) *$G := \mu A - I$ satisfies*

$$\|G(x) - G(y)\|^2 \leq \left[1 - \mu(2\beta - \mu L^2)\right] \|x - y\|^2, \quad \textit{for all } x, y \in T(H), \tag{8}$$

which implies that G is strictly contractive over $T(H)$, where $T(H)$ denotes the range of T. Moreover, the obvious relation $0 < \tau := 1 - \sqrt{1 - \mu(2\beta - \mu L^2)} \leq 1$ ensure that the closed ball

$$C_f := \left\{ x \in H : \|x - f\| \leq \frac{\|\mu A(f)\|}{\tau} \right\}$$

is well defined, for all $f \in \mathrm{Fix}(T)$.

(b) *$T^\lambda : H \to H$ satisfies $T^\lambda(C_f) \subset C_f$, for all $f \in \mathrm{Fix}(T)$ and $\lambda \in [0, 1]$. In particular, $T(C_f) \subset C_f$, for all $f \in \mathrm{Fix}(T)$.*

(c) *$T^\lambda : H \to H$, for all $\lambda \in (0, 1]$, is strictly contractive mapping having its unique fixed point $\xi_\lambda \in \bigcap_{f \in \mathrm{Fix}(T)} C_f$.*

(d) *Suppose that the sequence of parameters $\{\lambda_n\} \subset (0, 1]$ satisfies $\lim_{n\to\infty} \lambda_n = 0$. Let ξ_n be the unique fixed point of $T_{<n>} := T^{\lambda_n}$, that is,*

$$\xi_n := \xi_{\lambda_n} \in \mathrm{Fix}(T_{<n>}), \quad \textit{for all } n.$$

Then, the sequence $\{\xi_n\}$ *converges strongly to the unique solution* $u^* \in \text{Fix}(T)$ *of* VIP(Fix(T), A).

The conclusion (d) in Lemma 8 is a generalization of a pioneer result by Browder [6].

Yamada [80] presented the following hybrid steepest descent method for solving HVIP (6).

Algorithm 1 Let $T : H \to H$ be a nonexpansive mapping with $\text{Fix}(T) \neq \emptyset$. Suppose that a mapping $A : H \to H$ is L-Lipschitz continuous and β-strongly monotone over $T(H)$. Start with any initial choice $x_0 \in H$, any $\mu \in \left(0, \frac{2\beta}{L^2}\right)$ and generate a sequence $\{x_n\}$ by

$$x_{n+1} := T^{\lambda_{n+1}}(x_n) := T(x_n) - \lambda_{n+1}\mu A(T(x_n)). \tag{9}$$

The Yamada [80] established the following strong convergence result for the sequence generated by the Algorithm 1.

Theorem 1 *Let* $T : H \to H$ *be a nonexpansive mapping with* $\text{Fix}(T) \neq \phi$. *Suppose that a mapping* $A : H \to H$ *is* L*-Lipschitz continuous and* β*-strongly monotone over* $T(H)$. *Then with any* $x_0 \in H$, *any* $\mu \in \left(0, \frac{2\beta}{L^2}\right)$ *and any sequence* $\{\lambda_n\} \subset (0, 1]$ *satisfying*

(i) $\lim_{n\to\infty} \lambda_n = 0$,
(ii) $\sum_{n=1}^{\infty} \lambda_n = +\infty$,
(iii) $\lim_{n\to\infty} \frac{\lambda_n - \lambda_{n+1}}{\lambda_{n+1}^2} = 0$,

the sequence $\{x_n\}$ *generated by* (9) *converges strongly to the uniquely existing solution of* VIP(Fix(T), A).

An example of the sequence $\{\lambda_n\}$ of real numbers that satisfies conditions (i)–(iii) of Theorem 1 is $\lambda_n = \frac{1}{n^t}$ for $0 < t < 1$.

In 2004, Yamada and Ogura [81] extended Theorem 1 for quasi-nonexpansive mappings.

Recently, Iiduka [30] considered HVIP (6) in the setting of finite dimensional Euclidean space for fimrly nonexpansive mapping. He present an application of this problem to a power control problem for a direct-sequence code-division multiple-access data network. He considered the following problem.

Problem 1 Assume that

(A1) $C \subset \mathbb{R}^n$ is a nonempty, closed convex set and the explicit form of P_C is known;
(A2) $T : \mathbb{R}^n \to \mathbb{R}^n$ is a firmly nonexpansive mapping such that $\text{Fix}(T) \subset C$ and $\text{Fix}(T) \neq \emptyset$;
(A3) $A : \mathbb{R}^n \to \mathbb{R}^n$ is a continuous mapping.

The problem is to find $x^* \in$ VIP(Fix(T), A) ($\subset C$).

He presented the following, so-called, *fixed point optimization algorithm.*

Algorithm 2 [30] STEP 0. Choose $x_0 \in C$, $\lambda_1 \in (0, \infty)$ and $\alpha_1 \in [0, 1)$ arbitrary, and set $n := 1$.
STEP 1. Given $x_n \in C$, choose $\lambda_n \in (0, \infty)$ and $\alpha_n \in [0, 1)$, and compute $x_{n+1} \in C$ as follows:

$$\begin{aligned} y_n &:= T\left(x_n - \lambda_n A(x_n)\right), \\ x_{n+1} &:= P_C\left(\alpha_n x_n + (1-\alpha_n) y_n\right). \end{aligned} \tag{10}$$

Update $n := n + 1$, and go to Step 1.

Iiduka [30] established the following convergence theorem for Algorithm 2.

Theorem 2 [30, Theorem 6] *In addition to the assumptions of Problem* 1, *we further assume that* $\{A(x_n)\}$ *is bounded,* $\mathrm{VIP}(\mathrm{Fix}(T), A) \neq \emptyset$, *and there exists* $n_0 \in \mathbb{N}$ *such that* $\mathrm{VIP}(\mathrm{Fix}(T), A) \subset \Omega := \bigcap_{n=n_0}^{\infty} \{x \in \mathrm{Fix}(T) : \langle A(x_n), x_n - x\rangle \geq 0\}$. *If* $\{\alpha_n\} \subset [0, 1)$ *and* $\{\lambda_n\} \subset (0, \infty)$ *satisfy the following conditions:*

(i) $\limsup_{n\to\infty} \alpha_n < 1$,
(ii) $\sum_{n=1}^{\infty} \lambda_n^2 < \infty$,

then, the sequence $\{x_n\}$ *generated by the Algorithm* 2 *has the following properties:*

(a) *For every* $y \in \Omega$, $\lim_{n\to\infty} \|x_n - y\|$ *exists, and* $\{x_n\}$ *and* $\{y_n\}$ *are bounded.*
(b) $\lim_{n\to\infty} \|x_n - y_n\| = 0$ *and* $\lim_{n\to\infty} \|x_n - T(x_n)\| = 0$.
(c) *If* $\|x_n - y_n\| = o(\lambda_n)$, *then* $\{x_n\}$ *converges to a point in* $\mathrm{VIP}(\mathrm{Fix}(T), A)$.

Iiduka [30] also studied the conditions under which $\{A(x_n)\}$ is bounded, VIP $(\mathrm{Fix}(T), A) \neq \emptyset$, and the condition "there exists $n_0 \in \mathbb{N}$ such that $\mathrm{VIP}(\mathrm{Fix}(T), A) \subset \Omega := \bigcap_{n=n_0}^{\infty} \{x \in \mathrm{Fix}(T) : \langle A(x_n), x_n - x\rangle \geq 0\}$" holds (see, Remark 7 (a)–(c) in [30]).

Moudafi and Maingé [57] considered the following general form of hierarchical variational inequality problem (HVIP).

$$\text{Find } x^* \in \mathrm{Fix}(T) \text{ such that } \langle x^* - S(x^*), y - x^*\rangle, \quad \text{for all } y \in \mathrm{Fix}(T), \tag{11}$$

where $S, T : H \to H$ are nonexpansive mappings and $\mathrm{Fix}(T) = \{x \in C : x = T(x)\}$ is the set of all fixed points of T and C is a nonempty closed convex subset of a real Hilbert space H.

It can be easily seen that the HVIP (11) is equivalent to the fixed point problem of finding $x^* \in C$ such that

$$x^* = P_{\mathrm{Fix}(T)} \circ S(x^*), \tag{12}$$

where $P_{\mathrm{Fix}(T)}$ is the metric projection on the convex set $\mathrm{Fix}(T)$. By using the definition of a normal cone to $\mathrm{Fix}(T)$, that is,

$$N_{\text{Fix}(T)} : x \mapsto \begin{cases} \{d \in H : \langle d, y - x\rangle \leq 0, \ \forall y \in \text{Fix}(T)\}, & \text{if } x \in \text{Fix}(T), \\ \emptyset, & \text{otherwise,} \end{cases} \tag{13}$$

we can easily see that the HVIP (11) is equivalent to the following variational inequality:

$$0 \in (I - S)x^* + N_{\text{Fix}(T)}(x^*). \tag{14}$$

When the solution set of the HVIP (11) is a singleton set (which is the case, for example, when S is a contraction) the problem reduces to the viscosity fixed point solution introduced in [55] and further developed in [19, 76].

Example 1 [57] Let $A : H \to H$ be L-Lipschitz continuous and β-strongly monotone and $S = I - \gamma A$ where $\gamma \in \left(0, \frac{2L}{\beta^2}\right)$. Then, the HVIP (11) reduces to the HVIP (6).

On the other hand, if we take $C = H$, $T = J_\lambda^A$ and $S = J_\lambda^B$ with A, B are two maximal monotone operators and J_λ^A, J_λ^B are the corresponding resolvent mappings, then the HVIP (11) reduces to the problem of finding $x^* \in H$ such that

$$0 \in \left(I - J_\lambda^B\right)(x^*) + N_{A^{-1}(0)}(x^*), \tag{15}$$

where $N_{A^{-1}(0)}$ denotes the normal cone to $A^{-1}(0) = \text{Fix}(J_\lambda^A)$, the set of zeros of A. The inclusion (15) can be written as to find x^* such that

$$0 \in B_\lambda(x^*) + N_{A^{-1}(0)}(x^*),$$

where $B_\lambda := \left(\lambda I + B^{-1}\right)^{-1}$ is the the Yosida approximation of B.

Example 2 [57] Let $\psi : H \to \mathbb{R}$ be a convex function such that $\nabla\psi$ is a β-strongly monotone and L-Lipschitz continuous (which is equivalent to the fact that $\nabla\psi$ is L^{-1} cocoercive), $\phi : H \to \mathbb{R}$ be a lower semicontinuous convex function and $Q = I - \gamma\nabla\psi$ with $\gamma \in (0, 2/L)$. Setting

$$T = \text{prox}_{\lambda\phi} := \arg\min\left\{\phi(y) + \frac{1}{2\lambda}\|. - y\|^2\right\}. \tag{16}$$

Then by the fact that $\text{Fix}(\text{prox}_{\lambda\psi}) = (\partial\phi)^{-1}(0) = \arg\min\phi$, the HVIP (11) reduces to the following *hierarchical minimization problem:*

$$\min_{x \in \arg\min\phi} \psi(x). \tag{17}$$

We denote by $\delta\phi$ and $\delta\psi$ the subdifferential operators of lower secontinuous convex functions ϕ and ψ, respectively. On the other hand, if we consider $A = \delta\phi$ and $B = \delta\psi$ in (15), the HVIP (11) reduces to the following *hierarchical minimization problem:*

$$\min_{x \in \arg\min \phi} \psi_\lambda(x),$$

where $\psi_\lambda(x) = \inf_y \{\psi(y) + (1/2\lambda)\|x - y\|^2\}$ is the Moreau-Yosida approximate of ψ.

Example 3 [57] Let $\psi : H \to \mathbb{R}$ be a convex function such that $\nabla\psi$ is β-strongly monotone and L-Lipschitz continuous (which is equivalent to the fact that $\nabla\psi$ is L^{-1} cocoercive). Setting $S = I - \gamma\nabla\phi$ with $\gamma \in (0, 2/L)$. The HVIP (11) reduces to the following hierarchical minimization problem studied by Yamada [80].

$$\min_{x \in \text{Fix}(T)} \phi(x).$$

On the other hand, when T is a nonexpansive mapping and $S = I - \tilde{\gamma}$ $(A - \gamma f)$, where A is a bounded linear $\tilde{\gamma}$-strongly monotone operator, f is a given α-contraction, and $\gamma > 0$ with $\tilde{\gamma} \in (0, \|A\| + \bar{\gamma})$. The HVIP (11) reduces to the problem of minimizing a quadratic function over the set of fixed points of a nonexpansive mapping studied by Marino and Xu [50], namely

$$\langle (A - \gamma f)(x^*), y - x^* \rangle \geq 0, \quad \text{for all } y \in \text{Fix}(T), \tag{18}$$

which is optimality condition for the following minimization problem:

$$\min_{x \in \text{Fix}(T)} \frac{1}{2}\langle Ax, x \rangle - h(x), \tag{19}$$

where h is the potential function for γf, that is, $h'(x) = \gamma f(x)$, for $x \in H$.

Moudafi [56] extended Krasnoselski–Mann (KM) iterative method for HVIP (11) and proposed the following algorithm:

Algorithm 3 [56]

$$x_{n+1} = (1 - \alpha_n)x_n + \alpha_n (\lambda_n S(x_n) + (1 - \lambda_n)T(x_n)), \quad \text{for all } n \geq 0, \tag{20}$$

where $x_0 \in C$, $\{\lambda_n\}$ and $\{\alpha_n\} \subset (0, 1)$.

Moudafi [56] and Yao and Liou [83] studied the weak convergence of the sequence generated by the Algorithm 3.

Theorem 3 [83] *Let C be a nonempty closed convex subset of a real Hilbert space H and $T, S : C \to C$ be two nonexpansive mappings such that* $\text{Fix}(T) \neq \emptyset$. *Let $\{x_n\}$ be a sequence generated by the Algorithm* 3. *Let* $\{\alpha_n\} \subset (0, 1)$ *and* $\{\lambda_n\} \subset (0, 1)$ *be two sequences of real numbers satisfying the following conditions:*

(i) $\sum_{n=0}^{\infty} \lambda_n < \infty$;
(ii) $\lim_{n\to\infty} \frac{\|x_{n+1} - x_n\|}{\alpha_n \lambda_n} = 0$.

Then,

(a) $\{x_n\}$ *converges weakly to a fixed point of* T*;*
(b) $\{x_n\}$ *is asymptotically regular, namely* $\lim_{n\to\infty} \|x_{n+1} - x_n\| = 0$*;*
(c) $\{x_n\}$ *converges weakly to a solution of* HVIP (11).

Yao and Liou [83] also showed that the conclusion of the Theorem 3 still holds if we replace the condition (i) in Theorem 3 by the following condition.

(i) $\sum_{n=0}^{\infty} \alpha_n \lambda_n < \infty$, and $\lim_{n\to\infty} \lambda_n = 0$.

When S is a contraction mapping, Yao and Liou [83] proved the following strong convergence theorem under different restriction on parameters.

Theorem 4 [83, Theorem 3.3] *Let C be a nonempty closed convex subset of a real Hilbert space H and $T, S : C \to C$ be nonexpansive and contraction mappings, respectively, such that* $\mathrm{Fix}(T) \neq \emptyset$*. Let $\{x_n\}$ be a sequence generated by Algorithm 3. Let $\{\alpha_n\} \subset (0, 1)$ and $\{\lambda_n\} \subset (0, 1)$ be two sequences of real numbers satisfying the following conditions:*

(i) $0 < \liminf_{n\to\infty} \alpha_n \leq \limsup_{n\to\infty} \alpha_n < 1$*;*
(ii) $\lim_{n\to\infty} \lambda_n = 0$ *and* $\sum_{n=0}^{\infty} \lambda_n = \infty$.

Then,

(a) $\{x_n\}$ *converges strongly to a fixed point of* T*;*
(b) $\{x_n\}$ *is asymptotically regular, namely* $\lim_{n\to\infty} \|x_{n+1} - x_n\| = 0$*;*
(c) $\{x_n\}$ *converges strongly to a solution of* HVIP (11).

In 2007, Mainge and Moudafi [48] proposed the following viscosity-like method for approximating a specific solution of HVIP (11).

$$x_{n+1} = \lambda_n f(x_n) + (1 - \alpha_n)\left(\alpha_n S(x_n) + (1 - \lambda_n) T(x_n)\right), \quad \text{for } n \geq 0, \tag{21}$$

where the initial guess $x_0 \in C$, $f : C \to C$ is a contraction mapping and $\{\lambda_n\}$ and $\{\alpha_n\}$ are sequences in $(0, 1)$ satisfying certain conditions. Note that this method can be regarded as a generalization of Halpern's algorithm. The strong convergence of the sequence generated by this method is also studied in [48].

Lu et al. [47] proposed the following generalization of iterative scheme (21).

$$x_{n+1} = \lambda_n \left[\alpha_n f(x_n) + (1 - \alpha_n) S(x_n)\right] + (1 - \lambda_n) T(x_n), \quad \text{for } n \geq 0, \tag{22}$$

where the initial guess $x_0 \in C$, $f : C \to C$ is a contraction mapping and $\{\lambda_n\}$ and $\{\alpha_n\}$ are sequences in $(0, 1)$ satisfying certain conditions. The strong convergence of the sequence generated by the above method is also studied. An application of HVIP to hierarchical minimization problem is also studied.

A fairly common method in solving some nonlinear problems is to replace the original problem by a family of regularized (perturbed) problems and each of these regularized problems has a unique solution. A particular (viscosity) solution of the original problem is obtained as a limit of these unique solutions of the regularized problems. This idea is used by Moudafi and Mainge [57] and considered the viscosity method for hierarchical fixed point problems of nonexpansive mappings as follows:

Let C be a nonempty closed convex subset of a real Hilbert space H. Given a contraction mapping $f : C \to C$ and two nonexpansive mappings $S, T : C \to C$. Then for $s, t \in (0, 1)$, the mapping

$$x \mapsto sf(x) + (1 - s)\,[tS(x) + (1 - t)T(x)]$$

is a contraction mapping on C. So it has a unique fixed point, denoted by $x_{s,t} \in C$; thus

$$x_{s,t} = sf(x_{s,t}) + (1 - s)\left[tS(x_{s,t}) + (1 - t)T(x_{s,t})\right]. \tag{23}$$

It is interesting to know the behavior of $\{x_{s,t}\}$ when $s, t \to 0$ separately or jointly. Moudafi and Mainge [57] initiated the investigation of the behavior of the net $\{x_{s,t}\}$ as $s \to 0$ first and then as $t \to 0$. It is further studied by Cianciaruso et al. [17], Marino and Xu [52], Xu [78], Yao et al. [82], and Zeng et al. [92].

Yamada [80] considered the following variational inequality problem defined over the intersection of the fixed point sets of a family of N nonexpansive mappings $T_i : H \to H$, where $N \geq 1$ an integer.

Problem 2 For all $i = 1, 2, \ldots, N$, assume that

(B1) $T_i : H \to H$ is a nonexpansive mapping with $\bigcap_{i=1}^{N} \mathrm{Fix}(T_i) \neq \emptyset$,
(B2) $A : H \to H$ is β-strongly monotone and L-Lipschitz continuous.

The hierarchical variational inequality problem defined over the set $\bigcap_{i=1}^{N} \mathrm{Fix}(T_i)$ is to find $x^* \in \bigcap_{i=1}^{N} \mathrm{Fix}(T_i)$ such that

$$\langle A(x^*), y - x^* \rangle \geq 0, \quad \text{for all } y \in \bigcap_{i=1}^{N} \mathrm{Fix}(T_i), \tag{24}$$

that is, to find

$$\begin{aligned} x^* &\in \mathrm{VIP}\left(\bigcap_{i=1}^{N} \mathrm{Fix}(T_i), A\right) \\ &:= \left\{x^* \in \bigcap_{i=1}^{N} \mathrm{Fix}(T_i) : \langle A(x^*), y - x^* \rangle \geq 0, \text{ for all } y \in \bigcap_{i=1}^{N} \mathrm{Fix}(T_i)\right\}. \end{aligned}$$

Yamada [80] proposed the following iterative method to compute the approximate solutions of HVIP (Problem 2).

Algorithm 4 Start with any initial choice $x_0 \in H$, any $\mu \in \left(0, \frac{2\beta}{L^2}\right)$ and generate a sequence $\{x_n\}$ by

$$x_{n+1} := T_{[n+1]}^{\lambda_{n+1}}(x_n) := T_{[n+1]}(x_n) - \lambda_{n+1}\mu A\left(T_{[n+1]}(x_n)\right), \tag{25}$$

where $[\cdot]$ is the modulo N function defined by $[i] := [i]_N := \{i - kN : k = 0, 1, 2, \dots\} \cap \{1, 2, \dots, N\}$.

The Yamada [80] established the following strong convergence result for the sequence generated by the Algorithm 4.

Theorem 5 [80, Theorem 3.3] *For $i = 1, 2, \dots$, let $T_i : H \to H$ be a nonexpansive mapping with $\bigcap_{i=1}^N \mathrm{Fix}(T_i) \neq \emptyset$ and*

$$\begin{aligned}\bigcap_{i=1}^N \mathrm{Fix}(T_i) &= \mathrm{Fix}\,(T_1 \circ T_2 \circ T_3 \circ \cdots \circ T_N)\\ &= \mathrm{Fix}\,(T_N \circ T_2 \circ T_3 \circ \cdots \circ T_{N-1})\\ &\vdots\\ &= \mathrm{Fix}\,(T_{N-1} \circ T_{N-2} \circ \cdots \circ T_N \circ T_1).\end{aligned}$$

Suppose that a mapping $A : H \to H$ is L-Lipschitz continuous and β-strongly monotone over $\Delta = \bigcap_{i=1}^N Fix(T_i)$. Then with any $x_0 \in H$, any $\mu \in \left(0, \frac{2\beta}{L^2}\right)$ and any sequence $\{\lambda_n\} \subset [0, 1]$ satisfying

(i) $\lim_{n\to\infty} \lambda_n = 0$,
(ii) $\sum_{n=1}^\infty \lambda_n = +\infty$,
(iii) $\sum_{n=1}^\infty |\lambda_n - \lambda_{n+N}| < +\infty$,

the sequence $\{x_n\}$ generated by (25) *converges strongly to the uniquely existing solution of* HVIP 2.

An example of the sequence $\{\lambda_n\}$ of real numbers that satisfies conditions (i)–(iii) of Theorem 5 is $\lambda_n = \frac{1}{n}$.

Very recently, Ceng et al. [10] proposed the following iterative method to find the approximate solutions of HVIP (Problem 2).

Algorithm 5 Start with any initial choice $x_0 \in H$, any $\mu \in \left(0, \frac{2\beta}{L^2}\right)$, $\{\lambda_n\} \subset [0, 1)$, $\{\alpha_n\} \subset (0, 1)$, $\{\beta_n\} \subset (0, 2\alpha)$ and generate a sequence $\{x_n\}$ by

$$\begin{aligned}x_{n+1} &:= \alpha_n x_{n-1} + (1-\alpha_n) T_{[n]}^{\lambda_n}(\bar{x}_n)\\ &:= \alpha_n x_{n-1} + (1-\alpha_n)\left[T_{[n]}(x_n - \gamma_n F(x_n)) - \lambda_n \mu A \circ T_{[n]}\,(x_n - \gamma_n F(x_n))\right],\end{aligned} \tag{26}$$

for $n \geq 1$, where $F : H \to H$ is an α-inverse strongly monotone mapping.

Ceng et al. [10] establish the following weak convergence result for the sequence generated by (26).

Theorem 6 [10, Theorem 3.1] *Let $F : H \to H$ be an α-inverse strongly monotone mapping, $A : H \to H$ is L-Lipschitz continuous and β-strongly monotone for some constants $L, \beta > 0$. For each $i = 1, 2, \dots, N$, let $T_i : H \to H$ be nonexpansive with $\bigcap_{i=1}^N \mathrm{Fix}(T_i) \neq \emptyset$. Suppose that* VIP $\left(\bigcap_{i=1}^N \mathrm{Fix}(T_i), A\right) \neq \emptyset$. *Let $\mu \in \left(0, \frac{2\beta}{L^2}\right)$, $\{\lambda_n\} \subset [0, 1)$, $\{\alpha_n\} \subset (0, 1)$, $\{\beta_n\} \subset (0, 2\alpha)$ be such that $\sum_{n=1}^\infty \lambda_n < \infty$, $\beta_n \leq \lambda_n$*

and $a \leq \alpha_n \leq b$ *for all* $n \geq 1$, *for some* $a, b \in (0, 1)$. *Then, the sequence* $\{x_n\}$ *generated by* (26) *converges weakly to an element of* $\bigcap_{i=1}^{N} \text{Fix}(T_i) \neq \emptyset$.

If, in addition, $\|x_n - T_{[n]}(\bar{x}_n)\| = o(\beta_n)$, *then the sequence* $\{x_n\}$ *converges weakly to an element of* VIP $\left(\bigcap_{i=1}^{N} \text{Fix}(T_i), A\right)$.

They also provided the following strong convergence result.

Theorem 7 [10, Theorem 3.4] *Under the hypothesis of Theorem* 6, *the sequence* $\{x_n\}$ *generated by* (26) *converges strongly to an element of* VIP $\left(\bigcap_{i=1}^{N} \text{Fix}(T_i), A\right)$ *if and only if* $\liminf_{n\to\infty} d(x_n, C) = 0$, *where* $C :=$ VIP $\left(\bigcap_{i=1}^{N} \text{Fix}(T_i), A\right)$.

5 Triple Hierarchical Variational Inequalities

A variational inequality problem defined over the set of solutions of hierarchical variational inequality problem is called a *triple hierarchical variational inequality problem.*

Problem 3 Let H be a real Hilbert space. Assume that

(B1) $T : H \to H$ is a nonexpansive mapping with $\text{Fix}(T) \neq \emptyset$;
(B2) $A_1 : H \to H$ is an α-inverse strongly monotone mapping;
(B3) $A_2 : H \to H$ is a β-strongly monotone and L-Lipschtiz continuous mapping;
(B4) $\text{VIP}(\text{Fix}(T), A_1) \neq \emptyset$.

The *triple hierarchical variational inequality problem* (in short, THVIP) is to find $x^* \in \text{VIP}\,(\text{Fix}(T), A_1)$ such that

$$\langle A_2(x^*), v - x^* \rangle \geq 0, \quad \text{for all } v \in \text{VIP}\,(\text{Fix}(T), A_1), \tag{27}$$

that is, to find

$$\begin{aligned} x^* &\in \text{VIP}\,(\text{VIP}\,(\text{Fix}(T), A_1)\,, A_2) \\ &= \{x^* \in \text{VIP}(\text{Fix}(T), A_1) : \langle A_2(x^*), v - x^* \rangle \geq 0,\ \forall v \in \text{VIP}(\text{Fix}(T), A_1)\}. \end{aligned}$$

Example 4 [29] Let C be a nonempty closed convex subset of a real Hilbert space H, $f_0 : H \to \mathbb{R}$ be a convex function with $\frac{1}{\gamma}$-Lipschitz continuous gradient, $f_1 : H \to \mathbb{R}$ be a convex function with $\frac{1}{\alpha}$-Lipschitz continuous gradient, and $f_2 : H \to H$ be a strongly convex function with a Lipschitz continuous gradient. Define

$$T := P_C(I - \lambda \nabla f_0), \quad \text{for } \lambda \in (0, 2\gamma],$$

and

$$A_i := \nabla f_i, \quad \text{for } i = 1, 2.$$

Then, by Proposition 2 and Proposition 3 (c) and (e), T is nonexpansive with $\text{Fix}(T) = \arg\min_{z\in C} f_0(z)$. Then, THVIP (Problem 3) with $T := P_C(I - \lambda\nabla f_0)$, for $\lambda \in (0, 2\gamma]$ and $A_i := \nabla f_i$, for $i = 1, 2$, reduces to the following *triple hierarchical constrained convex optimization problem* (in short, THCCOP):

$$\text{Minimize } f_2(x) \quad \text{subject to} \quad x \in \arg \min_{y \in \arg\min\limits_{z\in C} f_0(z)} f_1(y).$$

This problem is called *double-hierarchical constrained convex optimization* in [8].

Example 5 [29] Let $A_0 : H \to H$ be a γ-inverse strongly monotone and $f_1, f_2 : H \to \mathbb{R}$ be the same as in Example 4. Consider

$$T := P_{\mathbb{R}^n_+}(I - \lambda A_0), \quad \text{for } \lambda \in (0, 2\gamma],$$

and

$$A_i := \nabla f_i, \quad \text{for } i = 1, 2.$$

Then, the set VIP(Fix(T), A_1) coincides with the solution set of the following *mathematical program with the complementarity constraint* (MPCC) [46, 62]:

$$\text{Minimize } f_1(x) \quad \text{subject to } x \in \mathbb{R}^n_+, \quad A_0(x) \in \mathbb{R}^n_+, \quad \langle x, A_0(x)\rangle = 0.$$

Then THVIP (Problem 3) reduces to the following *convex optimization problem over solution set of MPCC*:

$$\text{Minimize } f_2(x), \quad \text{subject to } x \in \text{Sol(MPCC)},$$

where Sol(MPCC) denotes the set of solutions of MPCC.

Example 6 [29] Let H be a real Hilbert space and $S, T : H \to H$ be nonexpansive mappings with nonempty fixed point set. Let $f_2 : H \to \mathbb{R}$ be defined as in Example 4. Define

$$A_1 := \frac{1}{\delta}(I - S), \text{ for } \delta > 0, \quad \text{and} \quad A_2 := \nabla f_2,$$

where I denotes the identity mapping. Then, A_1 is $\frac{\delta^2}{2}$-inverse strongly monotone and VIP(Fix(T), A_1) can be represented as the solution set of the following *hierarchical fixed point problem* (HFPP) for nonexpansive mappings [48, 56]:

$$\text{find } \bar{x} \in \text{HFPP}(S, T) := \left\{\bar{x} \in \text{Fix}(T) : \bar{x} = P_{\text{Fix}(T)}S(\bar{x})\right\}.$$

Therefore, THVIP (Problem 3) reduces to the following *minimization problem over hierarchical fixed point problem*:

$$\text{Minimize } f_2(x), \quad \text{subject to } x \in \text{HFPP}(S, T).$$

Iiduka [27, 29] proposed the following iterative method to compute the solutions of THVIP (Problem 3).

Algorithm 6 [27, 29] STEP 0. Take $\mu > 0$, $\{\alpha_n\} \subset (0, 1]$, $\{\lambda_n\} \subset (0, \infty)$, choose $x_0 \in H$ arbitrarily, and let $n := 0$.
STEP 1. Given $x_n \in H$, compute $x_{n+1} \in H$ as

$$\begin{aligned} y_n &:= T(x_n - \lambda_n A_1(x_n)), \\ x_{n+1} &:= y_n - \mu\alpha_n A_2(y_n). \end{aligned} \tag{28}$$

Update $n := n + 1$ and go to Step 1.

Iiduka [27] established the following convergence result for Algorithm 6:

Theorem 8 [27, Theorem 3.2] *Assume that the Assumptions* (B1)–(B4) *are satisfied and $\{y_n\}$ in Algorithm 6 is bounded. Let $\mu \in (0, \frac{2\beta}{L^2})$, $\{\alpha_n\} \subset (0, 1]$ and $\{\lambda_n\} \subset (0, 2\alpha]$ be such that the following conditions hold:*

(i) $\sum_{n=1}^{\infty} \alpha_n = \infty$;
(ii) $\lim_{n\to\infty} \frac{1}{\alpha_{n+1}} \left| \frac{1}{\lambda_{n+1}} - \frac{1}{\lambda_n} \right| = 0$;
(iii) $\lim_{n\to\infty} \frac{1}{\lambda_{n+1}} \left| 1 - \frac{\alpha_n}{\alpha_{n+1}} \right| = 0$;
(iv) $\lim_{n\to\infty} \lambda_n = 0$;
(v) $\lim_{n\to\infty} \frac{\alpha_n}{\lambda_n} = 0$;
(vi) $\lim_{n\to\infty} \frac{\lambda_n^2}{\alpha_n} = 0$.

Then, the sequence $\{x_n\}$ generated by Algorithm 6 satisfies the following properties:

(a) *$\{x_n\}$, $\{A_1(x_n)\}$ and $\{A_2(y_n)\}$ are bounded.*
(b) $\lim_{n\to\infty} \frac{\|x_{n+1} - x_n\|}{\lambda_n} = 0$, $\lim_{n\to\infty} \frac{\|x_n - y_n\|}{\lambda_n} = 0$ *and* $\lim_{n\to\infty} \|x_n - T(x_n)\| = 0$.
(c) *If there exists $k > 0$ such that $\|x - T(x)\| \geq k \inf_{y\in \text{Fix}(T)} \|x - y\|$ for all $x \in H$, then the sequence $\{x_n\}$ converges strongly to the unique existing solution of* THVIP *(Problem* 3*).*

Iiduka [29] further studied the convergence of the sequence generated by the Algorithm 6 by making some changes in the parametric sequences. He established the following convergence result.

Theorem 9 [29, Theorem 4.1] *Assume that the assumptions* (B1)–(B4) *are satisfied and $\{y_n\}$ in Algorithm 6 is bounded. Let $\mu \in \left(0, \frac{2\beta}{L^2}\right)$, $\{\alpha_n\} \subset (0, 1]$ and $\{\lambda_n\} \subset (0, 2\alpha]$ be sequences of real numbers such that the following conditions hold:*

(i) $\lim_{n\to\infty} \alpha_n = 0$;
(ii) $\sum_{n=1}^{\infty} \alpha_n = \infty$;
(iii) $\sum_{n=1}^{\infty} |\alpha_{n+1} - \alpha_n| < \infty$;
(iv) $\sum_{n=1}^{\infty} |\lambda_{n+1} - \lambda_n| < \infty$;
(v) $\lambda_n \leq \alpha_n$ *for all* $n \in \mathbb{N}$.

Then, the sequence $\{x_n\}$ generated by Algorithm 6 satisfies the following properties:

(a) $\{x_n\}$ *is bounded.*
(b) $\lim_{n\to\infty} \|x_n - y_n\| = 0$ *and* $\lim_{n\to\infty} \|x_n - T(x_n)\| = 0$.
(c) *If* $\|x_n - y_n\| = o(\lambda_n)$, *then* $\{x_n\}$ *converges strongly to the unique solution of* THVIP *(Problem 3).*

Recently, Ceng et al. [12] extended Algorithm 6 by taking variable parameters and proposed the following iterative method.

Algorithm 7 [12] STEP 0. Take $\{\alpha_n\} \subset (0, 1]$, $\{\lambda_n\} \subset (0, 2\alpha)$ and $\{\mu_n\} \subset \left(0, \frac{2\beta}{L^2}\right)$, choose $x_0 \in H$ arbitrarily, and let $n := 0$,
STEP 1. Given $x_n \in H$, compute $x_{n+1} \in H$ as

$$\begin{aligned} y_n &:= T(x_n - \lambda_n A_1(x_n)), \\ x_{n+1} &:= y_n - \mu_n \alpha_n A_2(y_n). \end{aligned} \tag{29}$$

Update $n := n + 1$ and go to Step 1.

In Algorithm 7, a sequence $\{\mu_n\}$ of positive parameters is introduced so as to take into account possible inexact computation. For $\mu \in \left(0, \frac{2\beta}{L^2}\right)$ whenever $\mu_n = \mu$ for all $n \geq 0$, then Algorithm 7 reduces to Algorithm 6.

The following convergence result for the sequence generated by Algorithm 7 is established by Ceng et al. [12].

Theorem 10 [12, Theorem 3.1] *Assume that the Assumptions* (B1)–(B4) *are satisfied and the sequence* $\{y_n\}$ *generated by Algorithm 7 is bounded. Let* $\{\alpha_n\} \subset (0, 1]$, $\{\lambda_n\} \subset (0, 2\alpha)$ *and* $\{\mu_n\} \subset \left(0, \frac{2\beta}{L^2}\right)$ *such that*

(i) $\sum_{n=0}^{\infty} \alpha_n = \infty$ *and* $\lim_{n\to\infty} \alpha_n = 0$,
(ii) $\left|\mu_n - \frac{\beta}{L^2}\right| \leq \frac{\sqrt{\beta^2 - cL^2}}{L^2}$, *for some* $c \in \left(0, \frac{\beta^2}{L^2}\right)$,
(iii) $\lim_{n\to\infty} \left(\mu_{n+1} - \left(\frac{\alpha_n}{\alpha_{n+1}}\right)\mu_n\right) = 0$,
(iv) $\sum_{n=0}^{\infty} |\lambda_{n+1} - \lambda_n| < \infty$ *and* $\lambda_n \leq \alpha_n$, *for all* $n \geq 0$.

Then, the sequence generated by Algorithm 7 satisfies the following properties:

(a) $\{x_n\}$ *is bounded.*
(b) $\lim_{n\to\infty} \|x_n - y_n\| = 0$ *and* $\lim_{n\to\infty} \|x_n - T(x_n)\| = 0$.
(c) $\{x_n\}$ *converges strongly to a unique solution of* THVIP *(Problem 3) provided* $\|x_n - y_n\| = o(\lambda_n)$.

Remark 1 For $\mu \in \left(0, \frac{2\beta}{L^2}\right)$, whenever $\mu_n = \mu$ for all $n \geq 0$, then the condition (ii) in Theorem 10 holds.

Indeed, since

$$\lim_{t\to 0^+} \frac{\beta - \sqrt{\beta^2 - tL^2}}{L^2} = 0 < \mu,$$

and

$$\lim_{t\to 0^+} \frac{\beta + \sqrt{\beta^2 - tL^2}}{L^2} = \frac{2\beta}{L^2} > \mu,$$

there exist some $\delta_1, \delta_2 \in \left(0, \frac{\beta^2}{L^2}\right)$ such that

$$\frac{\beta - \sqrt{\beta^2 - tL^2}}{L^2} < \mu, \quad \text{for all } t \in (0, \delta_1),$$

and

$$\frac{\beta + \sqrt{\beta^2 - tL^2}}{L^2} > \mu, \quad \text{for all } t \in (0, \delta_2).$$

Therefore, it is obvious that we can pick a number $c \in \left(0, \frac{\beta^2}{L^2}\right)$ such that

$$\frac{\beta - \sqrt{\beta^2 - cL^2}}{L^2} < \mu < \frac{\beta + \sqrt{\beta^2 - cL^2}}{L^2},$$

that is,

$$\left|\mu - \frac{\beta}{L^2}\right| < \frac{\sqrt{\beta^2 - cL^2}}{L^2}.$$

Also, for $\mu \in \left(0, \frac{2\beta}{L^2}\right)$ whenever $\mu_n = \mu$ for all $n \geq 0$, condition (iii) in Theorem 10 is equivalent to

$$\lim_{n\to\infty} \frac{\alpha_n}{\alpha_{n+1}} = 1.$$

This condition is different from the condition (iii) in Theorem 9.

The conclusion of Theorem 10 is also proved by Zeng et al. [89, Theorem 3.1] by assuming the following parametric conditions.

(i) $\sum_{n=0}^{\infty} \alpha_n = \infty$ and $\lim_{n\to\infty} \alpha_n = 0$,
(ii) $\lim_{n\to\infty} \frac{\alpha_n - \alpha_{n+1}}{\alpha_{n+1}} = 0$ or $\sum_{n=0}^{\infty} |\alpha_{n+1} - \alpha_n| < \infty$,
(iii) $\lim_{n\to\infty} \frac{\lambda_n - \lambda_{n+1}}{\lambda_{n+1}} = 0$ or $\sum_{n=0}^{\infty} |\lambda_{n+1} - \lambda_n| < \infty$,
(iv) $\lambda_n \leq \alpha_n$ for all $n \geq 0$.

They presented an application of THVIP (Problem 3) to constrained pseudoinverse problem.

We now present an example to illustrate the Algorithm 7 and Theorem 10.

Example 7 Let $H = \mathbb{R}^2$ with inner product $\langle ., . \rangle$ and norm $\| \cdot \|$ are defined by

$$\langle x, y \rangle = ac + bd \quad \text{and} \quad \|x\| = \sqrt{a^2 + b^2},$$

for all $x, y \in \mathbb{R}^2$ with $x = (a, b)$ and $y = (c, d)$. Let $C = \{x \in \mathbb{R}^2 : \|x\| \leq 1\}$. Clearly, C is a nonempty, bounded, closed and convex subset of $\mathbb{R}^2$. Let

$$A_1 = \begin{Bmatrix} \frac{1}{3} & -\frac{1}{3} \\ -\frac{1}{3} & \frac{1}{3} \end{Bmatrix}.$$

Then, A_1 is α-inverse strongly monotone with $\alpha = \frac{1}{2}$.

Let

$$A_2 = \frac{1}{2}I = \begin{Bmatrix} \frac{1}{2} & 0 \\ 0 & \frac{1}{2} \end{Bmatrix}.$$

Then, $A_2 : C \to C$ is a L-Lipschitz continuous and β-strongly monotone operator with constants $L = \frac{1}{2}$ and $\beta = \frac{1}{2}$, respectively. Take $\mu = 2$ such that $0 < \mu < 2\beta/L^2$.

Let T be a 2×2 positive definite matric such that $\|T\| = 1$, for instance, putting

$$T = \begin{Bmatrix} \frac{3}{5} & \frac{2}{5} \\ \frac{2}{5} & \frac{3}{5} \end{Bmatrix}.$$

Then, $\|T\| = 1$ and $T : C \to C$ is a nonexpansive mapping with

$$\text{Fix}(T) = \{(a, a) : |a| \leq 1\} \neq \emptyset.$$

Further, we observe that the solution set HVIP (Fix(T), A_1) of the HVIP is the following:

$$\begin{aligned} \text{HVIP}\,(\text{Fix}(T), A_1) &= \{z^* \in \text{Fix}(T) : \langle A_1(z^*), z - z^* \rangle \geq 0, \ \text{for all } z \in \text{Fix}(T)\} \\ &= \{z^* \in \text{Fix}(T) : \langle 0, z - z^* \rangle \geq 0, \ \text{for all } z \in \text{Fix}(T)\} \\ &= \text{Fix}(T) = \{(a, a) : |a| \leq 1\} \neq \emptyset. \end{aligned}$$

It is easy to see that there exists a unique solution $x^* = (0, 0)$ to the following THVIP:

Find $x^* \in$ HVIP (Fix(T), A_1) (= Fix(T)) such that

$$\langle A_2(x^*), v - x^* \rangle \geq 0, \quad \text{for all } v \in \text{HVIP}\,(\text{Fix}(T), A_1).$$

By using Matlab programming, we analyze the convergence of the sequences generated by Algorithms 6 and 7.

For $\mu = 1 \in (0, 2\beta/L^2)$, we generate the sequence x_{n+1} by Algorithm 6.

No. of iterations	y_n	x_{n+1}
1	(0.001, 0.001)	(0.001, 0.001)
2	(0.001, 0.001)	(0.0095, 0.0095)
3	(0.0095, 0.0095)	(0.0086, 0.0086)
4	(0.0086, 0.0086)	(0.0073, 0.0073)
5	(0.0073, 0.0073)	(0.0058, 0.0058)
6	(0.0058, 0.0058)	(0.0044, 0.0044)
7	(0.0044, 0.0044)	(0.0031, 0.0031)
8	(0.0031, 0.0031)	(0.0020, 0.0020)
9	(0.0020, 0.0020)	(0.0012, 0.0012)
10	(0.0012, 0.0012)	(1.0e − 003)(0.6545, 0.6545)
11	(1.0e − 003)(0.6545, 0.6545)	(1.0e − 003)(0.3274, 0.3274)
12	(1.0e1.0e − 003)(0.3274, 0.3274)	(1.0e1.0e − 003)(0.1473, 0.1473)
13	(1.0e1.0e − 003)(0.1473, 0.1473)	(1.0e1.0e − 004)(0.5893, 0.5893)
14	(1.0e1.0e − 004)(0.5893, 0.5893)	(1.0e1.0e − 004)(0.2062, 0.2062)
15	(1.0e1.0e − 004)(0.2062, 0.2062)	(1.0e1.0e − 005)(0.6187, 0.6187)
16	(1.0e1.0e − 005)(0.6187, 0.6187)	(1.0e1.0e − 005)(0.1547, 0.1547)
17	(1.0e1.0e − 005)(0.1547, 0.1547)	(1.0e1.0e − 006)(0.3094, 0.3094)
18	(1.0e1.0e − 006)(0.3094, 0.3094)	(1.0e1.0e − 007)(0.4640, 0.4640)
19	(1.0e1.0e − 007)(0.4640, 0.4640)	(1.0e1.0e − 008)(0.4640, 0.4640)
20	(1.0e1.0e − 008)(0.4640, 0.4640)	(1.0e1.0e − 009)(0.2320, 0.2320)
21	(1.0e1.0e − 009)(0.2320, 0.2320)	(0, 0)

For $\mu_n \subset (0, 2\beta/L^2)$, we generate the sequence x_{n+1} by Algorithm 7

No. of iterations	y_n	x_{n+1}
1	(0.0100, 0.0100)	(0.0100, 0.0100)
2	(0.0100, 0.0100)	(0.0099, 0.0099)
3	(0.0099, 0.0099)	(0.0095, 0.0095)
4	(0.0095, 0.0095)	(0.0086, 0.0086)
5	(0.0086, 0.0086)	(0.0073, 0.0073)
6	(0.0073, 0.0073)	(0.0054, 0.0054)
7	(0.0054, 0.0054)	(0.0035, 0.0035)
8	(0.0035, 0.0035)	(0.0018, 0.0018)
9	(0.0018, 0.0018)	(1.0e1.0e − 003)(0.6402, 0.6402)
10	(1.0e1.0e − 003)(0.6402, 0.6402)	(1.0e1.0e − 003)(0.1216, 0.1216)
11	(1.0e1.0e − 003)(0.1216, 0.1216)	(0, 0)

Among classes of nonlinear mappings, the class of pseudo-contraction mappings is one of the most important classes because it is closely related to the class of monotone mappings. Up to now, considerable research efforts have been devoted to develop the iterative methods for computing the approximate fixed points of pseudo-contraction mappings; See for example [16, 51, 61, 72, 85, 86, 90, 94, 95] and the references therein.

We propose the following hybrid iterative algorithm for computing a fixed point of a pseudo-contraction mapping and finding a solution of THVIP (Problem 3) in the setting of real Hilbert spaces.

Algorithm 8 Suppose that the assumptions (B1)–(B4) in Problem 3 are satisfied.
STEP 1. Take $\mu > 0$. Put $C_1 = H$, choose $x_0 \in H$, $\lambda_1 \in (0, 2\alpha]$, $\alpha_1 \in (0, 1]$, $\beta_1 \in (0, 1)$ arbitrarily, and let $n := 1$.
STEP 2. Given $x_n \in C_n$, choose $\lambda_n \in (0, 2\alpha]$, $\alpha_n \in (0, 1]$ and $\beta_n \in (0, 1)$ and compute $x_{n+1} \in C_{n+1}$ as

$$\begin{cases} y_n := (1-\beta_n)x_n + \beta_n(I - \alpha_n\mu A_2)T_n(x_n), \\ C_{n+1} := \left\{ \begin{array}{l} z \in C_n : \|\beta_n(I - (I - \alpha_n\mu A_2)T_n)(y_n)\|^2 \\ \qquad \leq 2\beta_n[\langle x_n - z, (I - (I - \alpha_n\mu A_2)T_n)(y_n)\rangle \\ \qquad\quad - \langle \alpha_n\mu A_2 T_n(y_n) + \lambda_n A_1(T(y_n)), y_n - z\rangle] \end{array} \right\}, \\ x_{n+1} := P_{C_{n+1}}(x_0), \quad n \geq 0, \end{cases} \tag{30}$$

where $T_n := (I - \lambda_n A_1)T$ for all $n \geq 1$.
Update $n := n + 1$ and go to Step 2.

We now present the criteria for the strong convergence of the sequence generated by the Algorithm 8.

Theorem 11 *Let $T : H \to H$ be a L-Lipschitz continuous pseudo-contractive self-mapping defined on a real Hilbert space H such that* $\mathrm{Fix}(T) \neq \emptyset$. *Assume that $\{\beta_n\} \subset [a, b]$ for some $a, b \in \left(0, \frac{1}{L+1}\right)$ and $\{\alpha_n\} \subset (0, 1]$ and $\{\lambda_n\} \subset (0, 2\alpha]$ such that $\lim_{n\to\infty}\alpha_n = \lim_{n\to\infty}\lambda_n = 0$. Take a fixed number $\mu \in \left(0, \frac{2\beta}{L^2}\right)$. Then, the sequence $\{x_n\}$ generated by Algorithm 8 satisfies the following properties:*

(a) *$\{x_n\}$ is bounded;*
(b) $\lim_{n\to\infty}\|x_n - y_n\| = 0$ *and* $\lim_{n\to\infty}\|x_n - T(x_n)\| = 0$;
(c) *$\{x_n\}$ converges strongly to $P_{\mathrm{Fix}(T)}(x_0)$;*
(d) *If T is nonexpansive and A_1 is injective, $P_{\mathrm{Fix}(T)}(x_0)$ is a unique solution of* THVIP *(Problem 3) provided* $\lim_{n\to\infty}(\|x_n - y_n\| + \alpha_n)/\lambda_n = 0$.

Proof We first show that $P_{\mathrm{Fix}(T)}$ and $\{x_n\}$ are well defined.

From [51, 94], we note that $\mathrm{Fix}(T)$ is closed and convex. Indeed, by Zhou [94], we can define a mapping $g : H \to H$ by

$$g(x) = (2I - T)^{-1}(x), \quad \text{for all } x \in H.$$

It is clear that g is a nonexpansive self-mapping such that $\mathrm{Fix}(T) = \mathrm{Fix}(g)$. Hence, by Matinez-Yanes and Xu [53, Proposition 2.1 (iii)], we conclude that $\mathrm{Fix}(g) = \mathrm{Fix}(T)$ is a closed convex set. This implies that the projection $P_{\mathrm{Fix}(T)}$ is well defined. It is obvious that $\{C_n\}$ is closed and convex. Thus, $\{x_n\}$ is also well defined.

We now show that $\mathrm{Fix}(T) \subseteq C_n$ for all $n \geq 0$. Indeed, taking $p \in \mathrm{Fix}(T)$, we note that $(I - T)p = 0$ and (1) is equivalent to

$$\langle (I-T)(x)-(I-T)(y), x-y\rangle \geq 0, \quad \text{for all } x, y \in H. \tag{31}$$

By using Lemma 1 and the inequality (31), we obtain

$$\begin{aligned}
&\|x_n - p - \beta_n(I-(I-\alpha_n\mu A_2)T_n)(y_n)\|^2 \\
&= \|x_n - p\|^2 - \|\beta_n(I-(I-\alpha_n\mu A_2)T_n)(y_n)\|^2 \\
&\quad - 2\beta_n \langle (I-(I-\alpha_n\mu A_2)T_n)(y_n), x_n - p - \beta_n(I-(I-\alpha_n\mu A_2)T_n)(y_n)\rangle \\
&= \|x_n - p\|^2 - \|\beta_n(I-(I-\alpha_n\mu A_2)T_n)(y_n)\|^2 \\
&\quad - 2\beta_n \langle (I-T)(y_n) - (I-T)(p) + \lambda_n A_1(T(y_n)), y_n - p\rangle \\
&\quad - 2\beta_n \langle T_n(y_n) - (I-\alpha_n\mu A_2)(T_n(y_n)), y_n - p\rangle \\
&\quad - 2\beta_n \langle (I-(I-\alpha_n\mu A_2)T_n)(y_n), x_n - y_n - \beta_n(I-(I-\alpha_n\mu A_2)T_n)(y_n)\rangle \\
&\leq \|x_n - p\|^2 - \|\beta_n(I-(I-\alpha_n\mu A_2)T_n)(y_n)\|^2 \\
&\quad - 2\beta_n \langle T_n(y_n) - (I-\alpha_n\mu A_2)T_n(y_n) + \lambda_n A_1(T(y_n)), y_n - p\rangle \\
&\quad - 2\beta_n \langle (I-(I-\alpha_n\mu A_2)T_n)(y_n), x_n - y_n - \beta_n(I-(I-\alpha_n\mu A_2)T_n)(y_n)\rangle \\
&= \|x_n - p\|^2 - \|x_n - y_n + y_n - x_n + \beta_n(I-(I-\alpha_n\mu A_2)T_n)(y_n)\|^2 \\
&\quad - 2\beta_n \langle \alpha_n\mu A_2 T_n y_n + \lambda_n A_1(T(y_n), y_n - p\rangle \\
&\quad - 2\beta_n \langle (I-(I-\alpha_n\mu A_2)T_n)(y_n), x_n - y_n - \beta_n(I-(I-\alpha_n\mu A_2)T_n)(y_n)\rangle \\
&= \|x_n - p\|^2 - \|x_n - y_n\|^2 - \|y_n - x_n + \beta_n(I-(I-\alpha_n\mu A_2)T_n)(y_n)\|^2 \\
&\quad - 2\langle x_n - y_n, y_n - x_n + \beta_n(I-(I-\alpha_n\mu A_2)T_n)(y_n)\rangle \\
&\quad + 2\beta_n \langle (I-(I-\alpha_n\mu A_2)T_n)(y_n), y_n - x_n + \beta_n(I-(I-\alpha_n\mu A_2)T_n)(y_n)\rangle \\
&\quad - 2\beta_n \langle \alpha_n\mu A_2(T_n(y_n)) + \lambda_n A_1(T(y_n)), y_n - p\rangle \\
&= \|x_n - p\|^2 - \|x_n - y_n\|^2 - \|y_n - x_n + \beta_n(I-(I-\alpha_n\mu A_2)T_n)(y_n)\|^2 \\
&\quad - 2\langle x_n - y_n - \beta_n(I-(I-\alpha_n\mu A_2)T_n)(y_n), y_n - x_n \\
&\quad + \beta_n(I-(I-\alpha_n\mu A_2)T_n)(y_n)\rangle \\
&\quad - 2\beta_n \langle \alpha_n\mu A_2 T_n(y_n) + \lambda_n(A_1(T(y_n)), y_n - p\rangle \\
&\leq \|x_n - p\|^2 - \|x_n - y_n\|^2 - \|y_n - x_n + \beta_n(I-(I-\alpha_n\mu A_2)T_n)(y_n)\|^2 \\
&\quad + 2\Big|\langle x_n - y_n - \beta_n(I-(I-\alpha_n\mu A_2)T_n)(y_n), y_n - x_n \\
&\quad + \beta_n(I-(I-\alpha_n\mu A_2)T_n)(y_n)\rangle\Big| \\
&\quad - 2\beta_n \langle \alpha_n\mu A_2 T_n(y_n) + \lambda_n A_1(T(y_n)), y_n - p\rangle.
\end{aligned} \tag{32}$$

Since T is a L-Lipschitz continuous mapping, by Lemma 4 and Proposition 5, we have

$$\begin{aligned}
&\|(I-(I-\alpha_n\mu A_2)T_n)(x_n) - (I-(I-\alpha_n\mu A_2)T_n)(y_n)\| \\
&\leq \|x_n - y_n\| + \|(I-\alpha_n\mu A_2)T_n(x_n) - (I-\alpha_n\mu A_2)T_n(y_n)\| \\
&\leq \|x_n - y_n\| + (1-\alpha_n\tau)\|T_n(x_n) - T_n(y_n)\| \\
&= \|x_n - y_n\| + (1-\alpha_n\tau)\|(I-\lambda_n A_1)T(x_n) - (I-\lambda_n A_1)T(y_n)\| \\
&\leq \|x_n - y_n\| + \|(I-\lambda_n A_1)T(x_n) - (I-\lambda_n A_1)T(y_n)\| \\
&\leq \|x_n - y_n\| + \|T(x_n) - T(y_n)\| \\
&\leq (L+1)\|x_n - y_n\|,
\end{aligned} \tag{33}$$

where $\tau = 1 - \sqrt{1-\mu(2\eta - \mu\kappa^2)}$. From (30), we observe that

$$x_n - y_n = \beta_n (I - (I - \alpha_n \mu A_2)T_n) x_n.$$

Hence, by utilizing (33), we obtain

$$\begin{aligned}
&|\langle x_n - y_n - \beta_n(I - (I - \alpha_n\mu A_2)T_n)(y_n), y_n - x_n + \beta_n(I - (I - \alpha_n\mu A_2)T_n)(y_n)\rangle| \\
&= \beta_n|\langle (I - (I - \alpha_n\mu A_2)T_n)(x_n) - (I - (I - \alpha_n\mu A_2)T_n)(y_n), y_n - x_n \\
&\quad + \beta_n(I - (I - \alpha_n\mu A_2)T_n)(y_n)\rangle| \\
&\le \beta_n \|(I - (I - \alpha_n\mu A_2)T_n)(x_n) - (I - (I - \alpha_n\mu A_2)T_n)y_n\| \cdot \\
&\quad \|y_n - x_n + \beta_n(I - (I - \alpha_n\mu A_2)T_n)(y_n)\| \\
&\le \beta_n(L+1)\|x_n - y_n\| \, \|y_n - x_n + \beta_n(I - (I - \alpha_n\mu A_2)T_n)(y_n)\| \\
&\le \frac{\beta_n(L+1)}{2}\left(\|x_n - y_n\|^2 + \|y_n - x_n + \beta_n(I - (I - \alpha_n\mu A_2)T_n)(y_n)\|^2\right).
\end{aligned} \tag{34}$$

Combining (32) and (34), we get

$$\begin{aligned}
&\|x_n - p - \beta_n(I - (I - \alpha_n\mu A_2)T_n)(y_n)\|^2 \\
&\le \|x_n - p\|^2 - \|x_n - y_n\|^2 - \|y_n - x_n + \beta_n(I - (I - \alpha_n\mu A_2)T_n)(y_n)\|^2 \\
&\quad + \beta_n(L+1)\left(\|x_n - y_n\|^2 + \|y_n - x_n + \beta_n(I - (I - \alpha_n\mu A_2)T_n)(y_n)\|^2\right) \\
&\quad - 2\beta_n\langle \alpha_n\mu A_2(T_n(y_n)) + \lambda_n A_1(T(y_n)), y_n - p\rangle \\
&= \|x_n - p\|^2 + [\beta_n(L+1) - 1](\|x_n - y_n\|^2 \\
&\quad + \|y_n - x_n + \beta_n(I - (I - \alpha_n\mu A_2)T_n)(y_n)\|^2) \\
&\quad - 2\beta_n\langle \alpha_n\mu A_2(T_n(y_n)) + \lambda_n A_1(T(y_n)), y_n - p\rangle \\
&\le \|x_n - p\|^2 - 2\beta_n\langle \alpha_n\mu A_2(T_n(y_n)) + \lambda_n A_1(T(y_n)), y_n - p\rangle.
\end{aligned} \tag{35}$$

We observe that

$$\begin{aligned}
&\|x_n - p - \beta_n(I - (I - \alpha_n\mu A_2)T_n)(y_n)\|^2 \\
&= \|x_n - p\|^2 - 2\beta_n\langle x_n - p, (I - (I - \alpha_n\mu A_2)T_n)(y_n)\rangle \\
&\quad + \|\beta_n(I - (I - \alpha_n\mu A_2)T_n)(y_n)\|^2.
\end{aligned} \tag{36}$$

Therefore, from (35) and (36), we have

$$\begin{aligned}
\|\beta_n(I - (I - \alpha_n\mu A_2)T_n)(y_n)\|^2 \le 2\beta_n[&\langle x_n - p, (I - (I - \alpha_n\mu A_2)T_n)(y_n)\rangle \\
&-\langle \alpha_n\mu A_2(T_n(y_n)) + \lambda_n A_1(T(y_n)), y_n - p\rangle],
\end{aligned}$$

which implies that

$$p \in C_n, \quad \text{that is,} \quad \mathrm{Fix}(T) \subseteq C_n, \quad \text{for all } n \ge 1.$$

Since $x_n = P_{C_n}(x_0)$, we have

$$\langle x_0 - x_n, x_n - y\rangle \ge 0, \quad \text{for all } y \in C_n.$$

Since $\mathrm{Fix}(T) \subseteq C_n$, we obtain

$$\langle x_0 - x_n, x_n - u\rangle \ge 0, \quad \text{for all } u \in \mathrm{Fix}(T).$$

Therefore, for all $u \in \text{Fix}(T)$, we have

$$\begin{aligned} 0 &\le \langle x_0 - x_n, x_n - u\rangle \\ &= \langle x_0 - x_n, x_n - x_0 + x_0 - u\rangle \\ &= -\|x_0 - x_n\|^2 + \langle x_0 - x_n, x_0 - u\rangle \\ &\le -\|x_0 - x_n\|^2 + \|x_0 - x_n\|\|x_0 - u\|, \end{aligned}$$

which implies that

$$\|x_0 - x_n\| \le \|x_0 - u\|, \quad \text{for all } u \in \text{Fix}(T). \tag{37}$$

Thus, $\{x_n\}$ is bounded and so are $\{y_n\}$, $\{T(y_n)\}$, $\{T_n(y_n)\}$.

From $x_n = P_{C_n}(x_0)$ and $x_{n+1} = P_{C_{n+1}}(x_0) \in C_{n+1} \subset C_n$, we have

$$\langle x_0 - x_n, x_n - x_{n+1}\rangle \ge 0. \tag{38}$$

Hence,

$$\begin{aligned} 0 &\le \langle x_0 - x_n, x_n - x_{n+1}\rangle \\ &= \langle x_0 - x_n, x_n - x_0 + x_0 - x_{n+1}\rangle \\ &= -\|x_0 - x_n\|^2 + \langle x_0 - x_n, x_0 - x_{n+1}\rangle \\ &\le -\|x_0 - x_n\|^2 + \|x_0 - x_n\|\|x_0 - x_{n+1}\|, \end{aligned}$$

and therefore,

$$\|x_0 - x_n\| \le \|x_0 - x_{n+1}\|.$$

Thus, $\lim_{n\to\infty} \|x_n - x_0\|$ exists.

From Lemma 2 and the inequality (38), we obtain

$$\begin{aligned} \|x_{n+1} - x_n\|^2 &= \|(x_{n+1} - x_0) - (x_n - x_0)\|^2 \\ &= \|x_{n+1} - x_0\|^2 - \|x_n - x_0\|^2 - 2\langle x_{n+1} - x_n, x_n - x_0\rangle \\ &\le \|x_{n+1} - x_0\|^2 - \|x_n - x_0\|^2 \to 0. \end{aligned}$$

Since $x_{n+1} \in C_{n+1} \subseteq C_n$, from $\|x_n - x_{n+1}\| \to 0$, $\lambda_n \to 0$ and $\alpha_n \to 0$, it follows that

$$\begin{aligned} &\|\beta_n(I - (I - \alpha_n\mu A_2)T_n)(y_n)\|^2 \\ &\le 2\beta_n[\langle x_n - x_{n+1}, (I - (I - \alpha_n\mu A_2)T_n)(y_n)\rangle \\ &\quad - \langle \alpha_n\mu A_2(T_n(y_n)) + \lambda_n A_1(T(y_n)), y_n - x_{n+1}\rangle] \\ &\le 2\beta_n[\|x_n - x_{n+1}\|\|(I - (I - \alpha_n\mu A_2)T_n)(y_n)\| \\ &\quad + \|\alpha_n\mu A_2(T_n(y_n)) + \lambda_n A_1(T(y_n))\|\|y_n - x_{n+1}\|] \\ &\le 2\beta_n[\|x_n - x_{n+1}\|(\|y_n\| + \|T_n(y_n)\| + \alpha_n\mu\|A_2(T_n(y_n))\|) \\ &\quad + (\alpha_n\mu\|A_2(T_n(y_n))\| + \lambda_n\|A_1(T(y_n))\|)\|y_n - x_{n+1}\|] \to 0. \end{aligned}$$

We note that $\beta_n \in [a, b]$ for some $a, b \in \left(0, \frac{1}{L+1}\right)$, we thus obtain

$$\|y_n - (I - \alpha_n \mu A_2)T_n(y_n)\| \to 0.$$

We also note that

$$\begin{aligned}\|T(y_n) - (I - \alpha_n \mu A_2)T_n(y_n)\| &= \|T(y_n) - T_n(y_n) + \alpha_n \mu A_2(T_n(y_n))\| \\ &\leq \|T(y_n) - (I - \lambda_n A_1)T(y_n)\| + \alpha_n \mu \|A_2(T_n(y_n))\| \\ &= \lambda_n \|A_1(T(y_n))\| + \alpha_n \mu \|A_2(T_n(y_n))\| \to 0.\end{aligned}$$

Therefore, we get

$$\|y_n - T(y_n)\| \leq \|y_n - (I - \alpha_n \mu A_2)T_n(y_n)\| + \|T(y_n) - (I - \alpha_n \mu A_2)T_n(y_n)\| \to 0.$$

On the other hand, by utilizing Lemma 4 and Proposition 5, we deduce that

$$\begin{aligned}&\|x_n - (I - \alpha_n \mu A_2)T_n(x_n)\| \\ &\leq \|x_n - y_n\| + \|y_n - (I - \alpha_n \mu A_2)T_n(y_n)\| \\ &\quad + \|(I - \alpha_n \mu A_2)T_n(y_n) - (I - \alpha_n \mu A_2)T_n(x_n)\| \\ &\leq \|x_n - y_n\| + \|y_n - (I - \alpha_n \mu A_2)T_n(y_n)\| \\ &\quad + (1 - \alpha_n \tau)\|T_n(y_n) - T_n(x_n)\| \\ &\leq \|x_n - y_n\| + \|y_n - (I - \alpha_n \mu A_2)T_n(y_n)\| \\ &\quad + \|(I - \lambda_n A_1)T(y_n) - (I - \lambda_n A_1)T(x_n)\| \\ &\leq \|x_n - y_n\| + \|y_n - (I - \alpha_n \mu A_2)T_n(y_n)\| + \|T(y_n) - T(x_n)\| \\ &\leq \|x_n - y_n\| + \|y_n - (I - \alpha_n \mu A_2)T_n(y_n)\| + L\|y_n - x_n\| \\ &= (L + 1)\|x_n - y_n\| + \|y_n - (I - \alpha_n \mu A_2)T_n(y_n)\| \\ &= \beta_n(L + 1)\|x_n - (I - \alpha_n \mu A_2)T_n(x_n)\| + \|y_n - (I - \alpha_n \mu A_2)T_n(y_n)\|,\end{aligned}$$

that is,

$$\|x_n - (I - \alpha_n \mu A_2)T_n(x_n)\| \leq \frac{1}{1 - \beta_n(L + 1)}\|y_n - (I - \alpha_n \mu A_2)T_n(y_n)\| \to 0.$$

We note that

$$\begin{aligned}\|T(x_n) - (I - \alpha_n \mu A_2)T_n(x_n)\| &= \|T(x_n) - T_n(x_n) + \alpha_n \mu A_2(T_n(x_n))\| \\ &\leq \|T(x_n) - (I - \lambda_n A_1)T(x_n)\| + \alpha_n \mu \|A_2(T_n(x_n))\| \\ &= \lambda_n \|A_1(T(x_n))\| + \alpha_n \mu \|A_2(T_n(x_n))\| \to 0.\end{aligned}$$

Consequently,

$$\|x_n - T(x_n)\| \leq \|x_n - (I - \alpha_n \mu A_2)T_n(x_n)\| + \|T(x_n) - (I - \alpha_n \mu A_2)T_n(x_n)\| \to 0. \tag{39}$$

Relation (39) and Lemma 6 guarantee that every weak limit point of the sequence $\{x_n\}$ is a fixed point of T, that is, $\omega_w(x_n) \subseteq \text{Fix}(T)$. This fact, the inequality (37) and

Lemma 1 ensure the strong convergence of $\{x_n\}$ to $P_{\mathrm{Fix}(T)}(x_0)$. Since $\|x_n - y_n\| = \|\beta_n(I-(I-\alpha_n\mu A_2)T_n)(x_n)\| \to 0$, it immediately follows that the sequence $\{y_n\}$ converges strongly to $P_{F(T)}(x_0)$.

Finally, we prove that whenever T is nonexpansive and A_1 is injective and $(\|x_n - y_n\| + \alpha_n)/\lambda_n \to 0$ (as $n \to \infty$), $P_{\mathrm{Fix}(T)}(x_0)$ is the unique solution of Problem 3.

Indeed, put $\hat{x} := P_{\mathrm{Fix}(T)}(x_0)$. By condition (B4), we can take an arbitrarily fixed element $y \in \mathrm{VI}(\mathrm{Fix}(T), A_1)$ and put $M := \sup\{\|x_n - y\| + \|y_n - y\| : n \geq 1\} < \infty$. Then, from the condition (B3) and Lemmas 4 and 6, it follows that for all $n \geq 1$,

$$
\begin{aligned}
&\|(I-\alpha_n\mu A_2)T_n(x_n) - y\|^2 \\
&= \|(I-\alpha_n\mu A_2)T_n(x_n) - (I-\alpha_n\mu A_2)T_n(y) + (I-\alpha_n\mu A_2)T_n(y) - y\|^2 \\
&\leq \|(I-\alpha_n\mu A_2)T_n(x_n) - (I-\alpha_n\mu A_2)T_n(y)\|^2 \\
&\quad + 2\langle (I-\alpha_n\mu A_2)T_n(x_n) - y, (I-\alpha_n\mu A_2)T_n(y) - y\rangle \\
&\leq (1-\alpha_n\tau)^2\|T_n(x_n) - T_n(y)\|^2 \\
&\quad + 2\langle y - (I-\alpha_n\mu A_2)T_n(x_n), \lambda_n A_1(y) + \alpha_n\mu A_2(T_n(y))\rangle \\
&\leq \|(I-\lambda_n A_1)T(x_n) - (I-\lambda_n A_1)T(y)\|^2 \\
&\quad + 2\langle y - (I-\alpha_n\mu A_2)T_n(x_n), \lambda_n A_1(y) + \alpha_n\mu A_2(T_n(y))\rangle \\
&\leq \|Tx_n - Ty\|^2 + 2\langle y - (I-\alpha_n\mu A_2)T_n(x_n), \lambda_n A_1(y) + \alpha_n\mu A_2(T_n(y))\rangle \\
&\leq \|x_n - y\|^2 + 2\langle y - (I-\alpha_n\mu A_2)T_n(x_n), \lambda_n A_1(y) + \alpha_n\mu A_2(T_n(y))\rangle \\
&= \|x_n - y\|^2 + 2\langle y - T_n(x_n), \lambda_n A_1(y) + \alpha_n\mu A_2(T_n(y))\rangle \\
&\quad + 2\alpha_n\mu\langle A_2(T_n(x_n)), \lambda_n A_1(y) + \alpha_n\mu A_2(T_n(y))\rangle \\
&\leq \|x_n - y\|^2 + 2\langle y - (I-\lambda_n A_1)T(x_n), \lambda_n A_1(y) + \alpha_n\mu A_2(T_n(y))\rangle \\
&\quad + 2\alpha_n\mu\|A_2(T_n(x_n))\|\|\lambda_n A_1(y) + \alpha_n\mu A_2(T_n(y))\| \\
&\leq \|x_n - y\|^2 + 2(\langle y - T(x_n), \lambda_n A_1(y) + \alpha_n\mu A_2(T_n(y))\rangle \\
&\quad + \lambda_n\langle A_1(T(x_n)), \lambda_n A_1(y) + \alpha_n\mu A_2(T_n(y))\rangle) \\
&\quad + 2\alpha_n\mu\|A_2(T_n(x_n))\|\|\lambda_n A_1(y) + \alpha_n\mu A_2(T_n(y))\| \\
&\leq \|x_n - y\|^2 + 2(\langle y - T(x_n), \lambda_n A_1(y) + \alpha_n\mu A_2(T_n(y))\rangle \\
&\quad + (\lambda_n\|A_1(T(x_n))\| + \alpha_n\mu\|A_2T_n(x_n)\|)\|\lambda_n A_1(y) + \alpha_n\mu A_2(T_n(y))\|),
\end{aligned}
$$

and hence,

$$
\begin{aligned}
\|y_n - y\|^2 &= \|(1-\beta_n)(x_n - y) + \beta_n[(I-\alpha_n\mu A_2)T_n(x_n) - y]\|^2 \\
&\leq (1-\beta_n)\|x_n - y\|^2 + \beta_n\|(I-\alpha_n\mu A_2)T_n(x_n) - y\|^2 \\
&\leq (1-\beta_n)\|x_n - y\|^2 + \beta_n[\|x_n - y\|^2 \\
&\quad + 2(\langle y - T(x_n), \lambda_n A_1(y) + \alpha_n\mu A_2(T_n(y))\rangle \\
&\quad + (\lambda_n\|A_1(T(x_n))\| + \alpha_n\mu\|A_2(T_n(x_n))\|)\|\lambda_n A_1(y) + \alpha_n\mu A_2(T_n(y))\|)] \\
&= \|x_n - y\|^2 + 2\beta_n[\langle y - T(x_n), \lambda_n A_1(y) + \alpha_n\mu A_2(T_n(y))\rangle \\
&\quad + (\lambda_n\|A_1(T(x_n))\| + \alpha_n\mu\|A_2(T_n(x_n))\|)\|\lambda_n A_1(y) + \alpha_n\mu A_2(T_n(y))\|].
\end{aligned}
$$

This implies that

$$\begin{aligned}0 &\le \tfrac{1}{\lambda_n}\{\|x_n - y\|^2 - \|y_n - y\|^2 + 2\beta_n[\langle y - T(x_n), \lambda_n A_1(y) + \alpha_n \mu A_2(T_n(y))\rangle \\ &\quad +(\lambda_n \|A_1(T(x_n))\| + \alpha_n \mu \|A_2(T_n(x_n))\|)\|\lambda_n A_1(y) + \alpha_n \mu A_2(T_n(y))\|]\} \\ &= (\|x_n - y\| + \|y_n - y\|)\tfrac{\|x_n - y\| - \|y_n - y\|}{\lambda_n} + 2\beta_n[\langle y - T(x_n), A_1(y) + \mu \tfrac{\alpha_n}{\lambda_n} A_2(T_n(y))\rangle \\ &\quad +(\|A_1(T(x_n))\| + \mu \tfrac{\alpha_n}{\lambda_n}\|A_2(T_n(x_n))\|)\|\lambda_n A_1(y) + \alpha_n \mu A_2(T_n(y)\|] \\ &\le M \tfrac{\|x_n - y_n\|}{\lambda_n} + 2\beta_n[\langle y - T(x_n), A_1(y) + \mu \tfrac{\alpha_n}{\lambda_n} A_2(T_n(y))\rangle \\ &\quad +(\|A_1(T(x_n))\| + \mu \tfrac{\alpha_n}{\lambda_n}\|A_2(T_n(x_n))\|)\|\lambda_n A_1(y) + \alpha_n \mu A_2(T_n(y))\|],\end{aligned}$$

that is,

$$\begin{aligned}0 &\le \tfrac{M}{2\beta_n} \cdot \tfrac{\|x_n - y_n\|}{\lambda_n} + \left\langle y - T(x_n), A_1(y) + \mu \tfrac{\alpha_n}{\lambda_n} A_2(T_n(y))\right\rangle \\ &\quad + \left(\|A_1(T(x_n))\| + \mu \tfrac{\alpha_n}{\lambda_n}\|A_2(T_n(x_n))\|\right)\|\lambda_n A_1(y) + \alpha_n \mu A_2(T_n(y))\|.\end{aligned} \tag{40}$$

Since T is nonexpansive, it is known that $L = 1$ and $\{\beta_n\} \subset [a, b]$ for some $a, b \in \left(0, \frac{1}{2}\right)$. In terms of the conditions that $\alpha_n \to 0, \lambda_n \to 0$ and $(\|x_n - y_n\| + \alpha_n)/\lambda_n \to 0$, we deduce from (40) and $x_n \to \hat{x}$ $(=: P_{\mathrm{Fix}(T)}(x_0))$ that

$$\langle y - \hat{x}, A_1(y)\rangle \ge 0, \quad \text{for all } y \in \mathrm{Fix}(T).$$

The condition (B1) ensures

$$\langle y - \hat{x}, A_1(\hat{x})\rangle \ge 0, \quad \text{for all } y \in \mathrm{Fix}(T),$$

that is, $\hat{x} \in \mathrm{VIP}(\mathrm{Fix}(T), A_1)$. Furthermore, from the conditions (A2) and (A4), we conclude that Problem 3 has a unique solution. Hence, VIP (VIP(Fix(T), A_1), A_2) is a singleton. Thus we may assume that VIP (VIP(Fix(T), A_1), A_2) $= \{x^*\}$. This implies that $x^* \in$ VIP (Fix(T), A_1).

Now we show that $\hat{x} = x^*$. Indeed, since $\hat{x}, x^* \in \mathrm{VIP}(\mathrm{Fix}(T), A_1)$, we have

$$\langle A_1(\hat{x}), y - \hat{x}\rangle \ge 0, \quad \text{for all } y \in \mathrm{Fix}(T), \tag{41}$$

and

$$\langle A_1(x^*), y - x^*\rangle \ge 0, \quad \text{for all } y \in \mathrm{Fix}(T). \tag{42}$$

Setting $y = x^*$ in inequality (41) and $y = \hat{x}$ in inequality (42), and then adding the resultant inequalities, we obtain

$$\langle A_1(\hat{x}) - A_1(x^*), \hat{x} - x^*\rangle \le 0.$$

Since A_1 is α-inverse-strongly monotone, we have

$$\alpha\|A_1(\hat{x}) - A_1(x^*)\|^2 \le \langle A_1(\hat{x}) - A_1(x^*), \hat{x} - x^*\rangle \le 0.$$

Consequently, $A_1(\hat{x}) = A_1(x^*)$. Since A_1 is injective, we have $\hat{x} = x^*$. □

Ceng et al. [11] considered the following more general problem than Problem 3.

Problem 4 Let C be a nonempty closed convex subset of a real Hilbert space H and $f : C \to H$ be a ρ-contraction mapping with constant $\rho \in [0, 1)$. Let $S, T : C \to C$ be two nonexpansive mappings with $\text{Fix}(T) \neq \emptyset$ and $F : C \to H$ be L-Lipschitz continuous and β-strongly monotone with

$$0 < \mu < \frac{2\beta}{L^2} \quad \text{and} \quad 0 < \gamma < \tau,$$

where, $\tau = 1 - \sqrt{1 - \mu(2\beta - \mu L^2)}$. The *triple hierarchical variational inequality problem* is to find $x^* \in \Xi$ such that

$$\langle (\mu F - \gamma f)(x^*), v - x^* \rangle \geq 0, \quad \text{for all } v \in \Xi, \tag{43}$$

where Ξ denotes the solution set of the following *hierarchical variational inequality problem:*

$$\begin{cases} \text{Find } z^* \in \text{Fix}(T) \text{ such that} \\ \langle (\mu F - \gamma S)(z^*), z - z^* \rangle \geq 0, \quad \text{for all } v \in \text{Fix}(T), \end{cases} \tag{44}$$

where we assume that the solution set of above hierarchical variational inequality is nonempty.

Now we present two examples of above mentioned triple hierarchical variational inequality problem.

Example 8 [11] Let $H = \mathbb{R}^2$ with inner product $\langle ., . \rangle$ and norm $\| \cdot \|$ are defined by

$$\langle x, y \rangle = ac + bd \quad \text{and} \quad \|x\| = \sqrt{a^2 + b^2},$$

for all $x, y \in \mathbb{R}^2$ with $x = (a, b)$ and $y = (c, d)$. Let $C = \left\{ x \in \mathbb{R}^2 : \|x\| \leq \sqrt{2} \right\}$. Clearly, C is a nonempty, bounded and closed convex subset of $\mathbb{R}^2$. Let f be a 2×2 positive semidefinite matrix such that $0 < \|f\| < 1$, for instance, letting

$$f = \begin{Bmatrix} \frac{1}{3} & -\frac{1}{3} \\ -\frac{1}{3} & \frac{1}{3} \end{Bmatrix}.$$

Then, $\|f\| = \frac{2}{3}$ and $f : C \to C$ is a ρ-contraction mapping with contractivity constant $\rho = \frac{2}{3}$. Let $F = \frac{1}{2}I = \begin{Bmatrix} \frac{1}{2} & 0 \\ 0 & \frac{1}{2} \end{Bmatrix}$. Then $F : C \to C$ is L-Lipschitz continuous and β-strongly monotone with constants $L = \frac{1}{2}$ and $\beta = \frac{1}{2}$, respectively. Take $\mu = 2$ and $\gamma = 1$ such that $0 < \mu < 2\beta/L^2$ and $0 < \gamma \leq \tau$, where $\tau = 1 - \sqrt{1 - \mu(2\beta - \mu L^2)} = 1$. Let T and S be two 2×2 positive definite matrices such that $\|T\| = \|S\| = 1$, for instance, letting

$$T = \left\{ \begin{matrix} \frac{2}{3} & \frac{1}{3} \\ \frac{1}{3} & \frac{2}{3} \end{matrix} \right\} \text{ and } S = \left\{ \begin{matrix} \frac{3}{5} & \frac{2}{5} \\ \frac{2}{5} & \frac{3}{5} \end{matrix} \right\}.$$

Then, $\|T\| = \|S\| = 1$ and $S, T : C \to C$ are nonexpansive with $\text{Fix}(T) = \{(a, a) : |a| \leq 1\} \neq \emptyset$.

We observe that the solution set Ξ of the HVIP (Problem 4) is the following:

$$\begin{aligned} \Xi &= \{z^* \in \text{Fix}(T) : (\mu F - \gamma S)z^*, z - z^*\rangle \geq 0, \quad \text{for all } z \in \text{Fix}(T)\} \\ &= \{z^* \in \text{Fix}(T) : \langle (I - S)(z^*), z - z^*\rangle \geq 0, \quad \text{for all } z \in \text{Fix}(T)\} \\ &= \{z^* \in \text{Fix}(T) : \langle 0, z - z^*\rangle \geq 0, \quad \text{for all } z \in \text{Fix}(T)\} \\ &= \text{Fix}(T) = \{(a, a) : |a| \leq 1\} \neq \emptyset. \end{aligned}$$

It is easy to see that there exists a unique solution $x^* = (0, 0)$ to the following THVIP: find $x^* \in \Xi$ $(= \text{Fix}(T))$ such that

$$\langle (\mu F - \gamma f)(x^*), x - x^*\rangle \geq 0, \quad \text{for all } x \in \Xi,$$

that is,

$$\langle (I - f)(x^*), x - x^*\rangle \geq 0, \quad \text{for all } x \in \Xi.$$

Example 9 [9] Let C be a nonempty closed convex subset of a real Hilbert space H and $f : C \to H$ be l-Lipschitz continuous with constant $l > 0$. Suppose that $g_0 : H \to \mathbb{R}$ is a convex function with a $1/\alpha_0$-Lipschitz continuous gradient, $g_1 : H \to \mathbb{R}$ is a convex function with a $1/\alpha_1$-Lipschitz continuous gradient, and $g_2 : H \to \mathbb{R}$ is an α-strongly convex function with a α_2-Lipschitz continuous gradient. Define $T := P_C(I - \lambda \nabla g_0)$ for $\lambda \in (0, 2\alpha_0]$, $V := P_C(I - \tilde{\lambda} \nabla g_1)$ for $\tilde{\lambda} \in (0, 2\alpha_1]$ and $F := \nabla g_2$. Then $T, V : C \to C$ are nonexpansive mappings with $\text{Fix}(T) = \text{argmin}_{z \in C}\, g_0(z)$ and $\text{Fix}(V) = \text{argmin}_{z \in C}\, g_1(z)$, and F is L-Lipschitzian and β-strongly monotone with $L = 1/\alpha_2$ and $\beta = \alpha$. Assume that $\text{argmin}_{z \in C}\, g_0(z) \cap \text{argmin}_{z \in C}\, g_1(z) \neq \emptyset$. Then, we have

$$\begin{aligned} \emptyset &\neq \text{argmin}_{z \in C}\, g_0(z) \cap \text{argmin}_{z \in C}\, g_1(z) \\ &= \text{Fix}(T) \cap \text{Fix}(V) \end{aligned}$$

$$\begin{aligned} &\subset \{z^* \in \text{Fix}(T) : \langle (I - V)(z^*), z - z^*\rangle \geq 0, \quad \text{for all } z \in \text{Fix}(T)\} \\ &= \text{HVIP}(\text{Fix}(T), (I - V)). \end{aligned}$$

When $0 < \mu < 2\alpha\alpha_2^2$ and $0 \leq \gamma l < \tau$, where $\tau = 1 - \sqrt{1 - \mu\left(2\alpha - \frac{\mu}{\alpha_2^2}\right)}$, we have

$$0 < \mu < 2\beta/L^2 \quad \text{and} \quad 0 \leq \gamma l < \tau,$$

where $\tau = 1 - \sqrt{1 - \mu(2\beta - \mu L^2)}$. In particular, when $\mu = \beta/L^2 = \alpha\ \alpha_2^2$, we have

$$0 \le \gamma l < \tau = 1 - \sqrt{1 - \mu\left(2\alpha - \frac{\mu}{\alpha_2^2}\right)} = 1 - \sqrt{1 - \alpha^2\ \alpha_2^2}.$$

In this case, when $\gamma = \frac{1}{2l}\alpha^2\ \alpha_2^2$ $\left(\text{obviously, } \sqrt{1 - \alpha^2\ \alpha_2^2} < 1 - \frac{1}{2}\alpha^2\ \alpha_2^2\right)$, the triple hierarchical variational inequality problem of finding $x^* \in$ HVIP (Fix(T), $(\mu F - \gamma f)$) such that

$$\langle(\mu F - \gamma f)(x^*), x - x^*\rangle \ge 0, \quad \text{for all } x \in \text{HVIP}(\text{Fix}(T), (\mu F - \gamma f)),$$

reduces to the following THVIP: find $x^* \in$ HVIP (Fix(T), $(\mu F - \gamma f)$) such that

$$\left\langle\left(\nabla g_2 - \frac{\alpha}{2l} f\right)(x^*), x - x^*\right\rangle \ge 0, \quad \text{for all } x \in \text{HVIP}(\text{Fix}(T), (\mu F - \gamma f)).$$

Ceng et al. [11] combined the regularization method, the hybrid steepest descent method, and the projection method to propose an implicit scheme that generates a net in an implicit way. They studied the strong convergence of the sequence generated by the proposed scheme to the unique solution of THVIP (Problem 4).

Further, Cent et al. [9] proposed an approximation method to compute the solutions of above mentioned THVIP (Problem 4). They combined the hybrid steepest descent method, viscosity method, and the projection method to propose their method. The strong convergence of the net generated by the proposed method is also studied.

We consider the following THVIP where the HVIP is defined over the intersection of fixed point set of a nonexpansive mapping and fixed point set of strictly pseudo-contractive mapping.

Problem 5 Let C a nonempty closed convex subset of a real Hilbert space H and $F : C \to H$ be L-Lipschitz continuous and β-strongly monotone, where $L > 0$ and $\beta > 0$ are constants. Let $V : C \to H$ be ρ-contraction with coefficient $\rho \in [0, 1)$, $S, T_1 : C \to C$ be nonexpansive mappings, and $T_2 : C \to C$ be ζ-strictly pseudo-contractive mapping with $\text{Fix}(T_1) \cap \text{Fix}(T_2) \neq \emptyset$. Let $0 < \mu < 2\beta/L^2$ and $0 < \gamma \le \tau$, where $\tau = 1 - \sqrt{1 - \mu(2\beta - \mu L^2)}$. The problem is to find $x^* \in \Xi$ such that

$$\langle(\mu F - \gamma V)(x^*), x - x^*\rangle \ge 0, \quad \text{for all } x \in \Xi, \tag{45}$$

where Ξ denotes the solution set of the following hierarchical variational inequality problem (HVIP) of finding $z^* \in \text{Fix}(T_1) \cap \text{Fix}(T_2)$ such that

$$\langle(\mu F - \gamma S)(z^*), z - z^*\rangle \ge 0, \quad \text{for all } z \in \text{Fix}(T_1) \cap \text{Fix}(T_2). \tag{46}$$

Whenever T_1 and T_2 are strictly pseudo-contraction, Problem 5 is considered in [41].

Whenever $T_1 \equiv T$ is a nonexpansive mapping and $T_2 \equiv I$ is an identity mapping, then Problem 5 reduce to Problem 3.

We also consider the following triple hierarchical variational inequality problem:

Problem 6 Let C a nonempty closed convex subset of a real Hilbert space H and $F : C \to H$ be L-Lipschitz continuous and β-strongly monotone, where $L > 0$ and $\beta > 0$ are constants. Let $A : C \to H$ be a monotone and κ-Lipschitz continuous mapping, $V : C \to H$ be ρ-contraction with coefficient $\rho \in [0, 1)$, $S, T : C \to C$ be nonexpansive mappings with $\mathrm{Fix}(T) \cap \Gamma \neq \emptyset$. Let $0 < \mu < 2\beta/L^2$ and $0 < \gamma \leq \tau$, where $\tau = 1 - \sqrt{1 - \mu(2\beta - \mu L^2)}$. The problem is to find $x^* \in \Xi$ such that

$$\langle (\mu F - \gamma V)(x^*), x - x^* \rangle \geq 0, \quad \text{for all } x \in \Xi, \tag{47}$$

where Ξ denotes the solution set of the following hierarchical variational inequality problem (HVIP) of finding $z^* \in \mathrm{Fix}(T) \cap \Gamma$ such that

$$\langle (\mu F - \gamma S)(z^*), z - z^* \rangle \geq 0, \quad \text{for all } z \in \mathrm{Fix}(T) \cap \Gamma, \tag{48}$$

where $\Gamma = \mathrm{VIP}(C, A)$.

We remark that Problem 6 is a generalization of Problem 5. Indeed, in Problem 6, if we put $T = T_1$ and $A = I - T_2$, where $T_1 : C \to C$ is a nonexpansive mapping and $T_2 : C \to C$ is a ζ_2-strictly pseudo-contractive mapping. Then from the definition of strictly pseudo-contractive mapping, we have

$$\langle T_2(x) - T_2(y), x - y \rangle \leq \|x - y\|^2 - \frac{1 - \zeta_2}{2} \|(I - T_2)(x) - (I - T_2)(y)\|^2, \quad \text{for all } x, y \in C.$$

It is clear that the mapping $A = I - T_2$ is $\frac{1-\zeta_2}{2}$-inverse strongly monotone. Letting $\kappa = \frac{2}{1-\zeta_2}$, then $A : C \to H$ is monotone and κ-Lipschitz continuous. In this case, $\Gamma = \mathrm{Fix}(T_2)$. Therefore, Problem 6 reduces to Problem 5.

Motivated and inspired by Korpelevich's extragradient method [73], the iterative method proposed in [11] and multistep hybrid extragradient method proposed in [41], we propose the following multistep explicit and implicit hybrid extragradient-like methods for solving Problem 6.

Algorithm 9 Let C be a nonempty closed convex subset of a real Hilbert space H and $F : C \to H$ be L-Lipschitz continuous and β-strongly monotone $A : C \to H$ be monotone and κ-Lipschitz continuous, $V : C \to H$ be ρ-contraction with coefficient $\rho \in [0, 1)$ and $S, T : C \to C$ be nonexpansive mappings. Suppose that $\{\alpha_n\} \subset [0, \infty)$, $\{\beta_n\}, \{\gamma_n\} \subset [0, 1]$ and $\{\lambda_n\}, \{\delta_n\} \subset (0, 1)$. Let $0 < \mu < 2\beta/L^2$, $0 < \gamma \leq \tau$ and $A_n = \alpha_n I + A$ for all $n \geq 0$, where $\tau = 1 - \sqrt{1 - \mu(2\beta - \mu L^2)}$. The sequence $\{x_n\}$ is generated by the following iterative scheme:

$$\begin{cases} x_0 = x \in C \quad \text{chosen arbitrarily}, \\ y_n = (1-\beta_n)x_n + \beta_n P_C(x_n - \lambda A_n(x_n)), \\ z_n = \gamma_n x_n + (1-\gamma_n)T(P_C(y_n - \lambda A_n(y_n))), \\ x_{n+1} = P_C[\lambda_n \gamma(\delta_n V(x_n) + (1-\delta_n)S(x_n)) + (I - \lambda_n \mu F)(T(z_n))], \quad \text{for all } n \geq 0. \end{cases} \tag{49}$$

In particular, if $V \equiv 0$, then (49) reduces to the following iterative scheme:

$$\begin{cases} x_0 = x \in C \quad \text{chosen arbitrarily}, \\ y_n = (1-\beta_n)x_n + \beta_n P_C(x_n - \lambda A_n(x_n)), \\ z_n = \gamma_n x_n + (1-\gamma_n)T(P_C(y_n - \lambda A_n(y_n))), \\ x_{n+1} = P_C[\lambda_n(1-\delta_n)\gamma S(x_n) + (I - \lambda_n \mu F)(T(z_n))], \quad \text{for all } n \geq 0. \end{cases} \tag{50}$$

If $S \equiv V$, then (49) reduces to the following iterative scheme:

$$\begin{cases} x_0 = x \in C \quad \text{chosen arbitrarily}, \\ y_n = (1-\beta_n)x_n + \beta_n P_C(x_n - \lambda A_n(x_n)), \\ z_n = \gamma_n x_n + (1-\gamma_n)T(P_C(y_n - \lambda A_n(y_n))), \\ x_{n+1} = P_C[\lambda_n \gamma V(x_n) + (I - \lambda_n \mu F)(T(z_n))], \quad \text{for all } n \geq 0. \end{cases} \tag{51}$$

Moreover, if $S \equiv V \equiv 0$ then (49) reduces to the following iterative scheme:

$$\begin{cases} x_0 = x \in C \quad \text{chosen arbitrarily}, \\ y_n = (1-\beta_n)x_n + \beta_n P_C(x_n - \lambda A_n(x_n)), \\ z_n = \gamma_n x_n + (1-\gamma_n)T(P_C(y_n - \lambda A_n(y_n))), \\ x_{n+1} = P_C[(I - \lambda_n \mu F)(T(z_n))], \quad \text{for all } n \geq 0. \end{cases} \tag{52}$$

We now present the convergence analysis of Algorithm 9 for solving Problem 6.

Theorem 12 *Let C be a nonempty closed convex subset of a real Hilbert space H, $F : C \to H$ be L-Lipschitz and β-strongly monotone with constants $L, \beta > 0$, $A : C \to H$ be $1/\kappa$-inverse strongly monotone, $V : C \to H$ be ρ-contraction with coefficient $\rho \in [0, 1)$ and $S, T : C \to C$ be nonexpansive mappings. Let $0 < \lambda < 2/\kappa$, $0 < \mu < 2\beta/L^2$ and $0 < \gamma \leq \tau$, where $\tau = 1 - \sqrt{1 - \mu(2\beta - \mu L^2)}$. Assume that the solution set Ξ of* HVIP (48) *is nonempty and the sequences $\{\alpha_n\} \subset [0, \infty)$, $\{\beta_n\}$, $\{\gamma_n\} \subset [0, 1]$ and $\{\lambda_n\}$, $\{\delta_n\} \subset (0, 1)$ satisfy the following condition.*

(i) $\sum_{n=0}^{\infty} \alpha_n < \infty$;
(ii) $0 < \liminf_{n\to\infty} \beta_n \leq \limsup_{n\to\infty} \beta_n < 1$ *and* $0 < \liminf_{n\to\infty} \gamma_n \leq \limsup_{n\to\infty} \gamma_n < 1$;
(iii) $\lim_{n\to\infty} \lambda_n = 0$, $\lim_{n\to\infty} \delta_n = 0$ *and* $\sum_{n=0}^{\infty} \delta_n \lambda_n = \infty$;
(iv) *there are constants $\bar{k}, \theta > 0$ such that*

$$\|x - Tx\| \geq \bar{k}[d(x, \mathrm{Fix}(T) \cap \Gamma)]^{\theta}, \quad \forall x \in C;$$

(v) $\lim_{n\to\infty} \dfrac{\lambda_n^{1/\theta}}{\delta_n} = 0.$

Then, the following assertions hold.

(a) *If* $\{x_n\}$ *is the sequence generated by the scheme* (49) *and* $\{Sx_n\}$ *is bounded, then* $\{x_n\}$ *converges strongly to a point* $x^* \in \text{Fix}(T) \cap \Gamma$ *which is a unique solution of Problem* 6 *provided* $\|x_{n+1} - x_n\| + \|x_n - z_n\| = o(\lambda_n)$.

(b) *If* $\{x_n\}$ *is a sequence generated by the scheme* (50) *and* $\{S(x_n)\}$ *is bounded, then* $\{x_n\}$ *converges strongly to a unique solution* x^* *of the following* VIP *provided* $\|x_{n+1} - x_n\| + \|x_n - z_n\| = o(\lambda_n)$*:*

$$\textit{find } x^* \in \Xi \textit{ such that } \quad \langle F(x^*), x - x^*\rangle \geq 0, \quad \textit{for all } x \in \Xi. \qquad (53)$$

6 Triple Hierarchical Variational Inequalities for a Family of Nonexpansive Mappings

Ceng et al. [12, 89] considered the following monotone variational inequality with the variational inequality constraint which is defined over the intersection of the fixed point sets of a family of N nonexpansive mappings $T_i : H \to H$, where $N \geq 1$ an integer.

Problem 7 For each $i = 1, 2, \ldots, N$, assume that

(C1) $T_i : H \to H$ is a nonexpansive mapping with $\bigcap_{i=1}^{N} \text{Fix}(T_i) \neq \emptyset$,
(C2) $A_1 : H \to H$ is an α-inverse strongly monotone mapping,
(C3) $A_2 : H \to H$ is a β-strongly monotone and L-Lipschitz continuous mapping,
(C4) $\text{VIP}\left(\bigcap_{i=1}^{N} \text{Fix}(T_i), A_1\right) \neq \emptyset$.

The problem is to

$$\begin{aligned}
&\text{find } x^* \in \text{VIP}\left(\text{VIP}\left(\bigcap_{i=1}^{N} \text{Fix}(T_i), A_1\right), A_2\right)\\
&:= \left\{x^* \in \text{VIP}\left(\bigcap_{i=1}^{N} \text{Fix}(T_i), A_1\right) : \langle A_2(x^*), v - x^*\rangle \geq 0,\ \forall v \in \text{VIP}\left(\bigcap_{i=1}^{N} \text{Fix}(T_i), A_1\right)\right\}.
\end{aligned}$$

We write $T_{[k]} := T_{k \bmod N}$ for integer $k \geq 1$ with the mod function taking values in the set $\{1, 2, \ldots, N\}$, that is, if $k = jN + q$ for some integers $j \geq 0$ and $0 \leq q < N$, then $T_{[k]} = T_N$ if $q = 0$ and $T_{[k]} = T_q$ if $0 < q < N$.

Zeng et al. [89] proposed the following relaxed hybrid steepest descent method for finding the solution of Problem 7.

Algorithm 10 [89] Step 0: Take $\{\alpha_n\} \subset (0, 1]$, $\{\lambda_n\} \subset (0, 2\alpha]$, $\mu \subset (0, 2\beta/L^2)$, choose $x_0 \in H$ arbitrarily, and let $n := 0$.

STEP 1: Given $x_n \in H$, compute $x_{n+1} \in H$ as

$$\begin{aligned} y_n &:= T_{[n+1]}(x_n - \lambda_n A_1(x_n)), \\ x_{n+1} &:= y_n - \mu \alpha_n A_2(y_n). \end{aligned} \tag{54}$$

Update $n := n + 1$ and go to Step 1.

Zeng et al. [89] proved that the sequence $\{x_n\}$ generated be Algorithm 10 converges strongly to a unique solution of THVIP (Problem 7).

Theorem 13 [89, Theorem 3.2] *An addition to the assumptions of Problem 7, assume that the sequence $\{y_n\}$ generated by the Algorithm 10, is bounded. Let $\{\alpha_n\} \subset (0, 1]$, $\{\lambda_n\} \subset (0, 2\alpha]$ and $\mu_n \subset (0, 2\beta/L^2)$ such that*

(i) $\sum_{n=0}^{\infty} \alpha_n = \infty$ *and* $\lim_{n\to\infty} \alpha_n = 0$*;*
(ii) $\lim_{n\to\infty} \frac{\alpha_n - \alpha_{n+N}}{\alpha_{n+N}} = 0$ *or* $\sum_{n=0}^{\infty} |\alpha_{n+N} - \alpha_n| < \infty$*;*
(iii) $\lim_{n\to\infty} \frac{\lambda_n - \lambda_{n+N}}{\lambda_{n+N}} = 0$ *or* $\sum_{n=0}^{\infty} |\lambda_{n+N} - \lambda_n| < \infty$*;*
(iv) $\lambda_n \le \alpha_n$ *for all* $n \ge 0$.

Assume, further, that

$$\begin{aligned} \bigcap_{i=1}^{N} \mathrm{Fix}(T_i) &= \mathrm{Fix}\,(T_1 \circ T_2 \circ T_3 \circ \cdots \circ T_N) \\ &= \mathrm{Fix}\,(T_N \circ T_2 \circ T_3 \circ \cdots \circ T_{N-1}) \\ &\;\;\vdots \\ &= \mathrm{Fix}\,(T_2 \circ T_3 \circ \cdots \circ T_N \circ T_1). \end{aligned}$$

Then, the sequence $\{x_n\}$ generated by Algorithm 10 satisfies the following properties:

(a) $\{x_n\}$ *is bounded.*
(b) $\lim_{n\to\infty} \|x_{n+N} - x_n\| = 0$ *and* $\lim_{n\to\infty} \left\|x_n - T_{[n+N]} \circ \cdots \circ T_{[n+1]}(x_n)\right\| = 0$.
(c) $\{x_n\}$ *converges strongly to a unique solution of* THVIP *(Problem 7) provided* $\|x_n - y_n\| = o(\lambda_n)$.

Ceng et al. [12] extended Algorithm 10 by considering the variable parameter. They proposed the following iterative method to find the solution of Problem 7.

Algorithm 11 [12] STEP 0: Take $\{\alpha_n\} \subset (0, 1]$, $\{\lambda_n\} \subset (0, 2\alpha]$, $\{\mu_n\} \subset (0, 2\beta/L^2)$, choose $x_0 \in H$ arbitrarily, and let $n := 0$.
STEP 1: Given $x_n \in H$, compute $x_{n+1} \in H$ as

$$\begin{aligned} y_n &:= T_{[n+1]}(x_n - \lambda_n A_1(x_n), \\ x_{n+1} &:= y_n - \mu_n \alpha_n A_2(y_n). \end{aligned} \tag{55}$$

Update $n := n + 1$ and go to Step 1.

Ceng et al. [12] proved that the sequence $\{x_n\}$ generated be Algorithm 10 converges strongly to a unique solution of THVIP (Problem 7).

Theorem 14 [12, Theorem 3.2] *In addition to the assumption of Problem 7, assume that the sequence* $\{y_n\}$ *generated by the Algorithm* 10*, is bounded. Let* $\{\alpha_n\} \subset (0, 1]$, $\{\lambda_n\} \subset (0, 2\alpha]$ *and* $\{\mu_n\} \subset \left(0, 2\beta/L^2\right)$ *such that*

(i) $\sum_{n=0}^{\infty} \alpha_n = \infty$ *and* $\lim_{n\to\infty} \alpha_n = 0$;
(ii) $\left|\mu_n - \frac{\beta}{L^2}\right| \leq \frac{\sqrt{\beta^2 - cL^2}}{L^2}$, *for some* $c \in \left(0, \frac{\beta^2}{L^2}\right)$;
(iii) $\lim_{n\to\infty} \left(\mu_{n+N} - \left(\frac{\alpha_n}{\alpha_{n+N}}\right)\mu_n\right) = 0$;
(iv) $\sum_{n=0}^{\infty} |\lambda_{n+N} - \lambda_n| < \infty$ *and* $\lambda_n \leq \alpha_n$ *for all* $n \geq 0$.

Assume, further, that

$$\begin{aligned}\bigcap_{i=1}^{N} \operatorname{Fix}(T_i) &= \operatorname{Fix}(T_1 \circ T_2 \circ T_3 \circ \cdots \circ T_N)\\ &= \operatorname{Fix}(T_N \circ T_2 \circ T_3 \circ \cdots \circ T_{N-1})\\ &\vdots\\ &= \operatorname{Fix}(T_2 \circ T_3 \circ \cdots \circ T_N \circ T_1).\end{aligned}$$

Then, the sequence $\{x_n\}$ *generated by Algorithm* 10 *satisfies the following properties:*

(a) $\{x_n\}$ *is bounded.*
(b) $\lim_{n\to\infty} \|x_{n+N} - x_n\| = 0$ *and* $\lim_{n\to\infty} \left\|x_n - T_{[n+N]} \circ \cdots \circ T_{[n+1]}(x_n)\right\| = 0$.
(c) $\{x_n\}$ *converges strongly to a unique solution of* THVIP *(Problem 7) provided* $\|x_n - y_n\| = o(\lambda_n)$.

Now we study the Problem 7, where T_i $(i = 1, 2, \ldots, N)$ is a Lipschitz continuous and pseudo-contraction mapping on H with $N \geq 1$ an integer. In this case, we propose a hybrid iterative algorithm for solving Problem 7 concerning a finite family of mappings $\{T_i\}_{i=1}^N$ with $\bigcap_{i=1}^N \operatorname{Fix}(T_i) \neq \emptyset$. Under some appropriate conditions, we derive the strong convergence of our algorithm to the unique solution of Problem 7.

Let Ω be the intersection of the fixed point sets of N pseudo-contractive mappings $T_i : H \to H$ with $N \geq 1$ an integer, that is,

$$\Omega = \bigcap_{i=1}^{N} \operatorname{Fix}(T_i).$$

We propose the following hybrid iterative algorithm for computing a common fixed point of a finite family $\{T_i\}_{i=1}^N$ of pseudo-contractive mappings and a solution of Problem 7 in the setting of a real Hilbert space H.

Algorithm 12 Suppose that the assumptions (C1)–(C4) in Problem 7 are satisfied.
STEP 1. Take $\mu > 0$. Put $C_1 = H$, choose $x_0 \in H$, $\lambda_1 \in (0, 2\alpha]$, $\alpha_1 \in (0, 1]$, $\beta_1 \in (0, 1)$ arbitrarily, and let $n := 1$.
STEP 2. Given $x_n \in C_n$, choose $\lambda_n \in (0, 2\alpha]$, $\alpha_n \in (0, 1]$ and $\beta_n \in (0, 1)$ and compute $x_{n+1} \in C_{n+1}$ as

$$\begin{cases} y_n := (1-\beta_n)x_n + \beta_n(I-\alpha_n\mu A_2)\widetilde{T}_n(x_n), \\ C_{n+1} := \left\{ \begin{aligned} z \in C_n : & \left\|\beta_n(I-(I-\alpha_n\mu A_2)\widetilde{T}_n)(y_n)\right\|^2 \\ & \leq 2\beta_n\Big[\langle x_n - z, (I-(I-\alpha_n\mu A_2)\widetilde{T}_n)(y_n)\rangle \\ & \quad - \langle \alpha_n\mu A_2\widetilde{T}_n(y_n) + \lambda_n(A_1(T_n(y_n)), y_n - z\rangle\Big] \end{aligned} \right\}, \\ x_{n+1} := P_{C_{n+1}}(x_0), \quad \text{for all } n \geq 0, \end{cases} \tag{56}$$

where

$$\widetilde{T}_n := (I-\lambda_n A_1)T_n \quad \text{and} \quad T_n := T_{n \bmod N}, \quad \text{for integer } n \geq 1, \tag{57}$$

with the mod function taking values in the set $\{1, 2, \ldots, N\}$.

Update $n := n + 1$ and go to Step 2.

Under quite appropriate conditions, we establish the following strong convergence theorem for the sequence $\{x_n\}$ generated by the Algorithm 12.

Theorem 15 *For each $i = 1, 2, \ldots, N$, let $T_i : H \to H$ be a L-Lipschitz continuous pseudo-contractive self-mapping defined on a real Hilbert space H such that $\Omega = \bigcap_{i=1}^N \mathrm{Fix}(T_i) \neq \emptyset$. Assume that $\{\beta_n\} \subset [a, b]$ for some $a, b \in \left(0, \frac{1}{L+1}\right)$ and $\{\alpha_n\} \subset (0, 1]$ and $\{\lambda_n\} \subset (0, 2\alpha]$ such that $\lim_{n\to\infty} \alpha_n = \lim_{n\to\infty} \lambda_n = 0$. Take a fixed number $\mu \in (0, 2\eta/\kappa^2)$. Then the sequence $\{x_n\}$ generated by the Algorithm 12 satisfies the following properties:*

(a) *$\{x_n\}$ is bounded.*
(b) $\lim_{n\to\infty} \|x_n - y_n\| = 0$ *and* $\lim_{n\to\infty} \|x_n - T_l(x_n)\| = 0$ *for all* $l = 1, 2, \ldots, N$.
(c) *$\{x_n\}$ converges strongly to $P_\Omega(x_0)$.*
(d) *If T_l is nonexpansive for each $1 \leq l \leq N$ and A_1 is injective, $P_\Omega x_0$ is the unique solution of Problem 7 provided* $\lim_{n\to\infty}(\|x_n - y_n\| + \alpha_n)/\lambda_n = 0$.

Proof As stated in the proof of Theorem 11, we can readily see that each $\mathrm{Fix}(T_i)$ is closed and convex for $i = 1, 2, \ldots, N$. Hence Ω is a closed and convex set. This implies that the projection P_Ω is well defined. It is clear that the sequence $\{C_n\}$ is closed and convex. Thus, $\{x_n\}$ is also well defined.

We show that $\Omega \subseteq C_n$ for all $n \geq 0$. Indeed, taking $p \in \Omega$, we note that $(I - T_n)(p) = 0$ and

$$\langle (I-T_n)(x) - (I-T_n)(y), x - y\rangle \geq 0, \quad \text{for all } x, y \in H. \tag{58}$$

By Lemma 1 and inequality (58), we obtain

$$
\begin{aligned}
&\|x_n - p - \beta_n(I - (I - \alpha_n\mu A_2)\widetilde{T}_n)(y_n)\|^2 \\
&= \|x_n - p\|^2 - \|\beta_n(I - (I - \alpha_n\mu A_2)\widetilde{T}_n)(y_n)\|^2 \\
&\quad - 2\beta_n\langle(I - (I - \alpha_n\mu A_2)\widetilde{T}_n)(y_n), x_n - p - \beta_n(I - (I - \alpha_n\mu A_2)\widetilde{T}_n)(y_n)\rangle \\
&= \|x_n - p\|^2 - \|\beta_n(I - (I - \alpha_n\mu A_2)\widetilde{T}_n)(y_n)\|^2 \\
&\quad - 2\beta_n\langle(I - T_n)(y_n) - (I - T_n)(p) + \lambda_n A_1(T_n(y_n)), y_n - p\rangle \\
&\quad - 2\beta_n\langle\widetilde{T}_n(y_n) - (I - \alpha_n\mu A_2)\widetilde{T}_n(y_n), y_n - p\rangle \\
&\quad - 2\beta_n\langle(I - (I - \alpha_n\mu A_2)\widetilde{T}_n)(y_n), x_n - y_n - \beta_n(I - (I - \alpha_n\mu A_2)\widetilde{T}_n)(y_n)\rangle \\
&\le \|x_n - p\|^2 - \|\beta_n(I - (I - \alpha_n\mu A_2)\widetilde{T}_n)(y_n)\|^2 \\
&\quad - 2\beta_n\langle\widetilde{T}_n(y_n) - (I - \alpha_n\mu A_2)\widetilde{T}_n(y_n) + \lambda_n A_1(T_n(y_n)), y_n - p\rangle \\
&\quad - 2\beta_n\langle(I - (I - \alpha_n\mu A_2)\widetilde{T}_n)(y_n), x_n - y_n - \beta_n(I - (I - \alpha_n\mu A_2)\widetilde{T}_n)(y_n)\rangle \\
&= \|x_n - p\|^2 - \|x_n - y_n + y_n - x_n + \beta_n(I - (I - \alpha_n\mu A_2)\widetilde{T}_n)(y_n)\|^2 \\
&\quad - 2\beta_n\langle\alpha_n\mu A_2\widetilde{T}_n(y_n) + \lambda_n A_1(T_n(y_n)), y_n - p\rangle \\
&\quad - 2\beta_n\langle(I - (I - \alpha_n\mu A_2)\widetilde{T}_n)(y_n), x_n - y_n - \beta_n(I - (I - \alpha_n\mu A_2)\widetilde{T}_n)(y_n)\rangle \\
&= \|x_n - p\|^2 - \|x_n - y_n\|^2 - \|y_n - x_n + \beta_n(I - (I - \alpha_n\mu A_2)\widetilde{T}_n)(y_n)\|^2 \\
&\quad - 2\langle x_n - y_n, y_n - x_n + \beta_n(I - (I - \alpha_n\mu A_2)\widetilde{T}_n)(y_n)\rangle \\
&\quad + 2\beta_n\langle(I - (I - \alpha_n\mu A_2)\widetilde{T}_n)(y_n), y_n - x_n + \beta_n(I - (I - \alpha_n\mu A_2)\widetilde{T}_n)(y_n)\rangle \\
&\quad - 2\beta_n\langle\alpha_n\mu A_2(\widetilde{T}_n(y_n)) + \lambda_n A_1(T_n(y_n)), y_n - p\rangle \\
&= \|x_n - p\|^2 - \|x_n - y_n\|^2 - \|y_n - x_n + \beta_n(I - (I - \alpha_n\mu A_2)\widetilde{T}_n)(y_n)\|^2 \\
&\quad - 2\langle x_n - y_n - \beta_n(I - (I - \alpha_n\mu A_2)\widetilde{T}_n)(y_n), y_n - x_n + \beta_n(I - (I - \alpha_n\mu A_2)\widetilde{T}_n)(y_n)\rangle \\
&\quad - 2\beta_n\langle\alpha_n\mu A_2(\widetilde{T}_n(y_n)) + \lambda_n A_1(T_n(y_n)), y_n - p\rangle \\
&\le \|x_n - p\|^2 - \|x_n - y_n\|^2 - \|y_n - x_n + \beta_n(I - (I - \alpha_n\mu A_2)\widetilde{T}_n)(y_n)\|^2 \\
&\quad + 2|\langle x_n - y_n - \beta_n(I - (I - \alpha_n\mu A_2)\widetilde{T}_n)(y_n), y_n - x_n + \beta_n(I - (I - \alpha_n\mu A_2)\widetilde{T}_n)(y_n)\rangle| \\
&\quad - 2\beta_n\langle\alpha_n\mu A_2(\widetilde{T}_n(y_n)) + \lambda_n A_1(T_n(y_n)), y_n - p\rangle.
\end{aligned} \tag{59}
$$

Since each T_i is L-Lipschitz continuous for $i = 1, 2, \ldots, N$, by Lemmas 4 and 5, we obtain

$$
\begin{aligned}
&\|(I - (I - \alpha_n\mu A_2)(\widetilde{T}_n)(x_n)) - (I - (I - \alpha_n\mu A_2)(\widetilde{T}_n)(y_n))\| \\
&\le \|x_n - y_n\| + \|(I - \alpha_n\mu A_2)(\widetilde{T}_n(x_n)) - (I - \alpha_n\mu A_2)(\widetilde{T}_n(y_n))\| \\
&\le \|x_n - y_n\| + (1 - \alpha_n\tau)\|\widetilde{T}_n(x_n) - \widetilde{T}_n(y_n)\| \\
&= \|x_n - y_n\| + (1 - \alpha_n\tau)\|(I - \lambda_n A_1)(T_n(x_n)) - (I - \lambda_n A_1)(T_n(y_n))\| \\
&\le \|x_n - y_n\| + \|(I - \lambda_n A_1)(T_n(x_n)) - (I - \lambda_n A_1)(T_n(y_n))\| \\
&\le \|x_n - y_n\| + \|T_n(x_n) - T_n(y_n)\| \\
&\le (L + 1)\|x_n - y_n\|,
\end{aligned} \tag{60}
$$

where $\tau = 1 - \sqrt{1 - \mu(2\eta - \mu\kappa^2)}$. From (56), we observe that

$$x_n - y_n = \beta_n(I - (I - \alpha_n\mu A_2)\widetilde{T}_n)(x_n).$$

Hence, by utilizing (60), we get

$$\begin{aligned}
&\big|\langle x_n - y_n - \beta_n(I-(I-\alpha_n\mu A_2)\widetilde{T}_n)(y_n), y_n - x_n + \beta_n(I-(I-\alpha_n\mu A_2)\widetilde{T}_n)(y_n)\rangle\big| \\
&= \beta_n\big|\langle (I-(I-\alpha_n\mu A_2)\widetilde{T}_n)(x_n) - (I-(I-\alpha_n\mu A_2)\widetilde{T}_n)(y_n), y_n - x_n \\
&\quad + \beta_n(I-(I-\alpha_n\mu A_2)\widetilde{T}_n)(y_n)\rangle\big| \\
&\leq \beta_n\|(I-(I-\alpha_n\mu A_2)\widetilde{T}_n)(x_n) - (I-(I-\alpha_n\mu A_2)\widetilde{T}_n)(y_n)\| \\
&\quad \|y_n - x_n + \beta_n(I-(I-\alpha_n\mu A_2)\widetilde{T}_n)(y_n)\| \\
&\leq \beta_n(L+1)\|x_n - y_n\|\|y_n - x_n + \beta_n(I-(I-\alpha_n\mu A_2)\widetilde{T}_n)(y_n)\| \\
&\leq \frac{\beta_n(L+1)}{2}\left(\|x_n - y_n\|^2 + \|y_n - x_n + \beta_n(I-(I-\alpha_n\mu A_2)\widetilde{T}_n)(y_n)\|^2\right).
\end{aligned} \tag{61}$$

Combining (59) and (61), we get

$$\begin{aligned}
&\|x_n - p - \beta_n(I-(I-\alpha_n\mu A_2)\widetilde{T}_n)(y_n)\|^2 \\
&\leq \|x_n - p\|^2 - \|x_n - y_n\|^2 - \|y_n - x_n + \beta_n(I-(I-\alpha_n\mu A_2)\widetilde{T}_n)(y_n)\|^2 \\
&\quad + \beta_n(L+1)(\|x_n - y_n\|^2 + \|y_n - x_n + \beta_n(I-(I-\alpha_n\mu A_2)\widetilde{T}_n)(y_n)\|^2) \\
&\quad - 2\beta_n\langle \alpha_n\mu A_2(\widetilde{T}_n(y_n)) + \lambda_n A_1(T_n(y_n)), y_n - p\rangle \\
&= \|x_n - p\|^2 + [\beta_n(L+1) - 1](\|x_n - y_n\|^2 \\
&\quad + \|y_n - x_n + \beta_n(I-(I-\alpha_n\mu A_2)\widetilde{T}_n)(y_n)\|^2) \\
&\quad - 2\beta_n\langle \alpha_n\mu A_2(\widetilde{T}_n(y_n)) + \lambda_n A_1(T_n(y_n)), y_n - p\rangle \\
&\leq \|x_n - p\|^2 - 2\beta_n\langle \alpha_n\mu A_2(\widetilde{T}_n(y_n)) + \lambda_n A_1(T_n(y_n)), y_n - p\rangle.
\end{aligned} \tag{62}$$

We observe that

$$\begin{aligned}
&\|x_n - p - \beta_n(I-(I-\alpha_n\mu A_2)\widetilde{T}_n)(y_n)\|^2 \\
&= \|x_n - p\|^2 - 2\beta_n\langle x_n - p, (I-(I-\alpha_n\mu A_2)\widetilde{T}_n)(y_n)\rangle \\
&\quad + \|\beta_n(I-(I-\alpha_n\mu A_2)\widetilde{T}_n)(y_n)\|^2.
\end{aligned} \tag{63}$$

Therefore, from (62) and (63), we have

$$\begin{aligned}
\|\beta_n(I-(I-\alpha_n\mu A_2)\widetilde{T}_n)(y_n)\|^2 \leq 2\beta_n[&\langle x_n - p, (I-(I-\alpha_n\mu A_2)\widetilde{T}_n)(y_n)\rangle \\
&- \langle \alpha_n\mu A_2(\widetilde{T}_n(y_n)) + \lambda_n A_1(T_n(y_n)), y_n - p\rangle],
\end{aligned}$$

which implies that

$$p \in C_n, \quad \text{that is,} \quad \Omega \subset C_n, \quad \text{for all } n \geq 1.$$

From $x_n = P_{C_n}(x_0)$, we have

$$\langle x_0 - x_n, x_n - y\rangle \geq 0, \quad \text{for all } y \in C_n.$$

Since $\Omega \subseteq C_n$, we have

$$\langle x_0 - x_n, x_n - u\rangle \geq 0, \quad \text{for all } u \in \Omega.$$

So, for all $u \in \Omega$ we have

$$\begin{aligned} 0 &\leq \langle x_0 - x_n, x_n - u\rangle \\ &= \langle x_0 - x_n, x_n - x_0 + x_0 - u\rangle \\ &= -\|x_0 - x_n\|^2 + \langle x_0 - x_n, x_0 - u\rangle \\ &\leq -\|x_0 - x_n\|^2 + \|x_0 - x_n\|\|x_0 - u\|, \end{aligned}$$

which implies that

$$\|x_0 - x_n\| \leq \|x_0 - u\|, \quad \text{for all } u \in \Omega. \tag{64}$$

Thus, $\{x_n\}$ is bounded and so are $\{y_n\}$ and $\{\widetilde{T}_n(y_n)\}$.

From $x_n = P_{C_n}(x_0)$ and $x_{n+1} = P_{C_{n+1}}(x_0) \in C_{n+1} \subset C_n$, we have

$$\langle x_0 - x_n, x_n - x_{n+1}\rangle \geq 0. \tag{65}$$

Hence,

$$\begin{aligned} 0 &\leq \langle x_0 - x_n, x_n - x_{n+1}\rangle \\ &= \langle x_0 - x_n, x_n - x_0 + x_0 - x_{n+1}\rangle \\ &= -\|x_0 - x_n\|^2 + \langle x_0 - x_n, x_0 - x_{n+1}\rangle \\ &\leq -\|x_0 - x_n\|^2 + \|x_0 - x_n\|\|x_0 - x_{n+1}\|, \end{aligned}$$

and therefore,

$$\|x_0 - x_n\| \leq \|x_0 - x_{n+1}\|.$$

This implies that the limit $\lim_{n\to\infty} \|x_n - x_0\|$ exists.

From Lemma 1 and (65), we obtain

$$\begin{aligned} \|x_{n+1} - x_n\|^2 &= \|(x_{n+1} - x_0) - (x_n - x_0)\|^2 \\ &= \|x_{n+1} - x_0\|^2 - \|x_n - x_0\|^2 - 2\langle x_{n+1} - x_n, x_n - x_0\rangle \\ &\leq \|x_{n+1} - x_0\|^2 - \|x_n - x_0\|^2 \to 0, \quad \text{as } n \to \infty. \end{aligned}$$

Thus,

$$\lim_{n\to\infty} \|x_{n+1} - x_n\| = 0.$$

It is easy to see that $\lim_{n\to\infty} \|x_n - x_{n+i}\| = 0$ for each $i = 1, 2, \ldots, N$. Since $x_{n+1} \in C_{n+1} \subseteq C_n$, $\|x_n - x_{n+1}\| \to 0$ and $\lambda_n \to 0$, we have

$$\begin{aligned}
&\|\beta_n(I-(I-\alpha_n\mu A_2)\widetilde{T}_n)(y_n)\|^2\\
&\le 2\beta_n[\langle x_n-x_{n+1},(I-(I-\alpha_n\mu A_2)\widetilde{T}_n)(y_n)\rangle\\
&\quad-\langle \alpha_n\mu A_2(\widetilde{T}_n(y_n))+\lambda_n A_1(T_n(y_n)), y_n-x_{n+1}\rangle]\\
&\le 2\beta_n[\|x_n-x_{n+1}\|\|(I-(I-\alpha_n\mu A_2)\widetilde{T}_n)(y_n)\|\\
&\quad+\|\alpha_n\mu A_2(\widetilde{T}_n(y_n))+\lambda_n A_1(T_n(y_n))\|\|y_n-x_{n+1}\|]\\
&\le 2\beta_n[\|x_n-x_{n+1}\|(\|y_n\|+\|\widetilde{T}_n(y_n)\|+\alpha_n\mu\|A_2(\widetilde{T}_n(y_n))\|)\\
&\quad+(\alpha_n\mu\|A_2(\widetilde{T}_n(y_n))\|+\lambda_n\|A_1(T_n(y_n))\|)\|y_n-x_{n+1}\|]\to 0.
\end{aligned}$$

Since $\beta_n \in [a,b]$ for some $a, b \in \left(0, \frac{1}{L+1}\right)$, we obtain

$$\left\|y_n-(I-\alpha_n\mu A_2)(\widetilde{T}_n(y_n))\right\|\to 0.$$

We also note that

$$\begin{aligned}
&\left\|T_n(y_n)-(I-\alpha_n\mu A_2)(\widetilde{T}_n(y_n))\right\|\\
&=\left\|T_n(y_n)-\widetilde{T}_n(y_n)+\alpha_n\mu A_2(\widetilde{T}_n(y_n))\right\|\\
&\le\|T_n(y_n)-(I-\lambda_n A_1)(T_n(y_n))\|+\alpha_n\mu\left\|A_2(\widetilde{T}_n(y_n))\right\|\\
&=\lambda_n\|A_1(T_n(y_n))\|+\alpha_n\mu\left\|A_2(\widetilde{T}_n(y_n))\right\|\to 0.
\end{aligned}$$

Therefore, we get

$$\begin{aligned}
\|y_n-T_n(y_n)\|\le&\left\|y_n-(I-\alpha_n\mu A_2)(\widetilde{T}_n(y_n))\right\|\\
&+\left\|T_n(y_n)-(I-\alpha_n\mu A_2)(\widetilde{T}_n(y_n))\right\|\to 0.
\end{aligned}$$

By Lemmas 4 and 5, we deduce that

$$\begin{aligned}
&\left\|x_n-(I-\alpha_n\mu A_2)(\widetilde{T}_n(x_n))\right\|\\
&\le\|x_n-y_n\|+\left\|y_n-(I-\alpha_n\mu A_2)(\widetilde{T}_n(y_n))\right\|\\
&\quad+\left\|(I-\alpha_n\mu A_2)(\widetilde{T}_n(y_n))-(I-\alpha_n\mu A_2)(\widetilde{T}_n(x_n))\right\|\\
&\le\|x_n-y_n\|+\left\|y_n-(I-\alpha_n\mu A_2)(\widetilde{T}_n(y_n))\right\|+(1-\alpha_n\tau)\left\|\widetilde{T}_n(y_n)-\widetilde{T}_n(x_n)\right\|\\
&\le\|x_n-y_n\|+\left\|y_n-(I-\alpha_n\mu A_2)(\widetilde{T}_n(y_n))\right\|\\
&\quad+\|(I-\lambda_n A_1)(T_n(y_n))-(I-\lambda_n A_1)(T_n(x_n))\|\\
&\le\|x_n-y_n\|+\left\|y_n-(I-\alpha_n\mu A_2)(\widetilde{T}_n(y_n))\right\|+\|T_n(y_n)-T_n(x_n)\|\\
&\le\|x_n-y_n\|+\left\|y_n-(I-\alpha_n\mu A_2)(\widetilde{T}_n(y_n))\right\|+L\|y_n-x_n\|\\
&=(L+1)\|x_n-y_n\|+\left\|y_n-(I-\alpha_n\mu A_2)(\widetilde{T}_n(y_n))\right\|\\
&=\beta_n(L+1)\left\|x_n-(I-\alpha_n\mu A_2)(\widetilde{T}_n(x_n))\right\|+\left\|y_n-(I-\alpha_n\mu A_2)(\widetilde{T}_n(y_n))\right\|,
\end{aligned}$$

that is,

$$\begin{aligned}&\|x_n - (I - \alpha_n\mu A_2)(\widetilde{T}_n(x_n))\| \\ &\qquad \le \frac{1}{1-\beta_n(L+1)} \|y_n - (I - \alpha_n\mu A_2)(\widetilde{T}_n(y_n))\| \to 0.\end{aligned}$$

Also,

$$\begin{aligned}&\|T_n(x_n) - (I - \alpha_n\mu A_2)(\widetilde{T}_n(x_n))\| \\ &= \|T_n(x_n) - \widetilde{T}_n(x_n) + \alpha_n\mu A_2(\widetilde{T}_n(x_n))\| \\ &\le \|T_n(x_n) - (I - \lambda_n A_1)(T_n(x_n))\| + \alpha_n\mu \|A_2(\widetilde{T}_n(x_n))\| \\ &= \lambda_n \|A_1(T_n(x_n))\| + \alpha_n\mu \|A_2(\widetilde{T}_n(x_n))\| \to 0.\end{aligned}$$

Consequently,

$$\|x_n - T_n(x_n)\| \le \|x_n - (I - \alpha_n\mu A_2)(\widetilde{T}_n(x_n))\| \tag{66}$$

$$\qquad + \|T_n(x_n) - (I - \alpha_n\mu A_2)(\widetilde{T}_n(x_n))\| \to 0, \tag{67}$$

and hence, for each $i = 1, 2, \ldots, N$,

$$\begin{aligned}\|x_n - T_{n+i}(x_n)\| &\le \|x_n - x_{n+i}\| + \|x_{n+i} - T_{n+i}(x_{n+i})\| + \|T_{n+i}(x_{n+i}) - T_{n+i}(x_n)\| \\ &\le (L+1)\|x_n - x_{n+i}\| + \|x_{n+i} - T_{n+i}(x_{n+i})\| \to 0 \quad \text{as } n \to \infty.\end{aligned}$$

So, we obtain $\lim_{n\to\infty} \|x_n - T_{n+i}(x_n)\| = 0$ for each $i = 1, 2, \ldots, N$. This implies that

$$\lim_{n\to\infty} \|x_n - T_l(x_n)\| = 0, \quad \text{for all } l = 1, 2, \ldots, N. \tag{68}$$

The relation (68) and Lemma 6 guarantee that every weak limit point of $\{x_n\}$ is a fixed point of T_l. Since l is an arbitrary element in the finite set $\{1, 2, \ldots, N\}$, every weak limit point of $\{x_n\}$ lies in Ω, that is, $\omega_w(x_n) \subset \Omega$. This fact, the inequality (64) and Lemma 3 ensure the strong convergence of $\{x_n\}$ to $P_\Omega x_0$. Since

$$\|x_n - y_n\| = \|\beta_n(I - (I - \alpha_n\mu A_2)\widetilde{T}_n)(x_n)\| \to 0,$$

it follows that $\{y_n\}$ converges strongly to $P_\Omega(x_0)$.

Finally, we prove that whenever T_i is nonexpansive for each $1 \le i \le N$, and A_1 is injective and $\|x_n - y_n\| + \alpha_n)/\lambda_n \to 0$ as $n \to \infty$, $P_\Omega(x_0)$ is the unique solution of Problem 7.

Indeed, put $\hat{x} := P_\Omega(x_0)$. By condition (C4), we can take an arbitrarily fixed $y \in \mathrm{VI}(\Omega, A_1)$ and put $M := \sup\{\|x_n - y\| + \|y_n - y\| : n \ge 1\} < \infty$. Then, from the condition (C3) and Lemmas 4 and 5, it follows that for all $n \ge 1$,

$$
\begin{aligned}
&\|(I-\alpha_n\mu A_2)(\widetilde{T}_n(x_n))-y\|^2\\
&=\|(I-\alpha_n\mu A_2)(\widetilde{T}_n(x_n))-(I-\alpha_n\mu A_2)(\widetilde{T}_n(y))+(I-\alpha_n\mu A_2)(\widetilde{T}_n(y))-y\|^2\\
&\le\|(I-\alpha_n\mu A_2)(\widetilde{T}_n(x_n))-(I-\alpha_n\mu A_2)(\widetilde{T}_n(y))\|^2\\
&\quad+2\langle(I-\alpha_n\mu A_2)(\widetilde{T}_n(x_n))-y,(I-\alpha_n\mu A_2)(\widetilde{T}_n(y))-y\rangle\\
&\le(1-\alpha_n\tau)^2\|\widetilde{T}_n(x_n)-\widetilde{T}_n(y)\|^2\\
&\quad+2\langle y-(I-\alpha_n\mu A_2)(\widetilde{T}_n(x_n)),\lambda_nA_1(y)+\alpha_n\mu A_2(\widetilde{T}_n(y))\rangle\\
&\le\|(I-\lambda_nA_1)(T_n(x_n))-(I-\lambda_nA_1)(T_n(y))\|^2\\
&\quad+2\langle y-(I-\alpha_n\mu A_2)(\widetilde{T}_n(x_n)),\lambda_nA_1(y)+\alpha_n\mu A_2(\widetilde{T}_n(y))\rangle\\
&\le\|T_n(x_n)-T_n(y)\|^2+2\langle y-(I-\alpha_n\mu A_2)(\widetilde{T}_n(x_n)),\lambda_nA_1(y)+\alpha_n\mu A_2(\widetilde{T}_n(y))\rangle\\
&\le\|x_n-y\|^2+2\langle y-(I-\alpha_n\mu A_2)(\widetilde{T}_n(x_n)),\lambda_nA_1(y)+\alpha_n\mu A_2(\widetilde{T}_n(y))\rangle\\
&=\|x_n-y\|^2+2\langle y-\widetilde{T}_n(x_n),\lambda_nA_1(y)+\alpha_n\mu A_2(\widetilde{T}_n(y))\rangle\\
&\quad+2\alpha_n\mu\langle A_2(\widetilde{T}_n(x_n)),\lambda_nA_1(y)+\alpha_n\mu A_2(\widetilde{T}_n(y))\rangle\\
&\le\|x_n-y\|^2+2\langle y-(I-\lambda_nA_1)(T_n(x_n)),\lambda_nA_1(y)+\alpha_n\mu A_2(\widetilde{T}_n(y))\rangle\\
&\quad+2\alpha_n\mu\|A_2(\widetilde{T}_n(x_n))\|\|\lambda_nA_1(y)+\alpha_n\mu A_2(\widetilde{T}_n(y))\|\\
&\le\|x_n-y\|^2+2(\langle y-T_n(x_n),\lambda_nA_1(y)+\alpha_n\mu A_2(\widetilde{T}_n(y))\rangle\\
&\quad+\lambda_n\langle A_1(T_n(x_n)),\lambda_nA_1(y)+\alpha_n\mu A_2(\widetilde{T}_n(y))\rangle)\\
&\quad+2\alpha_n\mu\|A_2(\widetilde{T}_n(x_n)\|\|\lambda_nA_1(y)+\alpha_n\mu A_2(\widetilde{T}_n(y))\|\\
&\le\|x_n-y\|^2+2(\langle y-T_n(x_n),\lambda_nA_1(y)+\alpha_n\mu A_2(\widetilde{T}_n(y))\rangle\\
&\quad+(\lambda_n\|A_1(T_n(x_n))\|+\alpha_n\mu\|A_2(\widetilde{T}_n(x_n))\|)\|\lambda_nA_1(y)+\alpha_n\mu A_2(\widetilde{T}_n(y))\|),
\end{aligned}
$$

and hence,

$$
\begin{aligned}
\|y_n-y\|^2&=\|(1-\beta_n)(x_n-y)+\beta_n[(I-\alpha_n\mu A_2)(\widetilde{T}_n(x_n))-y]\|^2\\
&\le(1-\beta_n)\|x_n-y\|^2+\beta_n\|(I-\alpha_n\mu A_2)(\widetilde{T}_n(x_n))-y\|^2\\
&\le(1-\beta_n)\|x_n-y\|^2+\beta_n[\|x_n-y\|^2\\
&\quad+2(\langle y-T_n(x_n),\lambda_nA_1(y)+\alpha_n\mu A_2(\widetilde{T}_n(y))\rangle\\
&\quad+(\lambda_n\|A_1(T_n(x_n))\|+\alpha_n\mu\|A_2(\widetilde{T}_n(x_n))\|)\|\lambda_nA_1(y)+\alpha_n\mu A_2(\widetilde{T}_n(y))\|)]\\
&=\|x_n-y\|^2+2\beta_n[\langle y-T_n(x_n),\lambda_nA_1(y)+\alpha_n\mu A_2(\widetilde{T}_n(y))\rangle\\
&\quad+(\lambda_n\|A_1(T_n(x_n))\|+\alpha_n\mu\|A_2(\widetilde{T}_n(x_n))\|)\|\lambda_nA_1(y)+\alpha_n\mu A_2(\widetilde{T}_n(y))\|].
\end{aligned}
$$

This implies that

$$
\begin{aligned}
0&\le\frac{1}{\lambda_n}\Big\{\|x_n-y\|^2-\|y_n-y\|^2+2\beta_n\Big[\Big\langle y-T_n(x_n),\lambda_nA_1(y)+\alpha_n\mu A_2(\widetilde{T}_n(y))\Big\rangle\\
&\quad+\Big(\lambda_n\|A_1(T_n(x_n))\|+\alpha_n\mu\left\|A_2(\widetilde{T}_n(x_n))\right\|\Big)\left\|\lambda_nA_1(y)+\alpha_n\mu A_2(\widetilde{T}_n(y))\right\|\Big]\Big\}\\
&=(\|x_n-y\|+\|y_n-y\|)\frac{\|x_n-y\|-\|y_n-y\|}{\lambda_n}\\
&\quad+2\beta_n\Big[\Big\langle y-T_n(x_n),A_1(y)+\mu\frac{\alpha_n}{\lambda_n}A_2(\widetilde{T}_n(y))\Big\rangle
\end{aligned}
$$

$$+ \left(\|A_1(T_n(x_n))\| + \mu\tfrac{\alpha_n}{\lambda_n}\left\|A_2(\widetilde{T}_n(x_n))\right\|\right)\left\|\lambda_n A_1(y) + \alpha_n\mu A_2(\widetilde{T}_n(y))\right\|\Big]$$

$$\le M\tfrac{\|x_n - y_n\|}{\lambda_n} + 2\beta_n\Big[\left\langle y - T_n(x_n), A_1(y) + \mu\tfrac{\alpha_n}{\lambda_n}A_2(\widetilde{T}_n(y))\right\rangle$$

$$+ \left(\|A_1(T_n(x_n))\| + \mu\tfrac{\alpha_n}{\lambda_n}\left\|A_2(\widetilde{T}_n(x_n))\right\|\right)\left\|\lambda_n A_1(y) + \alpha_n\mu A_2(\widetilde{T}_n(y))\right\|\Big],$$

that is,

$$\begin{aligned} 0 \le \tfrac{M}{2\beta_n}\cdot\tfrac{\|x_n - y_n\|}{\lambda_n} &+ \left\langle y - T_n(x_n), A_1(y) + \mu\tfrac{\alpha_n}{\lambda_n}A_2(\widetilde{T}_n(y))\right\rangle \\ &+ \left(\|A_1(T_n(x_n))\| + \mu\tfrac{\alpha_n}{\lambda_n}\left\|A_2(\widetilde{T}_n(x_n))\right\|\right)\left\|\lambda_n A_1(y) + \alpha_n\mu A_2(\widetilde{T}_n(y))\right\|. \end{aligned} \tag{69}$$

Since T_i is nonexpansive for each $1 \le i \le N$, it is known that $L = 1$ and $\{\beta_n\} \subset [a, b]$ for some $a, b \in \left(0, \frac{1}{2}\right)$. In terms of the conditions that $\alpha_n \to 0$, $\lambda_n \to 0$ and $(\|x_n - y_n\| + \alpha_n)/\lambda_n \to 0$, we deduce from (69) and $x_n \to \hat{x}$ $(=: P_\Omega(x_0))$ that

$$\langle y - \hat{x}, A_1(y)\rangle \ge 0, \quad \text{for all } y \in \Omega.$$

The condition (C1) ensures

$$\langle y - \hat{x}, A_1(\hat{x})\rangle \ge 0, \quad \text{for all } y \in \Omega,$$

that is, $\hat{x} \in \mathrm{VIP}(\Omega, A_1)$. Furthermore, from the conditions (C2) and (C4), we conclude that Problem 7 has a unique solution. Hence, $\mathrm{VIP}(\mathrm{VIP}(\Omega, A_1), A_2)$ is a singleton set. Thus, we may assume that $\mathrm{VIP}(\mathrm{VIP}(\Omega, A_1), A_2) = \{x^*\}$. This implies that $x^* \in \mathrm{VIP}(\Omega, A_1)$.

Now we show that $\hat{x} = x^*$. Indeed, since $\hat{x}, x^* \in \mathrm{VIP}(\Omega, A_1)$, we have

$$\langle A_1(\hat{x}), y - \hat{x}\rangle \ge 0, \quad \text{for all } y \in \Omega, \tag{70}$$

and

$$\langle A_1(x^*), y - x^*\rangle \ge 0, \quad \text{for all } y \in \Omega. \tag{71}$$

Setting $y = x^*$ in inequality (70) and $y = \hat{x}$ in inequality (71), and then adding the resultant inequalities, we get

$$\langle A_1(\hat{x}) - A_1(x^*), \hat{x} - x^*\rangle \le 0.$$

Since A_1 is α-inverse-strongly monotone, we have

$$\alpha\left\|A_1(\hat{x}) - A_1(x^*)\right\|^2 \le \left\langle A_1(\hat{x}) - A_1(x^*), \hat{x} - x^*\right\rangle \le 0.$$

Consequently, $A_1(\hat{x}) = A_1(x^*)$. Since A_1 is injective, we have $\hat{x} = x^*$. □

Remark 2 Algorithm 11 in [86] for a Lipschitz continuous pseudo-contraction is extended to develop our hybrid iterative algorithm for computing a common fixed point of N Lipschitz continuous pseudo-contractions, that is, Algorithm 11. Beyond question, our Theorem 14 is more general and more flexible than [86, Theorem 3.1] to a great extent. Meantime, the proof of Theorem 14 is very different from that of [86, Theorem 3.1] because our technique of argument depends on Lemma 5.

Acknowledgments In this research, the first author was support by a research project no. SR/S4/MS:719/11 of the Department of Science and Technology, Govt. of India. While, the second author was supported by the National Science Foundation of China (11071169), and Ph.D. Program Foundation of Ministry of Education of China (20123127110002).

References

1. Ansari, Q.H., Lalitha, C.S., Mehta, M.: Generalized Convexity, Nonsmooth Variational Inequalities, and Nonsmooth Optimization. CRC Press, Taylor & Francis Group, Boca Raton (2014)
2. Auslender, A., Teboulle, M.: Asymptotic Cones and Functions in Optimization and Variational Inequalities. Springer, New York (2003)
3. Baiocchi, C., Capelo, A.: Variational and Quasivariational Inequalities: Applications to Free Boundary Problems. Wiley, Chichester (1984)
4. Bauschke, H.: The approximation of fixed points of compositions of nonexpansive mappings in Hilbert spaces. J. Math. Anal. Appl. **202**, 150–159 (1996)
5. Browder, F.E.: Existence and approximation of solutions of nonlinear variational inequalities. Proc. Natl. Acad. Sci. U.S.A. **56**, 1080–1086 (1966)
6. Browder, F.E.: Convergence of approximates to fixed points of nonexpansive nonlinear mappings in Banach spaces. Arch. Ration. Mech. Anal. **24**, 82–90 (1967)
7. Browder, F.E., Petryshyn, W.V.: Construction of fixed points of nonlinear mappings in Hilbert spaces. J. Math. Anal. Appl. **20**, 197–228 (1967)
8. Cabot, A.: Proximal point algorithm controlled by a slowly vanishing term: applications to hierarchical minimization. SIAM J. Optim. **15**, 555–572 (2005)
9. Ceng, L.C., Ansari, Q.H., Wen, C.-F.: Hybrid steepest descent viscosity method for triple hierarchical variational inequalities. Abstr. Appl. Anal. **2012**, Article ID 907105 (2012)
10. Ceng, L.C., Ansari, Q.H., Wong, N.-C., Yao, Y.C.: Implicit iterative method for hierarchical variational inequalities. J. Appl. Math. **2012**, Article ID 472935 (2012)
11. Ceng, L.C., Ansari, Q.H., Yao, Y.C.: Iterative method for triple hierarchical variational inequalities in Hilbert spaces. J. Optim. Theory Appl. **151**, 482–512 (2011)
12. Ceng, L.C., Ansari, Q.H., Yao, J.C.: Relaxed hybrid steepest descent method with variable parameters for triple hierarchical variational inequality. Appl. Anal. **91**, 1793–1810 (2012)
13. Ceng, L.C., Wen, C.F.: Hybrid steepest-descent methods for triple hierarchical variational inequalities. Taiwan. J. Math. **17**, 1441–1472 (2013)
14. Ceng, L.C., Xu, H.K., Yao, J.C.: A hybrid steepest-descent method for variational inequalities in Hilbert spaces. Appl. Anal. **87**, 575–589 (2008)
15. Chang, S.S.: Viscosity approximation methods for a finite family of nonexpansive mappings in Banach spaces. J. Math. Anal. Appl. **323**, 1402–1416 (2006)
16. Chidume, C.E., Zegeye, H.: Approximate fixed point sequences and convergence theorems for Lipschitz pseudo-contractive maps. Proc. Am. Math. Soc. **132**, 831–840 (2003)
17. Cianciaruso, F., Colao, V., Muglia, L., Xu, H.K.: On an implicit hierarchical fixed point approach to variational inequalities. Bull. Austr. Math. Soc. **80**, 117–124 (2009)
18. Combettes, P.L.: A block-iterative surrogate constraint splitting method for quadratic signal recovery. IEEE Trans. Signal Process **51**, 1771–1782 (2003)

19. Combettes, P.L., Hirstoaga, S.A.: Approximating curves for nonexpansive and monotone operators. J. Convex Anal. **13**, 633–646 (2006)
20. Dafermos, S.: Traffic equilibrium and variational inequalities. Transp. Sci. **14**, 42–54 (1980)
21. Facchinei, F., Pang, J.-S.: Finite-Dimensional Variational Inequalities and Complementarity Problems, vol. I, II. Springer, New York (2003)
22. Fichera, G.: Problemi elettrostatici con vincoli unilaterali; il problema di Signorini con ambigue condizioni al contorno. Atti Acad. Naz. Lincei. Mem. Cl. Sci. Fis. Mat. Nat. Sez. I **7**, 91–140 (1964)
23. Geobel, K., Kirk, W.A.: Topics in Metric Fixed Point Theory. Cambridge Studies in Advanced Mathematics, vol. 28. Cambridge University Press, Cambridge (1990)
24. Glowinski, R., Lions, J.L., Trémolières, R.: Numerical Analysis of Variational Inequalities. North-Holland, Amsterdam (1981)
25. Goh, C.J., Yang, X.Q.: Duality in Optimization and Variational Inequalities. Taylor & Francis, London (2002)
26. Hartmann, P., Stampacchia, G.: On some nonlinear elliptic differential functional equations. Acta Math. **115**, 271–310 (1966)
27. Iiduka, H.: Strong convergence for an iterative method for the triple-hierarchical constrained optimization problem. Nonlinear Anal. **71**, e1292–e1297 (2009)
28. Iiduka, H.: A new iterative algorithm for the variational inequality problem over the fixed point set of a firmly nonexpansive mapping. Optimization **59**, 873–885 (2010)
29. Iiduka, H.: Iterative algorithm for solving triple-hierarchical constrained optimization problem. J. Optim. Theory Appl. **148**, 580–592 (2011)
30. Iiduka, H.: Fixed point optimization algorithmand its application to power control in CDMA data networks. Math. Prog. (2011). doi:10.1007/s10107-010-0427-x
31. Iiduka, H.: Iterative algorithm for triple-hierarchical constrained nonconvex optimization problem and its application to network bandwidth allocation. SIAM J. Optim. **22**, 862–878 (2012)
32. Iiduka, H., Takahashi, W., Toyoda, M.: Approximation of solutions of variational inequalities for monotone mappings. Panam. Math. J. **14**, 49–61 (2004)
33. Iiduka, H., Uchida, M.: Fixed point optimization algorithms for network bandwidth allocation problems with compoundable constraints. IEEE Commun. Lett. **15**, 596–598 (2011)
34. Iiduka, H., Yamada, I.: A use of conjugate gradient direction for the convex optimization problem over the fixed point set of a nonexpansive mapping. SIAM J. Optim. **19**, 1881–1893 (2009)
35. Isac, G.: Topological Methods in Complementarity Theory. Kluwer Academic Publishers, Dordrecht (2000)
36. Jitpeera, T., Kumam, P.: A new explicit triple hierarchical problem over the set of fixed points and generalized mixed equilibrium problems. J. Inequal. Appl. **2012**, Article ID 82 (2012)
37. Jitpeera, T., Kumam, P.: Algorithms for solving the variational inequality problem over the triple hierarchical problem. Abstr. Appl. Anal. **2013**, Article ID 827156 (2013)
38. Jung, J.S.: Iterative approaches to common fixed points of nonexpansive mappings in Banach spaces. J. Math. Anal. Appl. **302**, 509–520 (2005)
39. Kim, T.H., Xu, H.K.: Strong convergence of modified Mann iterations. Nonlinear Anal. **61**, 51–60 (2005)
40. Kinderlehrer, D., Stampacchia, G.: An Introduction to Variational Inequalities and their Applications. Academic Press, New York (1980)
41. Kong, Z.R., Ceng, L.C., Ansari, Q.H., Pang, C.T.: Multistep hybrid extragradient method for triple hierarchical variational inequalities. Abstr. Appl. Anal. **2013**, Article ID 718624 (2013)
42. Konnov, I.V.: Combined Relaxation Methods for Variational Inequalities. Springer, Berlin (2001)
43. Konnov, I.V.: Equilibrium Models and Variational Inequalities. Elsevier, Amsterdam (2007)
44. Lions, J.-L., Stampacchia, G.: Inéquations variationnelles non coercives. C.R. Math. Acad. Sci. Paris **261**, 25–27 (1965)
45. Lions, J.-L., Stampacchia, G.: Variational inequalities. Comm. Pure Appl. Math. **20**, 493–519 (1967)

46. Luo, Z.Q., Pang, J.S., Ralph, D.: Mathematical Programs with Equilibrium Constraints. Cambridge University Press, New York (1996)
47. Lu, X., Xu, H.-K., Yin, X.: Hybrid methods for a class of monotone variational inequalities. Nonlinear Anal. **71**, 1032–1041 (2009)
48. Mainge, P.E., Moudafi, A.: Strong convergence of an iterative method for hierarchical fixed-point problems. Pac. J. Optim. **3**, 529–538 (2007)
49. Mancio, O., Stampacchia, G.: Convex programming and variational inequalities. J. Optim. Theory Appl. **9**, 2–23 (1972)
50. Marino, G., Xu, H.K.: A general iterative methods for nonexpansive mappings in Hilbert spaces. J. Math. Anal. Appl. **318**, 43–52 (2006)
51. Marino, G., Xu, H.K.: Weak and strong convergence theorems for strict pseudo-contractions in Hilbert spaces. J. Math. Anal. Appl. **329**, 336–349 (2007)
52. Marino, G., Xu, H.K.: Explicit hierarchical fixed point approach to variational inequalities. J. Optim. Theory Appl. **149**, 61–78 (2011)
53. Matinez-Yanes, C., Xu, H.K.: Strong convergence of the CQ method for fixed point processes. Nonlinear Anal. **64**, 2400–2411 (2006)
54. Minty, G.J.: Monotone (nonlinear) operators in Hilbert space. Duke Math. J. **29**, 341–346 (1962)
55. Moudafi, A.: Viscosity approximation methods for the fixed point problems. J. Math. Anal. Appl. **241**, 46–55 (2000)
56. Moudafi, A.: Krasnoselski-Mann iteration for hierarchical fixed-point problems. Inverse Prob. **23**, 1635–1640 (2007)
57. Moudafi, A., Mainge, P.-E.: Towards viscosity approximations of hierarchical fixed-point problems. Fixed Point Theory Appl. **2006**, Article ID 95453 (2006)
58. Nagurney, A.: Network Economics: A Variational Inequality Approach. Academic Publishers, Dordrecht (1993)
59. Nakajo, K., Takahashi, W.: Strong convergence theorems for nonexpansive mappings and nonexpansive semigroups. J. Math. Anal. Appl. **279**, 372–379 (2003)
60. O'Hara, J.G., Pillay, P., Xu, H.K.: Iterative approaches to finding nearest common fixed points of nonexpansive mappings in Hilbert spaces. Nonlinear Anal. **54**, 1417–1426 (2003)
61. Osilike, M.O., Udomene, A.: Demiclosedness principle and convergence theorems for strictly pseudo-contractive mappings of Browder-Petryshyn type. J. Math. Anal. Appl. **256**, 431–445 (2001)
62. Outrata, J., Kočvara, M., Zowe, J.: Nonsmooth Approach to Optimization Problems with Equilibrium Constraints. Kluwer Academic Publishers, Dordrecht (1998)
63. Patriksson, M.: Nonlinear Programming and Variational Inequality Problems: A Unified Approach. Kluwer Academic Publishers, Dordrecht (1999)
64. Reich, S.: Weak convergence theorems for nonexpansive mappings in Banach spaces. J. Math. Anal. Appl. **67**, 274–276 (1979)
65. Sezan, M.I.: An overview of convex projections theory and its application to image recovery problems. Ultramicroscopy **40**, 55–67 (1992)
66. Shioji, N., Takahashi, W.: Strong convergence of approximated sequences for nonexpansive mappings in Banach spaces. Proc. Am. Math. Soc. **125**, 3641–3645 (1997)
67. Slavakis, K., Yamada, I.: Robust wideband beamforming by the hybrid steepest descent method. IEEE Trans. Signal Process **55**, 4511–4522 (2007)
68. Smith, M.J.: The existence, uniqueness and stability of traffic equilibria. Transp. Res. **13B**, 295–304 (1979)
69. Stampacchia, G.: Formes bilineaires coercivitives sur les ensembles convexes. C.R. Math. Acad. Sci. Paris **258**, 4413–4416 (1964)
70. Stampacchia, G.: Variational inequalities. In: Ghizzetti, A. (ed.) Theory and Applications of Monotone Operators. Proceedings of NATO Advanced Study Institute, Venice, Oderisi, Gubbio (1968)
71. Takahashi, W., Takeuchi, Y., Kubota, R.: Strong convergence theorems by hybrid methods for families of nonexpansive mappings in Hilbert spaces. J. Math. Anal. Appl. **341**, 276–286 (2008)

72. Udomene, A.: Path convergence, approximation of fixed points and variational solutions of Lipschitz pseudo-contractions in Banach spaces. Nonlinear Anal. **67**, 2403–2414 (2007)
73. Wairojjana, N., Jitpeera, T., Kumam, P.: The hybrid steepest descent method for solving variational inequality over triple hierarchical problems. J. Inequal. Appl. **2012**, Article ID 280 (2012)
74. Wu, D.P., Chang, S.S., Yuan, G.X.: Approximation of common fixed points for a family of finite nonexpansive mappings in Banach spaces. Nonlinear Anal. **63**, 987–999 (2005)
75. Xu, H.K.: Iterative algorithm for nonlinear operators. J. Lond. Math. Soc. **66**, 240–256 (2002)
76. Xu, H.-K.: Viscosity approximation methods for nonexpansive mappings. J. Math. Anal. Appl. **298**, 279–291 (2004)
77. Xu, H.K.: Strong convergence of an iterative method for nonexpansive mappings and accretive operators. J. Math. Anal. Appl. **314**, 631–643 (2006)
78. Xu, H.K.: Viscosity methods for hierarchical fixed point approach to variational inequalities. Taiwan. J. Math. **14**, 463–478 (2010)
79. Xu, H.K., Kim, T.H.: Convergence of hybrid steepest-descent methods for variational inequalities. J. Optim. Theory Appl. **119**, 185–201 (2003)
80. Yamada, I.: The hybrid steepest descent method for the variational inequality problems over the intersection of fixed point sets of nonexpansive mappings. In: Butnariu, D., Censor, Y., Reich, S. (eds.) Inherently Parallel Algorithms in Feasibility and Optimization and Their Applications, pp. 473–504. Elservier, New York (2001)
81. Yamada, I., Ogura, N.: Hybrid steepest descent method for the variational inequality problem over the fixed point set of certain quasi-nonexpansive mappings. Numer. Funct. Anal. Optim. **25**, 619–655 (2004)
82. Yao, Y., Chen, R., Xu, H.-K.: Schemes for finding minimum-norm solutions of variational inequalities. Nonlinear Anal. **72**, 3447–3456 (2010)
83. Yao, Y., Liou, Y.-C.: Weak and strong convergence of Krasnoselski–Mann iteration for hierarchical fixed point problems. Inverse Prob. **24**, Article ID 015015 (2008)
84. Yao, Y., Liou, Y.-C.: An implicit extragradient method for hierarchical variational inequalities. Fixed Point Theory Appl. **2011**, Article ID 697248 (2011)
85. Yao, Y., Liou, Y.C., Chen, R.: Strong convergence of an iterative algorithm for pseudo-contractive mapping in Banach spaces. Nonlinear Anal. **67**, 3311–3317 (2007)
86. Yao, Y., Liou, Y.C., Marino, G.: A hybrid algorithm for pseudo-contractive mappings. Nonlinear Anal. **71**, 4997–5002 (2009)
87. Yao, Y., Yao, J.C.: On modified iterative method for nonexpansive mappings and monotone mappings. Appl. Math. Comput. **186**, 1551–1558 (2007)
88. Zeng, L.C.: A note on approximating fixed points of nonexpansive mappings by the Ishikawa iteration process. J. Math. Anal. Appl. **226**, 245–250 (1998)
89. Zeng, L.C., Wong, M.M., Yao, J.C.: Strong convergence of relaxed hybrid steepestdescent methods for triple hierarchical constrained optimization. Fixed Point Theory Appl. **2012**, Article ID 29 (2012)
90. Zeng, L.C., Wong, N.C., Yao, J.C.: Strong convergence theorems for strictly pseudo-contractive mappings of Browder-Petryshyn type. Taiwan. J. Math. **10**, 837–849 (2006)
91. Zeng, L.C., Wong, N.C., Yao, J.C.: On the convergence analysis of modified hybrid steepest-descent methods with variable parameters for variational inequalities. J. Optim. Theory Appl. **132**, 51–69 (2007)
92. Zeng, L.C., Wen, C.-F., Yao, J.C.: An implicit hierarchical fixed-point approach to general variational inequalities in Hilbert spaces. Fixed Point Theory Appl. **2011**, Article ID 748918 (2007)
93. Zeng, L.C., Yao, J.C.: Implicit iteration scheme with perturbed mapping for common fixed points of a finite family of nonexpansive mappings. Nonlinear Anal. **64**, 2507–2515 (2006)
94. Zhou, H.Y.: Convergence theorems of fixed points for Lipschitz pseudo-contractions in Hilbert spaces. J. Math. Anal. Appl. **343**, 546–556 (2008)
95. Zhou, H.Y.: Convergence theorems of common fixed points for a finite family of Lipschitz pseudo-contractions in Banach spaces. Nonlinear Anal. **68**, 2977–2983 (2008)

Split Feasibility and Fixed Point Problems

Qamrul Hasan Ansari and Aisha Rehan

Abstract In this survey article, we present an introduction of split feasibility problems, multisets split feasibility problems and fixed point problems. The split feasibility problems and multisets split feasibility problems are described. Several solution methods, namely, CQ methods, relaxed CQ method, modified CQ method, modified relaxed CQ method, improved relaxed CQ method are presented for these two problems. Mann-type iterative methods are given for finding the common solution of a split feasibility problem and a fixed point problem. Some methods and results are illustrated by examples.

Keywords Split feasibility problems · Multisets split feasibility problems · Fixed point problems · Variational inequalities · Projection gradient method · Mann's iterative method · CQ methods · Relaxed CQ algorithm · Extragradient method · Relaxed extragradient method

1 Introduction

Let C and Q be nonempty closed convex sets in $\mathbb{R}^N$ and $\mathbb{R}^M$, respectively, and A be a given $M \times N$ real matrix. The *split feasibility problem* (in short, SFP) is to find x^* such that

$$x^* \in C \quad \text{and} \quad Ax^* \in Q. \tag{1}$$

Q. H. Ansari (✉) · A. Rehan
Department of Mathematics, Aligarh Muslim University, Aligarh 202002, India
e-mail: qhansari@gmail.com

A. Rehan
e-mail: aishashaheen370@gmail.com

Q. H. Ansari
Department of Mathematics and Statistics, King Fahd University of Petroleum and Minerals, Dhahran, Saudi Arabia

Q. H. Ansari (ed.), *Nonlinear Analysis*, Trends in Mathematics,
DOI: 10.1007/978-81-322-1883-8_9, © Springer India 2014

It was introduced by Censor and Elfving [14] for modeling inverse problems, which arise from phase retrievals and in medical image reconstruction [5]. Recently, it is found that SFP can also be used to model the intensity modulated radiation therapy [13, 15, 16, 20]. It has also several applications in various fields of science and technology.

If C and Q are nonempty closed convex subsets of real Hilbert spaces $\mathscr{H}_1$ and $\mathscr{H}_2$, respectively, and $A \in B(\mathscr{H}_1, \mathscr{H}_2)$, where $B(\mathscr{H}_1, \mathscr{H}_2)$ denotes the space of all bounded linear operators from $\mathscr{H}_1$ to $\mathscr{H}_2$, then the SFP is to find a point x^* such that

$$x^* \in C \quad \text{and} \quad Ax^* \in Q. \tag{2}$$

A special case of the SFP (2) is the following *convexly constrained linear inverse problem* (in short, CCLIP) [29] of finding x^* such that

$$x^* \in C \quad \text{and} \quad Ax^* = b. \tag{3}$$

It has extensively been investigated in the literature by using the projected Landweber iterative method [42]. However, SFP has received much less attention so far, due to the complexity resulted from the set Q.

The original algorithm introduced in [14] involves the computation of the inverse A^{-1} (assuming the existence of the inverse of A) and thus does not become popular. A more popular algorithm that solves SFP seems to be the CQ algorithm of Byrne [5, 6], which is found to be a gradient-projection method in convex minimization (it is also a special case of the proximal forward-backward splitting method [19, 21]).

Throughout the chapter, we denote by Γ the solution set of the SFP, that is,

$$\Gamma = \{x \in C : Ax \in Q\} = C \cap A^{-1}Q.$$

We also assume that the SFP is consistence, that is, the solution set Γ is nonempty, closed and convex.

For each $j = 1, 2, \ldots, J$, let K_j, be a nonempty closed convex subset of a M-dimensional Euclidean space $\mathbb{R}^M$ with $\cap_{j=1}^{J} K_j \neq \emptyset$. The *convex feasibility problem* (in short, CFP) is to find an element of $\cap_{j=1}^{J} K_j$. Solving the SFP is equivalent to find a member of the intersection of two sets Q and $A(C) = \{Ac : c \in C\}$ or of the intersection of two sets $A^{-1}(Q)$ and C, so the split feasibility problem can be viewed as a particular case of the CFP.

During the last decade, SFP has been extended and generalized in many directions. Several iterative methods have been proposed and analyzed; See, for example, references given in the bibliography.

1.1 Multiple-Sets Split Feasibility Problem

The *multiple-sets split feasibility problem* (in short, MSSFP) is to find a point closest to a family of closed convex sets in one space such that its image under a linear transformation will be closest to another family of closed convex sets in the image space. It can be a model for many inverse problems where constraints are imposed on the solutions in the domain of a linear operator as well as in the operator's range. It generalizes the *convex feasibility problems* and *split feasibility problems*. Formally, given nonempty closed convex sets $C_i \subseteq \mathbb{R}^N$, $i = 1, 2, \ldots, t$, and the nonempty closed convex sets $Q_j \subseteq \mathbb{R}^M$, $j = 1, 2, \ldots, r$, in the N and M dimensional Euclidean spaces, respectively, the *multiple-sets split feasibility problem* (in short, MSSFP) is to

$$\text{find } x^* \in C := \bigcap_{i=1}^{t} C_i \quad \text{such that} \quad Ax^* \in Q := \bigcap_{j=1}^{r} Q_j, \tag{4}$$

where A is given $M \times N$ real matrix. This can serve as a model for many inverse problems where constraints are imposed on the solutions in the domain of a linear operators as well as in the operator's range. The multiple-sets split feasibility problem extends the well-known convex feasibility problem, which is obtained from (4) when there are no matrix A and the set Q_j present at all.

The multiple split feasibility problems [15] arise in the field of intensity-modulated radiation therapy (in short, IMRT) when one attempts to describe physical dose constraints and equivalent uniform does (EUD) within a single model. The intensity-modulated radiation therapy is described in Sect. 1.1.1. For further details, see Censor et al. [13].

1.1.1 Intensity-Modulated Radiation Therapy

Intensity-modulated radiation therapy (in short, IMRT) [13] is an advanced mode of high-precision radiotherapy, that used computer-controlled linear accelerators to deliver precise radiation doses to specific areas within the tumor. IMRT allows for the radiation doses to confirm more precisely to the three-dimensional (3D) shape of the tumor by modulating-or controlling the intensity of the radiation beam in multiple small volumes. IMRT also allows higher radiation doses to be focused to regions within the tumor while minimizing the dose to surrounding normal critical structures. Treatment is carefully planned by using 3-D computed tomograpy (CT) or magnetic resonance (MRI) images of the patient in conjuction with computarized dose calculations to determine the dose intensity pattern that will best conform to the tumor shape. Typically, combinations of multiple intensity-modulated field coming from different beam directions produce a custom tailored radiation dose that maximizes tumor dose while also minimizing the dose to adjacent normal tissues. Because the ratio of normal tissue dose to tumor dose is reduced to a minimum with the IMRT approach higher and more effective radiation doses can safely delivered to tumor with fewer side effects compared with conventional radiotherapy techniques. IMRT also has the potential to reduce treatment toxicity, even when doses are not increased. Radiation therapy, including IMRT stops cancer cells from dividing and growing, thus slowing or stopping tumour growth. In many cases, radiation therapy is capable of killing all of the cancer cells, thus shrinking or eliminating tumors.

1.1.2 The Multiple-Sets Split Feasibility Problem in Intensity-Modulated Radiation Therapy

Let us first define the notations:

$\mathbb{R}^J$: The radiation intensity space, the J-dimensional Euclidean space.

$\mathbb{R}^I$: The dose space, the I-dimensional Euclidean space.

$x = (x_j)_{j=1}^J \in \mathbb{R}^J$: vector of beamlet intensity.

$h = (h_i)_{i=1}^I \in \mathbb{R}^I$: vector of doses absorbed in all voxels.

d_{ij} : doses absorbed in the voxel i due to radiation of unit intensity from the jth beamlet.

S_t : Set of all voxels indices in the structure t.

N_t : Number of voxel in the structure S_t.

We divide the entire volume of patient into I voxels, enumerated by $i = 1, 2, \ldots, I$. Assume that $T + Q$ anatomical structures have been outlined including planning target volumes (PTVs) and organ at risk (OAR). Let us count all PTVs and OARs sequentially by S_t, $t = 1, 2, \ldots, T, T + 1, \ldots, T + Q$, where the first T structure represents the planning target volume and the next Q structure represents the organ at risk.

Let us assume that the radiation is delivered independently from each of the J beamlet, which are arranged in certain geometry and indexed by $j = 1, 2, \ldots, J$. The intensities x_j of the beamlets are arranged in a J-dimensional vector $x = (x_j)_{j=1}^J \in \mathbb{R}^J$ in the J dimensional Euclidean space $\mathbb{R}^J$- the radiation intensity space.

The quantities $d_{ij} \geq 0$, which represent the dose absorbed in voxel i due to radiation of unit intensity from the jth beamlet are calculable by any forward program. Let h_i denote the total dose absorbed in the voxel i and let $h = (h_i)_{i=1}^I$ be the vector of doses absorbed in all voxels. We call the space $\mathbb{R}^I$-the dose space. we can calculate h_i as

$$h_i = \sum_{j=1}^{J} d_{ij} x_j. \tag{5}$$

The dose influence matrix $D = (d_{ij})$ is the $I \times J$ matrix whose elements are the $d_{ij}^{\prime}s$ mentioned above. Thus, (5) can be written as the vector equation

$$h = Dx. \tag{6}$$

The constraint are formulated in two different Euclidean vector space. The delivery constraints are formulated in the Euclidean vector space of radiation intensity vector (that is, vector whose component are radiation intensities). The equivalent uniform dose (in short, EUD) constraints are formulated in the Euclidean vector space of dose vectors (that is, vectors whose components are dose in each voxel).

Now, let us assume that M constraints in the dose space and N constraints in the intensity space. Let H_m be the set of dose vectors that fulfil the mth dose constraints and, let X_n be the set of beamlet intensity vectors that fulfil the nth intensity constraint. Each of the constraint sets H_m and X_n can be one of the specific H and X sets, respectively, described below.

In the dose space, a typical constraint is that given critical structure S_t, the dose should not exceed an upper bound u_t. The corresponding set $H_{\max,t}$ is

$$H_{\max,t} = \{h \in \mathbb{R}^I \mid h_i \leq u_t, \quad \text{for all } i \in S_t\}. \tag{7}$$

Similarly, in the target volumes (in short, TVs), the dose should not fall below a lower bound l_t. The set $H_{\min,t}$ of dose vectors that fulfil this constraint is

$$H_{\min,t} = \{h \in \mathbb{R}^I \mid l_t \leq h_i \quad \text{for all } i \in S_t\}. \tag{8}$$

To handle the equivalent uniform dose EUD constraint for each structure S_t, we define a real-valued function $E_t = \mathbb{R}^I \to \mathbb{R}$, called the EUD function, is defined by

$$E_t(h) = \left(\frac{1}{N_t} \sum_{i \in S_t} (h_i)^{\alpha_t}\right)^{1/\alpha_t}. \tag{9}$$

where N_t is the number of voxels in the structure S_t.

The parameter α_t is a tissue-specific number which is negative for target volumes TVs and positive for organ at risk OAR. For $\alpha_t = 1$,

$$E_t(h) = \frac{1}{N_t} \sum_{i \in S_t} (h_i), \tag{10}$$

that is, it is the mean dose of the organ for which it is calculated.

On the other hand, letting $\alpha_t \to \infty$ makes the equivalent uniform dose EUD function approach the maximal value, $\max\{h_i \mid i \in S_t\}$.

For each planning target volume PTVs structure $S_t, t = 1, 2, \ldots, T$, the parameter α_t is chosen negative and the equivalent uniform dose EUD constraint is described by the set

$$H_{\mathrm{EUD},t} = \{h \in \mathbb{R}^I \mid E^{\min} \leq E_t(h)\}, \tag{11}$$

where $E^{\min}$ is given, for each planning target volumes PTVs structure, by the treatment planner. For each organ at risk OAR, S_μ, $\mu = T+1, T+2, \ldots, T+Q$, the parameter is chosen $\alpha_t \geq 1$ and the equivalent uniform dose EUD constraint can be described by the set

$$H_{\mathrm{EUD},t} = \{h \in \mathbb{R}^I \mid E_t(h) \leq E^{\max}\}, \tag{12}$$

where $E^{\max}$ is given, for each organ at risk OAR, by the treatment planner. Due to the non-negativity of dose, $h \geq 0$ the equivalent uniform dose EUD function is convex for all $\alpha_t \geq 1$ and concave for all $\alpha_t < 1$. Therefore, the constraint sets $H_{\mathrm{EUD},t}$ are always convex sets in the dose vector space, since they are level sets of the convex functions $E_t(h)$ for organ at risk OAR (with $\alpha_t \geq 1$), or of the convex functions $-E_t(h)$ for the targets (with $\alpha_t < 0$).

In the radiation intensity space, the most prominent constraint is the non-negativity of the intensities, described by the set.

$$X_+ = \{x \in \mathbb{R}^J \mid \quad x_j \geq 0 \quad \forall j = 1, 2, \ldots, J\}. \tag{13}$$

Thus, our unified model for physical dose and equivalent uniform dose EUD constraints takes the form of multiple-sets split feasibility problem, where some constraints (the non-negativity of radiation intensities) are defined in the radiation intensity space $\mathbb{R}^J$ and other constraints (upper and lower bounds on dose and the equivalent uniform dose EUD constraints) are defined in the dose space $\mathbb{R}^I$, and the two spaces are related by a (known) linear transformation D (the dose matrix).

The unified problem can be formulated as follows:

$$\text{find } x^* \in X_+ \bigcap \left(\bigcap_{i=1}^{N} X_n \right) \text{ such that } h^* = Dx^* \text{ and } h^* \in \left(\bigcap_{m=1}^{M} H_m \right). \tag{14}$$

2 Preliminaries

This section provides the basic definitions and results, which will be used in the sequel.

Throughout the chapter, we adopt the following terminology and notations.

Let $\mathscr{H}$ be a real Hilbert space whose norm and inner product are denoted by $\|\cdot\|$ and $\langle .,. \rangle$, respectively. Let C be a nonempty subset of $\mathscr{H}$. The set of fixed points of a mapping $T : C \to C$ is denoted by $\mathrm{Fix}(T)$. Let $\{x_n\}$ be a sequence in $\mathscr{H}$ and $x \in \mathscr{H}$. We use $x_n \to x$ and $x_n \rightharpoonup x$ to denote the strong and weak convergence of the sequence $\{x_n\}$ to x, respectively. We also use $\omega_w(x_n)$ to denote the weak ω-limit sets of the sequence $\{x_n\}$, namely,

$$\omega_w(x_n) := \{x \in \mathscr{H} : x_{n_i} \rightharpoonup x \text{ for some subsequence } \{x_{n_i}\} \text{ of } \{x_n\}\}.$$

The following result provides the weak convergence of a bounded sequence.

Proposition 1 [65, Proposition 2.6] *Let C be a nonempty closed convex subset of a real Hilbert space $\mathscr{H}$ and $\{x_n\}$ be a bounded sequence such that the following conditions hold:*

(i) *Every weak limit point of $\{x_n\}$ lies in C;*
(ii) $\lim_{n \to \infty} \|x_n - x\|$ *exists for every $x \in C$.*

Then, the sequence $\{x_n\}$ converges weakly to a point in C.

Lemma 1 [59] *Let* $\{x_n\}$ *and* $\{y_n\}$ *be bounded sequences in a real Hilbert space* $\mathscr{H}$ *and* $\{\alpha_n\}$ *be a sequence in* $[0, 1]$ *with* $0 < \liminf_{n\to\infty} \alpha_n \leq \limsup_{n\to\infty} \alpha_n < 1$. *Suppose that* $x_{n+1} = (1-\alpha_n)y_n + \alpha_n x_n$ *for all* $n \geq 0$, *and* $\limsup_{n\to\infty}(\|y_{n+1}-y_n\| - \|x_{n+1}-x_n\|) \leq 0$. *Then,* $\lim_{n\to\infty} \|y_n - x_n\| = 0$.

Lemma 2 [33] *Let* $\mathscr{H}$ *be a real Hilbert space. Then, for all* $x, y \in \mathscr{H}$ *and* $\lambda \in [0, 1]$,

$$\|\lambda x + (1-\lambda)y\|^2 = \lambda\|x\|^2 + (1-\lambda)\|y\|^2 - \lambda(1-\lambda)\|x-y\|^2.$$

Definition 1 A mapping $T : \mathscr{H} \to \mathscr{H}$ is said to be

(a) *Lipschitz continuous* if there exists a constant $L > 0$ such that

$$\|Tx - Ty\| \leq L\|x - y\|, \quad \text{for all } x, y \in \mathscr{H}; \tag{15}$$

(b) *contraction* if there exists a constant $\alpha \in (0, 1)$ such that

$$\|Tx - Ty\| \leq \alpha\|x - y\|, \quad \text{for all } x, y \in \mathscr{H}; \tag{16}$$

If $\alpha = 1$, then T is said to be *nonexpansive*;

(c) *firmly nonexpansive* if $2T - I$ is nonexpansive, or equivalently,

$$\langle x - y, Tx - Ty\rangle \geq \|Tx - Ty\|^2, \quad \text{for all } x, y \in \mathscr{H}. \tag{17}$$

Alternatively, T is *firmly nonexpansive* if and only if T can be expressed as

$$T = \frac{1}{2}(I + S),$$

where $S : \mathscr{H} \to \mathscr{H}$ is nonexpansive;

(d) *averaged mapping* if it can be written as the average of the identity mapping I and a nonexpansive mapping, that is,

$$T = (1-\alpha)I + \alpha S, \tag{18}$$

where $\alpha \in (0, 1)$ and $S : \mathscr{H} \to \mathscr{H}$ is nonexpansive. More precisely, when Eq. (18) holds, we say that T is α-averaged.

The Cauchy-Schwartz inequality implies that every firmly nonexpansive mapping is nonexpansive but converse need not be true.

Proposition 2 *Let* $S, T, V : \mathscr{H} \to \mathscr{H}$ *be given mappings.*

(a) *If* $T = (1-\alpha)S + \alpha V$ *for some* $\alpha \in (0, 1)$, S *is averaged and* V *is nonexpansive, then* T *is averaged.*

(b) *T is firmly nonexpansive if and only if the complement $I - T$ is firmly nonexpansive.*
(c) *If $T = (1 - \alpha)S + \alpha V$ for some $\alpha \in (0, 1)$, S is firmly nonexpansive and V is nonexpansive, then T is averaged.*
(d) *The composite of finitely many averaged mappings is averaged. That is, if each of the mappings $\{T_i\}_{i=1}^N$ is averaged, then so is the composite $T_1 o \ldots o T_N$. In particular, if T_1 is α_1-averaged and T_2 is α_2-averaged, where $\alpha_1, \alpha_2 \in (0, 1)$, then the composite $T_1 \circ T_2$ is α-averaged, where $\alpha = \alpha_1 + \alpha_2 - \alpha_1\alpha_2$.*
(e) *If the mappings $\{T_i\}_{i=1}^N$ are averaged and have a common fixed point, then*

$$\bigcap_{i=1}^{N} Fix(T_i) = Fix(T_1 \circ \cdots \circ T_N).$$

The notion Fix(T) denotes the set of all fixed points of the mapping T, that is, $Fix(T) = \{x \in \mathscr{H} : Tx = x\}$.

Definition 2 Let T be a nonlinear operator whose domain $D(T) \subseteq \mathscr{H}$, and range is $R(T) \subseteq \mathscr{H}$. The operator T is said to be

(a) *monotone* if

$$\langle x - y, Tx - Ty \rangle \geq 0, \quad \text{for all } x, y \in D(T). \tag{19}$$

(b) *strongly monotone* (or *β-strongly monotone*) if there exists a constant $\beta > 0$ such that

$$\langle x - y, Tx - Ty \rangle \geq \beta \|x - y\|^2, \quad \text{for all } x, y \in D(T). \tag{20}$$

(c) *inverse strongly monotone* (or *ν-inverse strongly monotone*) (ν-ism) if there exists a constant $\nu > 0$ such that

$$\langle x - y, Tx - Ty \rangle \geq \nu \|Tx - Ty\|^2, \quad \text{for all } x, y \in D(T). \tag{21}$$

It can be easily seen that if T is nonexpansive, then $I - T$ is monotone.

It is well-known that if the function $f : \mathscr{H} \to \mathbb{R}$ is Lipschitz continuous, then its gradient ∇f is $\frac{1}{L}$-ism.

Lemma 3 *Let $f : \mathscr{H} \to \mathbb{R}$ be a Lipschitz continuous function with Lipschtiz constant $L > 0$. Then, the gradient operator $\nabla f : \mathscr{H} \to \mathscr{H}$ is $\frac{1}{L}$-ism, that is,*

$$\langle \nabla f(x) - \nabla f(y), x - y \rangle \geq \frac{1}{L} \|\nabla f(x) - \nabla f(y)\|^2, \quad \text{for all} x, y \in \mathscr{H}. \tag{22}$$

Proposition 3 *Let $T : \mathscr{H} \to \mathscr{H}$ be a mapping.*

(a) *T is nonexpansive if and only if the complement $I - T$ is $\frac{1}{2}$-ism.*
(b) *If T is ν-ism, then for $\gamma > 0$, γT is $\frac{\nu}{\gamma}$-ism.*

(c) *T is averaged if and only if the complement $I - T$ is ν-ism for some $\nu > \frac{1}{2}$. Indeed, for $\alpha \in (0, 1)$, T is α-averaged if and only if $I - T$ is $\frac{1}{2\alpha}$-ism.*

2.1 Metric Projection

Let C be a nonempty subset of a normed space X and $x \in X$. An element $y_0 \in C$ is said to be a *best approximation* of x if

$$\|x - y_0\| = d(x, C),$$

where $d(x, C) = \inf_{y \in C} \|x - y\|$. The number $d(x, C)$ is called *the distance from x to C*. The (possibly empty) set of all best approximations from x to C is denoted by

$$P_C(x) = \{y \in C : \|x - y\| = d(x, C)\}.$$

This defines a mapping P_C from X into 2^C and it is called the *metric projection* onto C. The metric projection mapping is also known as the *nearest point projection*, *proximity mapping* or *best approximation operator*.

Theorem 1 *Let C be a nonempty closed convex subset of a Hilbert space $\mathcal{H}$. Then, for each $x \in \mathcal{H}$, there exists a unique $y \in C$ such that*

$$\|x - y\| = \inf_{z \in C} \|x - z\|.$$

The above theorem says that $P_C(.)$ is a single-valued projection mapping from $\mathcal{H}$ onto C.

Some important properties of projections are gathered in the following proposition.

Proposition 4 *Let C be a nonempty closed convex subset of a real Hilbert space $\mathcal{H}$. Then,*

(a) *P_C is idempotent, that is, $P_C(P_C(x)) = P_C(x)$, for all $x \in \mathcal{H}$;*
(b) *P_C is firmly nonexpansive, that is, $\langle x - y, P_C(x) - P_C(y)\rangle \geq \|P_C(x) - P_C(y)\|^2$, for all $x, y \in \mathcal{H}$;*
(c) *P_C is nonexpansive, that is, $\|P_C(x) - P_C(y)\| \leq \|x - y\|$, for all $x, y \in \mathcal{H}$;*
(d) *P_C is monotone, that is, $\langle P_C(x) - P_C(y), x - y\rangle \geq 0$, for all $x, y \in \mathcal{H}$.*

2.2 Projection Gradient Method

Let C be a nonempty closed convex subset of a Hilbert space $\mathcal{H}$ and $f : C \to \mathbb{R}$ be a function. Consider the constrained minimization problem:

$$\min_{x \in C} f(x). \tag{23}$$

Assume that the minimization problem (23) is consistent.

If $f : C \to \mathbb{R}$ is Fréchet differentiable convex function, then it is well known (see, for example, [2, 39]) that the minimization problem (23) is equivalent to the following *variational inequality problem:*

$$\text{Find } x^* \in C \quad \text{such that} \quad \langle \nabla f(x^*), y - x^* \rangle \geq 0, \quad \text{for all } y \in C, \tag{24}$$

where $\nabla f : \mathscr{H} \to \mathscr{H}$ is the gradient of f. The following is the general form of the *variational inequality problem:*

$$\text{VIP}(F, C) \qquad \text{Find } x^* \in C \quad \text{such that} \quad \langle F(x^*), y - x^* \rangle \geq 0, \quad \text{for all } y \in C,$$

where $F : C \to \mathscr{H}$ be a nonlinear mapping. For further details and applications of variational inequalities, we refer to [2, 30, 39] and the references therein. The following result provides the equivalence between a variational inequality problem and a fixed point problem.

Proposition 5 *Let C be a nonempty closed convex subset of a real Hilbert space $\mathscr{H}$ and $F : C \to \mathscr{H}$ be an operator. Then, $x^* \in C$ is a solution of the VIP(F, C) if and only if for any $\gamma > 0$, x^* is a fixed point of the mapping $P_C(I - \gamma F) : C \to C$, that is,*

$$x^* = P_C(x^* - \gamma F(x^*)), \tag{25}$$

where $P_C(x^* - \gamma F(x^*))$ denotes the projection of $(x^* - \gamma F(x^*))$ onto C, and I is the identity mapping.

In view of the above proposition and discussion, we have the following proposition.

Proposition 6 *Let C be a nonempty closed convex subset of a real Hilbert space $\mathscr{H}$ and $F : C \to \mathscr{H}$ be a convex and Fréchet differential function. Then, the following statement are equivalent:*

(a) *$x^* \in C$ is a solution of the minimization problem* (23)*;*
(b) *$x^* \in C$ solves $VIP(F, C)$* (24)*;*
(c) *$x^* \in C$ is a solution of the fixed point* Eq. (25).

From the above equivalence, we have the following projection gradient method.

Theorem 2 (Projection Gradient Method) *Let C be a nonempty closed convex subset of a real Hilbert space $\mathscr{H}$ and $F : C \to \mathscr{H}$ be a Lipschitz continuous and strongly monotone mapping with constants $L > 0$ and $\beta > 0$, respectively. Let $\gamma > 0$ be a constant such that $\gamma < \frac{2\beta}{L^2}$. Then,*

(i) *$P_C(I - \gamma F) : C \to C$ is a contraction mapping and there exists a solution $x^* \in C$ of the VIP(F, C).*
(ii) *The sequence $\{x_n\}$ generated by the following iterative, process:*

$$x_{n+1} = P_C(I - \gamma F)x_n, \quad \text{for all } n \in \mathbb{N},$$

converges strongly to a solution x^* of the $VIP(F, C)$.

In view of Proposition 6 and Theorem 2, we have the following method for finding an approximate solution of a convex and differentiable minimization problem.

Theorem 3 *Let C be a nonempty closed convex subset of a real Hilbert space $\mathscr{H}$ and $f : C \to \mathbb{R}$ be a convex and Fréchet differentiable function such that the gradient ∇f is Lipschitz continuous and strongly monotone mapping with constants $L > 0$ and $\beta > 0$, respectively. Let $\{\gamma_n\}$ be a sequence of strictly positive real numbers such that*

$$0 < \liminf_{n\to\infty} \gamma_n \le \limsup_{n\to\infty} \gamma_n < \frac{2\beta}{L^2}. \tag{26}$$

Then, the sequence $\{x_n\}$ generated by the following projection gradient method

$$x_{n+1} = P_C(I - \gamma_n \nabla f)x_n, \quad \text{for all } n \in \mathbb{N}, \tag{27}$$

converges strongly to a unique solution of the minimization problem (23).

The sequence $\{x_n\}$ generated by the Eq. (27) converges weakly to the unique solution of the minimization problem (23) even when ∇f is not necessary strongly monotone.

We present an example to illustrate projection gradient method.

Example 1 Let $C = [0, 1]$ be a closed convex set in $\mathbb{R}$, $f(x) = x^2$ and $\gamma_n = 1/5$ for all n. Then, all the conditions of the Theorem 3 are satisfied and the sequence generated by the Eq. (27) converges to 0 with initial guess $x_1 = 0.01$. We have the following table of iterates:

From Table 1, it is clear that the solution $x = 0$ is obtained after 11th iteration. We performed the iterative scheme in Matlab R2010 (Fig. 1).

Table 1 Convergence of $\{x_n\}$ in Example 1

Number of iterations (n)	x_n	Number of iterations (n)	x_n	Number of iterations (n)	x_n
1	0.0100	6	0.0008	11	0.0001
2	0.0060	7	0.0005	12	0.0000
3	0.0036	8	0.0003	13	0.0000
4	0.0022	9	0.0002	14	0.0000
5	0.0013	10	0.0001	15	0.0000

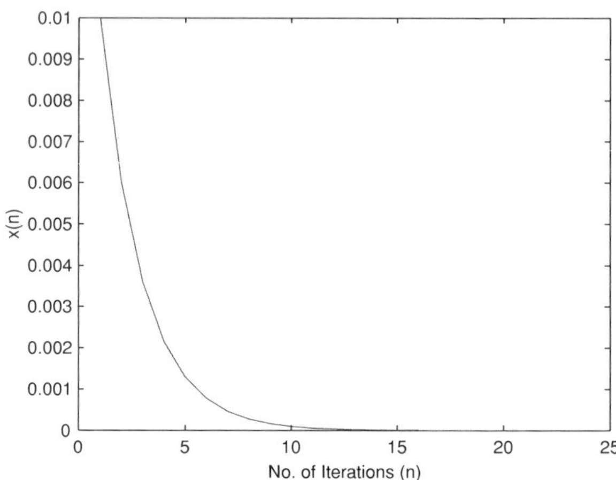

Fig. 1 Convergence of $\{x_n\}$ in Example 1

2.3 Mann's Iterative Method

Let C be a nonempty closed convex subset of a real Hilbert space $\mathscr{H}$ and $T : C \to C$ be a mapping. The well-known Mann's iterative algorithm is the following.

Algorithm 1 *(Mann's Iterative Algorithm)* For arbitrary $x_0 \in \mathscr{H}$, generate a sequence $\{x_n\}$ by the recursion:

$$x_{n+1} = (1 - \alpha_n)x_n + \alpha_n T x_n, \quad n \geq 0, \tag{28}$$

where $\{\alpha_n\}$ is (usually) assumed to be a sequence in $[0, 1]$.

Theorem 4 *Let C be a nonempty closed convex subset of a real Hilbert space $\mathscr{H}$ and $T : C \to C$ be a nonexpansive mapping with a fixed point. Assume that $\{\alpha_n\}$ is a sequence in* $[0, 1]$ *such that*

$$\sum_{n=1}^{\infty} \alpha_n(1 - \alpha_n) = \infty. \tag{29}$$

Table 2 Convergence of $\{x_n\}$ in Example 2

Iterations (n)	x_n	Iterations (n)	x_n	Iterations (n)	x_n
1	0.0100	9	0.1367	17	0.1714
2	0.4800	10	0.1919	18	0.1717
3	−0.1386	11	0.1604	19	0.1715
4	0.4784	12	0.1777	20	0.1716
5	−0.0346	13	0.1683	21	0.1716
6	0.3280	14	0.1733	22	0.1716
7	0.0770	15	0.1707	23	0.1716
8	0.2324	16	0.1720	24	0.1716

Then, the sequence $\{x_n\}$ generated by Mann's Algorithm 1 converges weakly to a fixed point of T.

Xu [65] studied the weak convergence of the sequence generated by the Mann's Algorithm 1 to a fixed point of an α-averaged mapping.

Theorem 5 [65, Theorem 3.5] *Let $\mathscr{H}$ be a real Hilbert space and $T : \mathscr{H} \to \mathscr{H}$ be an α-averaged mapping with a fixed point. Assume that $\{\alpha_n\}$ is a sequence in $[0,1/\alpha]$ such that*

$$\sum_{n=1}^{\infty} \alpha_n \left(\frac{1}{\alpha} - \alpha_n \right) = \infty. \tag{30}$$

Then, the sequence $\{x_n\}$ generated by Mann's Algorithm 1 converges weakly to a fixed point of T.

We illustrates Mann's Algorithm 1 with the help of the following examples:

Example 2 Let $T : [0, 1] \to [0, 1]$ be a mapping defined by

$$Tx = \frac{x^2}{4} - \frac{x}{2} + \frac{1}{4}, \quad \text{for all } x \in [0, 1].$$

Then, T is nonexpansive. Let $\{\alpha_n\} = \{\frac{1}{n+1}\}$. Then, all the conditions of Theorem 4 are satisfied and the sequence $\{x_n\}$ generated by Mann's Algorithm 1 converges to a fixed point of T, that is, to $x = 0.1716$. We take the initial guest $x_1 = .01$ and perform the Mann's Algorithm 1 by using Matlab R2010. We obtain the iterates in Table 2.

From Table 2, it is clear that the sequence generated by the Mann's Algorithm 1 converges to $x = 0.1716$ which is obtained after 19th iteration (Fig. 2).

Example 3 Let $T : [0, 1] \to [0, 1]$ be defined by

$$Tx = \frac{9}{10}x + \frac{1}{10}Sx, \quad \text{for all } x \in [0, 1].$$

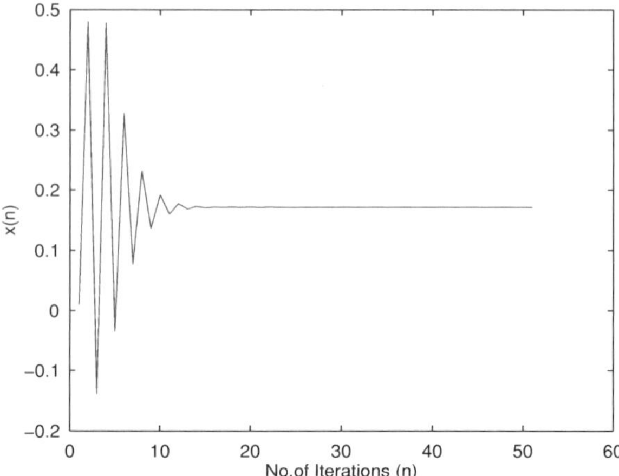

Fig. 2 Convergence of $\{x_n\}$ in Example 2

Table 3 Convergence of $\{x_n\}$ in Example 3

Number of iterations	x_n	Number of iterations	x_n	Number of iterations	$x(n)$
1	0.0100	5	0.1693	9	0.1715
2	0.2215	6	0.1725	10	0.1716
3	0.1550	7	0.1712	11	0.1716
4	0.17770	8	0.1717	12	0.1716

where $Sx = \frac{x^2}{4} - \frac{x}{2} + \frac{1}{4}$ is a nonexpansive map and $\{\alpha_n\} = 10 - \{\frac{1}{n}\}$. Then, T is a $\frac{1}{10}$-averaged mapping and all the conditions of Theorem 5 are satisfied. Hence, the sequence $\{x_n\}$ generated by Mann's Algorithm 1 converges to a fixed point of T, that is, to 0.1716 with initial guess $x_1 = 0.01$.

From Table 3, it is clear that the fixed point $x = 0.1716$ is obtained after 9th iteration. We performed the iterative scheme in Matlab R2010 (Fig. 3).

3 CQ-Methods for Split Feasibility Problems

In the pioneer paper [14], Censor and Elfving introduced the concept of a split feasibility problem (SFP) and used multidistance method to obtain the iterative algorithms for solving this problem. Their algorithms as well as others obtained later involves matrix inverses at each step. Byrne [5, 6] proposed a new iterative method called CQ-method that involves only the orthogonal projections onto C and Q and does not need to compute the matrix inverses, where C and Q are nonempty closed convex

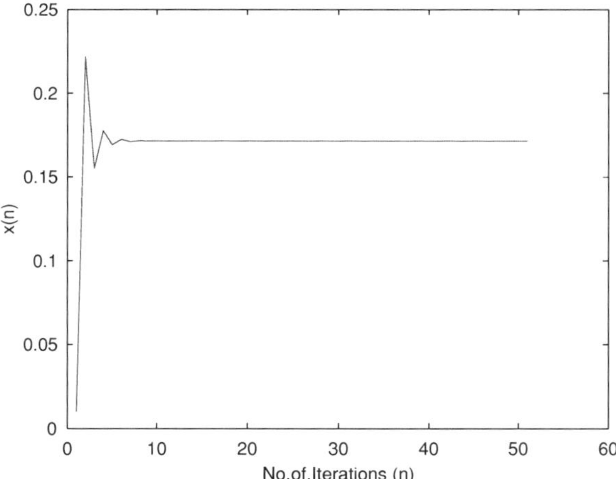

Fig. 3 Convergence of $\{x_n\}$ in Example 3

subsets of $\mathbb{R}^N$ and $\mathbb{R}^M$, respectively. It is one of the main advantages of this method compare to other methods. The CQ algorithm is as follows:

$$x_{n+1} = P_C\left(x_n - \gamma A^\top (I - P_Q)Ax_n\right), \quad n = 0, 1, \ldots,$$

where $\gamma \in (0, 2/L)$, A is a $M \times N$ matrix, $A^\top$ denotes the transpose of the matrix A, L is the largest eigenvalue of the matrix $A^\top A$, and P_C and P_Q denote the orthogonal projections onto C and Q, respectively. Byrne also studied the convergence of the CQ algorithm for arbitrary nonzero matrix A. Inspired by the work of Byrne [5, 6], Yang [68] proposed a modification of the CQ algorithm, called relaxed CQ algorithm in which he replaced P_C and P_Q by P_{C_n} and P_{Q_n}, respectively, where C_n and Q_n are half-spaces. One common advantage of the CQ algorithm and relaxed CQ algorithm is that the computation of the matrix inverses is not necessary. However, a fixed step-size related to the largest eigenvalue of the matrix $A^\top A$ is used. Computing the largest eigenvalue may be hard and conservative estimate of the step-size usually results in slow convergence. So, Qu and Xiu [53] modified the CQ algorithm and relaxed CQ algorithm by adopting Armijo-like searches. The modified algorithm need not compute the matrix inverses and the largest eigenvalue of the matrix $A^\top A$, and make a sufficient decrease of the objective function at each iteration. Zhao et al. [75] proposed a modified CQ algorithm by computing step-size adaptively and perform an additional projection step onto some simple closed convex set $X \subseteq \mathbb{R}^N$ in each iteration. Since all the algorithms have been introduced in finite-dimensional setting, Xu [65] proposed the relaxed CQ algorithm in infinite-dimensional setting, and also proved the weak convergence of the proposed algorithm. In 2011, Li [45] developed some improved relaxed CQ methods with the optimal step-length to solve the split feasibility problem based on the modified relaxed CQ algorithm [53].

In this section, we present different kinds of CQ algorithms, namely, CQ algorithm, relaxed CQ algorithm, modified CQ algorithm, modified relaxed CQ algorithm, modified projection type CQ algorithm, modified projection type relaxed CQ algorithm and improved relaxed CQ algorithm. We present the convergence results for these algorithms. We also present an example to illustrate CQ algorithm and its convergence result.

3.1 CQ Algorithm

Let C and Q be nonempty closed convex sets in $\mathbb{R}^N$ and $\mathbb{R}^M$, respectively, and A be an $M \times N$ real matrix with its transpose matrix $A^\top$. Let $\gamma > 0$ and assume that $x^* \in \Gamma$. Then, $Ax^* \in Q$ which implies the equation $(I - P_Q)Ax^* = 0$ which in turns implies the equation $\gamma A^\top(I - P_Q)Ax^* = 0$, hence the fixed point equation $(I - \gamma A^\top(I - P_Q)A)x^* = x^*$. Requiring that $x^* \in C$, Xu [65] considered the fixed point equation:

$$P_C(I - \gamma A^\top(I - P_Q)A)x^* = x^*. \tag{31}$$

and also observed that solutions of the fixed point Eq. (31) are exactly solutions of SFP.

Proposition 7 [65, Proposition 3.2] *Given $x^* \in \mathbb{R}^N$. Then, x^* solves the SFP if and only if x^* solves the fixed point* Eq. (31).

Byrne [5, 6] introduced the following CQ algorithm:

Algorithm 2 Let $x_0 \in \mathbb{R}^N$ be an initial guess. Generate a sequence $\{x_n\}$ by

$$x_{n+1} = P_C\left(x_n - \gamma A^\top(I - P_Q)Ax_n\right), \quad n = 0, 1, 2, \ldots, \tag{32}$$

where $\gamma \in (0, 2/L)$ and L is the largest eigenvalue of the matrix $A^\top A$.

It can be easily seen that the CQ algorithm does not require the computation of the inverse of any matrix. We need only to compute the projection onto the closed convex sets C and Q, respectively.

Byrne [5] studied the convergence of the above method and established the following convergence result.

Theorem 6 [5, Theorem 2.1] *Assume that the SFP is consistent. Then, the sequence $\{x_n\}$ generated by the CQ Algorithm 2 converges to a solution of the SFP.*

Remark 1 The particular cases of the CQ algorithm are the Landweber and projected Landweber methods [42]. These algorithms are discussed in detail in the book by

Table 4 Convergence of $\{x_n\}$ in Example 4

Number of iterations (n)	$x(n)$	Number of iterations (n)	$x(n)$
1	0.0100	10	0.4970
2	0.7940	11	0.5018
3	0.3236	12	0.4989
4	0.6058	13	0.5006
5	0.4365	14	0.4996
6	0.5381	15	0.5002
7	0.4771	16	0.4999
8	0.5137	17	0.5000
9	0.5049	18	0.5000

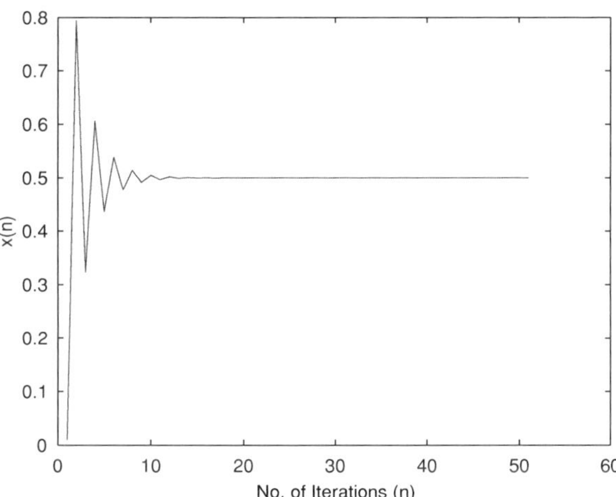

Fig. 4 Convergence of $\{x_n\}$ in Example 4

Bertero and Boccacci [8], primarily in the context of image restoration within infinite-dimensional spaces of functions (see also Landweber [41]). With $C = \mathbb{R}^N$ and $Q = \{b\}$, the CQ algorithm becomes the Landweber iterative method for solving the linear equations $Ax = b$.

The following example illustrates the CQ Algorithm 2 and its convergence result.

Example 4 Let $C = Q = [-1, 1]$ and $A(x) = 2x$. Then, A is a bounded linear operator with norm 2. Let $\gamma = 2/5$. Then, all the conditions of Theorem 6 are satisfied.

We perform the computation of the CQ Algorithm 2 by taking the initial guess $x_1 = 0.01$ and by using Matlab R2010. We obtain the iterates in Table 4.

From Table 4, it is clear that the sequence generated by the CQ Algorithm 2 converges to 0.5 after 16th iteration (Fig. 4).

Xu [65] extended Algorithm 2 in the setting of real Hilbert spaces to find the minimum-norm solution of the SFP with the help of regularization parameter. He considered $x_{\min}$ to be the minimum-norm solution of the SFP if $x_{\min} \in \Gamma$ has the property

$$\|x_{\min}\| = \min\{\|x^*\| : x^* \in \Gamma\}.$$

3.2 Relaxed CQ Algorithm

Let C and Q be nonempty closed convex sets in $\mathbb{R}^N$ and $\mathbb{R}^M$ respectively, and A be an $M \times N$ real matrix with its transpose matrix $A^\top$. Let P_{C_n} and P_{Q_n} denote the orthogonal projections onto the half-spaces C_n and Q_n, respectively. In some cases it is impossible or need too much time to compute the orthogonal projections [7, 32, 34]. Therefore, if this is the case, the efficiency of the projection type methods will be seriously affected, as would the CQ algorithm. Inexact technology plays an important role in designing efficient, easily implemented algorithms for the optimization problem, variational inequality problem and so on. The relaxed projection method may be viewed as one of the inexact methods. Fukushima [32] proposed a relaxed projection algorithm for solving variational inequalities and the theoretical analysis. The numerical experiment shows the efficiency of his method.

Inspired by the work of Fukushima [32], Yang [68] proposed the relaxed CQ algorithm. In order to describe relaxed CQ algorithm, he made some assumptions on C and Q, which are as follow:

- The solution set of the split feasibility problem is nonempty.

$$C = \{x \in \mathbb{R}^N : c(x) \leq 0\} \quad \text{and} \quad Q = \{y \in \mathbb{R}^M : q(y) \leq 0\}. \tag{33}$$

where c and q are the convex functionals on $\mathbb{R}^N$ and $\mathbb{R}^M$, respectively.
The subgradients $\partial c(x)$ and $\partial q(y)$ of c and q at x and y, respectively, are defined as follows:

$$\partial c(x) = \{z \in \mathbb{R}^N : c(u) \geq c(x) + \langle u - x, z\rangle,\ \forall u \in \mathbb{R}^N\} \neq \emptyset, \quad \text{for all } x \in C,$$

and

$$\partial q(y) = \{w \in \mathbb{R}^M : q(v) \geq q(y) + \langle v - y, w\rangle,\ \forall v \in \mathbb{R}^M\} \neq \emptyset, \quad \text{for all } y \in Q.$$

Note that the differentiability of $c(x)$ or $q(y)$ is not assumed. Therefore, both C and Q are general enough. For example, suppose any system of inequalities $c_i(x) \leq 0$, $i \in J$, where $c_i(x)$ are convex and J is an arbitrary index set, may be regarded as equivalent to the single inequality $c(x) \leq 0$ with $c(x) = \sup\{c_i(x) : i \in J\}$. One may easily get an element of $\partial c(x)$ by the expression of $\partial c(x)$ provided all $c_i(x)$ are differentiable.

With these assumptions, Yang [68] proposed the following relaxed CQ algorithm.

Algorithm 3 Let x_0 be arbitrary. For $n = 0, 1, 2, \ldots$, calculate

$$x_{n+1} = P_{C_n}\left(x_n - \gamma A^\top (I - P_{Q_n})Ax_n\right), \tag{34}$$

where $\{C_n\}$ and $\{Q_n\}$ are the sequences of closed convex sets defined as follows:

$$C_n = \{x \in \mathbb{R}^N : c(x_n) + \langle \xi_n, x - x_n \rangle \leq 0\}, \tag{35}$$

where $\xi_n \in \partial c(x_n)$, and

$$Q_n = \{y \in \mathbb{R}^M : q(A(x_n)) + \langle \eta_n, y - A(x_n) \rangle \leq 0\}, \tag{36}$$

where $\eta_n \in \partial q(A(x_n))$.

It can be easily seen that $C \subset C_n$ and $Q \subset Q_n$ for all n. Due to special form of C_n and Q_n, the orthogonal projections onto C_n and Q_n may be directly calculated [32]. Thus, the proposed algorithm can be easily implemented.

Yang [68] proved the following convergence result for Algorithm 3.

Theorem 7 [68, Theorem 1] *Let $\{x_n\}$ be the sequence generated by the Algorithm 3. Then, $\{x_n\}$ converges to a solution of the SFP.*

Xu [65] further studied the relaxed CQ algorithm in the setting of Hilbert spaces. He proposed the generalized form of the Algorithm 3 in the setting of Hilbert spaces and studied the weak convergence of the sequence generated by the proposed method.

3.3 Modified CQ Algorithm and Modified Relaxed CQ Algorithm

In CQ method and relaxed CQ method, we use a fixed step-size related to the largest eigenvalue of the matrix $A^\top A$, which sometimes affects convergence of the algorithms. Therefore, several modifications of these methods are proposed during the recent years. This section deals with such modified CQ method and relaxed CQ method.

By adopting Armijo-like searches, which are popular in iterative algorithms for solving nonlinear programming problems, variational inequality problems and so on [30, 67], Qu and Xiu [53] presented modification of CQ algorithm and relaxed CQ algorithm. In these modifications, it is not needed to compute the matrix inverses and the largest eigenvalue of the matrix $A^\top A$, and make a sufficient decrease of the objective functions at each iteration.

Let C, Q and A be the same as in Sect. 3.1. Qu and Xiu [53] proposed the following modified CQ algorithm:

Algorithm 4 Given constants $\beta > 0$, $\sigma \in (0, 1)$, $\gamma \in (0, 1)$. Let x_0 be arbitrary. For $n = 0, 1, 2, \ldots$, calculate

$$x_{n+1} = P_C\left(x_n - \alpha_n A^\top (I - P_Q) A x_n\right),$$

where $\alpha_n = \beta\gamma^{m_n}$ and m_n is the smallest nonnegative integer m such that

$$f(P_C(x_n - \beta\gamma^m A^\top (I - P_Q) A x_n)) \leq$$

$$f(x_n) - \sigma\left\langle A^\top (I - P_Q) A x_n, x_n - P_C\left(x_n - \beta\gamma^m A^\top (I - P_Q) A x_n\right)\right\rangle,$$

where $f(x) := \frac{1}{2}\|Ax - P_Q Ax\|^2$.

Algorithm 4 is in fact a special case of the standard gradient projection method with the Armijo-like search rule for solving convexly constrained optimization.

Qu and Xiu [53] established the following convergence of the modified CQ Algorithm 4.

Theorem 8 [53, Theorem 3.1] *Let $\{x_n\}$ be a sequence generated by the Algorithm 4. Then the following conclusions hold:*

(a) *$\{x_n\}$ is bounded if and only if the solution set S of minimization problem:*

$$\min_{x \in C} f(x) := \frac{1}{2}\|Ax - P_Q Ax\|^2,$$

is nonempty. In this case, $\{x_n\}$ must converge to an element of S.

(b) *$\{x_n\}$ is bounded and $\lim_{n\to\infty} f(x_n) = 0$ if and only if the SFP is solvable. In such a case, $\{x_n\}$ must converge to a solution of the SFP.*

Remark 2 Algorithm 4 is more applicable and it is easy to compute as compared to CQ Algorithm 2 proposed by Byrne [5], as it need not determine or estimate the largest eigenvalue of the matrix $A^\top A$. The step-size α_n is judiciously chosen so that the function value $f(x_{n+1})$ has a sufficient decrease. It can also be identified the existence of solution to the concerned problem by the iterative sequence.

Qu and Xiu [53] studied relaxed CQ algorithm proposed in Sect. 3.2 and proposed a modified relaxed CQ algorithm. Let C, Q, A, C_n and Q_n be the same as in Sect. 3.2.

For every n, let $F_n : \mathbb{R}^N \to \mathbb{R}^N$ be function defined as

$$F_n(x) = A^\top (I - P_{Q_n}) Ax, \quad \text{for all } x \in \mathbb{R}^N.$$

Modified relaxed CQ algorithm is the following:

Algorithm 5 Let x_0 be arbitrary and $\gamma > 0$, $l \in (0, 1)$, $\mu \in (0, 1)$ be given. For $n = 0, 1, 2, \ldots$, let

$$\overline{x}_n = P_{C_n}(x_n - \alpha_n F_n(x_n)),$$

where $\alpha_n = \gamma l^{m_n}$ and m_n is the smallest nonnegative integer m such that

$$\|F_n(\overline{x}_n) - F_n(x_n)\| \leq \mu \frac{\|\overline{x}_n - x_n\|}{\alpha_n}. \tag{37}$$

Set

$$x_{n+1} = P_{C_n}(x_n - \alpha_n F_n(\overline{x}_n)),$$

where C_n, Q_n are the sequences of closed convex sets defined as in (35) and (36).

Qu and Xiu [53] established the following convergence theorem.

Theorem 9 [53, Theorem 4.1] *Let $\{x_n\}$ be a sequence generated by Algorithm 5. If the solution set of SFP is nonempty, then $\{x_n\}$ converges to a solution of SFP.*

Inspired by Tseng's modified forward-backward splitting method for finding a zero of the sum of two maximal monotone mappings [60], Zhao et al. [75] proposed a modification of CQ algorithm, which computes the step-size adaptively, and performs an additional projection step onto some simple closed convex set $X \subseteq \mathbb{R}^N$ in each iteration. Let C, Q and A be the same as in Sect. 3.1.

Algorithm 6 [75] Let x_0 be arbitrary, $\sigma_0 > 0$, $\beta \in (0, 1)$, $\theta \in (0, 1)$, $\rho \in (0, 1)$. For $n = 0, 1, 2, \ldots$ compute

$$\bar{x}_n = P_C(x_n - \gamma_n F(x_n)), \tag{38}$$

where $F = A^\top(I - P_Q)A$, γ_n is chosen to be the largest $\gamma \in \{\sigma_n, \sigma_n\beta, \sigma_n\beta^2, \ldots\}$ satisfying

$$\gamma \|F(\bar{x}_n) - F(x_n)\| \leq \theta \|\bar{x}_n - x_n\|. \tag{39}$$

Let

$$x_{n+1} = P_X\left(\bar{x}_n - \gamma_n(F(\bar{x}_n) - F(x_n))\right). \tag{40}$$

If

$$\gamma_n \|F(x_{n+1}) - F(x_n)\| \leq \rho \|x_{n+1} - x_n\|, \tag{41}$$

then set $\sigma_n = \sigma_0$; otherwise, set $\sigma_n = \gamma_n$.

This algorithm involves projection onto a nonempty closed convex set X rather than onto the set C, which can be advantageous when X has a simpler structure than C. The set X can be chosen variously. It can be chosen to be a simple bounded subset of $\mathbb{R}^N$ that contains at least one solution of split feasibility problem, it can also be directly chosen as $X = \mathbb{R}^N$. In fact, it can be more generally chosen to be a dynamically changing set X_n, provided $\bigcap_{n=0}^{\infty} X_n$ contains a solution of the split feasibility problem. This does not affect the convergent result. The last step is used to reduce the inner iterations for searching the step-size γ_n.

For such algorithm, we usually take

$$\frac{1}{2}\|x_n - P_C(x_n)\|^2 + \frac{1}{2}\|Ax_n - P_Q(Ax_n)\|^2 < 0$$

or

$$\frac{1}{2}\|(I - P_Q)Ax_n\|^2 < \varepsilon$$

as the termination criterion, where $\varepsilon > 0$ is chosen to be sufficiently small.

Zhao et al. [75] established the following convergence result for the Algorithm 6.

Theorem 10 [75, Theorem 2.1] *Let $\{x_n\}$ be a sequence generated by Algorithm 6, X be a nonempty closed convex set in $\mathbb{R}^N$ with a simple structure. If $X \cap \Gamma$ is nonempty, then $\{x_n\}$ converges to a solution of SFP.*

Remark 3 This modified CQ algorithm differs from the extragradient-type method [38, 40, 53], whose second equation is

$$x_{n+1} = P_C(x_n - \gamma_n F(\overline{x}_n)).$$

It also differs from the modified projection-type method [54, 57], whose second equation is

$$x_{n+1} = x_n - \gamma_n(x_n - \overline{x}_n + \alpha_n(F(\overline{x}_n) - F(x_n)).$$

In Algorithm 6, the orthogonal projections P_C and P_Q had been calculated many times even in one iteration step, so they should be assumed to be easily calculated. However, sometimes it is difficult or even impossible to compute them. In order to overcome such situation turn to relaxed or inexact methods [31, 32, 34, 53, 68], which are more efficient and easily implemented. Zhao et al. [75] introduced relaxed modified CQ algorithm for split feasibility problem. Let C, Q, A, C_n and Q_n be the same as in the Sect. 3.2:

Algorithm 7 [75, Algorithm 3.1] Let x_0 be arbitrary, $\sigma_0 > 0, \beta \in (0, 1), \theta \in (0, 1)$, $\rho \in (0, 1)$ for $n = 0, 1, 2, \ldots$, compute

$$\bar{x}_n = P_{C_n}\left(x_n - \gamma_n F_n(x_n)\right), \tag{42}$$

where $F_n(x) = A^\top(I - P_{Q_n})Ax_n$ and γ_n is chosen to be the largest $\gamma \in \{\sigma_k, \sigma_k\beta, \sigma_k\beta^2 \ldots\}$ satisfying

$$\gamma\|F_n(\bar{x}_n) - F_n(x_n)\| \leq \theta\|\bar{x}_n - x_n\|. \tag{43}$$

Let

$$x_{n+1} = P_X\left(\bar{x}_n - \gamma_n(F_n(\bar{x}_n) - F_n(x_n))\right). \tag{44}$$

If

$$\gamma_n\|F_n(x_{n+1}) - F_n(x_n)\| \leq \rho\|x_{n+1} - x_n\|, \tag{45}$$

then set $\sigma_n = \sigma_0$; otherwise, set $\sigma_n = \gamma_n$, where $\{C_n\}$ and $\{Q_n\}$ are the sequences of closed convex sets defined as in (35) and (36), respectively.

Since projections onto half-spaces can be directly calculated, the relaxed algorithm is more practical and easily implemented than Algorithm 6 [31, 32, 34, 53, 68]. Here, we may take

$$\frac{1}{2}\|x_n - P_{C_n}(x_n)\|^2 + \frac{1}{2}\|Ax_n - P_{Q_n}(Ax_n)\|^2 < \epsilon,$$

or

$$\frac{1}{2}\|(I - P_{Q_n})Ax_n\|^2 < \epsilon$$

as the termination criterion.

We have the following convergence result for the Algorithm 7.

Theorem 11 [75, Theorem 3.1] *Let $\{x_n\}$ be a sequence generated by Algorithm 7, X be a nonempty closed convex set in $\mathbb{R}^N$ with a simple structure. if $X \cap \Gamma$ is nonempty, then $\{x_n\}$ converges to a solution of SFP.*

Remark 4 In Algorithm 7, the set X can be chosen to be any closed subset of $\mathbb{R}^N$ with a simple structure, provided it contains a solution of split feasibility problem. Dynamically changing it does not affect the convergence. For example, set $X_n = C_n$, then we get the following double-projection method:

$$\bar{x}_n = P_{C_n}\left(x_n - \gamma_n F_n(x_n)\right),$$

$$x_{n+1} = P_{C_n}\left(\bar{x}_n - \gamma_n(F_n(\bar{x}_n) - F_n(x_n))\right),$$

for $n = 0, 1, 2, \ldots$. This method differs from the modified relaxed CQ algorithm in [53]. Their method is in fact an extragradient method, with the second equation written as

$$x_{n+1} = P_{C_n}\left(x_n - \gamma_n F_n(\bar{x}_n)\right).$$

3.4 Improved Relaxed CQ Methods

Li [45] proposed the following two improved relaxed CQ methods and shown how to determine the optimal step length. The detailed procedure of the new methods is presented as follows:

Let C, Q, A, C_n and Q_n be the same as in the Sect. 3.2 and F_n be the same as defined in Sect. 3.3:

Algorithm 8 Initialization: choose $\mu \in (0, 1)$, $\epsilon > 0$, $x_0 \in \mathbb{R}^N$ and $n = 0$.

Step 1. Prediction: Choose an $\alpha_n > 0$ such that

$$\bar{x}_n = P_{C_n}(x_n - \alpha_n F_n(x_n)) \tag{46}$$

and

$$\alpha_n \|F_n(x_n) - F_n(\bar{x}_n)\| \leq \mu \|x_n - \bar{x}_n\| \tag{47}$$

Step 2. Stopping Criterion : compute

$$e_n(x_n, \alpha_n) = x_n - \bar{x}_n.$$

If $\|e_n(x_n, \alpha_n)\| \leq \epsilon$, terminate the iteration with the approximate solution x_n. Otherwise, go to step 3.
Step 3. Correction: The new iterate x_{n+1} is updated by

$$x_{n+1} = x_n^* = P_{C_n}(x_n - \beta_n \alpha_n F_n(\bar{x}_n)), \tag{48}$$

where

$$\beta_n = \delta_n \beta_n^*, \quad \beta_n^* = \frac{\langle x_n - \bar{x}_n, d_n(x_n, \bar{x}_n, \alpha_n)\rangle}{\|d_n(x_n, \bar{x}_n, \alpha_n)\|^2}, \quad \delta_n \in [\delta_L, \delta_V] \subseteq (0, 2), \tag{49}$$

and

$$d_n(x_n, \bar{x}_n, \alpha_n) = x_n - \bar{x}_n - \alpha_n(F_n(x_n) - F_n(\bar{x}_n)). \tag{50}$$

Set $n := n + 1$ and go to step 1.

Algorithm 9 : Initialization: Choose $\mu \in (0, 1)$, $\epsilon > 0$, $x_0 \in \mathbb{R}^{\mathbb{N}}$ and $n = 0$.
Step 1. Prediction: Choose an $\alpha_n > 0$ such that

$$\bar{x}_n = P_{C_n}(x_n - \alpha_n F_n(x_n)) \tag{51}$$

and

$$\alpha_n \|F_n(x_n) - F_n(\bar{x}_n)\| \leq \mu \|x_n - \bar{x}_n\|. \tag{52}$$

Step 2. Stopping Criteria : Compute

$$e_n(x_n, \alpha_n) = x_n - \overline{x}_n.$$

If $\|e_n(x_n, \alpha_n)\| \leq \epsilon$, terminate the iteration with the approximate solution x_n. Otherwise go to step 3.
Step 3. Correction: The corrector x_n^* is given by the following equation

$$x_n^* = P_{C_n}(x_n - \beta_n \alpha_n F_n(\bar{x}_n)), \tag{53}$$

where

$$\beta_n = \gamma_n \beta_n^*, \quad \beta_n^* = \frac{\langle x_n - \bar{x}_n, d_n(x_n, \bar{x}_n, \alpha_n)\rangle}{\|d_n(x_n, \bar{x}_n, \alpha_n)\|^2}, \quad \delta_n \in [\delta_L, \delta_U] \subseteq (0, 2), \tag{54}$$

and

$$d_n(x_n, \bar{x}_n, \alpha_n) = x_n - \bar{x}_n - \alpha_n(F_n(x_n) - F_n(\bar{x}_n)). \tag{55}$$

Step 4. Extension: The new iterate x_{n+1} is updated by

$$x_{n+1} = P_{C_n}(x_n - \rho_n(x_n - x_n^*)), \tag{56}$$

where

$$\rho_n = \gamma_n \rho_n^*, \quad \rho_n^* = \frac{\|x_n - x_n^*\|^2 + \beta_n \alpha_n \langle x_n^* - \bar{x}_n, F_n(\bar{x}_n)\rangle}{\|x_n - x_n^*\|^2}, \quad \gamma_n \in [\gamma_L, \gamma_U] \subseteq (0, 2). \tag{57}$$

Set $n := n + 1$ and go to step 1.

Remark 5 In the prediction step, if the selected α_n satisfies $0 < \alpha_n \leq \mu/L$ (L is the largest eigenvalue of the matrix $A^\top A$), from [45, Lemma 2.3], we have

$$\alpha_n \|F_n(x_n) - F_n(\bar{x}_n)\| \leq \alpha_n L \|x_n - \bar{x}_n\| \leq \mu \|x_n - \bar{x}_n\|, \tag{58}$$

and thus condition (47) or (52) is satisfied. Without loss of generality, we can assume that $\inf\{\alpha_n\} = \alpha_{\min} > 0$. Since we do not know the value of $L > 0$ but it exist, in practice, a self-adaptive scheme is adopted to find such a suitable $\alpha_n > 0$. For given x_n and a trial $\alpha_n > 0$, along with the value of $F_n(x_n)$, we set the trial $\overline{x}_n$ as follows:

$$\bar{x}_n = P_{C_n}(x_n - \alpha_n F_n(x_n)).$$

Then calculate

$$r_n := \frac{\alpha_n \|F_n(x_n) - F_n(\bar{x}_n)\|}{\|x_n - \bar{x}_n\|},$$

if $r_n \leq \mu$, the trial $\bar{x}_n$ is accepted as predictor; otherwise, reduce α_n by $\alpha_n := .9\mu\alpha_n^* \min(1, 1/r_n)$ to get a new smaller trial α_n and repeat this procedure. In the case that the predictor has been accepted, a good initial trial α_{n+1} for the next iteration is prepared by the following strategy:

$$\alpha_{n+1} = \begin{cases} \frac{0.9}{r_n}\alpha_n & \text{if } r_n \leq \nu, \\ \alpha_n & \text{otherwise}, \end{cases} \tag{59}$$

(usually $\nu \in [0.4, 0.5]$).

Condition (47) or (52) ensure that $\alpha_n \| F_n(x_n) - F_n(\bar{x}_n)\|$ is smaller than $\|x_n - \bar{x}_n\|$, however, too small $\alpha_n \| F_n(x_n) - F_n(\bar{x}_n)\|$ leads to slow convergence. The proposed adjusting strategy (59) is intended to avoid such a case as indicated in [35, 36]. Actually it is very important to balance the quantity of $\alpha_n \| F_n(x_n) - F_n(\bar{x}_n)\|$ and $\|x_n - \bar{x}_n\|$ in practical computation. Note that there are at least two times to utilize the value of function in the prediction step: one is $F_n(x_n)$, and the other is $F_n(\bar{x}_n)$ for testing whether the condition (47) or (52) holds. When α_n is selected well enough, $\bar{x}_n$ will be accepted after only one trial and in this case, the prediction step exactly utilizing the value of concerned function twice in one iteration.

It follow from [45, Relation (3.16)] and [45, Relation (3.27)] that for Algorithm 8, there exists a constant $\tau_1 > 0$ such that

$$\|x_{n+1} - x^*\|^2 \leq \|x_n - x^*\|^2 - \tau_1 \cdot \|x_n - \bar{x}_n\|^2. \tag{60}$$

From [45, Relation (3.38)], for Algorithm 9, there exist a constant $\tau_2 > 0$ such that

$$\|x_{n+1} - x^*\|^2 \leq \|x_n - x^*\|^2 - \tau_2 \cdot \{\|x_n - \bar{x}_n\|^2 + \|x_n - x_n^*\|^2\}. \tag{61}$$

Finally, we have the following convergence result of the proposed methods.

Theorem 12 [45] *Let $\{x_n\}$ be a sequence generated by the proposed methods (Algorithms 8 and 9), $\{\alpha_n\}$ be a positive sequence and* $\inf \{\alpha_n\} = \alpha_{\min} > 0$. *If the solution set of the SFP is nonempty, then $\{x_n\}$ converges to a solution of the SFP.*

4 Extragradient Methods for Common Solution of Split Feasibility and Fixed Point Problems

Korplevich [40] introduced the so-called *extragradient method* for finding a solution of a saddle point problem. She/He proved that the sequences generated by this algorithm converge to a saddle point. Motivated by the idea of an extragradient method,

Ceng et al. [10] introduced and analyzed an extragradient method with regularization for finding a common element of the solution set Γ of the split feasibility problem (SFP) and the set Fix(S) of the fixed points of a nonexpansive mapping S in the setting of Hilbert spaces. Combining the regularization method and extragradient method due to Nadezhkina and Takahashi [50], they proposed an iterative algorithm for finding an element of Fix$(S) \cap \Gamma$. They proved that the sequences generated by the proposed method converges weakly to an element $z \in$ Fix$(S) \cap \Gamma$.

On the other hand, Ceng et al. [11] introduced relaxed extragradient method for finding a common element of the solution set Γ of the SFP and the set Fix(S) of fixed points of a nonexpansive mapping S in the setting of Hilbert spaces. They combined Mann's iterative method and extragradient method to propose relaxed extragradient method. The weak convergence of the sequences generated by the proposed method is also studied. The relaxed extragradient method with regularization is studied by Deepho and Kumam [26]. They considered the set S of fixed points of a asymptotically quasi-nonexpansive and Lipschtiz continuous mapping in the setting of Hilbert spaces. They obtained the weak convergence result for their method.

Recently, Ceng et al. [12] proposed three different kinds of iterative methods for computing the common element of the solution set Γ of the split feasibility problem (SFP) and the set $\text{Fix}(S)$ of the fixed points of a nonexpansive mapping in the setting of Hilbert spaces. By combining Mann's iterative method and the extragradient method, they first proposed Mann-type extragradient-like algorithm for finding an element of the set $\text{Fix}(S) \cap \Gamma$. Moreover, they derived the weak convergence of the proposed algorithm under appropriate conditions. Second, they combined Mann's iterative method and the viscosity approximation method to introduce Mann-type viscosity algorithm for finding an element of the $\text{Fix}(S) \cap \Gamma$. The strong convergence of the sequences generated by the proposed algorithm to an element of the set $\text{Fix}(S) \cap \Gamma$ under mild conditions is also proved. Finally, by combining Mann's iterative method and the relaxed CQ methods, they introduced Mann-type relaxed CQ algorithm for finding an element of the set $\text{Fix}(S) \cap \Gamma$. They also established a weak convergence result for the sequences generated by the proposed Mann type relaxed CQ algorithm under appropriate assumptions.

Very recently, Li et al. [44] and Zhu et al. [76] developed iterative methods for finding the common solutions of a SFP and a fixed point problem.

In this section, we discuss extragradient method with regularization, relaxed extragradient method and relaxed extragradient method with regularization. We also mention the convergence results for these methods. Two examples are presented to illustrate these methods. We present Mann-type extragradient-like algorithm, Mann-type viscosity algorithm, and Mann-type relaxed CQ algorithm for computing an element of the set $\text{Fix}(S) \cap \Gamma$. The weak convergence results for these methods are presented. Some methods are illustrated by some examples.

4.1 An Extragradient Method

Throughout this section, we assume that $\Gamma \cap \text{Fix}(S) \neq \emptyset$.

We present the following extragradient method with regularization for finding a common element of the solution set Γ of the split feasibility problem and the set $\text{Fix}(S)$ of the fixed points of a nonexpansive mapping S. We also mention the weak convergence of this method.

Theorem 13 [10, Theorem 3.1] *Let C be a nonempty closed convex subset of a real Hilbert space $\mathscr{H}_1$ and $S : C \to C$ be a nonexpansive mapping such that Fix(S) $\cap \Gamma \neq \emptyset$. Let $\{x_n\}$ and $\{y_n\}$ be the sequences in C generated by the following extragradient algorithm:*

$$\begin{cases} x_0 = x \in C \quad \text{chosen arbitrarily}, \\ y_n = P_C(I - \lambda_n \nabla f_{\alpha_n})x_n, \\ x_{n+1} = \beta_n x_n + (1-\beta_n) S P_C(x_n - \lambda_n \nabla f_{\alpha_n}(y_n)), \quad \text{for all } n \geq 0, \end{cases} \tag{62}$$

where $\nabla f_{\alpha_n} = \alpha_n I + A^*(I - P_Q)A$, $\sum_{n=0}^{\infty} a_n < \infty$, $\{\lambda_n\} \subset [a, b]$ *for some* $a, b \in \left(0, \frac{1}{\|A\|^2}\right)$ *and* $\{\beta_n\} \subset [c, d]$ *for some* $c, d \in (0, 1)$. *Then, both the sequences* $\{x_n\}$ *and* $\{y_n\}$ *converge weakly to an element* $\bar{x} \in \text{Fix}(S) \cap \Gamma$.

Furthermore, by utilizing [50, Theorem 3.1], we can immediately obtain the following weak convergence result.

Theorem 14 [10, Theorem 3.2] *Let* $\mathcal{H}_1$, C *and* S *be the same as in Theorem 13. Let* $\{x_n\}$ *and* $\{y_n\}$ *be the sequences in* C *generated by the following Nadezhkina and Takahashi extragradient algorithm:*

$$\begin{cases} x_0 = x \in C \quad \text{chosen arbitrarily}, \\ y_n = P_C(I - \lambda_n \nabla f)x_n, \\ x_{n+1} = \beta_n x_n + (1-\beta_n) S P_C(x_n - \lambda_n \nabla f(y_n)), \quad \text{for all } n \geq 0, \end{cases} \tag{63}$$

where $\nabla f = A^*(I - P_Q)A$, $\{\lambda_n\} \subset [a, b]$ *for some* $a, b \in \left(0, \frac{1}{\|A\|^2}\right)$ *and* $\{\beta_n\} \subset [c, d]$ *for some* $c, d \in (0, 1)$. *Then, both the sequences* $\{x_n\}$ *and* $\{y_n\}$ *converge weakly to an element* $\bar{x} \in Fix(S) \cap \Gamma$.

By utilizing Theorem 13, we obtain the following results.

Corollary 1 [10, Corollary 3.2] *Let* $C = \mathcal{H}_1$ *be a Hilbert space and* $S : \mathcal{H}_1 \to \mathcal{H}_1$ *be a nonexpansive mapping such that* $Fix(S) \cap (\nabla f)^{-1}0 \neq \emptyset$. *Let* $\{x_n\}$ *be a sequence generated by*

$$\begin{cases} x_0 = x \in C \quad \text{chosen arbitrarily}, \\ x_{n+1} = \beta_n x_n + (1-\beta_n) S(x_n - \lambda_n \nabla_n f_{\alpha_n}(I - \lambda_n \nabla f_{\alpha_n})x_n), \quad \text{for all } n \geq 0, \end{cases} \tag{64}$$

where $\Sigma_{n=0}^{\infty} a_n < \infty$, $\{\lambda_n\} \subset [a, b]$ *for some* $a, b \in \left(0, \frac{1}{\|A\|^2}\right)$ *and* $\{\beta_n\} \subset [c, d]$ *for some* $c, d \in (0, 1)$. *Then, the sequence* $\{x_n\}$ *converges weakly to* $\bar{x} \in Fix(S) \cap (\nabla f)^{-1}$.

For the definition of maximal monotone operator and resolvent operator, see Chap. 6.

Corollary 2 [10, Corollary 3.3] *Let* $C = \mathcal{H}_1$ *be a Hilbert space and* $B : \mathcal{H}_1 \to 2^{\mathcal{H}_1}$ *be a maximal monotone mapping such that* $B^{-1}0 \cap (\nabla f)^{-1}0 \neq \emptyset$. *Let* j_r^B *be the resolvent of* B *for each* $r > 0$. *Let* $\{x_n\}$ *be a sequence generated by*

$$\begin{cases} x_0 = x \in C \quad \text{chosen arbitrarily}, \\ x_{n+1} = \beta_n x_n + (1-\beta_n) j_r^B(x_n - \lambda_n \nabla f_{\alpha_n}(I - \lambda_n \nabla f_{\alpha_n})x_n), \quad \forall n \geq 0, \end{cases} \tag{65}$$

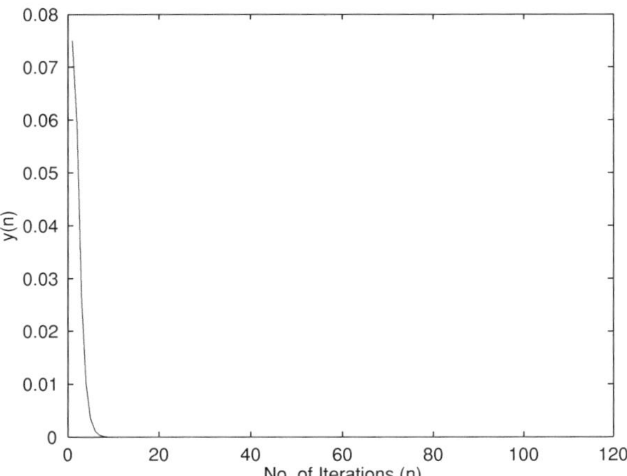

Fig. 5 Convergence of $\{y_n\}$ in Example 5

where $\Sigma_{n=0}^{\infty} a_n < \infty$, $\{\lambda_n\} \subset [a, b]$ *for some* $a, b \in \left(0, \frac{1}{\|A\|^2}\right)$ *and* $\{\beta_n\} \subset [c, d]$ *for some* $c, d \in (0, 1)$. *Then, the sequence* $\{x_n\}$ *converges weakly to* $\bar{x} \in B^{-1}0 \cap (\nabla f)^{-1}$.

Example 5 Let $C = Q = [0, 1]$ and $S : C \to C$ be defined as

$$Sx = \frac{x(x+1)}{4}, \quad \text{for all } x \in C.$$

Then, S is a nonexpansive mapping and $0 \in \text{Fix}(S)$. Let $Ax = x$ be a bounded linear operator. Let $\alpha_n = \frac{1}{n^2}$, $\beta_n = \frac{1}{2n}$ and $\lambda_n = \frac{1}{2(n+1)}$. All the conditions of Theorem 13 are satisfied. The sequences $\{x_n\}$ and $\{y_n\}$ generated by the scheme (62) starting with $x_1 = 0.1$. Then, we observe that these sequences converge to an element $0 \in \text{Fix}(S) \cap \Gamma$ (Figs. 5 and 6).

We did the computation in Matlab R2010 and got the solution 0 after 8th iterates (Figs. 5 and 6, Table 5).

4.2 Relaxed Extragredient Methods

In this section, we present a relaxed extragradiend method and study the weak convergence of the sequences generated by this method. We also present a relaxed extragradiend method with regularization for finding a common element of the solution set Γ of the SFP and the set $\text{Fix}(S)$ of fixed points of a asymptotically quasi-nonexpansive and Lipschtiz continuous mapping in the setting of Hilbert spaces. The weak convergence of the sequences generated by this method is also presented.

Theorem 15 [11, Theorem 3.2] *Let* C *be a nonempty closed and convex subset of a Hilbert space* $\mathscr{H}_1$ *and* $S : C \to C$ *be a nonexpansive mapping such that*

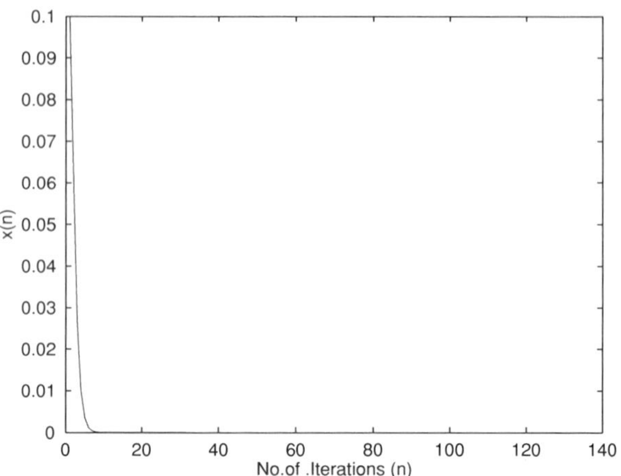

Fig. 6 Convergence of $\{x_n\}$ in Example 5

Table 5 Convergence of $\{x_n\}$ and $\{y_n\}$ in Example 5

Number of iterations (n)	y_n	x_n	Number of iterations (n)	$y(n)$	$x(n)$
1	0.0750	0.1000	6	0.0011	0.0011
2	0.0584	0.0610	7	0.0004	0.0004
3	0.0265	0.0269	8	0.0001	0.0001
4	0.0101	0.0101	9	0.0000	0.0000
5	0.0035	0.0035	10	0.0000	0.0000

Fix(S) $\cap \Gamma \neq \emptyset$. *Assume that* $0 < \lambda < \frac{2}{\|A\|^2}$, *and let* $\{x_n\}$ *and* $\{y_n\}$ *be the sequences in* C *generated by the following Mann-type extragradient-like algorithm:*

$$\begin{cases} x_0 = x \in C \quad \text{chosen arbitrarily,} \\ y_n = (1-\beta_n)x_n + \beta_n P_C(x_n - \lambda \nabla f_{\alpha_n}(x_n)), \\ x_{n+1} = \gamma_n x_n + (1-\gamma_n) S P_C(y_n - \lambda \nabla f_{\alpha_n}(y_n)), \quad \text{for all } n \geq 0, \end{cases} \tag{66}$$

where $\nabla f_{\alpha_n} = \nabla f + \alpha_n I = A^*(I - P_Q)A + \alpha_n I$ *and the sequences of parameters* $\{\alpha_n\}$, $\{\beta_n\}$, $\{\gamma_n\}$ *satisfy the following conditions:*

(i) $\sum_{n=0}^{\infty} \alpha_n < \infty$;
(ii) $\{\beta_n\} \subset [0, 1]$ *and* $0 < \liminf_{n\to\infty} \beta_n \leq \limsup_{n\to\infty} \beta_n < 1$;
(iii) $\{\gamma_n\} \subset [0, 1]$ *and* $0 < \liminf_{n\to\infty} \gamma_n \leq \limsup_{n\to\infty} \gamma_n < 1$.

Then, both the sequences $\{x_n\}$ *and* $\{y_n\}$ *converge weakly to an element* $z \in$ *Fix*$(S) \cap \Gamma$.

The following relaxed extragradiend method with regularization for finding a common element of the solution set Γ of the SFP and the set $\text{Fix}(S)$ of fixed points of a asymptotically quasi-nonexpansive and Lipschtiz continuous mapping in the setting of Hilbert spaces is proposed and studied by Deepho and Kumam [26]. They also studied the weak convergence of the sequences generated by this method.

Theorem 16 [26, Theorem 3.2] *Let C be a nonempty closed and convex subset of a Hilbert space $\mathscr{H}_1$ and $S : C \to C$ be a uniformly L-Lipschitz continuous and asymptotically quasi-nonexpansive mapping such that $Fix(S) \cap \Gamma \neq \emptyset$. Assume that $\{k_n\} \in [0, \infty)$ for all $n \in \mathbb{N}$ such that $\sum_{n=1}^{\infty}(k_n - 1) < \infty$. Let $\{x_n\}$ and $\{y_n\}$ be the sequences in C generated by the following algorithm:*

$$\begin{cases} x_0 = x \in C \quad chosen\ arbitrarily, \\ y_n = P_C(I - \lambda_n \nabla f_{\alpha_n}(x_n)), \\ x_{n+1} = \beta_n x_n + (1 - \beta_n) S^n(y_n), \quad for\ all\ n \geq 0, \end{cases} \tag{67}$$

where $\nabla f_{\alpha_n} = \nabla f + \alpha_n I = A^*(I - P_Q)A + \alpha_n I$, $S^n = \underbrace{S \circ S \circ \cdots \circ S}_{\text{n times}}$. The sequences of parameters $\{\alpha_n\}$, $\{\beta_n\}$, $\{\lambda_n\}$ satisfy the following conditions:

(i) $\sum_{n=1}^{\infty} \alpha_n < \infty$;
(ii) $\{\lambda_n\} \subset [a, b]$ *for some* $a, b \in \left(0, \frac{1}{\|A\|^2}\right)$ *and* $\sum_{i=1}^{\infty} |\lambda_{n+1} - \lambda_n| < \infty$;
(iii) $\{\beta_n\} \subset [c, d]$ *for some* $c, d \in (0, 1)$.

Then, both the sequences $\{x_n\}$ and $\{y_n\}$ converge weakly to an element $z \in Fix(S) \cap \Gamma$.

5 Mann-Type Iterative Methods for Common Solution of Split Feasibility and Fixed Point Problems

In this section, we present three different kinds of Mann-type iterative methods for finding a common element of the solution set Γ of the split feasibility problem and the set $\text{Fix}(S)$ of fixed points of a nonexpansive mapping S in the setting of infinite dimensional Hilbert spaces.

By combining Mann's iterative method and the extragradient method, we first propose Mann-type extragradient-like algorithm for finding an element of the set $\text{Fix}(S) \cap \Gamma$; moreover, we drive the weak convergence of the proposed algorithm under appropriate conditions. Second, we combine Mann's iterative method and the viscosity approximation method to introduce Mann-type viscosity algorithm for finding an element of the $\text{Fix}(S) \cap \Gamma$; moreover, we derive the strong convergence of the sequences generated by the proposed algorithm to an element of the set $\text{Fix}(S) \cap \Gamma$ under mild conditions. Finally, by combining Mann's iterative method and the relaxed CQ methods, we introduce Mann type relaxed CQ algorithm for finding

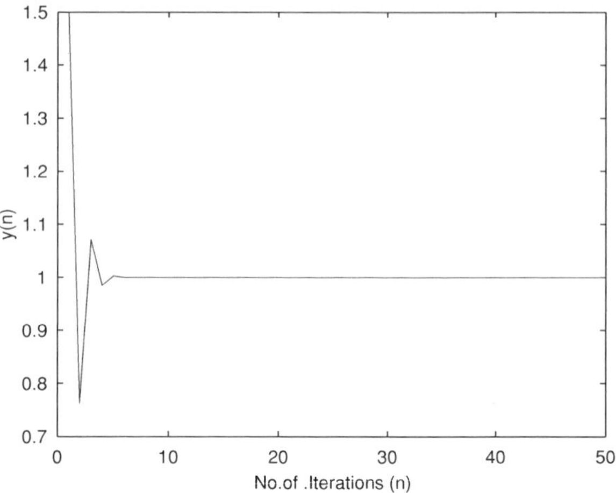

Fig. 7 Convergence of $\{y_n\}$ in Example 6

an element of the set $\text{Fix}(S) \cap \Gamma$. We also establish a weak convergence result for the sequences generated by the proposed Mann-type relaxed CQ algorithm under appropriate assumptions.

5.1 Mann-Type Extragradient-Like Algorithm

Let C and Q be nonempty closed convex subset of Hilbert spaces $\mathscr{H}_1$ and $\mathscr{H}_2$, respectively, and $A \in B(\mathscr{H}_1, \mathscr{H}_2)$. By combining Mann's iterative method and the extragradient method, Ceng et al. [12] proposed the following *Mann-type extragradient-like algorithm* for finding an element of the set $\text{Fix}(S) \cap \Gamma$ (Figs. 7 and 8):

The sequences $\{x_n\}$ and $\{y_n\}$ generated by the following iterative scheme:

$$\begin{cases} x_0 = x \in \mathscr{H}_1 \quad \text{chosen arbitrarily,} \\ y_n = (1 - \beta_n)x_n + \beta_n P_C(1 - \lambda_n A^*(I - P_Q)A)x_n, \\ x_{n+1} = \alpha_n x_n + (1 - \alpha_n) S P_C(I - \lambda_n A^*(I - P_Q)A)y_n, \quad \text{for all n} \geq 0, \end{cases} \tag{68}$$

where the sequences of parameters $\{\alpha_n\}$, $\{\beta_n\}$ and $\{\lambda_n\}$ satisfy some appropriate conditions.

The following result provides the weak convergence of the above scheme.

Theorem 17 [12, Theorem 3.2] *Let $S : C \to C$ be a nonexpansive mapping such that $\text{Fix}(S) \cap \Gamma \neq \emptyset$. Let $\{x_n\}$ and $\{y_n\}$ be the sequences by the Mann-type extragradient-like algorithm* (68)*, where the sequences of parameters $\{\alpha_n\}$, $\{\beta_n\}$ and $\{\lambda_n\}$ satisfies the following conditions:*

(i) $\{\alpha_n\} \subset [0, 1]$ *and* $0 < \liminf_{n\to\infty} \alpha_n \leq \limsup_{n\to\infty} < 1$;

(ii) $\{\beta_n\} \subset [0, 1]$ *and* $\liminf_{n\to\infty} \beta_n > 0$;

(iii) $\{\lambda_n\} \subset \left(0, \frac{2}{\|A\|^2}\right)$ *and* $0 < \liminf_{n\to\infty} \lambda_n \leq \limsup_{n\to\infty} \lambda_n < \frac{2}{\|A\|^2}$.

Then, both the sequences $\{x_n\}$ *and* $\{y_n\}$ *converges weakly to an element* $z \in Fix(S) \bigcap \Gamma$, *where*

$$z = \|\cdot\| - \lim_{n\to\infty} P_{Fix(S)\cap\Gamma} x_n.$$

We illustrate the above scheme and theorem by presenting the following example.

Example 6 Let $C = Q = [-1, 1]$ be closed convex set in $\mathbb{R}$. Let $S : C \to C$ be a mapping defined by

$$Sx = \frac{(x+1)^2}{4}, \quad \text{for all } x \in C.$$

Then, clearly S is a nonexpansive map and $1 \in \text{Fix}(S) \cap \Gamma$. Let $Ax = x$ be a bounded linear operator. If we choose $\alpha_n = \frac{1}{20} - \frac{1}{n}$ and $\beta_n = 1 - \frac{1}{2n}$, then all the conditions of Theorem 17 are satisfied. We choose the initial point $x_1 = 2$ and perform the iterative scheme in Matlab R2010. We obtain the solution after 6th iteration (Figs. 7 and 8, Table 6).

5.2 Mann-Type Viscosity Algorithm

Ceng et al. [12] modified the Mann-type extragradient-like algorithm, proposed in the last section, to obtain the strong convergence of the sequences. This modification is of viscosity approximation nature [9, 22, 48].

Theorem 18 [12, Theorem 4.1] *Let* $f : C \to C$ *be a* ρ*-contraction with* $\rho \in [0, 1)$ *and* $S : C \to C$ *be a nonexpansive mapping such that* $Fix(S) \cap \Gamma \neq \emptyset$. *Let* $\{x_n\}$ *and* $\{y_n\}$ *be the sequences generated by the following Mann-type viscosity algorithm:*

$$\begin{cases} x_0 = x_1 \in \mathscr{H}_1 \quad \text{chosen arbitrarily}, \\ y_n = P_C(I - \lambda_n A^*(I - P_Q)A)x_n, \\ z_n = P_C(I - \lambda_n A^*(I - P_Q)A)y_n, \\ x_{n+1} = \theta_n f(y_n) + \mu_n x_n + \nu_n z_n + \delta_n S z_n, \ \forall n \geq 0, \end{cases} \tag{69}$$

where the sequences of parameters $\{\theta_n\}$, $\{\mu_n\}$, $\{\nu_n\}$, $\{\delta_n\} \subset [0, 1]$ *and* $\{\lambda_n\} \subset \left(0, \frac{2}{\|A\|^2}\right)$ *satisfy the following conditions:*

(i) $\theta_n + \mu_n + \nu_n + \delta_n = 1$;

(ii) $\lim_{n\to\infty} \theta_n = 0$ *and* $\Sigma_{n=0}^{\infty} \theta_n = \infty$;

(iii) $\liminf_{n\to\infty} \delta_n > 0$;

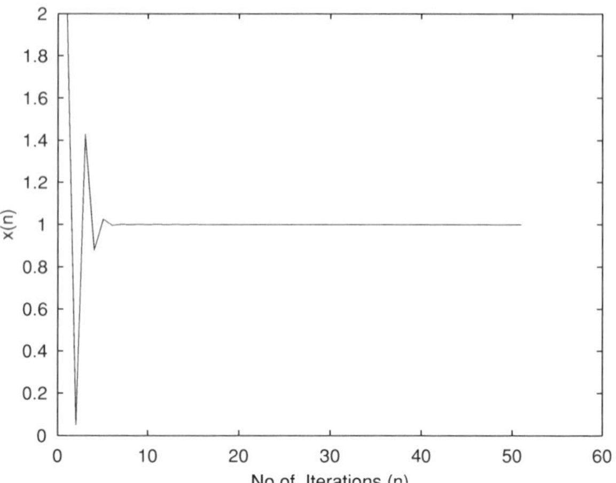

Fig. 8 Convergence of $\{x_n\}$ in Example 6

Table 6 Convergence of $\{x_n\}$ and $\{y_n\}$ in Example 6

Number of iterations (n)	y_n	x_n	Number of iterations (n)	$y(n)$	$x(n)$
1	1.500	2.000	7	1.000	1.000
2	0.7625	0.0500	8	1.000	1.000
3	1.0713	1.4275	9	1.000	1.000
4	0.9849	0.8789	10	1.000	1.000
5	1.0024	1.0242	11	1.000	1.000
6	0.9997	0.9964	12	1.000	1.000

(iv) $\lim_{n\to\infty}\left(\frac{\nu_{n+1}}{1-\mu_{n+1}} - \frac{\nu_n}{1-\mu_n}\right) = 0;$

(v) $0 < \liminf_{n\to\infty} \lambda_n \le \limsup_{n\to\infty} \lambda_n < \frac{2}{\|A\|^2}$ *and* $\lim_{n\to\infty}(\lambda_n - \lambda_{n+1}) = 0.$

Then, both the sequences $\{x_n\}$ *and* $\{y_n\}$ *converge strongly to* $x^* \in Fix(S) \cap \Gamma$ *which is also a unique solution of the variational inequality (VI):*

$$\langle (I-f)x^*, x - x^*\rangle \ge 0, \quad \textit{for all } x \in Fix(S) \cap \Gamma.$$

In other words, x^* *is a unique fixed point of the contraction* $P_{Fix(S)\cap\Gamma} f$, $x^* = (P_{Fix(S)\cap\Gamma} f)x^*$.

5.3 Mann-Type Relaxed CQ Algorithm

As pointed out earlier, the CQ algorithm (Algorithm 2) involves two projections P_C and P_Q and hence might hard to be implemented in the case where one of them fails to have closed-form expression. Thus, in [65] it was shown that if C and Q are level sets of convex functions, then the projections onto half-spaces are just needed to make the CQ algorithm implementable in this case. Inspired by relaxed CQ algorithm, Ceng et al. [12] proposed the following Mann-type relaxed CQ algorithm via projections onto half-spaces.

Define the closed convex sets C and Q as the level sets:

$$C = \{x \in \mathscr{H}_1 : c(x) \leq 0\} \quad \text{and} \quad Q = \{x \in \mathscr{H}_2 : q(x) \leq 0\}, \tag{70}$$

where $c : \mathscr{H}_1 \to \mathbb{R}$ and $q : \mathscr{H}_2 \to \mathbb{R}$ are convex functions. We assume that c and q are subdifferentiable on C and Q, respectively, namely, the subdifferentials

$$\partial c(x) = \{z \in \mathscr{H}_1 : c(u) \geq c(x) + \langle u - x, z\rangle, \ \forall u \in \mathscr{H}_1\} \neq \emptyset$$

for all $x \in C$, and

$$\partial q(x) = \{w \in \mathscr{H}_2 : q(v) \geq q(y) + \langle v - y, w\rangle, \ \forall v \in \mathscr{H}_1\} \neq \emptyset$$

for all $y \in Q$. We also assume that c and q are bounded on the bounded sets. Note that this condition is automatically satisfied when the Hilbert spaces are finite dimensional. This assumption guarantees that if $\{x_n\}$ is a bounded sequence in $\mathscr{H}_1$ (respectively, $\mathscr{H}_2$) and $\{x_n^*\}$ is another sequence in $\mathscr{H}_1$ (respectively, $\mathscr{H}_2$) such that $x_n^* \in \partial c(x_n)$ (respectively, $x_n^* \in \partial q(x_n)$) for each $n \geq 0$, then $\{x_n^*\}$ is bounded.

Let $S : \mathscr{H}_1 \to \mathscr{H}_1$ be a nonexpansive mapping. Assume that the sequences of parameters $\{\alpha_n\}$, $\{\beta_n\}$ and $\{\lambda_n\}$ satisfy the following conditions:

(i) $\{\alpha_n\} \subset [0, 1]$ and $0 < \liminf\limits_{n\to\infty} \alpha_n \leq \limsup\limits_{n\to\infty} \alpha_n < 1$;

(ii) $\{\beta_n\} \subset [0, 1]$ and $\liminf\limits_{n\to\infty} \beta_n > 0$;

(iii) $\{\lambda_n\} \subset \left(0, \frac{2}{\|A\|^2}\right)$ and $0 < \liminf\limits_{n\to\infty} \leq \limsup\limits_{n\to\infty} \lambda_n < \dfrac{2}{\|A\|^2}$.

Let $\{x_n\}$ and $\{y_n\}$ be the sequence defined by the following Mann-type relaxed CQ algorithm:

$$\begin{cases} x_0 = x \in \mathscr{H}_1 \quad \text{chosen arbitrarily,} \\ y_n = (1 - \beta_n)x_n + \beta_n P_{C_n}(I - \lambda_n A^*(I - P_{Q_n})A)x_n, \\ x_{n+1} = \alpha_n x_n + (1 - \alpha_n) S P_{C_n}(I - \lambda_n A^*(I - P_{Q_n})A)y_n, \quad \text{for all n} \geq 0, \end{cases} \tag{71}$$

where $\{C_n\}$ and $\{Q_n\}$ are the sequences of closed convex sets defined as follows:

$$C_n = \{x \in \mathscr{H}_1 : c(x_n) + \langle \xi_n, x - x_n \rangle \leq 0\}, \tag{72}$$

where $\xi_n \in \partial c(x_n)$, and

$$Q_n = \{y \in \mathscr{H}_2 : q(Ax_n) + \langle \eta_n, y - Ax_n \rangle \leq 0\}, \tag{73}$$

where $\eta_n \in \partial q(Ax_n)$.

It can be easily seen that $C \subset C_n$ and $Q \subset Q_n$ for all $n \geq 0$. Also, note that C_n and Q_n are half-spaces; thus, the projections P_{C_n} and P_{Q_n} have closed-form expressions.

Ceng et al. [12] established the following weak convergence theorem for the sequences generated by the scheme (71).

Theorem 19 [12, Theorem 5.1] *Suppose that* $Fix(S) \cap \Gamma \neq \emptyset$. *Then, the sequences* $\{x_n\}$ *and* $\{y_n\}$ *generated by the algorithm* (71) *converge weakly to an element* $z \in Fix(S) \cap \Gamma$, *where*

$$z = \|\cdot\| - \lim_{n\to\infty} P_{Fix(S)\bigcap\Gamma} x_n.$$

6 Solution Methods for Multiple-Sets Split Feasibility Problems

For each $i = 1, 2, \ldots, t$ and each $j = 1, 2, \ldots, r$, let $C_i \subseteq \mathscr{H}_1$ and $Q_j \subseteq \mathscr{H}_2$ be nonempty closed convex set in Hilbert spaces $\mathscr{H}_1$ and $\mathscr{H}_2$, respectively. Let $A \in B(\mathscr{H}_1, \mathscr{H}_2)$. The *convex feasibility problem* (CFP) is to find a vector x^* such that

$$x^* \in \bigcap_{i=1}^{t} C_i. \tag{74}$$

During the last decade, it received a lot of attention due to its applications in approximation theory, image recovery and signal processing, optimal control, biomedical engineering, communications, and geophysics, see, for example, [7, 17, 58] and the references therein.

Consider the *multiple-sets split feasibility problem* (MSSFP) of finding a vector x^* satisfying

$$x^* \in C := \bigcap_{i=1}^{t} C_i \quad \text{such that} \quad Ax^* \in Q := \bigcap_{j=1}^{r} Q_j. \tag{75}$$

As we have seen in the first section that this problem can be a unified model of several practical inverse problems, namely, image reconstruction, signal processing, an inverse problem of intensity-modulated radiation therapy, etc. Of course, when $i = j = 1$, MSSFP reduces to SFP.

The MSSFP (75) can be viewed as a special case of the CFP (74). In fact, (75) can be rewritten as

$$x^* \in \bigcap_{i=1}^{t+r} C_i, \quad \text{where } C_{t+j} := \{x \in \mathscr{H}_1 : A^{-1}x \in Q_j\}, \quad 1 \le j \le r. \tag{76}$$

However, the methodologies for studying the MSSFP (75) are different from those for the CFP (74) in order to avoid usage of the inverse A^{-1}. In other words, the methods for solving CFP (74) may not be applied to solve the MSSFP (75) without involving the inverse A^{-1}. The CQ Algorithm 2 is such an example where only the operator A (not the inverse A^{-1}) is relevant.

In view of Proposition 7, one can see that MSSFP (75) is equivalent to a common fixed point problem of finitely many nonexpansive mappings. Indeed, decompose MSSFP (75) into N subproblems ($1 \le i \le t$):

$$x_i^* \in C_i \quad \text{such that} \quad Ax_i^* \in Q := \bigcap_{j=1}^{r} Q_j. \tag{77}$$

For each $i = 1, 2, \ldots, t$, define a mapping T_i by

$$T_i(x) = P_{C_i}(I - \gamma_i \nabla f)x = P_{C_i}\left(I - \gamma_i \sum_{j=1}^{r} \beta_j A^*(I - P_{Q_j})A\right)x_i, \tag{78}$$

where f is defined by

$$f(x) = \frac{1}{2}\sum_{j=1}^{r} \beta_j \|Ax - P_{Q_j}Ax\|^2, \tag{79}$$

with $\beta_j > 0$ for all $j = 1, 2, \ldots, t$. Note that the gradient ∇f of f is

$$\nabla f(x) = \sum_{j=1}^{r} \beta_j A^*\left(I - P_{Q_j}\right)Ax, \tag{80}$$

which is L-Lipschitz continuous with constant $L = \sum_{j=1}^{r} \beta_j \|A\|^2$. If $\gamma_i \in (0, 2/L)$, then T_i is nonexpansive. Hence, fixed point algorithm for nonexpansive mappings can be applied to MSSFP (75)

Now we present the optimization method to solve MSSFP (75).

If x^* solves the MSSFP (75), then

(i) the distance from x^* to each C_i is zero, and
(ii) the distance from Ax^* to each Q_j is also zero.

This motivate us to consider the proximity function

$$p(x) := \frac{1}{2}\sum_{i=1}^{t}\alpha_i \|x - P_{C_i}(x)\|^2 + \frac{1}{2}\sum_{j=1}^{r}\beta_j \|Ax - P_{Q_j}(Ax)\|^2, \tag{81}$$

where $\alpha_i > 0$ for all i, $\beta_j > 0$ for all j. Then the proximity function is convex and differentiable with gradient

$$\nabla p(x) = \sum_{i=1}^{t}\alpha_i \left(I - P_{C_i}\right)(x) + \sum_{j=1}^{r}\beta_j A^* \left(I - P_{Q_j}\right) Ax, \tag{82}$$

where $*$ is the adjoint of A.

Proposition 8 [69] *x^* is a solution of MSSFP* (75) *if and only if $p(x^*) = 0$.*

Since the gradient $\nabla p(x)$ is L'-Lipschtiz continuous with constant

$$L' = \sum_{i=1}^{t}\alpha_i + \sum_{j=1}^{r}\beta_j \|A\|^2, \tag{83}$$

one can use the project gradient method to solve the

$$\min_{x\in\Omega} p(x), \tag{84}$$

where Ω is a closed convex subset of $\mathscr{H}_1$ whose intersection with the solution set of MSSFP (75) is nonempty, and get a solution of the so-called *constrained multiple-sets split feasibility problem* [15]:

$$\text{Find } x^* \in \Omega \quad \text{such that} \quad x^* \text{ solves (84)}. \tag{85}$$

In view of the above discussion, Censor et al. [15] proposed the following project gradient algorithm to find the solution of MSSFP (75) in the setting of finite-dimensional Hilbert spaces.

Algorithm 10 (*Projection Gradient Algorithm*) For any arbitrary $x_0 \in \mathscr{H}_1$, generates a sequence $\{x_n\}$ by

$$\begin{aligned} x_{n+1} &= P_\Omega\left(x_n - \gamma \nabla p(x_n)\right) \\ &= P_\Omega\left(x_n - \gamma\left(\textstyle\sum_{i=1}^{t}\alpha_i(I - P_{C_i})(x_n) + \Sigma_{j=1}^{r}\beta_j A^*(I - P_{Q_j})Ax_n\right)\right), \quad n \geq 0, \end{aligned} \tag{86}$$

where $\gamma \in (0, 2/L')$.

Censor et al. [15] established the convergence of the Algorithm 10. The following theorem is a version of their theorem in infinite dimensional Hilbert spaces established by Xu [64].

Theorem 20 *Let MSSFP* (75) *be consistent and a minimizer of the function p over Ω be inconsistent. Assume that $0 < \gamma < 2/L'$, where L' is given by* (83). *The sequence $\{x_n\}$ generated by Algorithm 10 converges weakly to a point z which is a solution of MSSFP* (75).

In this direction, several methods and results were obtained during the last decade. In [69], Yao et al. reviewed and presented some recent results on iterative approaches to MSSFP (75).

Zhao et al. [75] proposed the following modified projection algorithm for MSSFP (75) in finite dimensional Euclidean spaces.

Given closed convex sets $C_i \subseteq \mathbb{R}^N, i = 1, 2, \ldots, t$, and closed convex sets $Q_j \subseteq \mathbb{R}^M, j = 1, 2, \ldots, r$, in the N and M dimensional Euclidean spaces, respectively, and A an $M \times N$ real matrix. Let Ω be a closed convex subset of $\mathbb{R}^N$ whose intersection with the solution set of MSSFP (75) is nonempty,

Algorithm 11 For any arbitrary $x_0 \in \mathbb{R}^N, \sigma_0 > 0, \beta \in (0, 1), \theta \in (0, 1), \rho \in (0, 1)$. For $n = 0, 1, 2, \ldots$, compute

$$\bar{x}_n = P_\Omega(x_n - \gamma_n \nabla p(x_n)), \tag{87}$$

where γ_n is chosen to be the largest $\gamma \in \{\sigma_n, \sigma_n\beta, \sigma_n\beta^2, \ldots\}$ satisfying

$$\gamma \|\nabla p(\bar{x}_n) - \nabla p(x_n)\| \leq \theta \|\bar{x}_n - x_n\|. \tag{88}$$

Let

$$x_{n+1} = P_X(\bar{x}_n - \gamma_n(\nabla p(\bar{x}_n) - \nabla p(x_n))). \tag{89}$$

If

$$\gamma_n \|\nabla p(x_{n+1}) - \nabla p(x_n)\| \leq \rho \|x_{n+1} - x_n\|, \tag{90}$$

then set $\sigma_n = \sigma_0$; otherwise, set $\sigma_n = \gamma_n$, where $p(x)$ is proximity function as defined by (81).

We can take $p(x_n) < \epsilon$ or $\|\nabla p(x_n)\| < \epsilon$ as the stopping criteria in this algorithm.

We have the following result on the convergence of the sequence generated by Algorithm 11.

Theorem 21 [75, Theorem 4.1] *Let X be a nonempty closed convex set in $\mathbb{R}^N$ with a simple structure and $\{x_n\}$ be a sequence generated by Algorithm 11. If the set X contains at least one solution of the constrained multiple-sets split feasibility problem, then $\{x_n\}$ converges to a solution of the constrained multiple-sets split feasibility problem.*

A relaxed scheme of Algorithm 11 is also presented in [75].

Censor et al. [16] proposed a perturbed projection algorithm for multiple-sets split feasibility problem by applying the orthogonal projections onto a sequence of supersets of the original sets of the problem. Their work is based on the results of Santo and Scheimberg [55].

References

1. Ansari, Q.H.: Topics in Nonlinear Analysis and Optimization. World Education, Delhi (2012)
2. Ansari, Q.H., Lalitha, C.S., Mehta, M.: Generalized Convexity, Nonsmooth Variational Inequalities, and Nonsmooth Optimization. CRC Press, Boca Raton (2014)
3. Browder, F.E.: Fixed point theorems for noncompact mappings in Hilbert spaces. Proc. Natl. Acad. Sci. Soc. **43**, 1272–1276 (1965)
4. Byrne, C.: Block iterative methods for image reconstruction form projections. IEEE Trans. Image. Process. **5**, 96–103 (1996)
5. Byrne, C.: Iterative oblique projection onto convex subsets and the split feasibility problem. Inverse Prob. **18**, 441–453 (2002)
6. Byrne, C.: A unified treatment of some iterative algorithms in signal processing and image reconstruction. Inverse Prob. **20**, 103–120 (2004)
7. Bauschke, H.H., Borwien, J.M.: On projection algorithms for solving convex feasibility problems. SIAM Rev. **38**, 367–426 (1996)
8. Bertero, M., Boccacci, P.: Introduction to Inverse Problems in Imaging. Institue of Physics Publishing, Bristol (1988)
9. Ceng, L.-C., Ansari, Q.H., Yao, J.-C.: On relaxed viscosity iterative methods for variational inequalities in Banach spaces. J. Comput. Appl. Math. **230**, 813–822 (2009)
10. Ceng, L.-C., Ansari, Q.H., Yao, J.-C.: An extragradient method for solving split feasibility and fixed point problems. Comput. Math. Appl. **64**, 633–642 (2012)
11. Ceng, L.-C., Ansari, Q.H., Yao, J.-C.: Relaxed extragradient method for solving split feasibility and fixed point problem. Nonlinear Anal. **75**, 2116–2125 (2012)
12. Ceng, L.-C., Ansari, Q.H., Yao, J.-C.: Mann type iterative methods for finding a common solution of split feasibility and fixed point problems. Positivity **16**, 471–495 (2012)
13. Censor, Y., Bortfeld, T., Martin, B., Trofimov, A.: A unified approach for inversion problems in intensity-modulated radiation therapy. Phys. Med. Biol. **51**, 2353–2365 (2006)
14. Censor, Y., Elfving, T.: A mutliprojection algorithm using Bregman projections in a product space. Numer. Algorithms **8**, 221–239 (1994)
15. Censor, Y., Elfving, T., Kopf, N., Bortfeld, T.: The multiple-sets split feasibility problem and its applications for inverse problems. Inverse Prob. **21**, 2071–2084 (2005)
16. Censor, Y., Motova, A., Segal, A.: Perturbed projrections and subgradient projections for the multiple-sets split feasibility problem. J. Math. Anal. Appl. **327**, 1244–1256 (2007)
17. Combettes, P.L.: The convex feasibility problem in image recovery, In: Hawkes, P. (ed.) Advances in Image and Electron Physics, pp. 155–270. Academic Press, New York (1996)
18. Combettes, P.L., Bondon, P.: Hard-constrained inconsistent signal feasibility problems. IEEE Trans. Signal Process. **47**, 2460–2468 (1999)
19. Combettes, P.L., Pesquett, J.C.: A proximal decomposition method for solving convex variational inverse problems. Inverse Prob. **25**, Article ID 065014 (2008)
20. Chan, T., Shen, J.: Image Processing and Analysis: Variational, PDE, Wavelet, and Stochastic Methods. SIAM, Philadelphia, PA (2005)
21. Combettes, P.L., Wajs, R.: Signal recovery by proximal forward-backward splitting multiscale model. Multiscale Model. Simul. **4**, 1168–1200 (2005)
22. Ceng, L.-C., Xu, H.-K., Yao, J.-C.: The viscosity approximation method for asympotically nonexpansive mappings in Banach spaces. Nonlinear Anal. **69**, 1402–1412 (2008)
23. Daubechies, I., Fornasier, M., Loris, I.: Accelerated projected gradient method for linear inverse problems with sparsity constraits. J. Fourier Anal. Appl. **14**, 764–992 (2008)
24. Dang, Y., Gao, Y.: The strong convergence of a three-step algorithm for the split feasibility problem. Optim Lett. **7**, 1325–1339 (2013)
25. Davies, M., Mitianoudis, N.: A simple mixture model for sparse overcomplete ICA. IEEE Proc. Vis. Image Signal Process. **151**, 35–43 (2004)
26. Deepho, J., Kumam, P.: Split feasibility and fixed-point problems for asymptotically quasi-nonexpansive. Fixed Point Theory Appl. **2013**, Article ID 322 (2013)

27. Donoho, D.L.: De-noising by soft-thresholding. IEEE Trans. Inf. Theory **41**(3), 613–627 (1995)
28. Donoho, D.L., Vetterli, M., Devore, R.A., Daubechies, I.: Data compression and harmonic analysis. IEEE Trans. Inf. Theory **44**, 24352476 (1998)
29. Eicke, B.: Iteration methods for convexly constrained ill-posed problelms in Hilbert spaces. Numer. Funct. Anal. Optim. **13**, 413–429 (1992)
30. Facchinie, F., Pang, J.S.: Finite-Dimensional Variational Inequalities and Complementary Problems, Vol. I and II. Springer, New York (2003)
31. Fukushima, M.: On the convergence of a class of outer approximation algorithms for convex programs. J. Comput. Appl. Math. **10**, 147–156 (1984)
32. Fukushima, M.: A relaxed projection method for variational inequalities. Math. Program. **35**, 58–70 (1986)
33. Geobel, K., Kirk, W.A.: Topics in Metric Fixed Point Theory. Cambridge Studies in Advanced Mathematics, vol. 28. Cambridge University Press, Cambridge (1990)
34. He, B.: Inexact implicit methods for monotone general variational inequalities. Math. Program. **86**, 199–217 (1999)
35. He, B.S., Liao, L.Z.: Improvements of some projection methods for monotone nonlinear variational inequalities. J. Optim. Theory. Appl. **112**, 111–128 (2002)
36. He, B.S., Yuan, X.M., Zhang, J.Z.: Comparisons of two kinds of prediction methods for monotone variational inequalities. Comput. Optim. Appl. **27**, 247–267 (2004)
37. Hundal, H.: An alternating projection that does not converge in norm. Nonlinear Anal. **57**, 35–61 (2004)
38. Iusem, A.N.: An iterative algorithm for the variational inequality problem. Comput. Appl. Math. **13**, 103–114 (1994)
39. Kinderlehrar, D., Stampacchia, G.: An Introduction to Variational Inequality and Their Application. Academic Press, New York (1980)
40. Korpelevich, G.M.: The extragradient method for finding saddle points and other problems. Ekonomika Mat. Metody **12**, 747–756 (1976)
41. Lagendijk, R., Biemond, J.: Iterative Identification and Restoration of Images. Kluwer Academic Press, New York (1991)
42. Landweber, L.: An iterative formula for Fredholm integral equation of the first kind. Amer. J. Math. **73**, 615–624 (1951)
43. Levitin, E.S., Polyak, B.T.: Constrained minimization methods. Zh. Vychist. Mat. Mat. Fiz. **6**, 787–823 (1966)
44. Li, C.-L., Liou, Y.-C., Yao, Y.: A damped algorithm for split feasibility and fixed point problems. J. Inequal. Appl. **2013**, Article ID 379 (2013)
45. Li, M.: Improved relaxed CQ methods for solving the split feasibility problem. AMO - Adv. Model. Optim. **13**, 1244–1256 (2011)
46. Li, Y.: Iterative algorithm for a convex feasibility problem. An. St. Univ. Ovidius Constanta **18**, 205–218 (2010)
47. Mohammed, L.B., Auwalu, A., Afis, S.: Strong convergence for the split feasibility problem in real Hilbert Space. Math. Theo. Model. **3**, 2224–5804 (2013)
48. Moudafi, A.: Viscosity approximation methods for fixed-points problems. J. Math. Anal. Appl. **241**, 46–55 (2000)
49. Moudafi, A.: A relaxed alternating CQ-algorithm for convex feasibility problems. Nonlinear Anal. **79**, 117–121 (2013)
50. Nadezhkina, N., Takahashi, W.: Weak convergence theorem by an extragradient method for nonexpansive mappings and monotone mappings. J. Optim. Theory Appl. **128**, 191–201 (2006)
51. Noor, M.A., Noor, K.I.: Some new classes of quasi split feasibility problems. Appl. Math. Inf. Sci. **7**(4), 1547–1552 (2013)
52. Opial, Z.: Weak convergence of the sequence of successive approximations for nonexpansive mappings. Bull. Amer. Math. Soc. **73**, 595–597 (1967)
53. Qu, B., Xiu, N.: A note on the CQ algorithm for the split feasibility problem. Inverse Prob. **21**, 1655–1665 (2005)

54. Qu, B., Xiu, N.: A new halfspaces-relaxation projection method for the split feasibility problem. Linear Algebra Appl. **428**, 1218–1229 (2008)
55. Santos, P.S.M., Scheimberg, S.: A projection algorithm for general variational inequalities with perturbed constraint sets. Appl. Math. Comput. **181**, 649–661 (2006)
56. Sezan, M.T., Stark, H.: Application of convex projection theory to image recovery in tomography and related areas. In: H. Stark (ed.) Image Recovery Theory and Applications, pp. 415–462. Academic Press, Orlando (1987)
57. Solodov, M.V., Tseng, P.: Modified projection-type methods for monotone variational inequalities. SIAM J. Control Optim. **34**, 1814–1830 (1996)
58. Stark, H. (ed.): Image Recovery Theory and Applications. Academic Press, Orlando (1987)
59. Suzuki, T.: Strong convergence of Krasnoselskii and Mann's type sequence for one-parameter nonexpansive semigroups without Bochner integrals. J. Math. Anal. Appl. **305**, 227–239 (2005)
60. Tseng, P.: A modified forward-backward splitting method for maximal monotone mappings. SIAM J. Control Optim. **38**, 431–446 (2000)
61. Wang, C.Y., Xiu, N.H.: Convegence of the gradient projection method for generalized convex minimization. Comput. Optim. Appl. **16**, 111–120 (2000)
62. Wang, F., Xu, H.-K.: Cyclic algorithms for split feasibility problems in Hilbert spaces. Nonlinear Anal. **74**, 4105–4111 (2011)
63. Wang, W.-W., Gao, Y.: A modified algorithm for solving the split feasibility problem. Inter. Math. Forum **7**, 1389–1396 (2009)
64. Xu, H.-K.: A variable Krasnosel'skii-Mann algorithm and the multiple-set split feasibility problem. Inverse Prob. **22**, 2021–2034 (2006)
65. Xu, H.-K.: Iterative methods for the split feasibility problem in infinite-dimensional Hilbert spaces. Inverse Prob. **26**, Article ID 105018 (2010)
66. Xu, H.-K., Kim, T.H.: Convergence of hybrid steepest-decent methods for variational inequalities. J. Optim. Theory Appl. **119**, 185–201 (2003)
67. Xiu, N.H., Zhang, J.Z.: Some recent advances in projection-type methods for variational inequalities. J. Comput. Appl. Math. **152**, 559–585 (2003)
68. Yang, Q.: The relaxed CQ algorithm for solving split feasibility problem. Inverse Prob. **20**, 1261–1266 (2004)
69. Yao, Y., Chen, R., Matino, G., Liou, Y.C.: Appications of fixed-point and optimization methods to the multiple-set split feasibility problem. J. Appl. Math. **2012**, Article ID 927530 (2012)
70. Youla, D.: On deterministic convergence of iterations of relaxed projection operators. J. Vis. Commun. Image Represent. **1**, 12–20 (1990)
71. Youla, D.: Mathematical theory of image restoration by the method of convex projections. In: H. Stark (ed.) Image Recovery Theory and Applications, pp. 29–77. Academic Press, Orlando (1987)
72. Zhang, W., Han, D., Li, Z.: A self-adaptive projection method for solving the multiple-sets split feasibility problem. Inverse Prob. **25**, Article ID 115001 (2009)
73. Zhao, J., Yang, Q.: Several solutions methods for the split feasibility problems. Inverse Prob. **21**, 1791–1799 (2005)
74. Zhao, J., Yang, Q.: A simple projection method for solving the multiple-sets split feasibility problem. Inverse Prob. Sci. Eng. **21**(3), 537–546 (2013)
75. Zhao, J., Zhang, Y., Yang, Q.: Modified projection methods for the split feasibility problem and the multiple-sets split feasibility problem. J. Comput. Appl. Math. **219**, 1644–1653 (2012)
76. Zhu, L.-J., Liou, Y.-C., Chyu, C.-C.: Algorithmic and analytical approaches to the split feasibility problems and fixed point problems. Taiwanese J. Math. **17**, 1839–1853 (2013)

Isotone Projection Cones and Nonlinear Complementarity Problems

M. Abbas and S. Z. Németh

Abstract A brief introduction of complementarity problems is given. We discuss the notion of *-isotone projection cones and analyze how large is the class of these cones. We show that each generating *-isotone projection cone is superdual. We prove that a simplicial cone in R^m is *-isotone projection cone if and only if it is coisotone (i.e., it is the dual of an isotone projection cone. We consider the solvability of complementarity problems defined by *-isotone projection cones. The problem of finding nonzero solution of these problems is also presented.

Keywords Nonlinear complementarity problems · Cones · Isotone projection cones · *-isotone projection cones · Partial ordering · Fixed point problems · Variational inequalities · Projection mapping

1 Introduction

Let A be an $n \times n$ real matrix, $b \in \mathbb{R}^n$, and f be the function defined by

$$f(x) = \frac{1}{2}\langle x, Ax\rangle + \langle b, x\rangle.$$

In 1961, Dorn [12] considered the following optimization problem:

M. Abbas (✉)
Department of Mathematics and Applied Mathematics, University of Pretoria, Lynnwood road, Pretoria 0002, South Africa
e-mail: mujahid.abbas@up.ac.za

S. Z. Németh
School of Mathematics, The University of Birmingham, The Watson Building, Edgbaston, Birmingham B15 2TT, UK
e-mail: nemeths@for.mat.bham.ac.uk

Q. H. Ansari (ed.), *Nonlinear Analysis*, Trends in Mathematics,
DOI: 10.1007/978-81-322-1883-8_10,

$$\min_{x \in F} f(x), \tag{1}$$

where

$$F = \{x \in \mathbb{R}^n \mid x \geq 0,\ Ax + b \geq 0\}$$

is the feasibility region. It was shown that, if A is a positive definite (not necessarily symmetric) matrix, then problem (1) must have an optimal solution and

$$\min_{x \in F} f(x) = 0.$$

Dorn's paper was the first step in initiating the study of complementarity problems.

In 1963, Dantzig and Cottle [10] showed that, if A is a square (not necessarily symmetric) matrix whose all principal minors are positive, then problem (1) has an optimal solution x^* satisfying the following equation

$$\langle x^*, Ax^* + b \rangle = 0. \tag{2}$$

In 1964, Cottle [8] studied problem (1) under the assumption that A is a positive semidefinite matrix and observed that, in this case (1) may not have an optimal solution.

However, if A is positive semidefinite and $F \neq \emptyset$ then an optimal solution for (1) exists and again, $\min_{x \in F} f(x) = 0$. This result was generalized by Cottle [8, 9]. He considered the following nonlinear problem associated to a continuously differentiable mapping $h : \mathbb{R}^n \to \mathbb{R}^n$: The problem is as follows:

$$\begin{aligned} &\min_{x \in F} f(x) = \langle x, h(x) \rangle \\ &\text{where } F = \{x \in \mathbb{R}^n \mid x \geq 0,\ h(x) \geq 0\}. \end{aligned} \tag{3}$$

He showed that, if x_0 is an optimal solution of the above problem and the Jacobian matrix $J_h(x_0)$ has positive principal minors, then x_0 satisfies the following conditions:

$$x_0 \geq 0,\ h(x_0) \geq 0 \ \text{ and } \ \langle x_0, h(x_0) \rangle = 0.$$

If we take $h(x) = Ax + b$, then we obtain (1). This was the first nice result about the nonlinear complementarity problem.

In 1965, Lemke [27] contributed to the development of complementarity theory as a method for solving matrix games.

The complementarity problems are closely related to variational inequalities and to fixed point problems.

Concerning the complementarity problem, we distinguish two entirely distinct classes of problems:

1. The topological complementarity problem (T. C. P.)
2. The order complementarity problem (O. C. P.).

In the class of topological complementarity problem, we have the following categories:

1.1 Generalized Complementarity Problem

If $E(\tau)$ is a locally convex space, then E^* denotes the topological dual of E.

We say that $\langle E, F\rangle$ is a dual system if, E and F are vector spaces and $\langle ., .\rangle$ is bilinear functional on $E \times F$ such that

(i) $\langle x, y\rangle = 0$, for each $x \in E \Rightarrow y = 0$,
(ii) $\langle x, y\rangle = 0$, for each $y \in F \Rightarrow x = 0$.

If $E(\tau)$ is a locally convex space, we denote by $\langle E, E^*\rangle$ the dual system defined by the bilinear functional, $\langle x, u\rangle = u(x)$ for every $x \in E$ and $u \in E^*$.

If $\langle E, F\rangle$ is a dual system and $K \subset E$ is a convex cone, then the *dual cone* of K is $K^* = \{u \in F \mid \langle x, u\rangle \geq 0, \forall\, x \in K\}$. The *polar cone* of K is $K^{\perp} = \{u \in F \mid \langle x, u\rangle \leq 0, \forall\, x \in K\}$.

Let $\langle E, F\rangle$ be a dual system of locally convex spaces. For a given closed convex cone $K \subset E$ and a mapping $f : K \to F$, the *generalized complementarity problem* (G.C.P.) associated to K and f is the following problem:

$$\begin{aligned}&\text{Find } x_0 \in K \text{ such that}\\ &f(x_0) \in K^* \text{ and } \langle x_0, f(x_0)\rangle = 0.\end{aligned}$$

Note that if $f(x) = L(x) + b$, where $L : E \to F$ is a linear mapping and b an element of F then we have the *linear complementarity problem* (L.C.P.). If $f : K \to F$ is a nonlinear mapping, then we have the nonlinear complementarity problem (N.C.P).

If $E = F = \mathbb{R}^n$, $K = \mathbb{R}^n_+$, $\langle x, y\rangle = \sum_{i=1}^{n} x_i y_i$, where $x = (x_i)$, $y = (y_i) \in \mathbb{R}^n$, $A \in M_{n\times m}(\mathbb{R})$ and $b \in \mathbb{R}^n$, then we obtain the classical *linear complementarity problem* (L.C.P.):

$$\begin{aligned}&\text{Find } x_0 \geq 0 \text{ such that}\\ &Ax_0 + b \geq 0 \text{ and } \langle x_0, Ax_0 + b\rangle = 0.\end{aligned}$$

(In this case, $K = K^* = \mathbb{R}^n_+$).

If $\langle E, F\rangle$ is a dual system of locally convex spaces, $K \subset E$ a closed convex cone and $f : K \to F$ a multivalued mapping (that is $f : K \to 2^F$), then the *generalized multivalued complementarity problem* (G.M.C.P.) associated to f and K is:

$$\begin{aligned}&\text{Find } x_0 \in K \text{ and } y_0 \in F \text{ such that}\\ &y_0 \in f(x_0) \cap K^* \text{ and } \langle x_0, y_0\rangle = 0.\end{aligned}$$

Note that, if we take $f : K \to F$ a single valued mapping, then the (G.M.C.P.) becomes (G.C.P.) associated to K and f.

1.2 ε-Complementarity Problem

The origin of the definition of ε-complementarity problem is McLinden's paper [29].

Let $\langle E, F\rangle$ be a dual system of locally convex spaces and let $K \subset E$ be a closed convex cone.

Given mapping $f : K \to F$, an *ε-complementarity problem* (ε-C.P.) associated to f and K is:

$$\begin{aligned} &\text{for a give } \varepsilon > 0, \text{ find } x_0 \in K \\ &\text{such that } f(x_0) \in K^* \text{ and } \langle x_0, f(x_0)\rangle \leq \varepsilon. \end{aligned} \tag{4}$$

1.3 Order Complementarity Problem

Consider a vector lattice $(E, \leq)$, $K = \{x \in E \mid 0 \leq x\}$ and denote the Lattice operations sup (resp. inf) by $\vee$ (resp. $\wedge$).

Given a mapping $f : K \to E$, the *order complementarity problem* (O.C.P) is:

$$\begin{aligned} &\text{Find } x_0 \in K \\ &\text{such that } x_0 \wedge f(x_0) = 0. \end{aligned} \tag{5}$$

Since $x_0 \wedge f(x_0) = 0$, then it is clear that $f(x_0) \in K$. The order complementarity problem has interesting applications in Economics.

If $(H, \langle\cdot,\cdot\rangle, K)$ is a Hilbert lattice and $x, y \in K$, then $x \wedge y = 0$ if and only if $\langle x, y\rangle = 0$. Thus for Hilbert lattices, the problems (O.C.P.) and (G.C.P.) are equivalent.

There are many forms of complementarity problems based on the definition of the mapping and the structure of underlying space.

One of the most important problems in nonlinear analysis is the nonlinear complementarity problem, which can be stated as follows: Let K be a cone in a real Hilbert space $(H, \langle\cdot,\cdot\rangle)$, K^* the dual cone of K, and $f : K \to H$ a mapping; then the problem is to find an $x^* \in K$ such that $f(x^*) \in K^*$ and $\langle x^*, f(x^*)\rangle = 0$. The nonlinear complementarity problem defined by K and f will be denoted by *NCP*(f, K). The nonlinear complementarity problems can be viewed as particular fixed point problems, variational inequality problems, nonlinear optimization problems, convex optimization problems, nonlinear programing problems, etc. The complementarity problems are used to model several problems of economics, physics and engineering;

they can be constraints for an optimization problem; and they occur in the Karush-Kuhn-Tucker conditions for a nonlinear programming problem as well.

Motivated by solving complementarity problems, Isac and Németh have characterized a cone in the Euclidean space which admits an isotone projection onto it [15], where isotonicity is considered with respect to the order induced by the cone. They called such a cone isotone projection cone. The same authors [16] and Bernau [5] considered the similar problem for the Hilbert space. Bearing in mind the fixed point characterization of nonlinear complementarity problems, the isotonicity of the projection provides new existence results and iterative methods [17, 18, 20, 35] for these problems. Both the solvability and the approximation of solutions of nonlinear complementarity problems can be handled by using the metric projection onto the cone defining the problem, which emphasize the importance of studying the properties of projection mappings onto cones.

The aim of this chapter is to discuss the notion of *-isotone projection cones and its relationship to solvability of nonlinear complementarity problems. Our main references for this chapter are [1, 11, 13, 16, 32, 35, 36].

The structure of this chapter is as follows. In Sect. 2, we will fix the terminology and notations, and present some background results used in the chapter. In Sect. 3, we will discuss the notion of *-isotone projection cones and analyze how large is the class of these cones. We will show that each generating *-isotone projection cone is superdual. We will prove that a simplicial cone in R^m is *-isotone projection cone if and only if it is coisotone (i.e., it is the dual of an isotone projection cone). By using a more recent duality result from the preprint [37] of the second author it can be shown that infact the class of *-isotone and coisotone cones in Hilbert spaces is the same. However, the special terminology used in that paper is out of the scope of this chapter. In case of Euclidean spaces this result has been shown in [38]; but the proof uses the result for simplical cones presented here and it is also rather technical to be included in our chapter. We remark that the proofs in [37] are independent of the results of this chapter, but they are still subject to the scrutiny of reviewers. In Sect. 4, we will consider the solvability of complementarity problems defined by *-isotone projection cones.

2 Preliminaries

Let $(H, \langle\cdot,\cdot\rangle)$ be a real Hilbert space. All Hilbert spaces in this chapter are assumed to be real Hilbert spaces. Let $x, y \in H$, the line segment joining x and y is denoted by $[x, y]$ and defined by

$$[x, y] = \{\lambda x + (1-\lambda)y : 0 \leq \lambda \leq 1\}.$$

A subset $K \subset H$ is called is called *convex* if for every $x, y \in K$, $[x, y]$ is a subset of K. A nonempty subset $K \subset H$ is called a *cone* if $\lambda x \in K$, for all $x \in K$ and $\lambda \geq 0$. From this definition it can be seen that the zero element in $(H, \langle\cdot,\cdot\rangle)$

belongs to K. A cone K is called *pointed* if $K \cap (-K) = \{0\}$. For example, the set $\mathbb{R}^n_+ = \{(x_1, x_2, x_3, \ldots, x_n)^\top \in \mathbb{R}^n : x_i \geq 0$ for all $i = 1, 2, \ldots, n\}$ is a pointed cone. Note that a cone K in $(H, \langle\cdot,\cdot\rangle)$ is convex if and only if for every $x, y \in K$, $x + y \in K$. Indeed, if K is a convex, then for every $x, y \in K$, $\frac{1}{2}x + \frac{1}{2}y \in K$, that is $\frac{1}{2}(x + y) \in K$. Since K is a cone, $x + y \in K$. Conversely suppose that for every $x, y \in K$, $x + y \in K$. If $\lambda \in [0, 1]$, then $\lambda x \in K$ and $(1 - \lambda)y \in K$. Hence $\lambda x + (1 - \lambda)y \in K$. Hence, a subset $K \subset H$ is:

(i) a cone if $\lambda x \in K$, for all $x \in K$ and $\lambda > 0$ (non-negative homogenous)
(ii) convex cone if it is cone and $x + y \in K$, for all $x, y \in K$.
(iii) pointed cone if it is cone and $K \cap (-K) = \{0\}$.

For the simplicity of the terminology we shall call a pointed closed convex cone simply cone.

Note that convex cone need not to be a subspace. For example,

$$K = \{x \in C^2[a, b] : x(t) \geq 0 \text{ for all } t \in [a, b]\}$$

is a convex cone which is not a subspace of $C^2[a, b]$. $K - K$ is the called the *linear subspace* generated by K and it is the smallest linear subspace of H containing K. A cone $K \subset H$ is called *generating* if the linear subspace generated by K is H, that is, $K - K = H$. If $K \subset H$ is a cone, then

$$K^* = \{y \in H : \langle x, y\rangle \geq 0 \ \text{ for all } x \in K\}$$

is called the *dual* of K. The dual cone of a subspace of R^m is its orthogonal complement. Note that $x \in K^*$ if and only if $-x$ is the normal of a hyperplane that supports K at the origin. A cone $K \subset H$ is called *superdual* if $K^* \subset K$. If K is a cone, then

$$K^\perp = \{x \in H : \langle x, y\rangle \leq 0, \ \text{ for all } y \in K\}$$

is called the *polar* of K. Note that the polar of a cone consists of the origin and those nonzero vectors in H that make a nonacute angle with every nonzero vector in K. It is easy to see that $K^\perp = -K^*$. We say that the set A is generating the cone K if

$$K = \left\{\lambda_1 x^1 + \cdots + \lambda_\ell x^\ell : \ell \in \mathbb{N},\ \lambda_1, \ldots, \lambda_\ell \geq 0 \text{ and } x^1, \ldots, x^\ell \in A\right\}.$$

Note that if K is a generating cone, then K^* and $K^\perp$ are cones. This is the case for example if K is a simplicial cone in $\mathbb{R}^m$, that is, a cone generated by m linearly independent vectors.

The generating cones K and L are called *mutually polar* if $K = L^\perp$ (or equivalently $K^\perp = L$).

A relation ρ on H is called *reflexive* if $x\rho x$ for all $x \in H$. A relation ρ on H is called *transitive* if $x\rho y$ and $y\rho z$ imply $x\rho z$. A relation ρ on H is called *antisymmetric* if $x\rho y$ and $y\rho x$ imply $x = y$. A relation ρ on H is called an *order* if it is reflexive, transitive and antisymmetric. A relation ρ on H is called *translation invariant* if $x\rho y$ implies $(x+z)\rho(y+z)$ for any $z \in H$. A relation ρ on H is called *scale invariant* if $x\rho y$ implies $(\lambda x)\rho(\lambda y)$ for any $\lambda > 0$. A relation ρ on H is called *continuous* if for any two convergent sequences $\{x^n\}_{n\in\mathbb{N}}$ and $\{y^n\}_{n\in\mathbb{N}}$ with $x^n \rho y^n$ for all $n \in N$ we have $x^* \rho y^*$, where x^* and y^* are the limits of $\{x^n\}_{n\in\mathbb{N}}$ and $\{y^n\}_{n\in\mathbb{N}}$, respectively.

The relation ρ on H is a continuous, translation and scale invariant order if and only if it is induced by a cone $K \subset H$, that is, $\rho = \leq_K$, where $x \leq_K y$ if and only if $y - x \in K$. The cone K can be written as $K = \{x \in H : 0 \leq_K x\}$ and it is called the *positive cone* of the order $\leq_K$. The triplet $(H, \langle\cdot,\cdot\rangle, K)$ is called an *ordered vector space*. A cone $K \subset H$ is called *regular* if every decreasing sequence of elements in K is convergent. In $\mathbb{R}^m$ any cone is regular. The ordered vector space $(H, \langle\cdot,\cdot\rangle, K)$ is called a *vector lattice* if for every $x, y \in H$ there exist $x \wedge y := \inf\{x, y\}$ and $x \vee y := \sup\{x, y\}$. In this case we say that the cone K is latticial and for each $x \in H$ we denote $x^+ = 0 \vee x$, $x^- = 0 \vee (-x)$ and $|x| = x \vee (-x)$. Then, $x = x^+ - x^-$ and $|x| = x^+ + x^-$. If $H = \mathbb{R}^m$, then the latticial cones are exactly the simplicial cones.

Let C be a closed convex set and $P_C : H \to H$ be the projection mapping onto C defined by $P_C(x) \in C$ and $\|x - P_C(x)\| = \min\{\|x - y\| : y \in C\}$. If $x \in H$ and $y^0 \in C$, then $y^0 = P_C(x)$ if and only if $\langle y^0 - x, y - y^0\rangle \geq 0$ for all $y \in C$. That is, a hyperplane passing through y^0 with normal $x - y^0$ supports C at y^0. By using the definition of the metric projection and item (i) of the definition of a cone, it is easy to show that if K is a cone, then $P_K(\lambda x) = \lambda P_K(x)$, for any $x \in H$ and any $\lambda \geq 0$.

The following theorem is proved in [32].

Theorem 1 (Moreau) *Let H be a Hilbert space and $K, L \subset H$ two mutually polar generating cones in H. Then, the following statements are equivalent:*

(i) $z = x + y,\ x \in K,\ y \in L$ *and* $\langle x, y\rangle = 0$,
(ii) $x = P_K(z)$ *and* $y = P_L(z)$.

3 *-isotone Projection Cones in Hilbert Spaces

Definition 1 Let $(H, \langle\cdot,\cdot\rangle)$ be a Hilbert space and $K, L \subset H$ be cones. The mapping $\rho : H \to H$ is called *(L, K)-isotone* if $x \leq_L y$ implies $\rho(x) \leq_K \rho(y)$. If $L = K$, then an (L, K)-isotone mapping is called K-isotone, and if P_K is K-isotone, then K is called an *isotone projection cone*.

If K is a generating isotone projection cone, then it is latticial [16]. K is called *coisotone cone* if $K^\perp$ is a generating isotone projection cone [34]. If $H = \mathbb{R}^m$, the coisotone cones are exactly the simplicial cones generated by m linearly independent vectors which form pairwise nonacute angles [15, 19].

Definition 2 Let $(H, \langle .,. \rangle)$ be a Hilbert space and $K \subset H$ be a cone. If $P_K : H \to H$ is (K^*, K)-isotone, then the cone K is called **-isotone projection cone*.

The following proposition shows that the class of generating *-isotone projection cones is contained in the class of generating superdual cones.

Proposition 1 *Let $(H, \langle ., . \rangle)$ be a Hilbert space and $K \subset H$ be a generating cone. If the cone K is *-isotone projection cone, then it is superdual.*

Proof Let x be an arbitrary element of K^*. Since K is generating, there exist $u, v \in K$ such that $x = v - u$. Then, $u \leq_{K^*} v$. Since K is *-isotone projection cone, it follows that $u = P_K(u) \leq_K P_K(v) = v$ and consequently $x \in K$. Thus, $K^* \subset K$, that is, K is superdual. □

Theorem 2 *Let $(H, \langle ., . \rangle)$ be a Hilbert space and $K \subset H$ be a cone. The cone K is *-isotone projection cone, if and only if $P_K(u + v) \leq_K u$ for any $u \in K$ and any $v \in K^\perp$.*

Proof Suppose that K is an *-isotone projection cone. Let $u \in K$ and $v \in K^\perp$ be arbitrary. Then, $u + v \leq_{K^*} u$ implies that

$$P_K(u + v) \leq_K P_K(u) = u.$$

Conversely, suppose that

$$P_K(u + v) \leq_K u, \quad \text{for any } u \in K \text{ and any } v \in K^\perp. \tag{6}$$

Let $x, y \in H$ with $x \leq_{K^*} y$. Then, by Moreau's theorem $x \leq_{K^*} y \leq_{K^*} P_K(y)$. Thus,

$$x \leq_{K^*} P_K(y). \tag{7}$$

Let $u = P_K(y)$ and $v = x - P_K(y)$. Then, obviously $u \in K$ and, by Eq. (7), $v \in K^\perp$. Hence, we can use Eq. (6) to obtain

$$P_K(x) = P_K(u + v) \leq_K u = P_K(y).$$

Therefore, K is *-isotone projection cone. □

Corollary 1 *Let $(H, \langle ., . \rangle)$ be a Hilbert space and $K \subset H$ be a cone. The cone K is *-isotone projection cone, if and only if $P_K(x) \leq_K u$ for any $u \in K$ and any $x \in H$ with $x \leq_{K^*} u$.*

Proof Use Theorem 2 with $v = x - u \in K^\perp$. □

4 *-isotone Projection Cones in Euclidean Spaces

Let $\mathbb{R}^m$ be endowed with a Cartesian reference system. All matrices considered in this chapter will have real entries. We identify all vectors in $\mathbb{R}^m$ by column vectors. We denote the components of a vector in $\mathbb{R}^m$ by assigning lower indices to the letter which denotes the vector. In $\mathbb{R}^m$ the simplicial cones are of the form $L = A\mathbb{R}^m_+$, where A is a nonsingular $m \times m$ matrix. The generators of the cone L are the column vectors of A. Let $\langle .,. \rangle$ be the canonical scalar product of $\mathbb{R}^m$.

Lemma 1 *Let $K \subset \mathbb{R}^m$ be a cone and A a nonsingular matrix. Then,*

$$(AK)^* = (A^\top)^{-1} K^*, \tag{8}$$

and

$$(AK)^\perp = (A^\top)^{-1} K^\perp. \tag{9}$$

In particular,

$$(A\mathbb{R}^m_+)^* = (A^\top)^{-1} \mathbb{R}^m_+, \tag{10}$$

and

$$(A\mathbb{R}^m_+)^\perp = -(A^\top)^{-1} \mathbb{R}^m_+. \tag{11}$$

Proof Equation (9) follows easily from Eq. (8). Thus, it is enough to prove Eq. (8) only. $x \in (AK)^*$ if and only if $\langle x, Au \rangle \geq 0$ for any $u \in K$, which is equivalent to $\langle A^\top x, u \rangle \geq 0$ for any $u \in K$, or to $A^\top x \in K^*$. Hence, $x \in (AK)^*$ if and only if $x \in (A^\top)^{-1} K^*$. Therefore, $(AK)^* = (A^\top)^{-1} K^*$. □

For a vector $x \in \mathbb{R}^m$ denote $x^+ = \sup(x, 0)$ and $x^- = \sup(-x, 0)$, where the supremums are taken with respect to the order induced by $\mathbb{R}^m_+$. We write $x \geq 0$ if all components of x are non-negative.

The next proposition is a straightforward application of the Moreau's theorem. However, for the readers' convenience, we present all details of this proof.

Proposition 2 *Let A be a nonsingular matrix and $K = A\mathbb{R}^m_+$ the corresponding simplicial cone. Then, for any $y \in \mathbb{R}^m$ there exists a unique $x \in \mathbb{R}^m$ such that one of the following two equivalent statements hold:*

(a) $y = Ax^+ - (A^\top)^{-1} x^-$, $x \in \mathbb{R}^m$,
(b) $Ax^+ = P_K(y)$ *and* $-(A^\top)^{-1} x^- = P_{K^\perp}(y)$.

Proof Let us first prove that the statements (a) and (b) are equivalent.

Suppose that (a) holds. Then, by using Eq. (11), it follows that $Ax^+ \in K$, $-(A^\top)^{-1} x^- \in K^\perp$ and

$$\langle Ax^+, -(A^\top)^{-1} x^- \rangle = -\langle Ax^+, (A^{-1})^\top x^- \rangle = -\langle x^+, x^- \rangle = 0.$$

Thus, (b) follows from Moreau's theorem. The converse follows easily from the same theorem.

Next, we show that there exists a unique x such that item (a) holds. From Moreau's theorem, there exist a unique $p \in K$ and a unique $q \in K^\perp$ such that $y = p + q$ and $\langle p, q \rangle = 0$. Since $K = A\mathbb{R}^m_+$ and A is nonsingular, there exists a unique $u \in \mathbb{R}^m_+$

such that $p = Au$. By using Eq. (11), $K^{\perp} = -(A^{\top})^{-1}\mathbb{R}^m_+$, and hence there exists a unique $v \in \mathbb{R}^m_+$ such that $q = -(A^{\top})^{-1}v$. Thus,

$$0 = \langle p, q\rangle = \langle Au, -(A^{\top})^{-1}v\rangle = -\langle Au, (A^{-1})^{\top}v\rangle = -\langle u, v\rangle.$$

Therefore, $\langle u, v\rangle = 0$. Let $x = u - v$. Then, $u \in \mathbb{R}^m_+$, $v \in \mathbb{R}^m_+$ and $\langle u, v\rangle = 0$ implies $u = x^+$ and $v = x^-$. In conclusion,

$$y = p + q = Au - (A^{\top})^{-1}v = Ax^+ - (A^{\top})^{-1}x^-. \qquad \square$$

Recall that a square matrix is called *positive stable* if all its eigenvalues have positive real part. A real square matrix is called a *Z-matrix* if all of its off-diagonal entries are nonpositive. An M-matrix is a Z-matrix whose eigenvalues are positive. Therefore, a symmetric matrix is an M-matrix if and only if it is a positive definite Z-matrix. There are a large number of papers and books dealing with the properties and applications of the above classes of matrices. The reader can find more details about the special classes of M- and Z-matrices in [13, 14, 39]. A Stieltjes matrix is a symmetric positive definite Z-matrix [44]. It is easy to see that a symmetric matrix is an M-matrix if and only if it is a Stieltjes matrix. It is known that a Z matrix A is an M-matrix if and only if $Av \geq 0$ implies $v \geq 0$ [13, 14]. All square matrices A satisfying the property "$Av \geq 0$ implies $v \geq 0$" are called inverse positive or monotone [13]. Hence, all M-matrices (and in particular the Stieltjes matrices) are inverse positive. However, the converse of this statement is not true. It is known that a square matrix A is inverse positive if and only if $A^{-1} \geq 0$ [13].

Proposition 3 *An $m \times m$ positive definite matrix B is a Stieltjes matrix if and only if*

$$u \geq 0 \text{ and } x^- + B(u - x^+) \quad \Rightarrow \quad u - x^+ \geq 0. \tag{12}$$

Proof First, we show that any Stieltjes matrix B satisfies (12). We prove this by using induction on the dimension of the matrix. If B_1 is a one dimensional Stieltjes matrix, then $B_1 = (a)$, where $a > 0$. Let $u, x \in \mathbb{R}$. Thus, we have to show that $u \geq 0$ and $x^- + a(u - x^+) = x^- + B_1(u - x^+) \geq 0$ implies $u - x^+ \geq 0$. If $x \leq 0$ this is trivial because $u - x^+ = u \geq 0$. If $x \geq 0$, then $0 \leq x^- + a(u - x^+) = a(u - x^+)$ and hence $u - x^+ \geq 0$. Suppose, that the statement is true for m and prove it for $m + 1$.

We have to show that $\begin{pmatrix} u_1 \\ \vdots \\ u_{m+1} \end{pmatrix} \geq 0$ and

$$\begin{pmatrix} x_1^- \\ \vdots \\ x_{m+1}^- \end{pmatrix} + B_{m+1} \begin{pmatrix} u_1 - x_1^+ \\ \vdots \\ u_{m+1} - x_{m+1}^+ \end{pmatrix} \geq 0 \tag{13}$$

relation $P_K(y) \leq_K Au$ is equivalent to $Au - Ax^+ \in A\mathbb{R}^m_+$, or to $u - x^+ \geq 0$, where x is the element uniquely determined by the equation in the item (a) of Proposition 2. On the other hand, by Eq. (10), the relation $y \leq_{K^*} Au$ is equivalent to

$$A(u-x^+)+(A^\top)^{-1}x^- = Au - Ax^+ + (A^\top)^{-1}x^- = Au - y \in K^* = (A^\top)^{-1}\mathbb{R}^m_+,$$

or to, $x^- + A^\top A(u - x^+) \geq 0$. Thus, K is a *-isotone projection cone if and only if $x^- + A^\top A(u - x^+) \geq 0$ implies $u - x^+ \geq 0$. Therefore, by Proposition 3, K is a *-isotone projection cone if and only if $A^\top A$ is a Stieltjes matrix. □

In $\mathbb{R}^m$ the coisotone cones are simplicial cones with the generators forming pairwise nonacute angles. This means that the coisotone cones in $\mathbb{R}^m$ are the simplicial cones of the form $A\mathbb{R}^m_+$, where A is a nonsingular $m \times m$ matrix such that $A^\top A$ is a Stieltjes matrix.

Hence, we have the following theorem which shows that in $\mathbb{R}^m$ the class of coisotone cones is equal to the class of simplicial *-isotone projection cones.

Theorem 4 *Let A be an $m \times m$ nonsingular matrix and $K = A\mathbb{R}^m_+$ a simplicial cone. Then, K is a *-isotone projection cone if and only if it is coisotone.*

Proof K is coisotone if and only if $A^\top A$ is a Stieltjes matrix. Therefore, the result follows from Theorem 3. □

5 Connection Among Fixed Point Problems, Complementarity Problems and Variational Inequalities

In this section, first we study several equivalent forms of fixed point problems:

Definition 3 Let $K \subset H$ a closed convex set, and $f : H \to H$ a mapping, then the *variational inequality problem* defined by f and K is the problem of finding an $x^* \in K$ such that $\langle y - x^*, f(x^*)\rangle \geq 0$ for all y in K. We shall denote this problem by VI(f, K).

It is known that Fix(f, K) $\Leftrightarrow$ VI($I - f$, K) = VI(T, K), where $T = I - f$. Indeed, if x^* is the solution of Fix(f, K), then this implies that $T(x^*) = 0$. Thus x^* is the solution of VI(T, K). If x^* is the solution of VI(T, K), then $\langle y - x^*, T(x^*)\rangle \geq 0$ for all y in K. That is, $\langle y - x^*, x^* - f(x^*)\rangle \geq 0$ for all y in K. Taking $y = f(x^*)$, we have $\langle f(x^*) - x^*, x^* - f(x^*)\rangle \geq 0$. Thus

$$-\left\|x^* - f(x^*)\right\|^2 \geq 0$$

which implies that $f(x^*) = 0$. Hence x^* is the solution of Fix(f, K).

Also, we have: VI(f, K) $\Leftrightarrow$ Fix($P_K \circ (I - f)$, K). Indeed, x^* is the solution of Fix($P_K \circ (I - f)$, K) $\Leftrightarrow$ $P_K(x^* - fx^*) = x^*$. We know that $P_K(x^* - fx^*) =$

$x^* \Longleftrightarrow \langle x^* - fx^* - x^*, y - x^*)\rangle \le 0$ for all y in $K \Leftrightarrow \langle -fx^*, y - x^*)\rangle \le 0 \Leftrightarrow \langle fx^*, y - x^*)\rangle \ge 0$ for all y in K.

Next we show that variational inequality problem defined on a closed convex cone is equivalent to nonlinear complementarity problem.

VI$(f, K) \Leftrightarrow$ NCP(f, K), if K is closed convex cone:

If x^* is the solution of NCP(f, K) then $x^* \in K$, $f(x^*) \in K^*$ and $\langle fx^*, x^*\rangle = 0$. Consider

$$\langle y - x^*, f(x^*)\rangle = \langle y, f(x^*)\rangle - \langle x^*, f(x^*)\rangle = \langle y, f(x^*)\rangle \ge 0,$$

for all y in K as $f(x^*) \in K^*$. Suppose that x^* is the solution of $VI(f, K)$ which implies that $x^* \in K$ and $\langle y - x^*, f(x^*)\rangle \ge 0$ for all y in K. Take, $y = 0$ then $\langle -x^*, f(x^*)\rangle \ge 0$ gives that $\langle x^*, f(x^*)\rangle \le 0$. Take, $y = 2x^*$ then $\langle x^*, f(x^*)\rangle \ge 0$. Hence, $\langle x^*, f(x^*)\rangle = 0$. Also,

$$0 \le \langle y - x^*, f(x^*)\rangle = \langle y, f(x^*)\rangle - \langle x^*, f(x^*)\rangle = \langle y, f(x^*)\rangle,$$

for all y in K which implies that $f(x^*) \in K^*$.

Definition 4 Let $K \subset H$ a closed convex set, and $f : H \to R$ a mapping, then the *nonlinear optimization problem* defined by f and K is the problem of finding an $x^* \in K$ such that $f(x^*) \le f(y)$ for all y in K (that is, minimize $f(x)$ subject to $x \in K$). We shall denote this problem by NOPT(f, K).

If f is differentiable then the implication NOPT$(f, K) \Rightarrow$ VI$(\nabla f, K)$ is well known and it can be shown as follows: Let $f : H \to R$ a differentiable mapping. If x^* is the solution of NOPT(f, K) and y is an arbitrary point of K, then

$$x^* + t(y - x^*) \in K \text{ implies that } f(x^*) \le f(x^* + t(y - x^*))$$

which further implies that

$$\langle \nabla f(x^*), y - x^*\rangle = \lim_{t \searrow 0} \frac{f(x^* + t(y - x^*)) - f(x^*)}{t} \ge 0$$

for all y in K. Hence, x^* is the solution of VI$(\nabla f, K)$.

Now assume that f is convex and differentiable. It is straightforward to check that CNOPT$(f, K) \Leftrightarrow$ VI$(\nabla f, K)$. Indeed, if $f : H \to R$ a differentiable convex mapping, then NOPT$(f, K) \Rightarrow$ VI$(\nabla f, K)$. If x^* is the solution of VI$(\nabla f, K)$ and f is a convex mapping, then $x^* + t(y - x^*) \in K$ implies that

$$f(y) - f(x^*) \ge \langle \nabla f(x^*), y - x^*\rangle \ge 0,$$

for all y in K. Hence x^* is the solution of NOPT(f, K).

It is known that x^* is a solution of the nonlinear complementarity problem defined by K and f if and only if x^* is a fixed point of the mapping

$$K \ni x \mapsto P_K(x - f(x)), \tag{22}$$

where P_K is the projection mapping onto K. Indeed, for all x in H if we put $z = x - f(x)$ and $y = -f(x)$, then $z = x + y$. Suppose that x is a solution of nonlinear complementarity problem defined by K and f, that is $x \in K$ such that $-f(x) \in K^{\perp}$ and $\langle x, -f(x)\rangle = \langle x, y\rangle = 0$. Now this along with $z = x + y$ via Moreau's theorem gives $x = P_K(z)$. Therefore, x is a fixed point of the mapping $P_K(x - f(x))$.

Conversely suppose that x is a fixed point of the mapping $P_K(x - f(x))$, that is, $x = P_K(x - f(x))$. Thus $x \in K$. Now by Moreau's theorem we have

$$x - f(x) = P_K(x - f(x)) + P_{K^{\perp}}(x - f(x)) = x + P_{K^{\perp}}(x - f(x),$$

which further implies that $-f(x) \in K^{\perp}$, that is, $f(x) \in K^*$. Moreau's theorem also implies that $\langle x, P_{K^{\perp}}(x - f(x)\rangle = \langle x, -f(x)\rangle = 0 = \langle x, f(x)\rangle$. Hence, x is a solution of nonlinear complementarity problem defined by K and f.

Consider the following Picard iteration

$$x^{n+1} = P_K(x^n - f(x^n)) \tag{23}$$

for finding the fixed points of the mapping (22).

Note that if f is continuous and this iteration is convergent, then its limit is a fixed point of the mapping (22) and therefore a solution of the corresponding fixed point problem which in turn solves the nonlinear complementarity problem defined by K and f. Moreover, if f is continuous, the sequence $\{x^n\}_{n\in\mathbb{N}}$ is decreasing and the cone K regular, then the limit x^* of $\{x^n\}_{n\in\mathbb{N}}$ is a fixed point of the mapping (22), and therefore a solution of the corresponding complementarity problem. By using the ordering induced by the cone, it is interesting to study sufficient conditions for f such that $\{x^n\}_{n\in\mathbb{N}}$ to be decreasing. For this we introduce the notion of *-pseudomonotone decreasing mapping as follows:

The *-pseudomonotone decreasing mapping is a mapping which satisfies the following implication:

$$y - x \in K \text{ and } f(y) \in K^* \text{ implies } f(x) \in K^*.$$

This class of mappings extends the set of mappings which satisfy the following isotonicity property:

$$y - x \in K \Rightarrow f(x) - f(y) \in K^*.$$

For further details on equivalence among complementarity problems, fixed point problems and variational inequalities, see [2, 3] and the references therein.

In the next section, we show that if f is continuous and *-pseudomonotone decreasing, then $\{x^n\}_{n\in\mathbb{N}}$ is decreasing. By introducing other types of isotonicity properties for f, we will also analyze the problem of finding nonzero solutions for the complementarity problem.

6 Nonlinear Complementarity Problems on *-isotone Projection Cones

Recursions for complementarity problems, variational inequalities and optimization problems, similar to (23), were considered in several other works, for example [4, 6, 21, 25, 26, 28, 33, 40–43]. However, neither of these works used the order induced by the cone for analyzing the convergence. Instead, they used the Banach fixed point theorem based approach, assuming Kachurovskii-Minty-Browder type monotonicity (see [7, 22, 30, 31]) and global Lipschitz properties for f.

First we state two lemmas from [35] on which our main results are based.

Lemma 2 *Let H be a Hilbert space, $K \subset H$ a cone and $f : K \to H$ a continuous mapping. Consider the recursion (23). If the sequence $\{x^n\}_{n\in\mathbb{N}}$ is convergent and x^* is its limit, then x^* is a solution of the nonlinear complementarity problem NCP(f, K).*

Lemma 3 *Let H be a Hilbert space, $K \subset H$ a regular cone and $f : K \to H$ a continuous mapping. Consider the recursion (23). If the sequence $\{x^n\}_{n\in\mathbb{N}}$ is monotone decreasing, then it is convergent and its limit x^* is a solution of the nonlinear complementarity problem NCP(f, K).*

Definition 5 Let H be a Hilbert space, $K \subset H$ a cone. The mapping $f : K \to H$ is called a **-increasing* if f is (K, K^*)-isotone. The mapping f is called **-decreasing* if $-f$ is *-increasing.

The following notion is inspired by the notion of pseudomonotonicity defined by Karamardian and Schaible in [23].

Definition 6 Let H be a Hilbert space, $K \subset H$ a cone. The mapping $f : K \to H$ is called a **-pseudomonotone decreasing* if for every $x, y \in K$

$$y - x \in K \text{ and } f(y) \in K^* \quad \text{implies} \quad f(x) \in K^*.$$

Remark 1 (a) If f is *-decreasing, then it is *-pseudomonotone decreasing.
(b) If $f(K) \subset K^*$, then f is *-pseudomonotone decreasing.

Theorem 5 *Let H be a Hilbert space, $K \subset H$ a regular *-isotone projection cone and $f : K \to H$ a continuous mapping such that $f^{-1}(K^*) \neq \emptyset$. Consider the recursion (23) starting from an $x^0 \in f^{-1}(K^*)$. If f is *-pseudomonotone decreasing, then the sequence $\{x^n\}_{n\in\mathbb{N}}$ is convergent and its limit x^* is a solution of the nonlinear complementarity problem NCP(f, K).*

Proof Given that *-isotone projection cone is regular, by Lemma 3, it is enough to prove that the sequence $\{x^n\}_{n\in\mathbb{N}}$ is monotone decreasing. Moreover, it is enough to prove that $f(x^n) \in K^*$ for all $n \in \mathbb{N}$. Indeed, since K is a *-isotone projection cone, $f(x^n) \in K^*$ and $x^n \in K$ imply that

$$x^{n+1} = P_K(x^n - f(x^n)) \leq_K P_K(x^n) = x^n. \tag{24}$$

Hence, the sequence $\{x^n\}_{n\in\mathbb{N}}$ is monotone decreasing. We will prove the proposition

$$(\Gamma_n) \qquad f(x^n) \in K^*, \quad \text{for all } n \in \mathbb{N}$$

by induction. (Γ_0) is obviously true. We suppose that (Γ_n) is true and prove that (Γ_{n+1}) is also true. Since $f(x^n) \in K^*$, by relation (24) we have that $x^{n+1} \leq_K x^n$. Since f is *-pseudomonotone decreasing we have $f(x^{n+1}) \in K^*$; that is, (Γ_{n+1}) is true. □

Example 1 The monotone non-negative cone in $\mathbb{R}^m$ is defined in Example 2.13.9.4 of [11, pp. 198]. The monotone non-negative cone is an important cone in the isotonic regression and its applications (see [24] and the references therein). The monotone non-negative is also used in reconstruction problems (see [11], Sect. 5.13 and Remark 5.13.2.4). Suppose that K is the dual of the monotone non-negative cone in $\mathbb{R}^3$. Then, by Eqs. (435) and (429) of [11], we have

$$K = \left\{x \in \mathbb{R}^3 : x_1 + x_2 + x_3 \geq 0,\ x_1 + x_2 \geq 0,\ x_1 \geq 0\right\}, \tag{25}$$

and

$$K^* = \left\{x \in \mathbb{R}^3 : x_1 \geq x_2 \geq x_3 \geq 0\right\}.$$

It is a straightforward exercise to check that $K^* = U\mathbb{R}^3_+$, where $U = \begin{pmatrix} 1 & 1 & 1 \\ 0 & 1 & 1 \\ 0 & 0 & 1 \end{pmatrix}$. Then, $(U^\top)^{-1} = \begin{pmatrix} 1 & 0 & 0 \\ -1 & 1 & 0 \\ 0 & -1 & 1 \end{pmatrix}$ and therefore, by using Eq. (10), we get $K = (K^*)^* = \begin{pmatrix} 1 & 0 & 0 \\ -1 & 1 & 0 \\ 0 & -1 & 1 \end{pmatrix} \mathbb{R}^3_+$. The generators of K are the column vectors of $\begin{pmatrix} 1 & 0 & 0 \\ -1 & 1 & 0 \\ 0 & -1 & 1 \end{pmatrix}$, which form pairwise nonacute angles. Thus, K is a coisotone cone, which by Theorem 4 is a *-isotone projection cone. Consider the mapping $f = (f_1, f_2, f_3)^\top : K \to H$ defined by

$$f(x) = \left(3000 - x_1^3 - 2x_1 - 2x_2 - x_3,\ 2000 - x_1^3 - x_1 - x_2,\ 1000 - x_1^3\right)^\top. \tag{26}$$

We will show that f is *-pseudomonotone decreasing. For this we have to show that for every $x, y \in K$

$$y - x \in K \text{ and } f(y) \in K^* \text{ implies } f(x) \in K^*,$$

or equivalently

$$\begin{cases} y_1 + y_2 + y_3 \geq x_1 + x_2 + x_3, \\ y_1 + y_2 \qquad\ \ \geq x_1 + x_2, \\ y_1 \qquad\qquad\quad\ \geq x_1 \end{cases} \tag{27}$$

and

$$f_1(y) \geq f_2(y) \geq f_3(y) \geq 0 \tag{28}$$

implies

$$f_1(x) \geq f_2(x) \geq f_3(x) \geq 0. \tag{29}$$

Obviously, inequalities $(27)_3$ and $(28)_3$ imply inequality $(29)_3$. It is easy to see that

$$f_2(y) - f_3(y) = 1000 - y_1 - y_2$$

and

$$f_2(x) - f_3(x) = 1000 - x_2 - x_3.$$

Therefore, inequalities $(27)_2$ and $(28)_2$ imply inequality $(29)_2$. It is straightforward to see that

$$f_1(y) - f_2(y) = 1000 - y_1 - y_2 - y_3$$

and

$$f_1(x) - f_2(x) = 1000 - x_1 - x_2 - x_3.$$

Therefore, inequalities $(27)_1$ and $(28)_1$ imply inequality $(29)_1$. In conclusion, inequalities (27) and (28) imply inequalities (29). Thus, f is *-pseudomonotone decreasing. Consider a point

$$\begin{aligned} x^0 &= \left(x_1^0, x_2^0, x_3^0\right)^\top \in f^{-1}(K^*) \\ &= \left\{(x_1, x_2, x_3)^\top \in \mathbb{R}^3 \mid x_1 + x_2 + x_3 \in [0, 1000],\ x_1 + x_2 \in [0, 1000],\ x_1 \in [0, 10]\right\}. \end{aligned}$$

Of course, there are infinitely many such points. For example, it is easy to see that the box $[0, 10] \times [0, 490] \times [0, 500]$ is contained in $f^{-1}(K^*)$, so one could choose x^0 from this box. Thus, if we consider the recursion (23) with f defined by (26) and starting from x^0, then, by Theorem 5, the sequence $\{x^n\}_{n\in\mathbb{N}}$ defined by this recursion is convergent to a solution of the complementarity problem defined by f and the cone K (given by (25)).

The next theorem gives a sufficient condition for the recursion (23) to be convergent to a nonzero solution.

Theorem 6 *Let H be a Hilbert space, $K \subset H$ a regular *-isotone projection cone and $f : K \to H$ a *-pseudomonotone decreasing, continuous mapping such that $f^{-1}(K^*) \neq \emptyset$. Let $J : K \to H$ be the inclusion mapping defined by $J(x) = x$ and $P_K : H \to K$ the projection mapping onto K. If there are $\hat{x} \in f^{-1}(K^*)$ and $u \in \hat{x} + K$ such that*

$$(P_K \circ (J - f))((\hat{x} + K) \cap (u - K) \cap f^{-1}(K^*)) \subset \hat{x} + K,$$

then $\hat{x}$ is a solution of the nonlinear complementarity problem NCP(f, K) and for any $x^0 \in (\hat{x} + K) \cap (u - K) \cap f^{-1}(K^*)$ the recursion (23) starting from x^0 is convergent and its limit x^* is a solution of the nonlinear complementarity problem NCP(f, K) such that $\hat{x} \leq_K x^* \leq_K u$. In particular, if $\hat{x} \neq 0$, then the recursion (23) is convergent to a nonzero solution.

Proof Since $\hat{x} \in (\hat{x} + K) \cap (u - K) \cap f^{-1}(K^*)$ and

$$(P_K \circ (J - f))((\hat{x} + K) \cap (u - K) \cap f^{-1}(K^*)) \subset \hat{x} + K,$$

we have $\hat{x} \leq_K (P_K \circ (J - f))(\hat{x}) = P_K(\hat{x} - f(\hat{x})) \leq_K \hat{x}$. Hence,

$$\hat{x} = P_K(\hat{x} - f(\hat{x})),$$

that is, $\hat{x}$ is a solution of the nonlinear complementarity problem $NCP(f, K)$. In the proof of Theorem 5 we have seen by induction that

$$x^n \in K \cap f^{-1}(K^*), \quad \text{for all } n \in \mathbb{N}. \tag{30}$$

We prove by induction the proposition

$$(\Omega_n) \qquad \hat{x} \leq_K x^n \leq_K u, \quad \text{for all } n \in \mathbb{N}. \tag{31}$$

Obviously, (Ω_0) is true. Suppose that (Ω_n) is true. Hence, by using relation (30), we have $x^n \in (\hat{x} + K) \cap (u - K) \cap f^{-1}(K^*)$. Thus,

$$\begin{aligned} x^{n+1} &= (P_K \circ (J - f))(x^n) \\ &\in (P_K \circ (J - f))((\hat{x} + K) \cap (u - K) \cap f^{-1}(K^*)) \subset \hat{x} + K. \end{aligned} \tag{32}$$

On the other hand, by using relation (30) and the (K^*, K)-isotonicity of P_K, we have

$$x^{n+1} = P_K(x^n - f(x^n)) \leq_K P_K(x^n) = x^n \leq_K u. \tag{33}$$

Relations (32) and (33) imply that (Ω_{n+1}) is also true. Taking the limit in relation (31), as n tends to infinity, we get $\hat{x} \leq_K x^* \leq_K u$. □

Definition 7 Let H be a Hilbert space, $K \subset H$ a cone, $f : K \to H$ a mapping and $L > 0$. The mapping f is called **-order weekly L-Lipschitz* if

$$f(x) - f(y) \leq_{K^*} L(x - y), \quad \text{for all } x, y \in K \text{ with } y \leq_K x.$$

If $L = 1$, then f is called **-order weekly nonexpansive*.

It is easy to see that the mapping f is *-order weekly L-Lipschitz if and only if the mapping $K \ni x \mapsto Lx - f(x)$ is *-increasing.

Definition 8 Let H be a Hilbert space, $K \subset H$ a cone, $f : K \to H$ a mapping and $L > 0$. Then, the mapping f is called *projection order weekly L-Lipschitz* if the mapping $K \ni x \mapsto P_K(Lx - f(x))$ is K-isotone where P_K is the projection mapping onto K. If $L = 1$ the mapping f is called *projection order weekly nonexpansive*.

If $K \subset H$ is a *-isotone projection cone, then it is easy to see that every *-order weekly L-Lipschitz mapping is projection order weekly L-Lipschitz. In particular, every *-order weekly nonexpansive mapping is projection order weekly nonexpansive. Therefore, the next theorem is also true if we replace the projection order weekly L-Lipschitz condition for f with the *-order weekly L-Lipschitz condition.

Theorem 7 *Let H be a Hilbert space, $K \subset H$ a regular *-isotone projection cone, $L > 0$ and $f : K \to H$ a *-pseudomonotone decreasing, projection order weekly L-Lipschitz, continuous mapping such that*

$$f^{-1}(K^*) \neq \emptyset.$$

Let $\hat{x}$ be a solution of the nonlinear complementarity problem NCP(f, K). Then, for any $x^0 \in (\hat{x} + K) \cap f^{-1}(K^*)$ the recursion

$$x^{n+1} = P_K\left(x^n - \frac{f(x^n)}{L}\right) \tag{34}$$

starting from x^0 is convergent and its limit x^* is a solution of the nonlinear complementarity problem NCP(f, K) such that $\hat{x} \leq_K x^*$. In particular, if $\hat{x} \neq 0$, then the recursion (34) is convergent to a nonzero solution.

Proof We will use the following well-known property of the projection mapping P_κ onto a cone κ: $P_\kappa(\lambda x) = \lambda P_\kappa(x)$ for all $x \in H$ and $\lambda > 0$. We remark that the nonlinear complementarity problem $NCP(f, K)$ is equivalent to the nonlinear complementarity problem $NCP(f/L, K)$. Denote $g = f/L$. Then, the recursion (34) can be written in the form

$$x^{n+1} = P_K(x^n - g(x^n)).$$

We will use Theorem 6 for the mapping g. Let $J : K \to H$ be the inclusion mapping defined by $J(x) = x$ and $u \in \hat{x} + K$ arbitrary. Since any solution of the nonlinear complementarity problem NCP(g, K) is a solution of the nonlinear complementarity problem NCP(f, K) too, it is enough to check the relation

$$(P_K \circ (J - g))((\hat{x} + K) \cap (u - K) \cap g^{-1}(K^*)) \subset \hat{x} + K. \tag{35}$$

We have

$$P_K(x - g(x)) = P_K\left(\frac{1}{L}(Lx - f(x))\right) = \frac{1}{L}P_K(Lx - f(x)), \quad \text{for all } x \in K. \tag{36}$$

Since the mapping f is projection order weakly L-Lipschitz, from relation (36) and the scale invariance of the ordering induced by K, it follows that the mapping g is projection order weekly nonexpansive. For each $x \in (\hat{x} + K) \cap (u - K) \cap g^{-1}(K^*)$ we have $\hat{x} \leq_K x$. Thus, since $K \ni x \mapsto P_K(x - g(x))$ is K-isotone and $\hat{x}$ is a solution of the nonlinear complementarity problem $NCP(g, K)$, it follows that

$$\hat{x} = P_K(\hat{x} - g(\hat{x})) \leq_K P_K(x - g(x)).$$

The previous relation can be rewritten as $(P_K \circ (J - g))(x) \in \hat{x} + K$. Therefore, relation (35) holds. □

Example 2 We will use the notations from Example 1. Let $L > 0$ be a constant and $f : K \to \mathbb{R}^3$ a *-decreasing mapping. We will analyze under which conditions is the mapping $x \mapsto f(x) - Lx$ also *-decreasing. Let $E = (U^\top)^{-1}$. Then, $K = E\mathbb{R}^3_+$, $K^* = U\mathbb{R}^3_+$, $E = [e^1, e^2, e^3]$ and $U = [u^1, u^2, u^3]$. The column vectors e_1, e_2, e_3 are the generators of K and the column vectors u^1, u^2, u^3 are the generators of K^*. Any element $x \in K$ can be uniquely written as

$$x = x_1^e e^1 + x_2^e e^2 + x_3^e e^3.$$

We also have the unique decomposition

$$f(x) = f_1^u(x)u^1 + f_2^u(x)u^2 + f_3^u(x)u^3.$$

Denote the components of x with respect to the canonical basis of $\mathbb{R}^3$ by x_1, x_2, x_3 and the components of $f(x)$ with respect to the canonical basis of $\mathbb{R}^3$ by $f_1(x)$, $f_2(x)$, $f_3(x)$. We will next use the terminology of a decreasing, increasing function in the classical sense. It is easy to see that f is *-decreasing if and only if f_1^u, f_2^u, f_3^u are decreasing with respect to each variable x_1^e ,x_2^e, x_3^e. In other words f is *-decreasing if and only if

$$f(x) = g_1(x)u^1 + g_2(x)u^2 + g_3(x)u^3, \tag{37}$$

where g_1, g_2, g_3 are decreasing with respect to each variable x_1^e, x_2^e, x_3^e. We have

$$\begin{pmatrix} x_1 \\ x_2 \\ x_3 \end{pmatrix} = E \begin{pmatrix} x_1^e \\ x_2^e \\ x_3^e \end{pmatrix},$$

from where we get

$$\begin{pmatrix} x_1^e \\ x_2^e \\ x_3^e \end{pmatrix} = U^\top \begin{pmatrix} x_1 \\ x_2 \\ x_3 \end{pmatrix} = \begin{pmatrix} 1 & 0 & 0 \\ 1 & 1 & 0 \\ 1 & 1 & 1 \end{pmatrix} \begin{pmatrix} x_1 \\ x_2 \\ x_3 \end{pmatrix},$$

or equivalently

$$\begin{cases} x_1^e = x_1, \\ x_2^e = x_1 + x_2, \\ x_3^e = x_1 + x_2 + x_3. \end{cases} \tag{38}$$

Let $f : K \to \mathbb{R}^3$ be a *-decreasing mapping. Then, by equations (37), there exists g_1, g_2, g_3 decreasing with respect to each variable x_1^e, x_2^e, x_3^e such that

$$\begin{pmatrix} f_1 \\ f_2 \\ f_3 \end{pmatrix} = U \begin{pmatrix} g_1 \\ g_2 \\ g_3 \end{pmatrix} = \begin{pmatrix} 1 & 1 & 1 \\ 0 & 1 & 1 \\ 0 & 0 & 1 \end{pmatrix} \begin{pmatrix} g_1 \\ g_2 \\ g_3 \end{pmatrix},$$

or equivalently

$$\begin{cases} f_1 = g_1 + g_2 + g_3, \\ f_2 = \qquad g_2 + g_3, \\ f_3 = \qquad\qquad g_3. \end{cases} \tag{39}$$

Let $h : K \to \mathbb{R}^3$ be defined by $h(x) = f(x) - Lx$, where $L > 0$ and h_1, h_2, h_3 the components of $h(x)$ with respect to the canonical basis and h_1^u, h_2^u, h_3^u the components of $h(x)$ with respect to the basis (u^1, u^2, u^3). We have to analyze when are h_1^u, h_2^u, h_3^u decreasing with respect to each variable x_1^e, x_2^e, x_3^e. From Eq. (39) we get

$$\begin{cases} h_1 = g_1 + g_2 + g_3 - Lx_1, \\ h_2 = \qquad g_2 + g_3 - Lx_2, \\ h_3 = \qquad\qquad g_3 - Lx_3. \end{cases} \tag{40}$$

We have

$$\begin{pmatrix} h_1 \\ h_2 \\ h_3 \end{pmatrix} = U \begin{pmatrix} h_1^u \\ h_2^u \\ h_3^u \end{pmatrix},$$

from which we get

$$\begin{pmatrix} h_1^u \\ h_2^u \\ h_3^u \end{pmatrix} = E^\top \begin{pmatrix} h_1 \\ h_2 \\ h_3 \end{pmatrix} = \begin{pmatrix} 1 & -1 & 0 \\ 0 & 1 & -1 \\ 0 & 0 & 1 \end{pmatrix} \begin{pmatrix} h_1 \\ h_2 \\ h_3 \end{pmatrix},$$

or equivalently

$$\begin{cases} h_1^u = h_1 - h_2 & , \\ h_2^u = \quad\quad h_2 - h_3, \\ h_3^u = \quad\quad\quad\quad h_3. \end{cases} \tag{41}$$

Combining Eqs. (40) and (41) we get

$$\begin{cases} h_1^u = g_1 - L(x_1 - x_2), \\ h_2^u = g_2 - L(x_2 - x_3), \\ h_3^u = g_3 - Lx_3, \end{cases} \tag{42}$$

or, by using Eq. (38)

$$\begin{cases} h_1^u = g_1 - L(2x_1^e - x_2^e), \\ h_2^u = g_2 - L(2x_2^e - x_1^e - x_3^e), \\ h_3^u = g_3 - L(x_3^e - x_2^e). \end{cases} \tag{43}$$

Since g_3 is decreasing with respect to each variable x_1^e, x_2^e and x_3^e, Eq. $(43)_3$ implies that h_3^u is decreasing with respect to x_1^e, x_3^e. Equation $(43)_1$ implies that h_1^u is decreasing with respect to x_1^e, x_3^e. Equation $(43)_2$ implies that h_2^u is decreasing with respect to x_2^e. It follows that $x \mapsto f(x) - Lx$ is *-decreasing if and only if h_1^u, h_3^u are decreasing with respect to x_2^e and h_2^u is decreasing with respect to x_1^e, x_3^e. Thus, $x \mapsto f(x) - Lx$ is *-decreasing if and only if $g_1 = k_1 - Lx_2^e$, $g_2 = k_2 - Lx_1^e - Lx_3^e$ and $g_3 = k_3 - Lx_2^e$, where $k_1 : K \to \mathbb{R}$, $k_2 : K \to \mathbb{R}$ and $k_3 : K \to \mathbb{R}$ are decreasing functions with respect to each variable x_1^e, x_2^e, x_3^e.

Thus, if we consider a mapping $f : K \to \mathbb{R}^3$ defined by Eq. (37) with $g_1 = k_1 - Lx_2^e$, $g_2 = k_2 - Lx_1^e - Lx_3^e$ and $g_3 = k_3 - Lx_2^e$, where $k_1 : K \to \mathbb{R}$, $k_2 : K \to \mathbb{R}$, $k_3 : K \to \mathbb{R}$ are decreasing functions with respect to each variable x_1^e, x_2^e, x_3^e, then f is *-order weekly L-Lipschitz and therefore projection order weekly L-Lipschitz. Moreover, f is *-decreasing and therefore *-pseudomonotone decreasing. Thus, we can use Theorem 7 to conclude that the nonlinear complementarity problem NCP(f, K) has a solution $\hat{x}$ and the recursion (23) starting from any $x^0 \in (\hat{x} + K) \cap f^{-1}(K^*)$ converges to a solution x^* of NCP(f, K) such that $\hat{x} \leq_K x^*$.

References

1. Abbas, M., Németh, S.Z.: Solving nonlinear complementarity problems by isotonicity of the metric projection. J. Math. Anal. Appl. **386**, 882–893 (2012)
2. Ansari, Q.H., Lalitha, C.S., Mehta, M.: Generalized Convexity, Nonsmooth Variational Inequalities and Nonsmooth Optimization. CRC Press, Taylor & Francis Group, Boca Raton (2014)
3. Ansari, Q.H., Yao, J.-C.: Some equivalences among nonlinear complementarity problems, least-element problems and variational inequality problems in ordered spaces. In: Mishra, S.K. (ed.) Topics in Nonconvex Optimization, Theory and Applications, pp. 1–25. Springer, New York (2011)
4. Auslender, A.: Optimization Méthodes Numériques. Masson, Paris (1976)

5. Bernau, S.J.: Isotone projection cones. In: Martinez, J. (ed.) Ordered Algebraic Structures, pp. 3–11. Springer, New York (1991)
6. Bertsekas, D.P., Tsitsiklis, J.N.: Parallel and Distributed Computation: Numerical Methods. Prentice-Hall, New Jersey (1989)
7. Browder, F.E.: Continuity properties of monotone non-linear operators in banach spaces. Bull. Amer. Math. Soc. **70**, 551–553 (1964)
8. Cottle, R.W.: Note on a fundamental theorem in quadratic programming. SIAM J. Appl. Math. **12**, 663–665 (1964)
9. Cottle, R.W.: Nonlinear programs with positively bounded jacobians. SIAM J. Appl. Math. **14**, 147–158 (1966)
10. Dantzig, G.B., Cottle, R.W.: Positive (semi-)definite programming. In: Abadie, J. (ed.) Nonlinear Programming (NATO Summer School, Menton, 1964), pp. 55–73. North-Holland, Amsterdam (1967)
11. Dattorro, J.: Convex Optimization and Euclidean Distance Geometry. $M\varepsilon\beta oo$, v2011.01.29 (2005)
12. Dorn, W.S.: Self-dual quadratic programs. SIAM J. Appl. Math. **9**, 51–54 (1961)
13. Fiedler, M.: Special Matrices and Their Applications in Numerical Mathematics. Martinus Nijhoff, Dordrecht (1986)
14. Horn, R.A., Johnson, C.R.: Topics in Matrix Analysis. Cambridge University Press, Cambridge (1991)
15. Isac, G., Németh, A.B.: Monotonicity of metric projections onto positive cones of ordered euclidean spaces. Arch. Math. **46**, 568–576 (1986)
16. Isac, G., Németh, A.B.: Every generating isotone projection cone is latticial and correct. J. Math. Anal. Appl. **147**, 53–62 (1990)
17. Isac, G., Németh, A.B.: Isotone projection cones in hilbert spaces and the complementarity problem. Boll. Un. Mat. Ital. B **7**, 773–802 (1990)
18. Isac, G., Németh, A.B.: Projection methods, isotone projection cones, and the complementarity problem. J. Math. Anal. Appl. **153**, 258–275 (1990)
19. Isac, G., Németh, A.B.: Isotone projection cones in eucliden spaces. Ann. Sci. Math Québec **16**, 35–52 (1992)
20. Isac, G., Németh, S.Z.: Regular exceptional family of elements with respect to isotone projection cones in hilbert spaces and complementarity problems. Optim. Lett. **2**, 567–576 (2008)
21. Iusem, A.N., Svaiter, B.F.: A variant of korpelevich's method for variational inequalities with a new search strategy. Optimization **42**, 309–321 (1997)
22. Kachurovskii, R.: On monotone operators and convex functionals. Uspeκhi. Mat. Nauk 15, 213–215 (1960)
23. Karamardian, S., Schaible, S.: Seven kinds of monotone maps. J. Optim. Theory Appl. **66**, 37–46 (1990)
24. Kearsley, A.J.: Projections onto order simplexes and isotonic regression. J. Res. Natl. Inst. Stand. Technol. **111**, 121–125 (2006)
25. Khobotov, E.N.: A modification of the extragradient method for solving variational inequalities and some optimization problems. Zhurnal Vychislitel'noi Matematiki Mat. Fiziki **27**, 1462–1473 (1987)
26. Korpelevich, G.M.: The extragradient method for finding saddle points and other problems. Matecon **12**, 747–756 (1976)
27. Lemke, C.E.: Bimatrix equilibrium points and mathematical programming. Management Sci. **11**, 681–689 (1965)
28. Marcotte, P.: Application of khobotov's algorithm to variational inequalities and network equilibrium problems. Information Syst. Oper. Res. **29**, 258–270 (1991)
29. McLinden, L.: An analogue of moreau's proximation theorem, with application to the nonlinear complementarity problem. Pacific J. Math. **88**, 101–161 (1980)
30. Minty, G.: Monotone operators in hilbert spaces. Duke Math. J. **29**, 341–346 (1962)
31. Minty, G.: On a "monotonicity" method for the solution of non-linear equations in banach spaces. Proc. Nat. Acad. Sci. USA **50**, 1038–1041 (1963)

32. Moreau, J.-J.: Décomposition orthogonale d'un espace hilbertien selon deux cônes mutuellement polaires. C. R. Acad. Sci. **255**, 238–240 (1962)
33. Nagurney, A.: Network Economics - A Variational Inequality Approach. Kluwer Academic Publishers, Dordrecht (1993)
34. Németh, S.Z.: Inequalities characterizing coisotone cones in euclidean spaces. Positivity **11**, 469–475 (2007)
35. Németh, S.Z.: Iterative methods for nonlinear complementarity problems on isotone projection cones. J. Math. Anal. Appl. **350**, 340–347 (2009)
36. Németh, S.Z.: An isotonicity property of the metric projection onto a wedge, preprint, Darmstadt (2011)
37. Németh, S.Z.: A duality between the metric projection onto a convex cone and the metric projection onto its dual in Hilbert spaces. arXiv:1212.5438 (2013)
38. Németh, A.B., Németh, S.Z.: A duality between the metric projection onto a convex cone and the metric projection onto its dual. J. Math. Anal. Appl. **392**, 103–238 (2012)
39. Saad, Y.: Matrices and Thier Applications in Numerical Mathematics. Martinus Nijhoff, Dordrecht (1986)
40. Sibony, M.: Méthodes itératives pour les équations et inéquations aux dérivées partielles non linéaires de type monotone. Calcolo **7**, 65–183 (1970)
41. Solodov, M.V., Svaiter, B.F.: A new projection method for variational inequality problems. SIAM J. Control Optim. **37**, 765–776 (1999)
42. Solodov, M.V., Tseng, P.: Modified projection-type methods for monotone variational inequalities. SIAM J. Control Optim. **34**, 1814–1830 (1996)
43. Sun, D.: A class of iterative methods for nonlinear projection equations. J. Optim. Theory Appl. **91**, 123–140 (1996)
44. Young, D.M.: Iterative Solution of Large Linear Systems. Academic Press, New York (1971)

Index

Q. H. Ansari (ed.), *Nonlinear Analysis*, Trends in Mathematics,
DOI: 10.1007/978-81-322-1883-8, © Springer India 2014

Printed by Printforce, the Netherlands